AUTOMOTIVE ENGINE REBUILDING

ABOUT THE AUTHOR

James G. Hughes has worked as an automotive machinist and has rebuilt and repaired a wide variety of domestic and import engines during the past twenty-five years. He has been an automotive training instructor at Rio Hondo College in Whittier, California for fifteen years, and recently served as the national service training instructor for Toyota Motor Sales, U.S.A., Inc.

Mr. Hughes is certified as a General Automobile Mechanic and as a General Heavy-Duty Truck Mechanic, and is a member of the Society of Automotive Engineers. He is the author of three study guides that prepare students for the Automotive Service Excellence (ASE) certification examinations.

AUTOMOTIVE ENGINE REBUILDING

James G. Hughes
Rio Hondo College

REGENTS/PRENTICE HALL
Englewood Cliffs, New Jersey 07632

The information contained herein has been developed from the best available sources. John Wiley and Sons, and the author cannot assume responsibility for accuracy of data or consequences of its application. Persons using this book should do so with due regard for their own safety.

Cover photos by James Worrel.

Library of Congress Cataloging-in-Publication Data

Hughes, James G.
    Automotive engine rebuilding / James G. Hughes.
        p.   cm.
    Reprint. Originally published: New York : Wiley, c1984.
    Includes index.
    ISBN 0-13-053786-1 :
    1. Automobiles--Motors--Maintenance and repair.   I. Title.
[TL210.H75 1991]
629.2'5'0288--dc20   90-28163
                                                             CIP

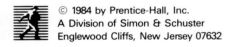

© 1984 by Prentice-Hall, Inc.
A Division of Simon & Schuster
Englewood Cliffs, New Jersey 07632

Printed in the United States of America

10 9 8 7 6 5 4 3

ISBN 0-13-053786-1

Prentice-Hall International (UK) Limited, *London*
Prentice-Hall of Australia Pty. Limited, *Sydney*
Prentice-Hall Canada Inc., *Toronto*
Prentice-Hall Hispanoamericana, S.A., *Mexico*
Prentice-Hall of India Private Limited, *New Delhi*
Prentice-Hall of Japan, Inc., *Tokyo*
Simon & Schuster Asia Pte. Ltd., *Singapore*
Editora Prentice-Hall do Brasil, Ltda., *Rio de Janeiro*

To my wife **Annetta** and family **Kelly** and **Melissa.** Without their support, understanding, and patience for four long years, this book would not have become a reality. Words are incapable of expressing my love and gratitude to them.

PREFACE

With rapid technological advances in the automobile, it is necessary that the automotive mechanic become a specialist. One of these specializations is in the area of engine overhaul. The overhaul may entail a complete rebuild or simply the repair or replacement of certain components. In any case, the engine mechanic today has to make a quick, accurate diagnosis, analyze failed parts, and perform work with attention to detail.

This book is intended to teach the concepts and procedures of engine work, many of which are learned only on-the-job. Often overlooked information and tips are made available, and the story of many failed and worn parts has been told, so as to show where to look for wear and how to recognize defective parts. The illustrations are worth years of practical experience in helping a mechanic interpret various engine conditions. In addition, a view of machine shop service procedures is covered because many dealership and independent garage mechanics "farm" out machining operations to automotive machine shops.

Most automotive textbooks focus on theoretical knowledge that is nice to know, but has little on-the-job application. In this book every attempt has been made to maintain balance between theory and the problems that will be encountered in the actual work.

The structure of this book gives the instructor flexibility in shop and classroom application. The chapters are presented following the typical sequence in which an engine would be rebuilt. However, the design allows material to be selected and covered in an order best suited to the individual situation.

Some additional features included in this book are:

1. True case history examples at the end of each chapter.
2. Review questions at the end of each chapter.
3. Realistic ASE sample questions.
4. Main and rod housing bore diameter specifications.
5. Cylinder head, block, and crankshaft casting number listings.
6. Addresses of core suppliers, equipment manufacturers, and other suppliers.

I am very grateful for the comments and suggestions of the many individuals who contributed to the development of this text. I extend special thanks to James E. Blackburn of Portland Community College, Clarence Bolin of Rio Hondo College, Robert D. Brunken of Portland Community College, Maurice W. Buffel of Southern Alberta Institute of Technology, Roland Granger of Alfred State College, Thomas G. Griesemer of Hudson Valley Community College, Ken Hendricks of San Jacinto College, Steve Hooper of Taylor Engine Rebuilding, David Lopez of Mike McCarthy Buick, George Moore of Aims Community College, the late Gene Morgado of Evergreen Valley College, Austin L. Morrill of Bailey Technical School, John C. Newland, John L. Ogborn of Rio Hondo College, Jay Ray of Pomona Adult School, John Remling of the Board of Cooperative Educational Services, Mike Slavich of S&W Auto Repair, Inc., Ray M. Southwick of Utah Technical College, George Spear of Los Angeles Trade & Technical College, Jay Steel of Taylor Engine Rebuilding, Louie Warren of Warren Automotive Inc. and David R. Whitcomb of Santa Ana College.

James G. Hughes
CGAM, CGTM

CONTENTS

CHAPTER 1

SAFETY AND SHOP PRACTICE

Shop safety should be everybody's business. Safe work habits prevent accidents. Six areas of shop safety are discussed in this chapter. These include personal safety, combustion and fire, safe work area, tool and equipment safety, safety in servicing, and nut and bolt safety.

An unsafe act, a hazardous condition, or a combination of both must occur before an accident can happen. Of these two conditions, approximately 90% of all accidents are a result of an unsafe act or work practice. Some of the most common factors that lead to worker errors are:

1. **Lack of job training** Each employee must know how to do the job safely and productively.

2. **Unaware of hazard** A lack of experience or knowledge or poor communication can result in hazards.

3. **Indifference** This can be caused by a lack of motivation, or a poor relationship with a supervisor or with fellow workers.

4. **Daring** Daring behavior cannot be tolerated. It blinds an individual to hazards that really do exist.

5. **Undue haste** Important steps or processes might be bypassed in order to get the job completed as quickly as possible.

6. **Following the example of others** If a person with poor work habits is never corrected, others will be tempted to imitate his or her conduct. This is especially true with newer, less experienced workers.

7. **Laziness** When a person is tired, lazy, or inclined to do as little as possible to get the job done, an accident is more apt to occur. An example is the employee who lifts an object by bending at the waist instead of by bending at the knees (since it usually requires less effort to lift by using your back).

PERSONAL SAFETY

How you look, dress, and act can help prevent accidents from occurring to yourself or to others.

Appearance

Take a look at your hair. If it is long, it may get entangled in rotating machinery. The following

1

story is a paraphrased newspaper account of a recent scalping.

A California man has just become the first person in the United States to lose his scalp, and then have it successfully replaced.

The man had crawled under an idling truck. The wind blew his 24-inch long hair into the spinning drive shaft. The drive shaft caught the hair and completely ripped off his scalp just above the eyebrows and ears.

A friend pulled the man from under the truck and called an ambulance. The man and his 10-inch by 16-inch scalp were rushed to the hospital where teams of doctors spent 17 hours in microsurgery. Blood vessels had to be sewed together from the scalp and the skull using thread thinner than an eye lash. (Source: San Gabriel Valley Tribune, March 5, 1977.)

If you do have long hair, it should either be tied back or contained by a cap to prevent it from getting into your eyes and work. You know how it feels to comb out a snarl in your hair. Think of how it must feel to be scalped!

Take a look at how you dress. Shirt tails, ties, or unbuttoned cuffs stand a good chance of getting wrapped up in machinery (Figure 1-1). Take the time to keep neat.

Never wear jewelry while performing automobile repair work. Jewelry can get tangled in operating machinery or become caught on ladder rungs, sheet metal protrusions, or accidentally dropped heavy parts. The result can be a severely lacerated or lost limb or finger. Metal jewelry is an

excellent conductor of electricity. For safety's sake, remove watches, rings, and long neck chains. Many severe burns and permanent scars have resulted from accidentally "shorting out" a live electrical connection (Figure 1-2).

Look at Figure 1-3 for an example of safe attire. The mechanic is wearing no loose clothing and

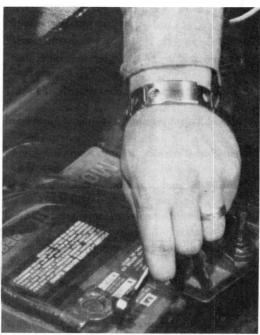

FIGURE 1-2
Leave jewelry off (Courtesy of Ford Parts and Service Division).

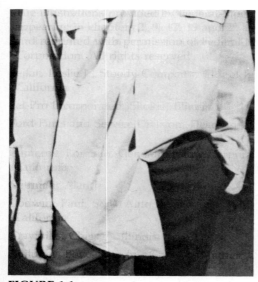

FIGURE 1-1
This attire is not safe (Courtesy of Ford Parts and Service Division).

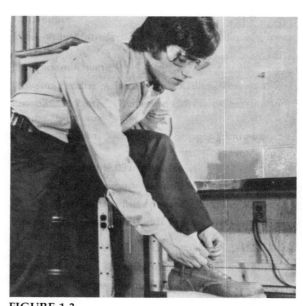

FIGURE 1-3
This is an example of proper shop attire (Courtesy of Ford Parts and Service Division).

no jewelry. He is wearing protective clothing, safety shoes, and safety glasses.

Protect Your Face and Eyes

Safety glasses or goggles are inexpensive insurance for one of the most common types of shop injuries. Occupational Safety and Health Act (OSHA) requirements state that

Protective eye and face equipment shall be required where there is reasonable probability of injury that can be prevented by such equipment. Suitable eye protec-

tors shall be provided where machines or operations present the hazard of flying objects, glare, liquids, injurious radiation, or a combination of these hazards.

Some common shop operations where eye injuries might occur are grinding, drilling, boring, welding, steam cleaning, using the hot tank, charging batteries, and discharging air conditioning refrigerant. Wear the proper eye or face protection every time such shop work is performed.

If you need to check whether or not a pair of glasses has safety lenses, a quick test can be made by using a polariscope (Figure 1-4).

Personal Conduct

While proper personal conduct can help prevent accidents, horseplay can send someone to the hospital. Air nozzle fights, creeper races, and practical jokes do not have any place in the shop (Figure 1-5). Pay attention to lifting rules. Bend with the knees when picking up an object (Figure 1-6). Do not twist at your waist when lifting. For moving heavy objects, get help, and use a hand-truck or dolly.

FIGURE 1-4
When an eye glass lens is placed in the polariscope, you can tell if it is safety glass by observing the pattern that forms as light passes through.

FIGURE 1-5
Horseplay is OUT! (Courtesy of Ford Parts and Service Division).

FIGURE 1-6
Lift the proper way (Courtesy of Ford Parts and Service Division).

Basic First Aid Awareness

In order to be able to quickly contact a doctor, the hospital, the fire department, and the police department in the event of an emergency, a list of the appropriate telephone numbers should be posted clearly by the telephone. *You should know the location of the first-aid kit.* This kit should include sterile gauze, bandages, scissors, and other related items for treating minor injuries. Know where the shop fire extinguishers are located. What facilities are in the shop area for flushing out the eyes?

COMBUSTION AND FIRE

Good work habits can help prevent fires (Figure 1-7). A clean and orderly shop helps to reduce fire danger. Use common sense when handling and storing combustible or flammable materials. This would include paint, thinner, cleaning solvent, kerosene, gasoline, and oil. To transfer highly flammable liquids from one metal container to another, you should bond them together and attach to a ground to prevent the possibility of a spark resulting from static electricity.

Do you know what to do in case of fire? Fire extinguishers may look alike, but they can be quite different in many ways. To be sure that you know about yours, **read the instruction plate on the body of the extinguisher. The use of a wrong extinguisher may spread the fire or cause an explosion.**

Flammable Liquids

Gasoline is a highly flammable liquid that vaporizes quickly when exposed to air. These vapors are heavier than air and will tend to collect in low places (floor, basement, or repair pit). A small spark, a lit cigarette, or a match can set off an explosion. *WARNING: Never use gasoline for*

FIGURE 1-7
This Volkswagen engine caught fire because a fuel hose was left loose.

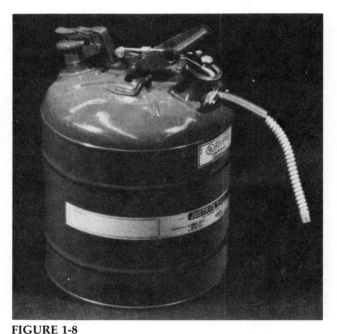

FIGURE 1-8
An approved gasoline safety can (Courtesy of Ford Parts and Service Division).

cleaning parts, yourself, or your clothing. Only store gasoline in an approved safety can (Figure 1-8). Make sure that all paint, thinner, cleaning solvent, kerosene, and oil is stored in approved storage cabinets. Ventilation should provide at least six complete air changes per hour.

WARNING: Do not ever attempt to siphon gasoline with your mouth. Death can result if the fluid is swallowed or the fumes inhaled. If gasoline is swallowed, the victim should drink milk, induce vomiting, and call a physician immediately.

Shop Rags

Used shop rags should be stored in an approved metal container. When oily, greasy, or paint-soaked shop rags are left lying about in piles, *spontaneous combustion* can occur. Heat is generated from within the pile during evaporation, and a fire starts by itself.

Batteries

Serious eye injuries from car batteries have more than doubled in the last three years according to one leading eye doctor. The most common injuries are chemical burns of the eye surfaces from splashed acid or from crust deposits on the terminal posts. The majority of these eye injuries were caused by explosions.

Automobile wet-cell storage batteries are filled with a mixture of water and sulfuric acid. Under certain conditions (particularly when being charged) the battery produces hydrogen and oxygen gas. This very explosive mixture is ignited easily by a cigarette, a match or spark, or by incorrect use of a battery charger. Occasionally a spontaneous explosion will occur in old or defective batteries when the conductors between the cells crack. Internal sparking then results when the electrolyte level falls below such a crack.

Never smoke near a charging battery or risk bridging battery terminals while you are wearing jewelry. *Do not test a battery by shorting across the posts with a pair of pliers.* If possible, disconnect the battery before starting to work on an engine. *NOTE: The ground cable should always be disconnected first when removing a battery from a car. When installing a battery in a car, you should connect the ground cable last.*

Be cautious when using jumper cables to start a disabled car (Figure 1-9). Here are the safety rules to follow when "jump" starting.

1. Match the voltage. Do not use a 24-volt battery to "jump" a 12-volt system.
2. Attach the positive cable from the charged battery to the disabled vehicle's positive battery post.
3. Attach the negative cable from the charged battery to the engine block on the disabled vehicle. The sparks that occur when this "jumper" cable is disconnected could cause an explosion if connected directly to the negative battery post.
4. Before "jump" starting, turn the heater blower motor switch on. Leave the blower motor running after the engine starts. *Do not turn off until after the "jumper" cables are removed.*

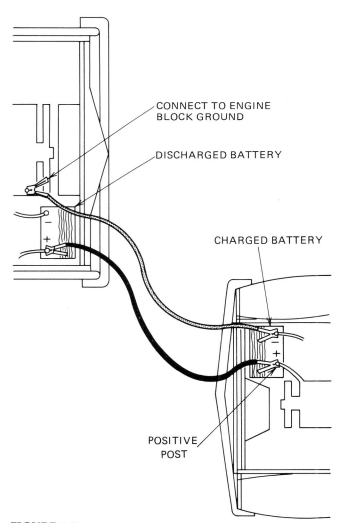

CONNECT TO ENGINE BLOCK GROUND

DISCHARGED BATTERY

CHARGED BATTERY

POSITIVE POST

FIGURE 1-9
Proper connections when "jump" starting.

This procedure will prevent voltage surge that could damage certain electrical components while disconnecting the cables.

5. If you are "jump" starting a computer-equipped vehicle, make sure the ignition is turned off before connecting or disconnecting the jumper cables. Otherwise, you run the risk of damaging the computer.

An understanding of the potential danger of a battery will lessen the risk of injury.

NOTE: Should battery acid ever get into the eyes, immediately flush with clean water for 15 minutes and then obtain medical attention.

Oxyacetylene Welding Equipment

When you use oxyacetylene equipment, keep away from anything combustible. Never use the torch next to the gas tank or lines. Do not use oxyacetylene equipment too close to concrete floors. The intense heat of the torch can cause concrete to explode violently.

Compressed gas cylinders should be stored with protective caps screwed over the valves and should be securely chained or strapped in an upright position (Figure 1-10). Where a special wrench is required, it must be left in position so the gas flow can be quickly turned off in case of an emergency.

Always wear approved goggles when oxyacetylene welding. Your eyes must be protected against hot flying particles, and against the infrared and ultraviolet light rays produced.

FIGURE 1-10

Welding "tanks" must be stored securely in an upright position.

Fires

If fire strikes, it is generally wise to call the fire department first and then attempt to extinguish the fire. *WARNING: If it gets hot or smoky, get out!*

What kind of a fire is it? Figure 1-11 defines the four classes of fire.

Which fire extinguisher should you use? Figure 1-12 indicates which type of fire extinguisher to use for the four classes of fire.

SAFE WORK AREA

Current insurance company statistics show that twice as many disabling accidents occur on the job than on the nation's streets and highways.

An automobile mechanic faces a number of potentially hazardous situations every day. In order to reduce work-related accidents, each person must develop an attitude of "safety consciousness." A good place to begin safety habits is in the work and storage areas.

Floors and Walkways

All floor and walkway surfaces should be kept clean, dry, and orderly. Oil or grease on floor surfaces can cause a person to slip and be injured. Wipe up any spilled oil or grease immediately. A commercial oil absorbent will help in cleaning up large oil spots (Figure 1-13). Make sure that aisles are kept clean and are wide enough for safe clearance. Uncluttered walking areas reduce the chances for an accident. It is a good safety habit to discard old parts and other junk by getting rid of it as soon as possible.

Is Cleaning the Shop Worth the Time and Effort?

One of the best reasons ever given for keeping a shop facility clean is the one by Henry Ford.

Class A Fire (regular combustible materials such as paper, wood, textiles, and rubbish).
Class B Fire (flammable liquids such as grease, oil, paint, and gasoline).
Class C Fire (electrical).
Class D Fire (certain pyrophoric metals such as magnesium, titanium, zirconium, and potassium).

FIGURE 1-11

There are four classes of fires.

FIGURE 1-12
This chart shows the type of fire extinguisher to use.

Class of Fire	Type of Fire Extinguisher								
	Soda-Acid	Pump Tank	Pressurized Water	Gas Water	Loaded Stream	Foam	Carbon Dioxide	Dry Chemical	Multipurpose Dry Chemical
A	Yes	Yes	Yes	Yes	Yes	Yes	No	No	Yes
B	No	No	No	No	Limited	Yes	Yes	Yes	Yes
C	No	No	No	No	No	No	Yes	Yes	Yes
D*	No	No	No	No	No	No	No	No	No

*A number of fire extinguishers are available for Class D fires. Because they vary widely in all respects, they cannot be covered here.

FIGURE 1-13
Oil absorbent helps clean up oil spills.

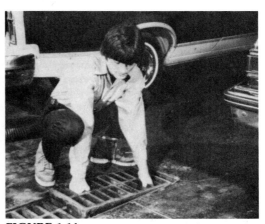

FIGURE 1-14
Cover any open drains (Courtesy of Ford Parts and Service Division).

When asked what he would do if he suddenly found himself the head of a business that was failing because of excessive production costs, he quickly answered: "The first thing would be to see if the plant was clean. It is 100 to 1 that I would find it dirty. I would clean it up. There is nothing so demoralizing to personnel as a dirty shop. Such a place drives away good men and attracts bad ones."

This statement was made years ago, but it is just as true today. Given a choice, no one would elect to work in a dirty shop. Every shop owner, every supervisor wants individuals to do the job well. However, if in filthy, unkept surroundings, bad worker traits, rather than good ones, will usually surface. No one can be expected to give a best effort for a company that does not care about personnel working conditons.

Your Work Area

Keep work benchtops clean and orderly. Clutter on top of benches and tables can fall off and cause an accident.

Check to see that all drain covers are in place (Figure 1-14). Open drains and poorly positioned covers have caused many toe, ankle, and leg injuries.

A good automobile mechanic has every tool in its proper place. A well-organized tool chest will save a lot of wasted time spent in looking for needed items. Floor jacks, jack stands, and creepers should be kept in one area, out of aisles.

Carbon Monoxide

More than 2000 people are killed and 10,000 more injured in this country each year due to carbon monoxide poisoning. *WARNING: Carbon monoxide fumes are deadly.* They are odorless, colorless, give very little warning, and can kill within minutes. Carbon monoxide gas attacks the hemoglobin in the blood and literally starves the body of oxygen.

FIGURE 1-15
Never start a car in the shop until the tailpipe is connected to an exhaust ventilation system (Courtesy of Ford Parts and Service Division).

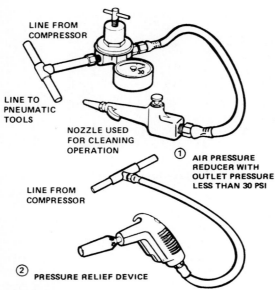

FIGURE 1-16
Air pressure reducing devices.

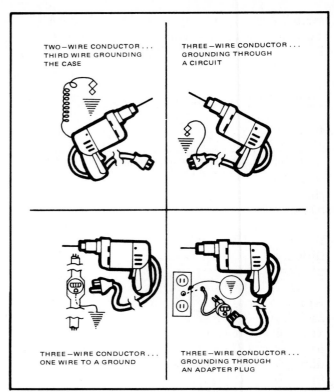

FIGURE 1-17
All portable electric tools must be properly grounded or U.L. double-insulated.

Early warning symptoms of carbon monoxide poisoning may include headaches, nausea, tightness across the chest, tiredness, drowsiness, and inattention. With continued exposure, a worker will become uncoordinated, faint, and die.

Make sure the shop work area is well ventilated. Never start an engine in the shop without proper exhaust ventilation (Figure 1-15).

Compressed Air

Compressed air can be dangerous, even fatal. At 40 psi, a blast of air from a distance of four inches can rupture an eardrum. At 40 psi, metal chips and other debris can be driven across the shop at 70 mph. Directed into the mouth, air at this pressure can rupture lungs and intestines. Aimed at the eyes, it can produce blindness. Air at pressures as low as 4 psi can rupture the bowel. Compressed air directed against the skin can cause tissue damage similar to burns. Air can be driven into scratches or punctures in the skin, causing pain and swelling, and perhaps even bubbles of air in the blood. *The utmost caution should accompany the use of compressed air.*

OSHA says: *The downstream pressure of compressed air used for cleaning purposes must remain at a pressure level below 30 psi whenever the nozzle is dead-ended.* Figure 1-16 shows two acceptable methods of meeting this safety requirement.

Electrical Grounding

Any electrically operated tool or piece of equipment must be properly grounded or U.L. double-insulated (Figure 1-17).

Never stand on a wet or damp floor when using electric tools. Insulate yourself by standing on a rubber mat or wooden blocks, if you cannot properly dry the area. Always make sure the switch is *off* before you plug in any electrical tool. Turn the switch *off* when you are through using it.

TOOL AND EQUIPMENT SAFETY

Section 1910.242 of the OSHA regulations states:

Each employer shall be responsible for the safe condition of tools and equipment used by employees, including tools and equipment which may be furnished by employees.

It is important that shop owners know the condition of the tools and equipment used by employees—shop-owned or not.

Hand tools should always be kept in clean, workable condition. Greasy, oily, or worn hand tools can slip out of your hand, causing skinned knuckles or broken fingers. Check all hand tools for cracks, chips, burrs, broken parts, or other dangerous conditions before you use them (Figures 1-18 and 1-19).

Box-end, Open-End, and Combination Wrenches

A good quality wrench, regardless of the type, is designed to keep leverage and load in safe balance. The mechanic does not have to improvise. There's a correct wrench design for almost every operation. Find the proper one of the correct size, and use it.

Here are some safety tips on the proper use of box-end, open-end and combination wrenches.

1. Never use a pipe extension or other form of "cheater bar" to increase the leverage of these wrenches.

2. Always use inch wrenches on inch fasteners and metric wrenches on metric fasteners (Figure 1-20).

3. Never expose any wrench to excessive heat that may draw the temper and ruin the tool.

4. If possible, always pull on a wrench handle. Adjust your stance to prevent a fall.

5. Do not use an open-end wrench on a "frozen" nut or to final tighten a nut. Use a box-end or socket wrench.

6. Never cock an open-end wrench. Make certain the nut or bolt head is fully seated in

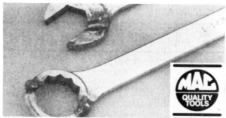
FIGURE 1-18
Never attempt to weld broken wrenches. The final product could be hazardous to your health (Courtesy of Mac Tools, Inc., Washington Court House, Ohio 43160).

FIGURE 1-19
This standard thin-wall socket has been abused. It was used on an impact wrench (Courtesy of Snap-on Tools Corporation).

the jaw opening. Position the jaw opening in the direction of pull (Figure 1-21).

7. Replace open-end wrenches with spread, nicked, or battered jaws, and box-end wrenches with rounded or damaged points.

8. Ratcheting box-end wrenches should not be used in heavy duty applications.

Adjustable Wrenches

This versatile tool is designed to furnish more capacity in a single tool. *Adjustable wrenches are not intended to replace fixed-opening wrenches for production or general service work.* Here are some tips on the proper use of an adjustable wrench.

1. Adjustable wrenches must be tightly adjusted to the nut or bolt head and pulled so the force is on the side of the fixed jaw (Figure 1-22).

2. Never use a hammer on this type of wrench.

3. Periodic inspection should be made to detect damaged jaw knurls, and pins and springs that should be replaced.

Socket Wrenches

Socket wrenches have more uses than conventional wrenches. Like all tools, they must be used

REGULAR SOCKET SIZES	DECIMAL EQUIVALENTS		METRIC SOCKET SIZES
1/8"	.125	.118	3mm.
3/16"	.187	.157	4mm.
1/4"	.250	.236	6mm.
5/16"	.312	.354	9mm.
3/8"	.375	.394	10mm.
7/16"	.437	.472	12mm.
1/2"	.500	.512	13mm.
9/16"	.562	.590	15mm.

REGULAR SOCKET SIZES	DECIMAL EQUIVALENTS		METRIC SOCKET SIZES
5/8"	.625	.630	16mm.
11/16"	.687	.709	18mm.
3/4"	.750	.748	19mm.
13/16"	.812	.787	20mm.
7/8"	.875	.866	22mm.
15/16"	.937	.945	24mm.
1"	1.00	.984	25mm.

FIGURE 1-20
Use this chart to familiarize yourself with the sizes of metric sockets as compared to regular ones. In some cases, a regular socket will fit on a metric bolt. However, the fit is not correct, and you can get hurt (Courtesy of Dana Corporation-Service Parts Group).

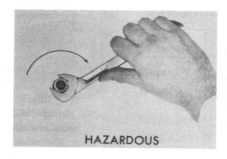

HAZARDOUS   SAFER

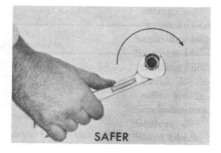

FIGURE 1-21

Have you ever pulled an open-end wrench right off the nut or bolt? This danger can be minimized if the wrench is positioned so the jaw opening faces in the direction of pull (Courtesy of Snap-on Tools Corporation).

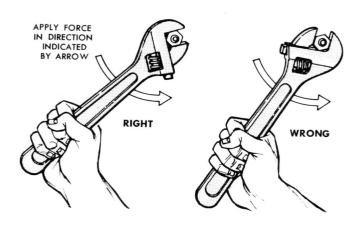

APPLY FORCE IN DIRECTION INDICATED BY ARROW

RIGHT

WRONG

FIGURE 1-22

Using an adjustable wrench (Courtesy of Ford Parts and Service Division).

properly. Here are some tips on socket wrench use.

1. Never use a hand socket on power drive or impact wrenches.
2. Do not hold a spinning impact universal socket in your hand.
3. Do not abuse adapters (Figure 1-23). Remember, when adapting "down" (big handle, small socket) you are building tremendous torque potential, When adapting "up" (small handle, big socket) you are losing turnpower. Always stay within safe limits.
4. Periodically clean and inspect your sockets to get better and longer service.
5. Sockets with cracked walls, breaks, or battered points must be replaced.

Socket Attachments

To select the correct socket handle (ratchet, breaker bar, or speed handle) and extension is vital if the job is to be done properly and in less time. There are nondetachable and detachable handles with innumerable combinations of attachments and

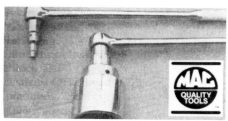

FIGURE 1-23

Do not abuse adapters. The top ratchet could damage the fastener while the bottom ratchet loses a lot of turning power (Courtesy of Mac Tools, Inc., Washington Court House, Ohio 43160).

sockets. Some tips on the proper use of socket attachments include:

1. Do not use a socket handle that is too big or too small for the job (Figure 1-24).
2. It is safer and easier to use one socket handle extension that is the correct length, instead of many shorter ones.
3. Periodically clean and inspect socket handles.

CHECKING TORQUE

Each square drive tool has a recommended maximum working torque that should be kept in mind to assure safe use. The table below shows recommended maximum working torque for standard tools by square drive, plus recommended maximum working torque for the male square drive which is usually the limitation for special tools.

Square Drive Size	Standard Tools Line						Specials
	Recommended Maximum Working Torque in Lb. In.						Special Tools Design Limit
	Ratchets	Breaker Bars	Extensions	Sliding Tee	Wobble Extensions	Speeders	
1/8"	NA	NA	NA	NA	NA	47	47
1/4"	340	340	340	340	240	240	340
3/8"	950	950	950	700	650	400	950
1/2"	3,400	2,800	2,800	2,800	NA	1,000	3,950
5/8"	NA	NA	NA	NA	NA	NA	6,000
3/4"	8,500	8,500	8,500	6,000	NA	NA	13,500
1"	14,000	12,000	14,000	14,000	NA	NA	26,000
1 1/2"	23,000	NA	40,000	22,000	NA	NA	80,000
2 1/2"	NA	NA	NA	NA	NA	NA	300,000

NOTES: NA — Not available in standard line. To convert above figures to lb. ft., divide by 12; to convert to newton metres, multiply by .113. For specials, torque value is based on 40 Rc minimum hardness. Torque values are based on obtaining reasonable service life when hand applied.

FIGURE 1-24

Here is a breakdown of government minimum loads for ratchets and other socket wrench handles. These figures can be used as a guide for the safe use of socket wrenches (Courtesy of Snap-on Tools Corporation).

FIGURE 1-25

Here is what can happen when pliers are used to remove a fuel line.

Pliers

Pliers are one of the most versatile tools in a tool box. There are many types and sizes; each is designed for a specific purpose. Here are some safety tips on the use of pliers.

1. Other than vise grips, pliers are not normally built to withstand pressure that is greater than normal hand squeezing.

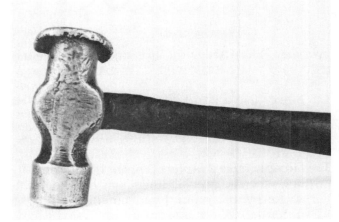

FIGURE 1-26

This hammer was found in a tool box. Why is it dangerous?

2. Never use pliers as a hammer. They may crack or break.

3. Do not use pliers on nuts or bolts. A wrench will do the job better, with less risk of damage or an accident (Figure 1-25).

Hammers

Hammers vary in configuration and in head hardness. Here are some safety tips on the proper use of hammers.

1. Keep your hammer clean at all times.

FIGURE 1-27
A dull chisel is unsafe.

FIGURE 1-28
Is this drift punch safe to use?

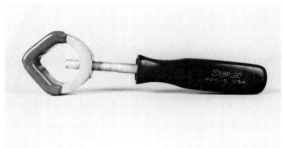

FIGURE 1-29
This tool can be useful when using a punch or chisel.

2. *Always wear safety glasses when striking any object.*
3. Never strike an object with the side of a hammer.
4. Never use one hammer to strike another hammer.
5. Do not use a hammer with a loose head. Wedges are available to tighten.
6. Always be sure the hammer face strikes the object squarely.
7. Never use a hammer with a mushroomed head (Figure 1-26).

Chisels and Punches

Chisels and punches come in many sizes and shapes, each with a specific function. Here are some tips on the proper use of chisels and punches.

1. Never use a dull chisel (Figure 1-27).
2. *Always wear safety glasses when using a punch or chisel.*
3. The diameter of the chisel you are using should be approximately 1/2 that of the hammer's face.
4. Never use a chisel or punch that has a mushroomed head. Correct it by grinding. Likewise, when the point of a punch becomes dull, chipped, or out-of-square. (Figure 1-28).
5. A punch and chisel holder is useful for certain jobs (Figure 1-29). It saves bruised knuckles and will take up the hammering shock.

Screwdrivers

Screwdrivers are, undoubtedly, the most abused of all hand tools. Use screwdrivers only for what they were intended to do—drive screws. Some tips on screwdriver use include:

1. Always use a screwdriver with a tip equal to the width of the screw head (Figure 1-30).
2. Always use the longest screwdriver that fits the screw slot and permits easy movements within the work space.
3. You can apply more power with less effort with a long screwdriver than with a short one.
4. Badly worn screwdriver blades can be returned to their original shape by careful grinding.

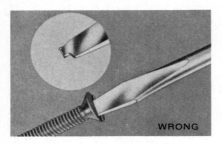

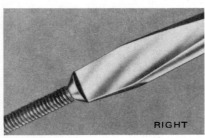

FIGURE 1-30
The screwdriver tip and the width of the screw should be equal (Courtesy of Snap-on Tools Corporation).

Files

There are many different kinds of files. Select the right file for the job to be done, hold the file correctly, and use the right stroke. Here are several tips on the proper use of a file.

1. Never use a file without a handle.
2. Always store files in a dry place. Rust affects the cutting edge of the file teeth.

Hacksaws

There are many uses for a hacksaw, and like any other tool, it must be used properly. Here are some tips on hacksaw use.

1. The teeth of a hacksaw blade should face away from the handle (Figure 1-31). A hacksaw is designed to cut only on the forward stroke. Apply slight downward pressure on the forward stroke, none on the return.
2. Do not rush when hacksawing. Use a long and steady stroke and keep the blade in a straight line.
3. A light oil or cutting fluid can be used for smoother sawing.
4. Hacksaw blades are made of several different types of material. A *high-speed steel blade* can be used to cut most machinable metal. *Tungsten alloy steel blades* have tougher, faster cutting teeth.

FIGURE 1-31
The proper way to install a hacksaw blade in frame (Courtesy of Ford Parts and Service Division).

Vises

A vise generally takes a lot of punishment. Some tips on the proper care and use of a vise are:

1. Never hammer directly on a vise unless there is an anvil built on the vise for that purpose.
2. Keep the screw mechanism clean and well oiled.
3. When jaw facings wear out or lose their gripping power, replace them.
4. Never tighten a vise handle with an extension bar or by hammering on the handle.
5. To avoid damage to parts clamped in the vise, install a soft cover (sheet brass or aluminum) over the regular jaws.

Grinding Wheels

Abrasive wheels, as manufactured today, offer a high level of product quality. Developments to increase grinding and finishing abilities have been made in the last several years. Some economy and safety hints on the use of grinding wheels include:

1. Always check the maximum operating speed established for the wheel against machine speed.
2. Check mounting flanges for equal and correct diameter. They should be at least 1/3 the diameter of the wheel and relieved around the hole. Always use the mounting blotters supplied with the wheel.
3. Make sure the work rest is properly adjusted at the center of wheel or above, and no more than 1/8″ away from the wheel.

4. Always use a guard covering at least 1/2 of the grinding wheel.

5. Allow newly mounted wheels to run at operating speed, with guard in place, for at least one minute before grinding.

6. *Always wear safety glasses or some type of eye protection when grinding.*

7. Turn off coolant before stopping the wheel to avoid creating an out-of-balance condition.

8. Do not ever use a wheel that has been dropped.

9. Do not grind on the side of the wheel, unless the wheel is specifically designed for that purpose.

10. Do not stand directly in front of a grinding wheel whenever a grinder is started.

Tips on the Proper Use of Air-Operated Tools

1. Make certain there is an adequate supply of clean, dry air at the right pressure and volume.

2. The shop air compressor must be drained (automatically or manually) to avoid excessive moisture.

3. Do not use an air ratchet as a breaker bar. It is primarily a nut runner.

4. Do not use an air drill as a grinder. It is not built to withstand the side thrust.

5. Do not overload an air drill chuck with cut down drill shanks.

6. Do not exceed the maximum bolt diameters specified for air impact wrenches.

7. Never use hand sockets on air impact wrenches.

8. Be sure that impact mechanisms in air tools are properly and frequently lubricated.

Tips on the Proper Use of Electric Tools

1. Keep all electrical cords dry. Replace all frayed and broken wiring immediately.

2. Ordinary plastic-dipped tool handles are designed for comfort, not electrical insulation. Tools having high dielectric insulation are available and are so identified. Do not confuse the two.

3. Do not overload electrical tools. Use a tool that is rated for the job being done.

FIGURE 1-32
Proper blocking after jacking up the other end of the vehicle (Courtesy of Durston Manufacturing Company).

4. Make certain electric tools and extension cords are properly grounded and always operate within the recommended voltage range.

5. Inspect motor brushes about every 400 hours. Replace them when they are worn to less than 1/3 of their original length.

6. Keep the inlet and outlet air passages clear of dirt and accumulated dust to ensure the motor runs cool.

Maintenance and Handling Tips for Tool Chests

1. Keep the tool storage unit clean.

2. Lubricate hinges, drawer slides, and casters periodically.

3. Do not try to move a rollaway unit over a rough or sloping surface alone. You could lose control.

4. Do not leave several top drawers open at one time. This will make the unit top heavy and easy to tip over.

5. Always lock the brakes on the swivel casters after moving a rollaway unit.

6. The specified weight limits for different rollaway models must not be exceeded.

Floor Jacks

When using a floor jack to raise a car, you must first check to see that the jacks wheels are not binding. The car should be parked on a reasonably level surface, the parking brake off, and the transmission in neutral during raising. After jacking the vehicle up, place wheel chocks on both sides of the free wheels (Figure 1-32). Then,

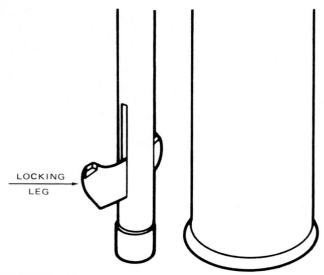

FIGURE 1-33
The leg must be out when the lift is up (Courtesy of Ford Parts and Service Division).

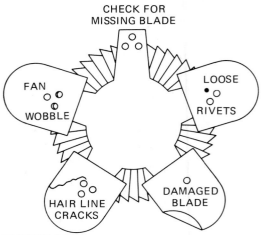

FIGURE 1-34
Check these items if you are required to work near the fan with the engine running.

before working on the car, correctly place jack stands under the car. *WARNING: Never crawl under a car held up only by a jack.*

Hydraulic Lift

When you use a hydraulic lift, *be sure the contact pads are in correct position.* Cracked windshields, misaligned hoods, and incompletely closed doors can result when a vehicle is lifted incorrectly. This is especially true with unitized frame vehicles. After the vehicle has been raised about six inches, shake it to make sure it is well balanced on the lift. If there are any "clunking" noises or scraping sounds, the vehicle is not secure.

Make certain that the locking leg or safety bar is in place when the lift is up (Figure 1-33). Never use the lift if it

1. Jerks or jumps during raising.
2. Settles after being raised.
3. Blows oil out the exhaust line.
4. Leaks oil at the packing gland.
5. Slowly rises after being raised.

Hydraulic Press

With a hydraulic press, always make sure that hydraulic pressure is being applied in a safe manner. Stand to the side when operating the press and *wear safety glasses.*

SAFETY IN SERVICING

This section covers some potentially dangerous service situations that a shop mechanic might encounter.

Fan Safety

A mechanic working under the hood of any car, while the engine is running, should be extremely cautious about the fan. Weakened fan blades are a potential hazard.

There is a recent case of a mechanic who died as a result of injuries received while working over a car's running engine. One of the flex-fan blades broke loose and fatally hit the man in the throat and chest.

Here are some safety tips to follow when required to work near the fan with the engine running.

1. Check each blade for cracks, looseness, and bend damage (Figure 1-34). If there are any defects, replace the fan before proceeding with work. *Never attempt to straighten a bent fan blade.* No matter how carefully you try to balance it, you could be setting up an accident that might injure or kill someone.
2. *Do not stand directly in line with a running fan.* Position yourself as far away from the fan as possible.
3. Be especially careful about flex-fans.
4. Keep electric cords and test equipment leads clear of the fan.
5. Do not wear neckties or other loose clothing.

Driving a Customer's Car

Before driving a customer's car, test the brakes and inspect the tires. Whenever you drive in the shop, watch for other cars and people. It often helps to have someone guide you. *NOTE: Make sure your hands and clothing are clean when getting into the car.*

Using Starting Spray

Caution should be exercised when using starting spray (ether) to help initiate firing in an engine (Figure 1-35). Overloading the intake manifold with this highly volatile fluid can result in a severe explosion. Blown seals, snapped piston rings, cracked pistons, damaged valves, burned through head gaskets, and bent connecting rods can result (Figure 1-36). More seriously, a flame can shoot out the top of the carburetor and cause personal injury.

Starting ether is used to lower the ignition point of the air-fuel mixture. When a controlled amount of ether is used, the firing will not exceed the normal combustion explosion force. When an excessive amount of ether is injected, a violent explosion will occur. This can be dangerous and

FIGURE 1-35
Starting ether can be dangerous.

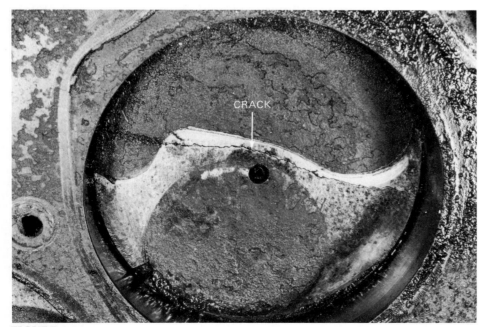

FIGURE 1-36
This cracked piston was caused by using too much starting fluid.

FIGURE 1-37
When refrigerant-12 is released to the atmosphere, it starts to "boil" rapidly at a temperature of minus 22° F.

possibly damage several engine parts as just mentioned. To avoid the danger of injecting excessive starting ether into the engine, carefully follow the instructions printed on the container.

Air Conditioning Servicing

When you are discharging, purging, evacuating, or recharging an air conditioning system, *always be sure to wear eye protection.* Refrigerant-12 vaporizes rapidly and quickly freezes virtually anything it contacts (Figure 1-37). If any Refrigerant-12 does get into your eyes, DO NOT RUB THEM. SEE A DOCTOR IMMEDIATELY. As a temporary measure, splash cold water on your eyes to bring the temperature above freezing while waiting for medical attention.

A Refrigerant-12 container can explode with heat. Therefore, keep it away from sunlight, steam heat, and welding equipment.

NUT AND BOLT SAFETY

Injuries, costly damage, and lost production result when a fastener failure occurs. Yet, nuts and bolts are probably the most misused and taken-for-granted objects in the shop. Consider the following:

1. Although most mechanics know a bolt or nut can be overtightened to the point of breaking, few mechanics realize bolts or nuts can break in service if not tightened enough.
2. A loose nut can result from using a bolt of the wrong grade.
3. A flat washer if installed upside down can cause bolt failure.
4. Each time a nut is reused, it becomes deformed and causes increased assembly friction and decreased loading strength, even though a torque wrench shows proper tightness.
5. Most torque wrench charts are valid only for clean, dry threads on new nuts and bolts.

Let us now look at some of the basics of nuts and bolts.

How to Identify Fastener Quality

Cap screws and bolts literally hold together the automobile. The quality of any cap screw or bolt can be identified by the head markings, which indicate the type of material used in their manufacture (Figure 1-38). Each grade is manufactured according to standards established by the Society of Automotive Engineers (SAE) and the American Society for Testing Materials (ASTM).

What Is Meant by the Term Torque?

Torque is twist and is usually measured in foot-pounds. For example, imagine yourself pulling with a force of ten pounds at a right angle to the handle of a wrench at a distance of $1\frac{1}{2}$ feet from the bolt. The torque or twist exerted would be 15 foot-pounds (ten pounds multiplied by $1\frac{1}{2}$ feet). *NOTE: Occasionally the terms ounces-inch, pounds-inch, and pounds-foot are used. From a purely technical standpoint this is the correct way to express torque. However, these terms will not be used in this book, because they are not generally used in the trade.*

What Is Tensile Strength?

Tensile strength can be defined as the amount of load required to fracture or break a material by *pulling longitudinally* on it. Tensile strength is usually expressed in pounds per square inch (psi). For instance, a 1/2–20 cap screw would contain .1597 square inch of cross-sectional stress area. If made from a 150,000 psi material, the cap screw would have a tensile strength of 150,000 × .1597, or 23,950 lb.

Grade Marking	Specification	Material	Physical Properties		
			Bolt Size [inches]	Proof Load Min. [psi]	Tensile Strength Min. [psi]
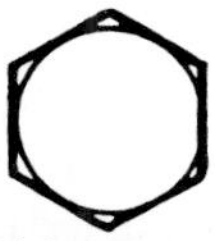 NO MARK	SAE—Grade 0	Steel	All Sizes	. . . .	. . . .
	SAE—Grade 1 ASTM—A 307	Low Carbon Steel	All Sizes	. . . .	55,000
	SAE—Grade 2	Low Carbon Steel	Up to 1/2'' Over 1/2'' to 3/4'' Over 3/4'' to 1-1/2''	55,000 52,000 28,000	60,000 64,000 55,000
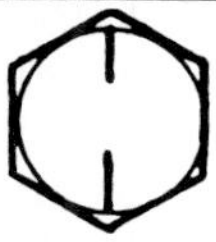	SAE—Grade 3	Medium Carbon Steel, Cold Worked	Up to 1/2'' Over 1/2'' to 5/8''	85,000 80,000	110,000 100,000
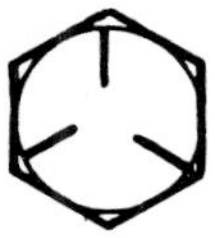	SAE—Grade 5	Medium Carbon Steel, Quenched and Tempered	Up to 3/4'' Over 3/4'' to 1'' Over 1'' to 1-1/2''	85,000 78,000 74,000	120,000 115,000 105,000
	ASTM—A 325		Up to 3/4'' Over 3/4'' to 1'' Over 1'' to 1-1/2'' Over 1-1/2'' to 3''	85,000 78,000 74,000 55,000	120,000 115,000 105,000 90,000
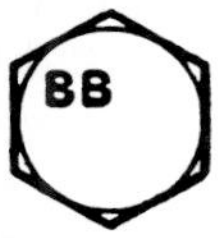 BB	ASTM—A 354 Grade BB	Low Alloy Steel Quenched and Tempered (Medium Carbon Steel, Quenched and Tempered may be substituted where possible)	Up to 2-1/2'' Over 2-1/2'' to 4''	80,000 75,000	105,000 100,000
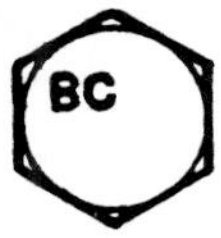 BC	ASTM—A 354 Grade BC	Low Alloy Steel Quenched and Tempered (Medium Carbon Steel, Quenched and Tempered may be substituted where possible	Up to 2-1/2'' Over 2-1/2'' to 4''	105,000 95,000	125,000 115,000
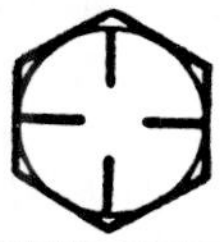	SAE—Grade 6	Medium Carbon Steel, Quenched and Tempered	Up to 5/8'' Over 5/8'' to 3/4''	110,000 105,000	140,000 123,000
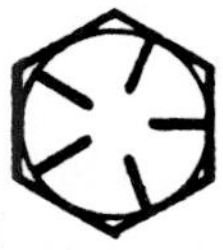	SAE—Grade 7	Medium Carbon Alloy Steel Quenched and tempered. Roll Threaded after heat treatment	Up to 1-1/2''	105,000	133,000
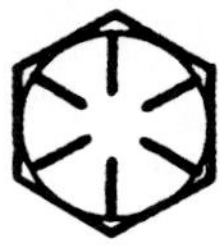	SAE—Grade 8 ASTM—A 354 Grade BD	Medium Carbon Alloy Steel, Quenched and Tempered	Up to 1-1/2''	120,000	150,000

ASTM Specifications
 A 307—Low Carbon Steel Externally and Internally Threaded Standard Fasteners.
 A 325—Quenched and Tempered Steel Bolts and Studs with Suitable Nuts and Plain Hardened Washers.
 A 354—Quenched and Tempered Alloy Steel Bolts and Studs with Suitable Nuts.

FIGURE 1-38

FIGURE 1-39
Here are actual photographs of some fatigue breaks (Courtesy of Premier Industrial Corporation).

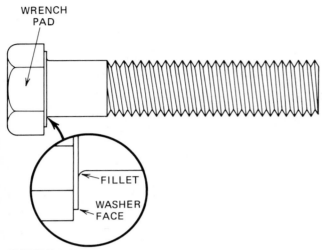

FIGURE 1-40
The fillet is a critical part of a bolt and must be protected when tension is applied (Courtesy of Bowman Distribution, Division of Barnes Group Inc.).

What Is Shear Strength?

Shear is opposite from tensile strength. It can be defined as the *push or pull at 90°* from the bolt axis. Shear strength is roughly 60% of a bolt's tensile strength.

What Is Meant by the Yield Point of a Bolt?

There is a point when a bolt begins to stretch without increasing the load on the assembly. In some alloy fasteners this may be 90% of the tensile strength rating.

What Is Fatigue?

Fatigue occurs when metal is bent back and forth by stress until it breaks. Bolt fatigue failures often occur considerably below the minimum tensile strength. This is normally the result of not tightening the bolt enough (Figure 1-39).

Why Do Bolts Break?

The fillet where the shank joins the head and the washer face are two very important parts of a bolt (Figure 1-40).

The purpose of the fillet is to reduce stress concentrations where the bolt head and shank meet. Scratches on the fillet will weaken the bolt and can cause the head to break off under tension.

Have you ever had this experience? You drill a hole in a piece of metal, install a bolt to fit the hole, and then while tightening to correct torque, the bolt head breaks off. This problem is usually caused by burrs and sharp edges from the drilled hole digging into the bolt fillet. As tension is applied, the scratch marks open up into small cracks. The bolt head then snaps off as full torque is applied. The solution is to protect the fillet. This can be done by slightly chamfering the edge of the hole, or by installing a flat washer under the bolt head. *Always install a flat washer correctly.* Every flat washer has sharp and rounded sides (because of the stamping process used during manufacturing). *Always place the rounded side next to the bolt head.*

Bolt failure is often caused by using a grade that is too low for the application. If a bolt's yield strength is too low for the torque being applied, it will stretch beyond its elastic limit. The nut will begin to work loose because the load on it is relaxed, which soon causes the bolt to break. Besides using the right grade bolt, it is very important for the bolt, nut, and washer to be matched. *Low-grade nuts and soft flat washers used with high-grade bolts usually result in failure.* Figure 1-41 shows high-grade nut markings.

What Is the Correct Torque for a Bolt?

A bolt takes two stresses when it is tightened: (1) twist and (2) tension. Tension is what you want. Twist is the necessary evil due to friction. About 90% of the applied torque goes to overcome friction, yet only tension remains after tightening. About 50% of the torque is absorbed in overcoming friction on the working faces of the bolt and nut, while 40% is taken up by thread friction. This leaves only 10% to apply bolt tension.

The ideal bolt tightening point is just below the yield point. This not only provides a cushion for working load variations but also prevents loosening. Standard torque charts (Figures 1-42 and 1-43) have been set up for average dry un-

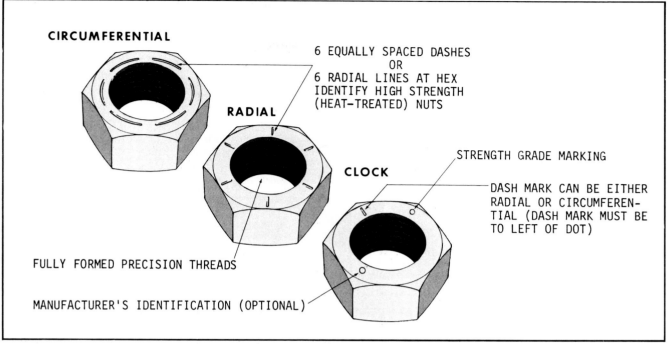

FIGURE 1-41

Hex nut grade 8 strength identification marking (Courtesy of Lawson Products, Inc.).

The following rules apply to the chart:
1. Consult manufacturers specific recommendations when available.
2. The chart may be used directly when any of the following lubricants are used:
   Never-Seez Compound, Molykote, Fel-Pro C-5, Graphite and Oil or similiar mixtures.
3. Increase the torque by 20% when using engine oil or chassis grease as a lubricant.
(These lubricants are not generally recommended for fasteners)
4. Reduce torque by 20% when cadmium plated bolts are used.
CAUTION: Tightening into aluminum usually will require less torque.

U.S. STANDARD

GRADE OF BOLT	S.A.E. 1 & 2	S.A.E. 5	S.A.E. 6	S.A.E. 8		
MINIMUM TENSILE STRENGTH	64,000 P.S.I.	105,000 P.S.I.	133,000 P.S.I.	150,000 P.S.I.		
GRADE MARKINGS ON HEAD →	⬡	⬡	⬡	⬡	SOCKET OR WRENCH SIZE	
U.S. STANDARD		TORQUE			U.S. REGULAR	
BOLT DIAMETER	DEC. EQUIV.	(IN FOOT POUNDS)			BOLT HEAD	NUT
1/4	.250	5	7	10	10.5	3/8	7/16
5/16	.3125	9	14	19	22	1/2	9/16
3/8	.375	15	25	34	37	9/16	5/8
7/16	.4375	24	40	55	60	5/8	3/4
1/2	.500	37	60	85	92	3/4	13/16
9/16	.5625	53	88	120	132	7/8	7/8
5/8	.625	74	120	167	180	15/16	1
3/4	.750	120	200	280	296	1-1/8	1-1/8
7/8	.875	190	302	440	473	1-5/16	1-5/16
1	1.000	282	466	660	714	1-1/2	1-1/2

FIGURE 1-42

(Courtesy of Dresser Industries, Inc.).

METRIC STANDARD

GRADE OF BOLT	5D	8G	10K	12K		
MINIMUM TENSILE STRENGTH	71,160 P.S.I.	113,800 P.S.I.	142,000 P.S.I.	170,674 P.S.I.		
GRADE MARKING ON HEAD →	5D	8G	10K	12K	SOCKET OR WRENCH SIZE	
METRIC		TORQUE			METRIC	
BOLT DIAMETER	U.S. DEC EQUIV.	(IN FOOT POUNDS)			BOLT HEAD	NUT
6mm	2362	5	6	8	10	10mm	10mm
8mm	3150	10	16	22	27	14mm	14mm
10mm	3937	19	31	40	49	17mm	17mm
12mm	4720	34	54	70	86	19mm	19mm
14mm	5512	55	89	117	137	22mm	22mm
16mm	6299	83	132	175	208	24mm	24mm
18mm	709	111	182	236	283	27mm	27mm
22mm	8661	182	284	394	464	32mm	32mm
24mm	945	261	419	570	689	36mm	36mm

FIGURE 1-43

(Courtesy of Dresser Industries, Inc.).

To compensate for these surface variations, this simple formula may be used to approximate the required torque:

$$T = FDC \div 12$$

T = Torque in ft-lb
F = Friction Factor (torque coefficient)
D = Cap Screw diameter in in.
C = Cap Screw tension required in lb.

Friction Factor (F) For Various Surfaces		Percentage Of Torque Change Required
Dry-unplated steel	.20	Use standard torque value shown
Cadmium plating	.15	Reduce standard torque 25%
Zinc plating	.21	Increase standard torque 10%
Aluminum	.15	Reduce standard torque 25%
Stainless Steel	.30	Increase torque 50%
Supertanium		
Special alloy steel	.20	

FIGURE 1-44

Friction Factors (F) and Torque Reductions for Lubricated Surfaces on Alloy Steel Cap Screws

Common Lubricants	Friction Factor	Percentage Of Torque Reduction Required
Collodial copper (Fel-pro C-5)	.11	Reduce standard torque 45%
Never-seize	.11	Reduce standard torque 45%
Grease	.12	Reduce standard torque 40%
Moly-cote (molybdenum disulphite)	.12	Reduce standard torque 40%
Heavy oils	.12	Reduce standard torque 40%
Graphite	.14	Reduce standard torque 30%
White lead	.15	Reduce standard torque 25%

FIGURE 1-45

plated conditions. However, surface variations such as thread roughness, scale, paint, oil, grease, and plating may alter these values considerably (Figure 1-44). Heavy grease, oil, and special thread lubricants may decrease required torque by almost 50% (Figures 1-45 and 1-46).

The Torque-Turn Method of Tightening

Instead of a torque wrench, some vehicle manufacturers recommend using the torque-turn method of tightening. *NOTE: This method only applies to Unified National Course (UNC) bolts with nuts.* Here is the procedure.

1. If the bolt is shorter than eight times its diameter, tighten until the assembly is ''snug.'' Hold the bolt from turning and tighten the nut an additional 1/2 turn.

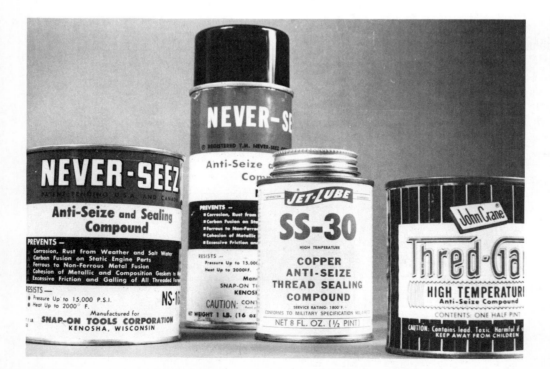

FIGURE 1-46
Antiseize compound prevents galling of threaded fasteners.

FIGURE 1-47

Closeup of a fatigue break. A crack formed in the upper right region and progressed towards the lower left (Courtesy of Bowman Distribution, Division of Barnes Group Inc.).

FIGURE 1-48

This is where fatigue breaks typically occur (Courtesy of Premier Industrial Corporation).

FIGURE 1-49

Notice the off-center pattern in this shear break (Courtesy of Premier Industrial Corporation).

2. If a bolt is longer than eight times its diameter, proceed as above, but instead tighten the nut an additional 2/3 turn.

This method of bolt tightening is based on thread geometry, and is considered to be quite accurate.

Fastener Failure Analysis

Over 75% of all fastener failures are caused by fatigue. As previously mentioned, fatigue breaks can be caused by the bolt not being tightened enough. A back-and-forth bending movement (cyclic stressing) of the bolt then results. Tiny cracks form at the stress points. Additional small cracks start as the bending movement continues. The fatiguing process continues across the bolt until it can no longer support the load. Then the bolt pulls apart, that is, breaks in tension. This is what causes the darkened area usually present in fatigue breaks (Figure 1-47).

Fatigue breaks most often happen at the fillet, at the thread runout point, or at the inner face of the nut on the assembly (Figure 1-48). The break usually happens across one thread and is flat and smooth except for the tension section.

Shear or twist break patterns are circular, and usually off-center (Figure 1-49), and the break will show no dark area. Shear or twist breaks occur for several basic reasons.

1. Improper thread fit (too much friction).
2. The bolt is not strong enough for the particular application.

Tension breaks are characterized by jagged lines in the break area (Figure 1-50). The breakage is caused by using a bolt that cannot carry the work load.

FIGURE 1-50

Jagged lines in the break area indicate a tension failure (Courtesy of Premier Industrial Corporation).

Rolled Threads Versus Cut Threads

Cut threads are made by lathe cutting, or by using a thread cutting die. The thread is formed by severing the metal to produce the grooves. The cut threads grain flow does not offer high resistance to shearing forces (Figure 1-51a).

Rolled threads are formed by rolling the bolt blank between a stationary die and a reciprocating die. Tremendous pressure is put on these dies to squeeze the metal into threads. The grain pattern in rolled threads offers a high resistance to shear (Figure 1-51b).

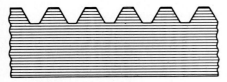

CUT THREADS

(a)

FIGURE 1-51a

Cut threads (Courtesy of Bowman Distribution, Division of Barnes Group Inc.).

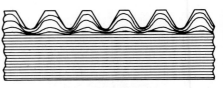

ROLLED THREADS

(b)

FIGURE 1-51b

Rolled threads (Courtesy of Bowman Distribution, Division of Barnes Group Inc.).

TRUE CASE HISTORY

Vehicle	Plymouth Roadrunner
Engine	8 cylinder, 383 CID (cubic inch displacement)
Mileage	Just a few miles on a new clutch assembly installed by the car owner.
Problem	While road testing the car, the clutch assembly exploded (Figure 1-52).

In the explosion, the back of the block had chunks of metal torn off and several holes were ripped in the floor pan. The owner suffered lacerations on his right leg that required stitches.

FIGURE 1-52
When a clutch explodes, metal fragments can hit the driver.

Cause	In the process of doing the clutch job, the clutch cover bolts were lost. The car owner went to the local hardware store and purchased some "no-mark" bolts that were threaded to the head. He did not realize that the lost bolts were shouldered "6-line" (grade 8). The purchased bolts were the wrong design and lacked the required strength.

TRUE CASE HISTORY

Vehicle	Ford Truck
Engine	8 cylinder, 390 CID
Mileage	Unknown. Truck was brought into the shop for a tune-up.
Problem	While standing in front of the truck, the tune-up mechanic was revving up the engine. All of a sudden, the truck lunged forward and pinned the mechanic against a workbench in his service stall. He suffered a crushed femur (thigh bone).
Cause	The engine mounts were torn (separated). This allowed the engine to change position when revved up. As a result, the transmission shift linkage moved and dropped into gear.
Comment	Never stand in front of a vehicle while revving the engine. Instead, block the wheels and stand off to the side out of line with the fan. Determine the condition of the engine mounts before beginning.

CHAPTER 1 SELF-TEST

1. What causes 90% of all accidents?
2. List seven most common factors of worker error.
3. Why can long hair be dangerous?
4. What should be done to prevent long hair from getting into your eyes and work?
5. What is the best practice to follow in regard to wearing a ring when working around electrical circuits?
6. What protective equipment is necessary when grinding?
7. What is a polariscope?
8. How should your back be held to keep from using the back muscles when lifting?
9. Is there a first-aid kit located in your shop? If so, where?
10. What danger exists when a highly flammable liquid is transferred from one metal container to another?
11. Why is it safer to wash dirty parts in cleaning solvent than gasoline?
12. Explain spontaneous combustion.
13. Why is it dangerous to test a battery by shorting across the posts with a pair of pliers?
14. State two reasons for disconnecting the battery before starting to work on the engine.
15. When removing a battery, why should the ground cable be removed first?
16. Before connecting or disconnecting the battery from the charger, always turn the switch _____ .
17. Why is it dangerous to blow off the top of a battery with compressed air?
18. Outline the procedure for "jump" starting a disabled car.
19. How should oxygen and acetylene cylinders be stored?
20. What is considered a Class A fire? Class B fire? Class C fire? Class D fire?
21. Are there fire extinguishers located in your shop?
22. Anytime you see spilled oil or grease present, what should you do?
23. Why should you get into the habit of picking up any article lying on the floor if it does not belong there?
24. What is the best neutralizer for spilled or splashed battery acid?
25. What are some of the early warning symptoms of carbon monoxide poisoning? Under what condition does carbon monoxide form?
26. Why is compressed air dangerous?
27. An electrically operated tool must be properly grounded or U.L._____ .
28. Why should hand tools be kept clean?
29. Why can pushing on a wrench be dangerous?
30. What is the disadvantage of an adjustable wrench?
31. How should an adjustable wrench be placed on a nut in regard to the pulling force?

32. When additional leverage is needed to loosen a nut, why is it dangerous to add a pipe extension to a breaker bar?
33. Why should a wrench with cracked or worn jaws not be used?
34. What can happen if a pair of pliers is used to remove a nut or bolt?
35. What is the purpose of the wedge in the hammer?
36. Why is it dangerous to use one hammer to strike another hammer?
37. Why should a mushroomed head be ground off a chisel?
38. Why is it so important to select the correct size screwdriver?
39. What danger exists when using a badly worn screwdriver blade?
40. Why is it dangerous to use a file without a handle?
41. How should files be stored?
42. The teeth of a hacksaw blade should face _____ from the handle.
43. What often results when a hacksaw is worked too fast?
44. Where should you stand when starting the grinder?
45. The tool rest on the grinder should be no more than _____ " away from the grinding wheel.
46. Why is it dangerous to use a hand socket on an air impact wrench?
47. What can happen if an electrical cord is allowed to get wet? Become frayed?
48. Why isn't it a good idea to leave several top drawers open at one time on a rollaway tool chest unit?
49. What precaution must be taken when you work under a car held up only by a floor jack?
50. List three things that can happen if a unitized frame vehicle is raised incorrectly with a hydraulic lift.
51. What should be done to the vehicle just as soon as it is raised off the ground with a hydraulic lift?
52. When working with a hydraulic press, always wear _____ .
53. Why are you cautioned not to stand directly in line with a running fan?
54. What can happen if an excessive amount of starting spray is used?
55. Why is an improperly tuned engine a fire hazard?
56. What should you do if air conditioning refrigerant gets into your eyes?
57. How can the quality of a bolt be easily identified?
58. Define torque as related to a bolt.
59. What is the difference between tensile strength and shear strength?
60. What is meant by the yield point of a bolt?
61. Explain how a bolt fatigue failure can be caused by insufficient tightening.
62. When installing a flat washer under a bolt head, what is the correct position?
63. The torque-turn method of tightening only applies to what type of thread?
64. What causes most fastener failures?
65. Describe the difference between rolled threads and cut threads.

CHAPTER 2

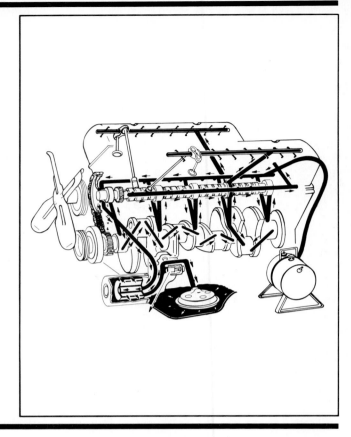

TOOLS
AND
EQUIPMENT

In the repair or rebuilding of today's engines, the mechanic generally is expected to perform the work in the shortest time and in a quality manner. This requires many different tools and pieces of equipment, all requiring knowledge on how to use them safely and efficiently.

A mechanic who is required to perform engine work should be familiar with all the common tools and equipment necessary to do the job. The classifications of these tools and equipment are:

1. General hand tools.
2. Special hand tools and fixtures.
3. Measuring instruments.
4. Testing devices.
5. Machining and power equipment.
6. Cleaning equipment.

NOTE: Much material in this chapter is elaborated on in later chapters. For this reason, it is very im- *portant to develop a basic understanding of each tool or piece of equipment presented here.*

GENERAL HAND TOOLS

An experienced mechanic is expected to know the general hand tools common to the trade— hand driven socket wrenches, box wrenches, end wrenches, screwdrivers, hammers, pliers, and a multitude of other devices classified as general hand tools. If you need a review, consult an introductory auto mechanic text.

SPECIAL HAND TOOLS
AND FIXTURES

While many engine jobs can be done using only general hand tools, some operations require certain special hand tools and fixtures. These special items are designed to ensure greater assembly accuracy and/or to reduce labor time. A discussion of certain special hand tools and fixtures follows.

Bearing Roll-Out Pin

There are times when the removal and replacement of main bearings is attempted while the engine remains in place in the vehicle chassis. With the crankshaft in place, the mechanic may find it difficult to get at the upper main bearing. The roll-out pin is a very simple and effective tool used to solve this problem.

The pins, generally sold in package assortments of several sizes (Figure 2-1), are made of a strong aluminum alloy material. To use the roll-out pin, place the pin in the crankshaft oil hole with the head laying as flat as possible on the journal (Figure 2-2). The edge of the head should butt up against the parting face of the old upper main bearing. By rotating the crankshaft in the correct direction, you can roll out the bearing. The new bearing can be installed by using the roll-out pin to push on the side that has the locking lip (Figure 2-3).

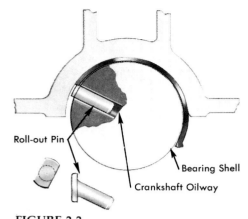

FIGURE 2-2
How the roll-out pin works (Courtesy of Federal-Mogul Corporation).

FIGURE 2-3
Rolling a new upper main bearing into place (Courtesy of Federal-Mogul Corporation).

FIGURE 2-1
Upper main bearing roll-out pins.

The roll-out pin helps eliminate problems that may result if other methods of removal are tried. Chiseling or pounding with a screwdriver is not recommended. This can damage the bearing seat in the saddle and the crankshaft surface finish. Do not try to substitute a cotter pin or screw for a roll-out pin. In one case where a sheet metal screw was being used for a roll-out pin, the head broke off. The rest of the screw was left jammed tight in the oil hole. The mechanic had to remove the crankshaft in order to get the piece out.

Piston Knurler

Piston skirts are designed to fit into the cylinder with a certain amount of clearance. When rings are replaced in an engine with little cylinder wear, the walls can be prepared by glaze breaking or flexible honing, and the old pistons used. When rings are replaced in worn cylinders that are tapered (up to .012"), the bore needs to be made parallel using a rigid hone. This truing up process will make the bore oversize. In order to maintain the correct skirt clearance, the old pistons can be expanded by knurling and reused. This procedure is generally done when the customer cannot be sold on a rebore job.

Knurling is a process of increasing the piston skirt diameter by "upsetting" the metal (Figure 2-4). When the knurler tool roller is brought into contact with the piston skirt, a controlled pressure is applied (Figure 2-5). This causes the teeth on the roller to displace the metal, thus increasing the piston diameter.

Ridge Reamers

If an engine has run for considerable mileage, a ridge will be formed at the upper end of ring travel (Figure 2-6a). It is especially important to remove this ridge if you are going to remove a piston and plan to reinstall it. Failure to cut the ridge away can cause the rings to hang up and catch during piston removal (Figure 2-6b). This can break the ring lands. Also, if a new square-cornered ring is installed in the top groove, and the cylinder ridge is not removed, the ring can hit the ridge (Figure 2-6c). This will cause a "clicking" noise. In some cases, the second ring land will bend and lock the second ring in its groove.

There are several types of ridge cutters on the market. The *lathe type* ridge reamer (Figure 2-7) positions on the inside of the cylinder wall and cuts from the bottom up. It is a very versatile ridge reamer (will accommodate slant bore blocks). The lathe action causes the cutter to closely follow the cylinder wall contour with no possibility of undercutting. The disadvantage is slowness of operation.

The *flexible blade type* ridge reamer (Figure 2-8) is extremely fast to operate. The cutter removes the ridge with a scraping action as it is turned. This type of ridge reamer requires extreme care to avoid undercutting and gouging.

When using any ridge reamer, observe the following tips for better results.

1. Use a spray lubricant on the cutter.
2. As the tool is turned by hand with a wrench

FIGURE 2-4
A knurled piston (Courtesy of Dana Corporation—Service Parts Group).

KNURLING ROLLER

FIGURE 2-5
One type of piston knurler.

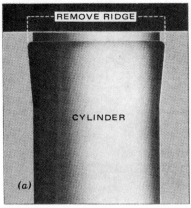
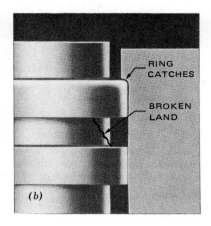
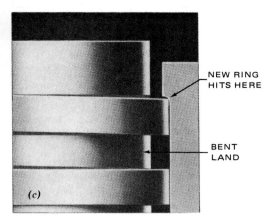

FIGURE 2-6

(a) Ridge should be cut (as shown by the dotted lines) before the piston is removed.
(b) Failure to remove the ridge can cause broken rings or lands when pushing out the piston.
(c) A new ring has a square corner. Failure to remove the ridge will cause noise and possible damage (Courtesy of Imperial Clevite, Inc., Engine Parts Division).

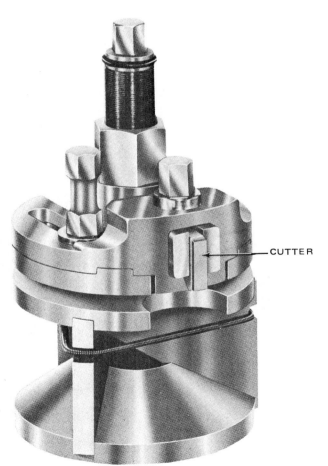

FIGURE 2-7
Lathe type ridge reamer (Courtesy of Lisle Corporation).

FIGURE 2-8
Flexible blade type ridge reamer.

or ratchet, "lead" the cutter by a few degrees (Figure 2-9) to minimize chatter and gouging.

3. Be careful to *never* cut more than 1/32″ into the ring travel area. The reamed surface should blend smoothly into the cylinder wall.

4. *Stop when all the ridge is just cut.* Check that there are no cuttings in the cylinder before removing the piston.

FIGURE 2-9

Always "lead" the cutter when using a ridge reamer. This ridge reamer is being used wrong. The wrench should be repositioned 180°.

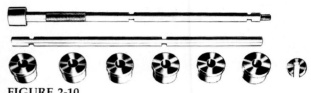

FIGURE 2-10

Camshaft bearing driver with solid driving plugs.

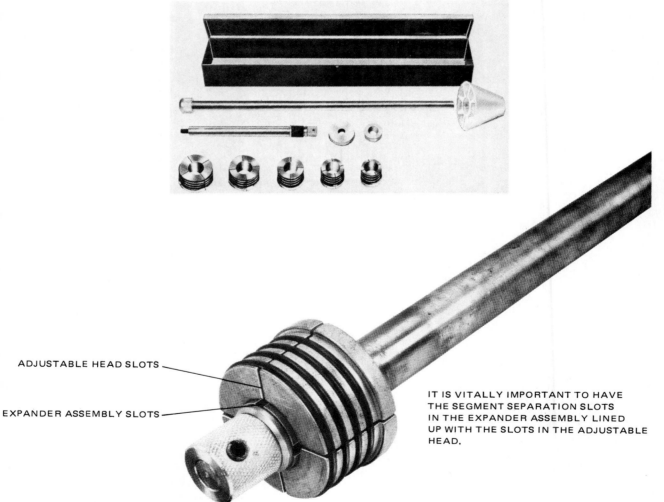

ADJUSTABLE HEAD SLOTS

EXPANDER ASSEMBLY SLOTS

IT IS VITALLY IMPORTANT TO HAVE THE SEGMENT SEPARATION SLOTS IN THE EXPANDER ASSEMBLY LINED UP WITH THE SLOTS IN THE ADJUSTABLE HEAD.

FIGURE 2-11

Camshaft bearing driver with split (expandable) driving plugs.

Camshaft Bearing Tool

Camshaft bearings are generally press-fitted into housing bores in the block. The bearings are installed and removed with a tool (Figures 2-10 and 2-11).

Ring Groove Cleaner

A ring groove cleaner tool often is used to get the carbon out of the ring grooves (Figure 2-12).

Select the correct cutter blade. Standard ring groove width sizes are 5/64", 3/32", 1/8", 5/32", and 3/16". Some model engines use a 1/16" wide groove. *Be careful not to make the groove wider or*

deeper. Make sure the oil ring holes or slots are clean.

Piston Ring Expander

Piston ring expander tools (Figure 2-13) are designed to expand the ring so it can be slipped over the piston into the groove. A ring expander tool can also be used to remove the rings without breaking them.

Never hand spiral on compression rings. This can cause the ring to take a "set" and bind in the groove (Figure 2-14).

Hand Scrapers

Hand scrapers (Figure 2-15) are used to remove carbon, old gasket material, adhesives, sludge, and other hard-to-clean substances from blocks, heads, and so on.

Piston Ring Compressor

When you install the piston assembly (Figures 2-16 and 2-17), compress the rings into their grooves to prevent them from hanging up on the top of the cylinder bore. Never try to install a piston assembly without a good ring compres-

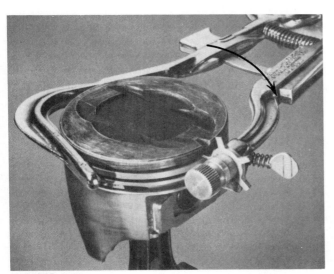

FIGURE 2-12
Using a ring groove cleaner (Courtesy of Ford Parts and Service Division).

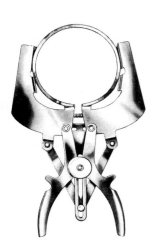

FIGURE 2-13
Piston ring expander.

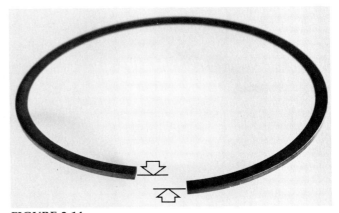

FIGURE 2-14
This hand installed compression ring will act like a lock washer in the piston groove.

FIGURE 2-15
A flexible and a rigid type scraper.

sor. Many ring jobs are ruined because of a bad compressor. To use a ring compressor, follow these recommendations.

FIGURE 2-16
Ratchet type piston ring compressor.

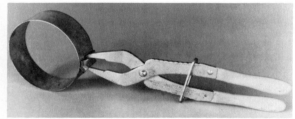

FIGURE 2-17
Clamp type piston ring compressor.

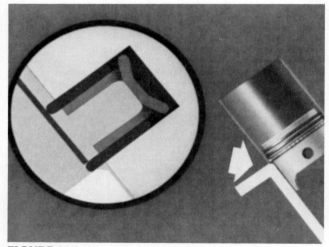

FIGURE 2-18
Here is an example of what can result from a poorly operating ring compressor tool (Courtesy of Ford Parts and Service Division).

1. *Dip the piston head in oil,* covering all the rings. Use the oil that will be used during the break-in period.
2. Tighten the compressor tool around the rings. *Make sure it is square and tight.* When the rings are not fully compressed, the oil control ring rails can catch on the cylinder bore edge and become distorted (Figure 2-18). Also, the compression rings can hit this same bore edge and ruin the scraping edge (Figure 2-19).
3. Place the piston assembly on top of the cylinder with the piston skirt in the bore. *Make sure the edge of the compressor is flush on top of the cylinder.*
4. Gently tap the piston through the compressor with a hammer handle. *Do not force the piston into the bore.*

NOTE: Sometimes, incorrect chamfer width at the top of the cylinder bore will cause ring installation problems. This chamfer should be machined at 45° and about 1/16" wide. Otherwise, the piston rings may be damaged as they enter the bore.

Connecting Rod Boots

Protective rubber or plastic boots, slipped over the rod bolts, will prevent nicking the crankshaft throw by the threads as the rod is removed or installed (Figure 2-20).

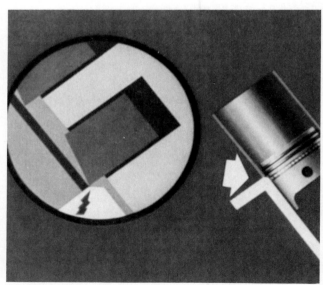

FIGURE 2-19
If compression rings cannot enter the bore properly, the scraping edge can be knocked off (Courtesy of Ford Parts and Service Division).

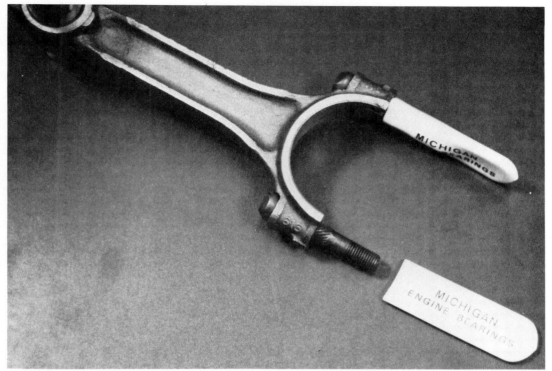

FIGURE 2-20
Rod bolt boots will protect the crankshaft.

Valve Spring Compressor

Removal or replacement of valves once the cylinder head is off the engine is best accomplished

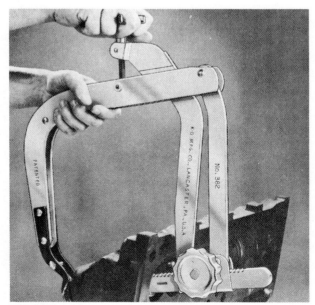

FIGURE 2-21
Valve spring compressor (Courtesy of K-D Tools).

by using a "C" type spring compressor (Figure 2-21). Here is the procedure.

1. Adjust the jaw width of the compressor to snugly fit the valve spring retainer (Figure 2-22).
2. Before compressing the valve spring, take a plastic faced hammer and strike the retainer several times. Apply straight downward

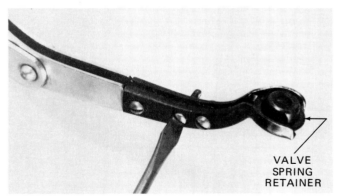

FIGURE 2-22
Adjusting compressor jaw.

blows around the retainer close to the valve stem. This procedure will help release the retainer from the keepers (retainer locks), and allow the spring to be compressed.

3. Compress the spring just enough to remove the keepers (Figure 2-23). Then relieve the compression and remove the retainer and spring.

VALVE SPRING COMPRESSOR

KEEPERS

FIGURE 2-23
Compressing valve spring.

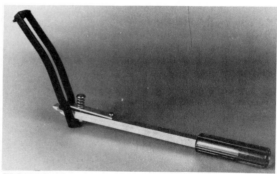

FIGURE 2-24
Flywheel turning tool.

FIGURE 2-25
Parts cleaning brush.

COMMENT: *Occasionally, the retainer and keepers will still not release and allow the spring to be compressed.*
Here are several suggestions.

1. Place the cylinder head face down on a board. Select a socket to fit the outer edge of the retainer. Spray a gum and varnish remover on the valve stem keeper area. Hit the socket with a soft-faced hammer. Some mechanics try to get the keepers to "fly" out by striking heavy blows to the socket with a ball-peen hammer. This is poor shop practice, and can crack valve guides.

2. While the compressor pushes on the spring, take a small brass punch and strike a blow near the center of the retainer.

Flywheel Turner

The flywheel turner (Figure 2-24) is engineered to grip and turn the flywheel so the crankshaft can be rotated to different positions.

Parts Cleaning Brushes

A typical parts cleaning brush (Figure 2-25) has a long handle and tapered nylon bristles that easily fit into tight areas. Figure 2-26 shows a rifle cleaning brush. It can be used for cleaning crankshaft oil holes and blocked oil galleries (passages).

Valve Guide Cleaners

It is an absolute must for valve guides to be clean before measurements can be taken. Varnish and carbon buildup must be removed. Several types of commercially manufactured valve guide cleaners (Figure 2-27) are available in the common guide sizes of 5/16", 11/32", and 3/8".

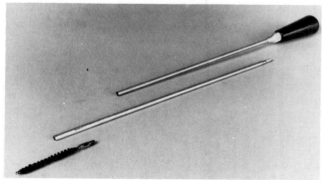

FIGURE 2-26
A rifle brush is excellent for cleaning small oil holes.

Valve Guide Drivers

Insert type valve guides can be removed by using a correctly sized guide driver (Figure 2-28). Select a driver that just fits the guide bore inside diameter (ID), making sure that the driving step is not wider than the guide outside diameter (OD).

Harmonic Balancer Puller

Most crankshaft harmonic balancers can be removed by using a puller as shown in Figure 2-29. This puller is equipped with a "duck's foot" that is attached to the balancer with two or three bolts. Removal of the balancer is accomplished by tightening the jackscrew against the end of the crankshaft, which then pulls the balancer straight off. *CAUTION: Use a step plate on the*

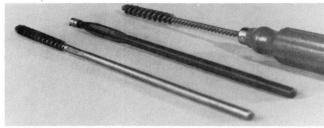

FIGURE 2-27
Three types of valve guide cleaners.

FIGURE 2-28
Valve guide driver set (Courtesy of K-D Tools).

FIGURE 2-29
Pulling off a harmonic balancer (Courtesy of Ford Parts and Service Division).

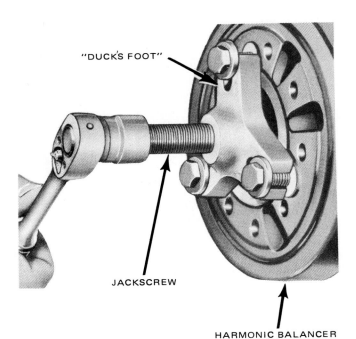

end of the jackscrew (Figure 2-30) to protect the threads in the end of the crankshaft snout.

Rear Main Seal Leak Repair Kit

This repair kit (Figure 2-31) uses a push rod device to drive a pointed steel spacer pin between the upper rear main seal and block. This procedure tightens the worn seal and stops the oil leak. The repair is best suited to engines that use a wick type seal, and in cases where the customer wants to spend less money to stop the leak.

Rear Main Seal Replacement Tool Kit

This replacement tool kit (Figure 2-32) removes and installs upper rear main seals (wick and lip type) without dropping the crankshaft. A cork-screw puller or special driver is used to remove the old seal. A pulling wire grips the wick type seal as it is pulled into position. The lip type seal is installed with the special driver.

Lifter Removal Tools

Sometimes valve lifters are hard to remove from their bores. Varnish film deposits form on the bottom part of the lifter that does not slide in the bore. The bottom diameter of the lifter then becomes larger than the lifter bore. A number of special tools are available to help remove stuck lifters (Figure 2-33). In extremely difficult cases it may first be necessary to soften or dissolve the varnish film with chemicals. Carburetor cleaner, lacquer thinner, or special chemical products (Figure 2-34) can be squirted around the lifter.

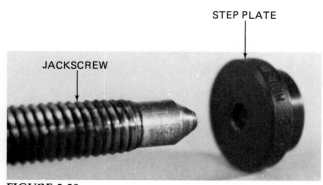

FIGURE 2-30
A step plate will protect the crankshaft snout threads.

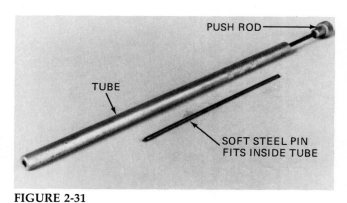

FIGURE 2-31
This tool can be used to stop rear main oil leaks without dropping the crankshaft.

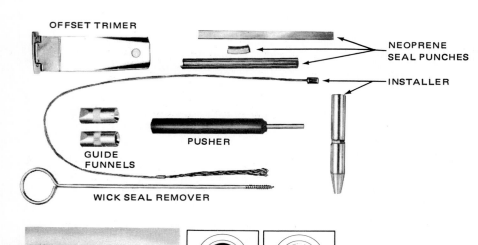

FIGURE 2-32

Rear main oil seal installer/remover kit (Courtesy of K-D Tools).

Once the varnish film has softened or dissolved, the lifter can be removed.

If a lifter still refuses to come out, it may be due to "mushroom" wear on the edge of the lifter bottom (Figure 2-35). *If a "mushroom" worn lifter is pulled out from the top, the lifter bore will be severely scored.* This author knows of a shop where a me-chanic who was working on a Buick with a 400 CID engine became aggravated trying to remove the lifters and decided to use a pry bar against the bottom of one lifter. A large section of the lifter bore broke off and ruined the block.

If the lifters are "mushroomed" from wear, it will be necessary to remove the camshaft and pull the lifters out from the bottom.

Rocker Arm Oil Stoppers

On engines with oil-fed push rods, to adjust the valves with the engine running can be messy. Squirting hot oil can splash on the exhaust manifold. Figure 2-36 shows a manufactured tool designed to eliminate this problem.

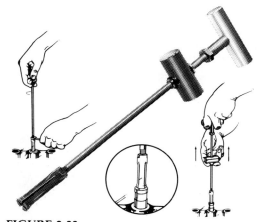

FIGURE 2-33

A tool for removing stuck lifters (Courtesy of Lisle Corporation).

FIGURE 2-34

This chemical product will remove varnish film on the outside of a lifter.

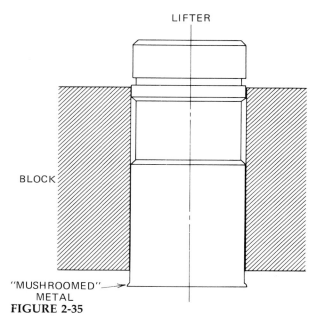

FIGURE 2-35

"Mushroom" lifter wear.

FIGURE 2-36

This type of rocker stopper deflects the oil while adjusting valves.

Cylinder Head Stands

When working on a cylinder head that has been removed from the engine, securely support the head. Figure 2-37 shows a popular pair of support stands.

Piston Vise

A piston vise (Figure 2-38) can be a very useful holding tool when machining pistons for groove inserts, when cleaning ring grooves, for making measurements, and so on.

Cylinder Head Gasket Installation Tool

The special tool shown in Figure 2-39 uses hollow dowel pins to hold and maintain gasket alignment while the cylinder head is being installed.

A special retracting tool then recovers the dowel pins through the head bolt holes.

Engine Rebuilding Stand

Figure 2-40 shows an engine mounted on a rebuilding stand. The engine can be moved easily to any position and safely locked. The engine is completely accessible and all work can be performed while it is on the rebuilding stand. The repaired engine can be test run and checked on the stand if it is properly mounted.

Cylinder Block Boring Stand

These particular boring stands (Figure 2-41) permit boring both banks of cylinders without removing the engine block from the stand. The block can be rolled into several stable positions.

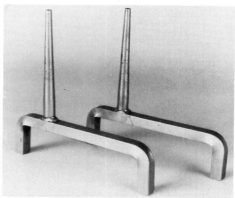

FIGURE 2-37
Cylinder head stands.

FIGURE 2-38
Piston vise.

FIGURE 2-39
Head gasket installation tool.

FIGURE 2-40
Engine rebuilding stand (Courtesy of Kent-Moore
Tool Division).

FIGURE 2-41
Boring stands (Courtesy of Kwik-Way).

MEASURING INSTRUMENTS

Vehicle manufacturers are salvaging more new engine components than ever before. For instance, original equipment manufacturer (OEM) production lines might save a block by machining all the main bearing bores oversize and installing thick wall bearings to compensate. By doing this, an otherwise rejected part can be restored. Some similar situations to know are:

1. One or more of the valve lifter bores is machined oversize. The lifter body is then made oversize to match.
2. One or more replaceable valve guides with an oversize outside diameter (OD).
3. Integral valve guides are machined with an oversize inside diameter (ID) and valves with oversize stems are installed.
4. Cylinder bores are honed slightly oversize and special high limit standard size pistons are installed.
5. One or more of the connecting rod or main bearing journals is ground undersize.
6. Push rods of varied lengths are installed.
7. Cylinder blocks are machined with the deck height lower than the designed specification. A special thick head gasket is used to compensate.

Do not assume when you rebuild an engine that parts are going to be standard. The above situations, though infrequent, do exist. Thorough dimensional checks, exact measurement, and precision fitting are absolutely necessary when rebuilding an engine today.

The more common measurement tools used in the industry are described next.

Micrometers (English)

Such clearance fits as the piston into the cylinder bore, valve stem into the valve guide, connecting rod onto the crankpin, and the piston pin into the connecting rod eye all require precision measurement. For these, the use of internal or external micrometers is essential. In the automotive trade, micrometers are often called *mikes*.

Mikes that measure the size of holes or bores are called *inside mikes*, while those that indicate the external measurement of a part are *outside mikes*.

Outside mikes are usually built with a measuring range of one inch (Figure 2-42). They are available in a range of sizes—zero to one inch, one inch to two inches, two inches to three inches, and so on. A standard outside mike is designed to be read in thousandths of an inch. One-thousandth of an inch (.001″) is the approximate thickness of the cellophane used to enclose a pack of cigarettes.

To read a mike, multiply the number of vertical divisions visible on the sleeve by 25. Then add the number of divisions on the bevel of the thimble counting from zero to the mark that coincides with the long horizontal line on the sleeve. Try reading the mikes that are illustrated (Figure 2-43a and b).

In some cases, the mechanic may want to make a measurement discrimination within .0001″. For this, the mike must have a *vernier scale* (Figure 2-44). The vernier scale lines are all mismatched with the thimble scale lines *except for one*. The number that matches is the ten-thousandths number (the fourth place decimal number). *NOTE: For a specified tolerance within .001″, a vernier mike should be used. For a ± .001″ tolerance, a standard mike can be used.* Try reading the vernier mike illustrated in Figure 2-45.

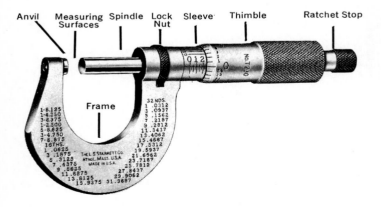

FIGURE 2-42

Outside micrometer (Courtesy of The L. S. Starrett Company).

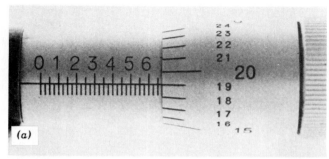

FIGURE 2-43a
(Courtesy of John Wiley and Sons, Inc.).

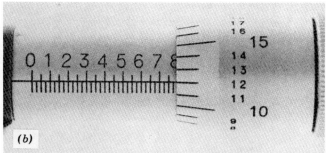

FIGURE 2-43b
(Courtesy of John Wiley and Sons, Inc.).

VERNIER
SCALE

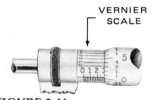

FIGURE 2-44
Vernier scale on a micrometer
(Courtesy of The L. S. Starrett
Company).

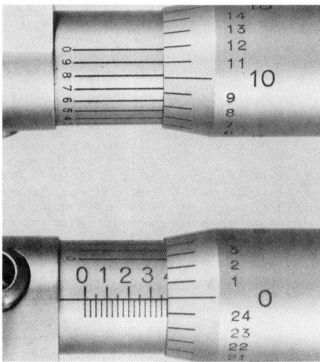

FIGURE 2-45
(Courtesy of John Wiley and Sons, Inc.).

A mike will give reliable results if it is not used beyond its discrimination, if it is properly calibrated, if it is properly cared for, and if the correct sense of "touch" is developed.

To calibrate an outside mike, the instrument is compared to a known *standard* (Figure 2-46). Place the standard between the anvil and spindle. Close the mike and rotate the standard until there is very light pressure on the standard. If the measurement reading you obtain deviates from the standard, then the mike needs adjusting. Some important factors in the care of a mike are:

1. Wipe clean of dust and oil after each use.
2. Do not open or close by holding onto the thimble and spinning the frame around the spindle axis (Figure 2-47).
3. *Never drop the mike.* This can spring the frame and cause misalignment between the anvil and spindle faces.
4. Make sure that the spindle face does not touch the anvil when storing the mike; temperature changes can spring it.
5. Do not touch the anvil and spindle faces with your fingers. Moisture and oils from your skin promote corrosion.
6. Prior to using a mike, clean the measuring surfaces by sliding a piece of paper held between the anvil and spindle faces.

By gripping the frame of an outside mike (Figure 2-48) your thumb and forefinger are free to operate the thimble. Very light pressure of the thumb and forefinger is all that is required when contacting the item to be measured. The correct sense of "touch" or "feel" will come from experience. *NOTE: Some mikes are built with a ratchet stop or*

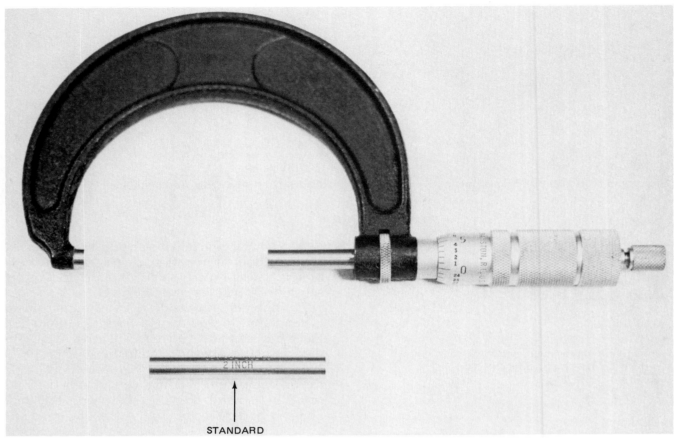

FIGURE 2-46
An outside micrometer and a standard.

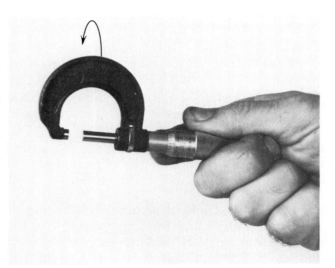

FIGURE 2-47
This is not the proper way to open a mike.

friction thimble that is designed to slip when the mike is turned to the correct tightness (Figure 2-49).

An inside mike is shown in Figure 2-50. Different length extension rods extend the range of the head and allow larger bore measurements. To lessen the chance of errors, it is a good idea to check inside mike readings by measuring across the contact surfaces with an outside mike. It is very easy to make an error when reading directly off an inside mike because of the various length extensions.

Micrometers (Metric)

The metric mike is used and read the same way as the English mike, except that the graduations are metric.

Each number on the metric mike sleeve represents 5 millimeters (mm) (Figure 2-51a). Each

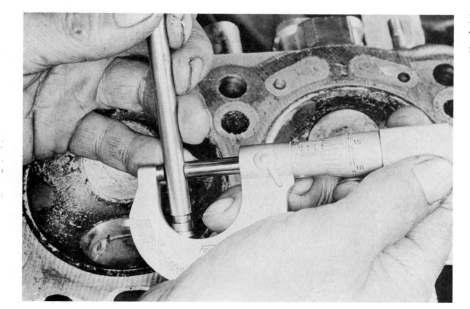

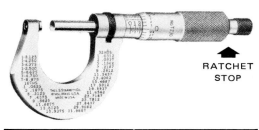

RATCHET STOP

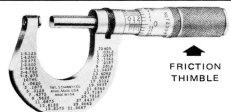

FRICTION THIMBLE

FIGURE 2-49

These mikes ensure consistent gauging pressure (Courtesy of The L. S. Starrett Company).

FIGURE 2-50

Measuring with an inside micrometer (Courtesy of John Wiley and Sons, Inc.).

of the ten equal spaces between each sleeve number represents .5 mm (Figure 2-51*b*). One revolution of the thimble changes the reading one space. The edge of the thimble is marked into 50 equal divisions (Figure 2-52). Every fifth line is numbered. Each thimble graduation represents .01 mm.

As with the English micrometer, add the sleeve and thimble readings together to obtain the total reading. See if you can read the measurement in Figure 2-53.

Dial Indicators

To measure the runout of a flywheel, the end play of a crankshaft, or the lift of a camshaft lobe,

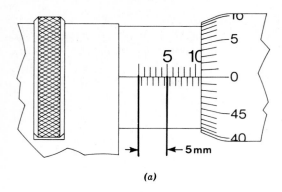

(a)

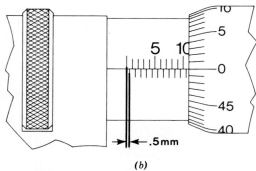

(b)

FIGURE 2-51

(a) Five millimeters (5 mm). (b) Five-tenths of a millimeter (.5 mm) (Courtesy of Imperial Clevite Inc., Engine Parts Division).

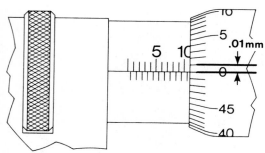

FIGURE 2-52

One-hundredth of a millimeter (.01 mm) (Courtesy of Imperial Clevite Inc., Engine Parts Division).

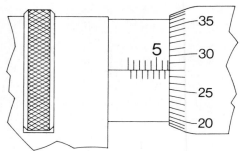

FIGURE 2-53

Micrometer set at 7.28 mm (Courtesy of Imperial Clevite Inc., Engine Parts Division).

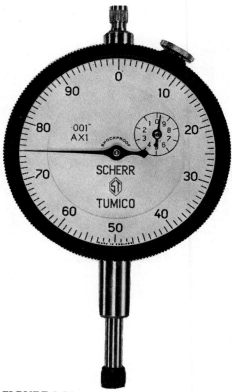

FIGURE 2-54

A travel indicating dial indicator with a continuous dial (Courtesy of Rank Scherr-Tumico, Inc.).

a travelling dial indicator would be the best tool to use. Usually the dial face has discrimination of .001″ or .0005″. Metric travel indicators are also available.

Many indicator faces are marked from zero and read continuously around to zero again (Figure 2-54). The face may have a revolution counter that counts the revolutions made by the sweep hand. Some dial indicator faces read in both di-

rections starting from zero and ending at the six o'clock position (Figure 2-55).

A magnetic base mounting attachement (Figure 2-56) or a clamp type fixture (Figure 2-57) is generally used to mount the dial indicator. *CAUTION: Always mount the dial indicator so that the plunger is at right angles to the object being measured.* Failure to do this can result in a cosine error and an inaccurate reading.

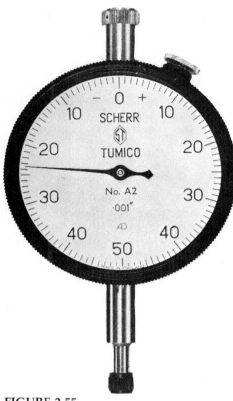

FIGURE 2-55

A travel indicating dial indicator with a balanced dial (Courtesy of Rank Scherr-Tumico, Inc.).

FIGURE 2-56

Magnetic base indicator holder (Courtesy of The L. S. Starrett Company).

FIGURE 2-57

A back plunger dial indicator with various attachment clamps (Courtesy of Rank Scherr-Tumico, Inc.).

Vernier Calipers

A *vernier caliper* is a precision measuring tool for inside, outside, and depth measurements (Figure 2-58). Vernier calipers have a maximum discrimination of .001". The most commonly used metric caliper discriminates to .02 mm. Vernier scales are marked with 25 or 50 divisions. A 50 division vernier presents a less cluttered appearance and is generally easier to read. The 50 division vernier caliper shown in Figure 2-59 is read as follows:

Main scale whole inch reading	1.000"
Additional main scale major divisions	.400"
Additional main scale minor divisions	.050"
Vernier and main scale point of coincidence	.009"
Total reading	1.459"

Dial Caliper

The dial caliper is a popular and accurate measuring instrument (Figure 2-60). It has a discrimination of .001" and can be used for inside, outside, and depth measurements. The dial caliper is easy to read because a *coincident line* does not have to be determined.

Machinist's Rule

Sometimes, certain engine dimensions are expressed in inches and fractions of inches (valve spring free length and valve head diameter, for example). A fractional inch machinist's rule often is used to measure these dimensions. A typical

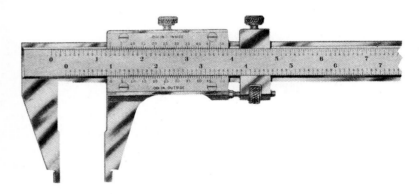

FIGURE 2-58
Vernier calipers (Courtesy of The L. S. Starrett Company).

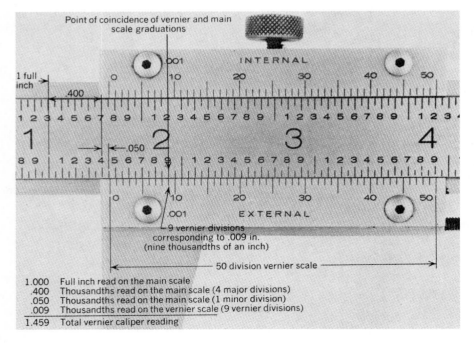

FIGURE 2-59
Reading a vernier caliper (Courtesy of John Wiley and Sons, Inc.).

FIGURE 2-60
Dial calipers (Courtesy of Rank Scherr-Tumico, Inc.).

machinist's rule would have the 1", 1/2", 1/4", 1/8", and 1/16" graduations on the front side (Figure 2-61a). The reverse side has one edge marked in 1/32" increments, and the other edge marked in 1/64" increments (Figure 2-61b). On the 1/32" side, every fourth mark generally is numbered. On the 1/64" side, every eighth mark is numbered. This is done to eliminate the need to count each graduation.

Many engine dimensions are specified in decimals. In some of these cases, a decimal rule could be used (Figure 2-62). Decimal rules are typically marked in increments of 1/10", 1/50" and 1/100". Decimal rules cannot discriminate to the individual thousandth. However, the width of the *etched division on the rule is approximately .003"*.

A typical metric rule has 1 mm and 1/2 mm graduations. Read the dimensions on the rule pictured in Figure 2-63.

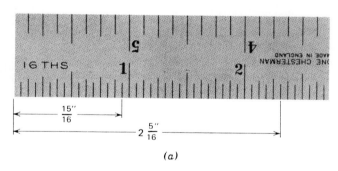

(a)

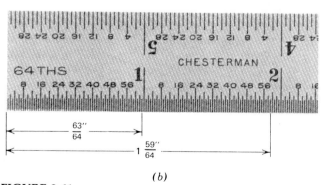

(b)

FIGURE 2-61
Fractional rules (Courtesy of John Wiley and Sons, Inc.).

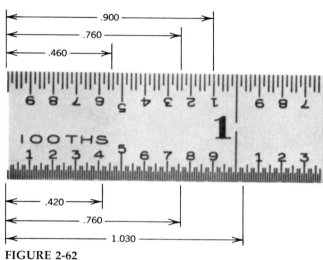

FIGURE 2-62
Decimal rule (Courtesy of John Wiley and Sons, Inc.).

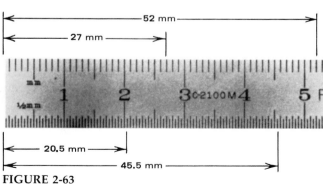

FIGURE 2-63
Metric rule (Courtesy of John Wiley and Sons, Inc.).

Telescoping Gauges

The telescoping gauge (sometimes incorrectly called a snap gauge) is a comparison measuring instrument. Telescoping gauges are widely used in automotive repair shops to measure cylinder bores, main bearing bores, cam bearing clearances, lifter to lifter bore clearances, and installed spring heights.

Telescoping gauges commonly come in a set of six gauges ranging from 5/16" to 6" (Figure 2-64). The gauge consists of two spring-loaded telescoping plungers with a locking handle. A telescoping gauge can be a very reliable tool if the following procedure is used.

1. Select the proper gauge.
2. Place the gauge into the bore being measured.
3. Release the locking handle and tilt the gauge slightly up so that the plungers may expand to a point larger than the bore diameter.
4. Lightly tighten the locking handle and move the gauge in an arc motion through the bore diameter (the plungers will telescope to the bore diameter).
5. Firmly tighten the locking handle.
6. Move the gauge back and forth through the bore (a light drag should be felt).
7. Remove the gauge and measure across the plungers with an outside micrometer (Figure 2-65). *CAUTION: Excessive micrometer pressure will depress the gauge plungers and result in an incorrect reading.*

8. Take two more measurements with the gauge in order to verify the original reading you obtained.

Small-Hole Gauges

The small-hole gauge, like the telescoping gauge, is a comparison measuring instrument often used to help obtain valve stem-to-guide clearance.

Small-hole gauges generally come in a set of four gauges ranging from 1/8" to 1/2". A very common type of small-hole gauge is the split ball type (Figure 2-66). A tapered rod is connected to

FIGURE 2-65
Transfering a telescoping gauge measurement to an outside micrometer.

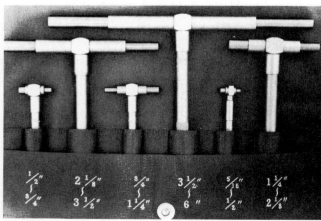

FIGURE 2-64
Set of telescoping gauges (Courtesy of John Wiley and Sons, Inc.).

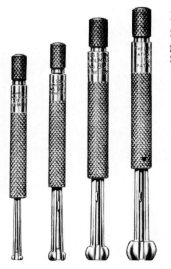

FIGURE 2-66
Set of split ball small-hole gauges (Courtesy of The L. S. Starrett Company).

the handle. When the handle is turned, this tapered rod is drawn between the split ball halves causing them to move outward into contact with the surface being measured. When measuring with the small-hole gauge, move it back and forth until the correct "feel" is obtained. Then remove the gauge and measure across the balls with an outside micrometer.

Radius Gauges

A typical radius gauge set is shown in Figure 2-67. This set ranges in size from 1/32" to 1/2". Radius gauges are useful for measuring the radii of grooves and external or internal fillets (rounded corners). For automotive application, they are used to check the crankshaft fillet radii at the end of the rod and main bearing journals to make sure they are the same size as originally specified.

This check should always be done during the crankshaft grinding process when the wheel corners are dressed. *The radius should never be less than 1/64" below the original radius.* It tends to weaken the crankshaft if smaller.

Conversely, rod and main bearing journal fillet radii should never be greater than specified. Larger than specified fillet radii on the crankshaft can cause bearing interference, that is, radii ride (Figure 2-68). A load concentration on the bearing edge will then occur, causing bearing breakdown.

Connecting Rod Big-End Gauge

If connecting rod bores are out-of-ground, the ID of the assembled rod bearings cannot be round. Oil clearance will be decreased at some points, while increased at others.

A special gauge is available for checking the out-of-roundness of the connecting rod big-end (Figure 2-69).

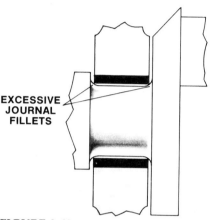

FIGURE 2-68
Fillets that are too large can cause bearing edge wear (Courtesy of Imperial Clevite Inc., Engine Parts Division).

FIGURE 2-69
Gauge for measuring connecting rod big-end concentricity.

FIGURE 2-67
Set of radius gauges.

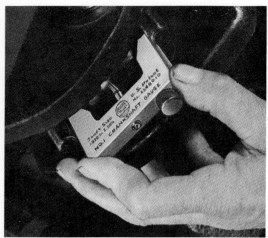

FIGURE 2-70

Positioning the crankshaft gauge against a journal (Courtesy of Federal-Mogul Corporation).

Crankshaft Gauge

The crankshaft gauge measures crankshaft journal diameters without removing the crankshaft from the engine (Figure 2-70). By taking multiple readings, you can determine journal out-of-roundness and taper.

Feeler Gauges

A feeler or thickness gauge (Figure 2-71) is most often used by the engine rebuilding mechanic to measure connecting rod side clearance, to check piston to cylinder fit, and to adjust valves. Frequently it is used with a precision straightedge to check for warpage on the block, cylinder heads, and manifolds.

A good steel feeler pack usually will contain leafs of .002″ to .010″ thickness (in steps of .001″), and leafs of .012″ to .024″ thickness (in steps of .002″).

Torque Indicating Wrenches

Excessive or unequal tightening of studs, nuts, and cap screws in an engine will distort parts. The cylinders, cylinder heads, connecting rod bores, main bearing bores, camshaft bearing bores, intake and exhaust manifolds, timing chain cover, and other engine components can become damaged with excessive or unequal tightening. Excessive tightening can also strain and stretch studs and cap screws to the point where failure will occur soon. It is a known fact that a single 1/2″ bolt may exert a leverage force as high as 16,000 pounds. This is enough to lift 4 or 5 automobiles!

COMMENT: On certain model Ford engines of a few years ago, you had to be very careful about over-tightening the oil filter canister through bolt. If this bolt was overtightened it would distort the block, causing the number 1 cam bearing to spin in its bore. This would block main gallery oil flow and ruin the engine.

There are several different designs of torque wrenches. With the *sensory or click type* torque wrench (Figure 2-72), torque is indicated mechanically when an audible click sound is heard. These torque wrenches are ideal for production line jobs where the same amount of torque is applied many times.

The *dial type* torque wrench indicates the amount of torque applied at any given moment (Figure 2-73). Dial type wrenches are especially

FIGURE 2-71

Thickness gauge set (Courtesy of The L. S. Starrett Company).

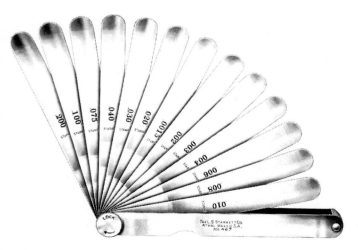

FIGURE 2-72

Sensory or click type torque wrench.

FIGURE 2-73

Dial type torque wrench.

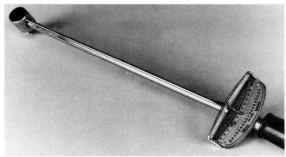

FIGURE 2-74

Beam type torque wrench.

suited to jobs requiring many different torque values because the dial does not have to be reset each time. To measure bearing or shaft preload use a dial type torque wrench.

A *beam type* torque wrench uses a flexible alloy steel beam which deflects when torque is applied (Figure 2-74). A pointer attached to the end of the beam indicates the torque reading on a scale. This type of torque wrench has the advantage of simple design. It does not use a multiplying device or a cam mechanism device that can be easily ruined.

Torque wrench testers are available to determine the accuracy of a torque wrench (Figure 2-75). There are many different sizes of torque wrenches with ranges in inch-ounces, inch-pounds, and foot-pounds. They are available with English, Metric, and newton-meter calibrations. See Figure 2-76.

The general recognized procedure for tightening most nut and bolt sets is as follows:

1. Start at the center of the assembly that is being attached. Draw up each nut or bolt until just snug, working alternately on each side.

2. Using a torque wrench, tighten each nut or bolt to about 1/2 the specified torque. Observe correct tightening sequence.

3. Final tighten to the specified torque observing correct sequence.

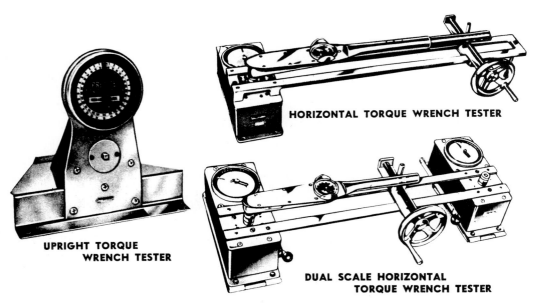

FIGURE 2-75

Torque wrench testers (Courtesy of Snap-on Tools Corporation).

HORIZONTAL TORQUE WRENCH TESTER

UPRIGHT TORQUE
WRENCH TESTER

DUAL SCALE HORIZONTAL
TORQUE WRENCH TESTER

1 meter kilogram (m kg)	= 7.23 foot-pounds
1 meter kilogram (m kg)	= 86.8 inch-pounds
1 centimeter kilogram (cm kg)	= .868 inch-pound
1 foot-pound	= 0.138 meter kilogram (m kg)
1 inch-pound	= 1.15 centimeter kilograms (cm kg)

FIGURE 2-76

NOTE: The newton-meter is a measurement of torque used by many countries using the metric system. One newton-meter is equivalent to .7375 foot-pounds.

Vee Blocks

The use of vee blocks (Figures 2-77 and 2-78) and a dial indicator is the recommended method for checking crankshaft straightness (alignment). *CAUTION: Mike the crankshaft main bearing journals for out-of-roundness prior to checking straightness.* Failure to do so might make you think the crankshaft is bent, when it really is not. With the front and rear main journals in the vee blocks, rotate the crankshaft and first check the center main journal with a dial indicator. Place a strip of paper in each ''v'' to prevent scratching the journals. Check the other main journals. *Journal-*

FIGURE 2-77
Vee blocks being used.

FIGURE 2-78
Vee blocks and dial indicator setup for checking crankshaft straightness.

to-journal alignment should be within .001" tir (total indicated reading), or overall alignment within .002" tir. Crankshaft flange runout, crankshaft snout runout, and a number of camshaft checks also can be made using vee blocks and a dial indicator.

Precision Straightedge

To check for surface flatness on the block deck, the cylinder head(s), the intake manifold, and the exhaust manifold(s), a precision straightedge and feeler gauge are used.

Cylinder head warpage is checked with a precision straightedge and feeler gauge (Figure 2-79).

Main bearing saddle bore alignment is sometimes checked by using a precision straightedge. The straightedge is placed in three alternate but parallel positions in the saddles. Using a feeler gauge that is half the maximum specified oil clearance, try to slip the feeler leaf under the straightedge.

Plastigage

Plastigage (Figure 2-80) is the trademark name of a product made by Perfect Circle. By using Plastigage, main and connecting rod bearing clearances can be quickly checked.

Plastigage consists of a soft, threadlike plastic material that is enclosed in an envelope (Figure 2-81). A thousandths scale is printed on one side of the envelope, and a millimeter scale on the other side.

To measure bearing clearance with Plastigage, proceed as follows:

1. Remove the bearing cap, and *wipe all oil from the crankshaft and bearing shell*. Rotate the crankshaft about 30° from bottom dead center (BDC).

2. Lay a piece of Plastigage across the full width of the bearing shell. Position the Plastigage so it is straight and about 1/4" off center.

3. Install the bearing cap to the torque recommended by the engine manufacturer. The Plastigage is flattened when the bearing cap is tightened. The less bearing clearance there is, the wider the Plastigage becomes.

4. Remove the bearing cap and match the width of the flattened Plastigage with the graduated scale on the envelope (Figure 2-82). The numbers on the scale indicate the bearing

FIGURE 2-79
Checking for cylinder head warpage with a precision straightedge and feeler gauge.

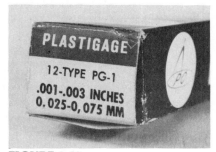

FIGURE 2-80
Plastigage.

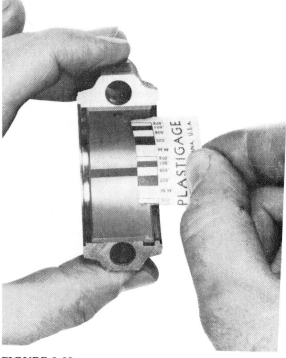

FIGURE 2-82
Measuring bearing clearance with Plastigage.

FIGURE 2-81
Plastigage is soft threadlike material.

clearance. ***NOTE:*** *If the width of the flattened Plastigage varies from end to end, it indicates that the bearing journal is tapered.* Also, examine the extreme ends for any indication of inadequate fillet radii.

Plastigage is available in three different ranges.

PG-1 (Green)   clearance range of from .001" to .003"

PR-1 (Red)   clearance range of from .002" to .006"

PB-1 (Blue)   clearance range of from .004" to .009"

CAUTION: *Observe special procedure when checking main bearing clearance with the engine mounted in the vehicle. An incorrect reading will be obtained unless the weight of the crankshaft and flywheel is relieved. This is accomplished by supporting the weight of the crankshaft with a jack placed under the counterweight adjoining the bearing being checked, or by installing a strip of a business card (1/2" wide) on the main bearing shell on each side of the one you are checking.*

CAUTION: *Exposing Plastigage to temperatures above 100°F, or below 32°F, can result in an erroneous reading.*

Connecting Rod Alignment Fixture

The piston and rod assembly must be in alignment to allow for minimum friction. The piston must move at exact right angles to the crankshaft centerline.

There are several good connecting rod alignment checking fixtures on the market. A popular rod alignment fixture used in many engine rebuilding shops is shown in Figure 2-83. It checks the connecting rods for bend, twist, and offset. The rods can be checked with or without the pistons attached.

Screw Pitch Gauges

There are times when it becomes necessary to determine the pitch (threads per inch) of a threaded fastener.

The screw pitch gauge (Figure 2-84) quickly determines the pitch of various threads. These gauges are housed in a steel case with a number of folding leaves at both ends. Each leaf has teeth corresponding to a certain thread pitch. If you match the teeth with the threads in question, the correct pitch can be read directly from the leaf.

Screw pitch gauges are available in various ranges to gauge American National Course and Fine threads, Whitworth threads, Metric threads, and International Standard threads.

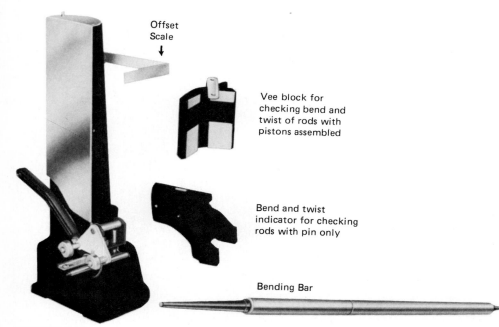

Offset Scale

Vee block for checking bend and twist of rods with pistons assembled

Bend and twist indicator for checking rods with pin only

Bending Bar

FIGURE 2-83
Rod aligner (Courtesy of Sunnen Products Company).

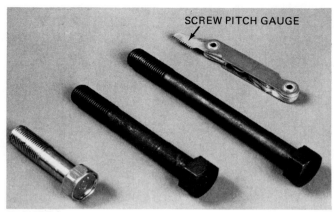

FIGURE 2-84
Screw pitch gauge.

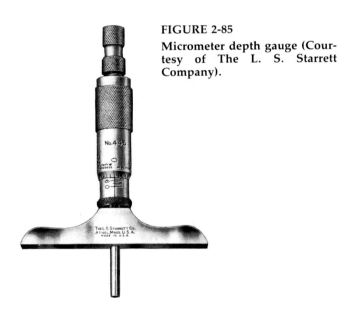

FIGURE 2-85

Micrometer depth gauge (Courtesy of The L. S. Starrett Company).

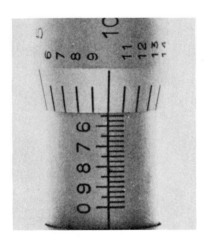

FIGURE 2-86

What is the reading on this depth micrometer? (Courtesy of John Wiley and Sons, Inc.).

Micrometer Depth Gauge

A micrometer depth gauge, as the name implies, is designed to measure the depth of holes, recesses, keyways, and so on. The tool consists of a base with a micrometer head (Figure 2-85). Various length measuring rods can be inserted through a hole in the micrometer screw. The rod protrudes through the base and moves as the thimble is rotated.

The reading is taken exactly the same as with an outside micrometer, *except that sleeve graduations run in the opposite direction and the covered number are the ones to be "counted"*. It is necessary to consider the rod length if using a rod other than the 0" to 1". For example, if the 1" to 2" rod is used, one inch must be added to the sleeve and thimble reading. Read the depth micrometer reading in Figure 2-86.

Cylinder Dial Gauge

The cylinder dial gauge is the ideal tool to quickly measure bore diameter, taper, and out-of-roundness. The gauge shows instantly the condition of the cylinder to one-thousandth of an inch as it is moved to various points on the wall. Various diameter cylinder bores can be measured by means of different length attachment rods.

The cylinder gauge shown in Figure 2-87 has the dial mounted on a block that moves at right angles to a sled. The sled has two straight line contact points which maintain alignment with the wall in the cylinder. A locking mechanism places the plunger in a clamped position before removing the gauge from the cylinders. This allows the cylinder diameter to then be read directly with an outside mike.

See Figure 2-88 for another type of bore gauge.

Dividers

Dividers (Figure 2-89) are used for measuring dimensions between lines or points. They can be used for transfering lengths taken from a rule and for scribing reference marks. At each end of the straight legs are sharp, hardened contact points. Measurements with a pair of dividers are made by visual comparison rather than by feel. Dividers are restricted in range by the span opening of the legs. They become less effective when the points are at a wide open incline to the working surface.

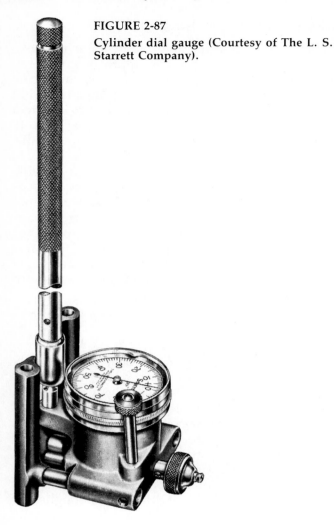

FIGURE 2-87
Cylinder dial gauge (Courtesy of The L. S. Starrett Company).

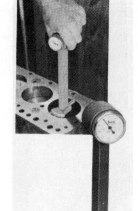

FIGURE 2-88
Cylinder dial gauge. Ridge removal is unnecessary when using this type (Courtesy of Central, Division of K-D Manufacturing Company).

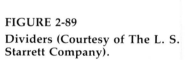

FIGURE 2-89
Dividers (Courtesy of The L. S. Starrett Company).

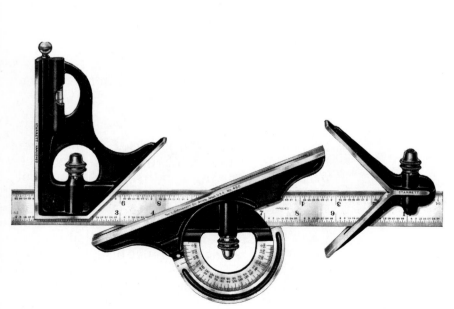

FIGURE 2-90
Combination square set (Courtesy of The L. S. Starrett Company).

Combination Square Set

The combination square set (Figure 2-90) has a wide range of applications and is one of the most versatile tools a mechanic can have. It can be used as a square, height gauge, depth gauge, level, steel rule, plumb, marking gauge, center gauge, scriber, and bevel protractor.

Precision Gauge

No pin fit should ever be made without precise gauging (Figure 2-91). Exact fits are required on today's high-horsepower, high-speed engines. A great deal of the pin fitting and rod reconditioning work done today is gauged with this instrument. The dial reads directly in ten-thousandths of an inch.

The precision gauge is a transfer or comparator type gauge. Any desired diameter within the range can be transfered to the gauge. The gauge then reads in tenths over or under the desired diameter.

Valve Seat Runout (Concentricity) Gauge

The valve seats in a cylinder head or block can be checked for runout by means of a dial gauge (Figure 2-92). In production, the factories grind valve seats not to exceed .001" off center. When doing a valve job, *the mechanic should reject anything over .002" tir.*

To check concentricity of a valve seat, place the gauge on a valve grinding pilot that has been installed in the guide. Rotate the lower section of the indicator keeping the adjustable bar trailing. Read the plus or minus variation on the dial. *CAUTION: When you measure valve seat runout, the seat must be clean before a valid measurement can be obtained.*

Ring Groove Wear Gauge

When new piston rings are installed in excessively worn grooves, the seal formed at the sides is inadequate and excessive "blow-by" and compression loss results. Also, new rings in worn grooves induce further wear, and frequently result in ring flutter and ultimate ring breakage.

The Perfect Circle top groove wear gauge provides a fast and accurate means of determining when ring grooves are worn too much (Figure 2-93). Steps on this gauge are made .006" thicker than the common standard ring widths. These steps are used as go, no-go gauges. If the appro-

FIGURE 2-91
Precision gauge (Courtesy of Sunnen Products Company).

FIGURE 2-92
Valve seat concentricity gauge.

priate step of the wear gauge will enter the ring groove at any point, either the piston should be replaced or groove inserts installed. ***NOTE:*** *A piston groove wears unevenly. Usually the greatest*

FIGURE 2-93
Ring groove wear gauge.

amount of groove wear on the piston will be at the point nearest the exhaust valve. Second grooves should be checked also.

TESTING DEVICES

Certain kinds of engine rebuilding equipment can be classified as special testing devices. There are times when these devices can be very valuable in the engine repair shop.

Engine Prelubricator

The engine prelubricator contains an oil reservoir tank that is attached to an air supply line. The oil supply line is hooked to an oil pressure port

hole on the outside of the block. Oil can then be supplied to the engine oil system at a predetermined and maintained pressure, even though the engine is not operating. The engine prelubricator shown in Figure 2-94 is manufactured in five or ten quart sizes.

The engine prelubricator has two important uses. First, a rebuilt engine can be charged with oil prior to being started. *This is very important to the life of engine bearings.* Many times a mechanic builds an engine, installs the oil pan, and thinks the oil pump will immediately be delivering oil under pressure. The time lag between first starting the engine and delivering oil to the bearings can be, in some cases, a minute or more (especially with oil pumps that do not prime easily).

When using the prelubricator to charge the lubrication system, set the air regulating valve on the tank to obtain an air gauge pressure of 25 psi. Higher pressures than this tend to aerate the oil. Hook up a direct reading oil pressure gauge to the engine, and have the valve cover(s) removed. Fill the prelubricator tank with engine oil. Use enough oil to fill the crankcase to correct level. Turn on the air valve. There should be a reading indicated on the oil pressure gauge, and oil should be dripping from the rocker arm assembly during charging. *NOTE:* *Turn the crankshaft slowly by hand while charging is taking place.* This will make sure that oil hole indexing occurs. When the prelubricator tank is empty (a "spitting" sound will be heard) immediately shut off the air supply.

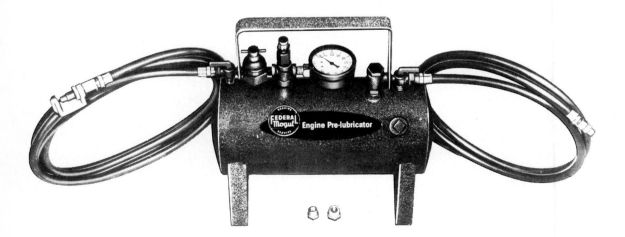

FIGURE 2-94
Engine prelubricator (Courtesy of Federal-Mogul Corporation).

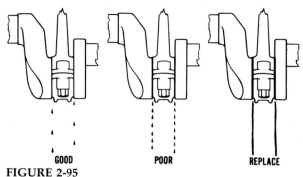

FIGURE 2-95

Oil leakage from connecting rod bearing ends.

Second, the engine prelubricator can be used to check bearing wear and to locate internal oil leaks. With the oil pan removed, oil leakage at the bearings and other points can be observed. The relative oil clearance can be checked by the flow of oil from the bearings. Leakage of 20 to 150 drops per minute indicates a satisfactory condition (Figure 2-95). *This oil flow specification is based on using SAE 20 or 30 oil at normal room temperature with 25 psi air pressure.*

If leakage continues where a time interval can hardly be noted between drops, that bearing should be removed and checked. If no oil leakage is observed, the bearing is fit too tight or an obstruction exists.

Valve Spring Tester

Valve spring tension is checked on a testing fixture. Figure 2-96 shows a spring tester that uses a direct reading scale to read the tension in pounds. This tester is used as follows:

1. Place the spring to be tested on the base plate.
2. Turn the handle until the height pointer is at the testing height, and then set the stop mechanism so that the rest of the springs can be quickly checked. *Never set the stop without using a spring,* because the base plate depresses as the spring is tested. Failure to observe this procedure can cause good springs to fail the test.
3. The tester is now ready for use. Compress the springs to be tested and read the scale.

Figure 2-97 shows a valve spring tester that uses a torque wrench to check tension. The spring testing height is set by turning the platform wheel on the graduated scale. The spring is then placed

FIGURE 2-96

A direct reading type valve spring tester.

FIGURE 2-97

A valve spring tester that uses a beam type torque wrench.

on the platform wheel, and the lever arm is pulled down with a beam type torque wrench. Pull the wrench until a clicking sound is heard from the tester. *Read the torque wrench at this exact moment and multiply the reading by two.* This is the spring tension at the specified height set on the platform wheel.

Ideally, a valve spring should be checked for both seated and open pressure. Correct seated pressure helps the valve maintain a tight seal. Correct spring pressure while the valve is open helps to overcome valve train inertia and prevent valve floating. *Spring pressures are always checked at a specific height.* If the manufacturer happens to provide a seated pressure specification, the checking height will be the same as the installed spring height specification. When the manufacturer provides an open pressure specification, the checking height is usually near the height of the spring at maximum valve opening. *All valve springs should test within 10% of tension specifications.*

Lifter Leakdown Tester

The hydraulic valve lifter is one of the most precision manufactured parts in an engine. The lifter body and plunger are measured electronically to .000033″ (33 millionths of an inch) and graded. When the lifter is assembled, .0002″ to .0003″ clearance is provided between the body ID and the plunger OD. The outside diameter of the lifter body is held to .0005″ tolerance. The lifter foot is contour ground to a slight radius, and hardened. Each lifter is tested for proper leakdown before leaving the factory on a testing machine. *Leakdown*, by definition, is the leakage of oil from the lifter lower chamber through the clearance between the body and plunger during opening and closing of the engine valve. Hydraulic valve lifters that *leak down too slowly* can drastically affect performance and power output since the valves will fail to open and close properly. Lifters that *leak down too fast* cause improper valve operation, resulting in excessive noise.

If possible, a leakdown test should be made on each hydraulic valve lifter before installing it in the engine. Do not assume that a new lifter has proper leakdown. It is better to take a few extra minutes and test the lifters.

The tester pictured in Figure 2-98 can be used to test the leakdown rate of a lifter prior to installation.

If you do not have a leakdown test fixture, an upright drill press can work as a substitute. Place the lifter to be tested in a small can. Cover the lifter with test oil (Figure 2-99). Place a push rod in the drill press chuck. Pull on the drill press handle with a *moderate* amount of force, using the push rod as a ram. The lifter plunger travel can be measured by observing the graduations on the drill press stop adjustment rod. The leakdown time can be noted by checking the second hand on a watch. *As a rule of thumb, most lifters will have a plunger travel of approximately .100″ in a 10 to 60-second time period.*

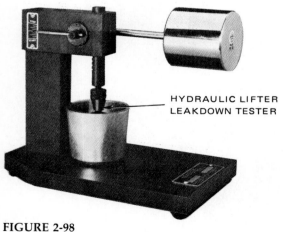

HYDRAULIC LIFTER
LEAKDOWN TESTER

FIGURE 2-98
Hydraulic valve lifter leakdown tester (Courtesy of Ford Parts and Service Division).

FIGURE 2-99
Special oil for testing hydraulic valve lifters.

Should new lifters be primed with engine oil? Aftermarket lifter manufacturers say positively no; use them as received in the box. When hydraulic valve lifters are manufactured, each lifter is tested for correct leakdown. The thin leakdown test oil is left in the lifter. This test oil is quickly displaced by pressurized engine oil when the engine starts, and the lifter plunger adjusts to its correct operating length. If the lifter is primed with engine oil before installation, the leakdown process to the correct operating length can be too slow. Hard starting then occurs, because the engine valves will be held open. Plus, there is the possibilty of damage if a piston hits an open valve.

Crack Testing Devices

Here are the testing device methods of locating cracks in general industry use.

1. The head or block is mounted in a special testing fixture (Figure 2-100). *Air pressure* is used to pressurize the water jackets after all openings have been sealed with rubber plugs. A soap solution is applied to the suspected areas. The formation of bubbles will indicate a crack.

2. *Water pressure* testing a block or head requires a special mounting fixture and sealing plates (Figure 2-101). Hot water is then circulated through the water jackets. Check visually for wetness on metal surfaces.

3. *Magnetic particle inspection* is a method of inspecting ferrous parts (Figure 2-102). A magnetic field is set up on the part being inspected. Fine iron powder is then sprayed on the part. Any cracks will show up as a highly visible gray line.

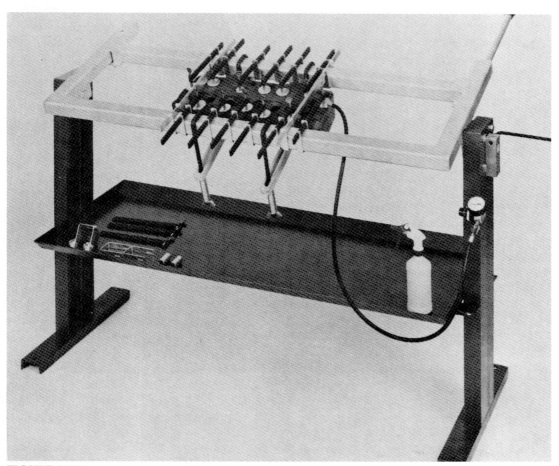

FIGURE 2-100

Using an air pressure tester to check a cylinder head for cracks (Courtesy of Irontite Products Inc.).

FIGURE 2-101
Pressure testing an engine block at normal operating temperature. This test should always be performed after reboring to detect possible porosity due to casting flaws.

FIGURE 2-102
Using a magnet to locate a cylinder head crack (Courtesy of Irontite Products, Inc.).

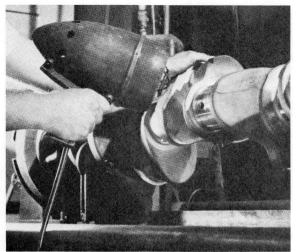

FIGURE 2-103
Inspecting a crankshaft under black light.

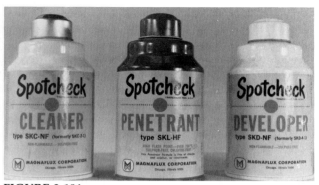

FIGURE 2-104
A dye penetrant crack detection kit.

4. *Magnaglo fluorescent inspection* (Figure 2-103) shows cracks as streaks of white against a dark blue background. Special fluorescent paste particles are mixed with oil to form a spray solution. This solution is then sprayed on the part to be inspected, *while the part is magnetized*. Any cracks will attract and hold the spray solution particles, which will be clearly seen under black light.

5. The *Zyglo* (a trademark of the Magnaflux Corporation) method of inspection uses a sprayed on fluorescent crack penetrant that is viewed under black light. This inspection method can be used on magnetic and non-magnetic materials. A brilliant fluorescent glow will indicate a crack.

6. The *dye penetrant* crack detection method (Figure 2-104) uses a three-step spray solution process. A dark red stain against a white background indicates any cracks. This method will locate cracks in ferrous and nonferrous materials under ordinary light.

MACHINING AND POWER EQUIPMENT

The task of describing and explaining the operation of engine rebuilding machine and power equipment is not a goal of this chapter. The commonly used machining and power equipment items will be covered in other chapters.

CLEANING EQUIPMENT

There are many methods used to clean engines and the related parts. Cleaning machines and equipment used in the trade include:

1. Steam cleaners.
2. Hot spray tanks.
3. Cold soak tanks.
4. Hot soak tanks.
5. Glass bead cleaning machines.
6. Airless shot blasting machines.
7. Pyrolytic ovens.

The general details and the advantages and disadvantages of each cleaning method will be discussed in Chapter 8.

TRUE CASE HISTORY

Vehicle   Ford Mustang

Engine   8 cylinder, 302 CID

Mileage   The engine was removed from the vehicle for a complete rebuild. The rebuilt engine was installed in the vehicle, and started without prelubricating. A direct reading oil pressure gauge was not hooked up.

Problem

1. The oil warning light stayed *on* at idle and low speeds. It stayed *off* at speeds above 20 mph.
2. A direct reading oil gauge showed the engine to have zero psi at idle. At all speeds above 20 mph (when the oil light went *off*) the oil gauge read about ten psi.

Cause   One of the oil gallery plugs behind the timing chain cover was left out when the engine was rebuilt (Figure 2-105). This was discovered by removing the oil pan and hooking up a prelubricator tank. A large flow of oil was observed running off the bottom of the timing chain.

FIGURE 2-105
Main oil gallery plug.

Comment   With one gallery plug missing the lubrication circuit still had enough restriction for the pump to produce ten psi; yet the oil light didn't indicate a problem above 20 mph. This example should point out the importance of installing a direct reading oil pressure gauge, and not to rely on the red "idiot" light.

TRUE CASE HISTORY

Vehicle   Plymouth Duster

Engine   8 cylinder, 318 CID

Mileage   67,500. The vehicle was brought into the shop for a minor overhaul (valve job, rings, and bearings).

Problem   All of the rod bearings wiped and failed shortly after the overhaul.

Cause   In this case, the mechanic measured the crankpins with a dial caliper that was inaccurate due to measuring face wear. He thus selected .002" undersize rod bearings, when standard size should have been used. Inadequate oil clearance caused the bearing failure.

Comment   A simple test can be made to determine if a caliper has measuring face wear. Open the jaws to exactly .001" and hold the caliper up to an outside light source. A band of colors will form in the .001" slit only if the measuring faces are parallel and not worn. This spectrum band contains the colors of the rainbow (violet, blue, green, yellow, and red).

The decision of going to any bearing undersize always requires careful study and exact measurement.

CHAPTER 2 SELF-TEST

1. When would a bearing roll-out pin be used?
2. What is the purpose of piston knurling?
3. State two reasons for removing the cylinder ridge when you overhaul an engine.
4. How are camshaft bearings generally fastened in the block?
5. What precaution must be observed when using a ring groove cleaner?
6. Never hand spiral on compression rings. Why?
7. When installing a piston assembly, dip the piston head in _____.
8. What is the reason for using connecting rod boots?
9. If the retainer and keepers will not release when using a spring compressor, what are two procedures that will help?
10. Why is it poor shop practice to try and release the keepers with ball-peen hammer blows on a socket wrench?
11. What are the three most common valve guide sizes for domestic engines?
12. Valve guide drivers are used to remove _____ type valve guides.
13. What is a step plate?
14. Describe the procedure that is sometimes used to tighten a wick type rear main seal to stop it from leaking oil.
15. Describe the procedure for removing and installing a lip type upper rear main seal without dropping the crankshaft.
16. What is generally the cause of difficult valve lifter removal?
17. What would be the reason for using rocker arm oil stoppers?
18. A piston vise can be a very useful holding tool when doing what jobs?
19. How do you use the cylinder head gasket installation tool mentioned in the chapter?
20. What are the advantages of working on an engine mounted on a rebuilding stand?
21. List five situations that are the result of vehicle manufacturers salvaging components on the production line.
22. How is the accuracy of a mike checked?
23. What would be the reading on a one-inch mike that has been turned out exactly $12\frac{1}{2}$ turns from zero position?
24. Why is it important to make sure the spindle face does not touch the anvil when storing the mike?
25. One revolution of the thimble on a metric mike changes the reading how much?
26. List three engine measurements that could be made with a travel indicating dial indicator.
27. Why should a dial indicator plunger be mounted at right angles to the object being measured?
28. Vernier calipers have a maximum discrimination of how much?
29. Why is a dial caliper easier to read than a vernier caliper?

30. How is a typical fractional inch machinist's rule graduated?

31. Decimal rules are typically marked in what increments?

32. When using a telescoping gauge to measure cylinder bore diameter, how can you be sure you are telescoped to the true bore diameter?

33. The small-hole gauge is often used to help obtain what cylinder head measurement?

34. Radius gauges are used for what engine application?

35. What can result if crankshaft fillet radii are greater than specified?

36. What will happen if an engine is assembled with out-of-round connecting rod big-end bores?

37. The crankshaft gauge is a special measuring tool designed for what purpose?

38. List four engine measurements that could be made with a feeler gauge.

39. Give several reasons why it is important not to overtighten cap screws in an engine.

40. What are the three different designs of torque wrenches discussed in the chapter?

41. Generally, where do you start tightening a set of nuts and bolts that attach an engine part?

42. List three crankshaft measurements that can be made using vee blocks and a dial indicator.

43. What is the procedure for checking main bearing saddle bore alignment with a precision straightedge?

44. What is the procedure for measuring bearing clearance with Plastigage?

45. Plastigage is available in what ranges?

46. When using Plastigage to check main bearing clearance on an engine mounted in the vehicle, what special procedure must be observed?

47. What three alignment conditions are encountered in connecting rods?

48. What is a screw pitch gauge?

49. The sleeve graduations on a micrometer depth gauge are different than an outside micrometer in what respect?

50. The cylinder dial gauge can be used to determine what three cylinder measurements?

51. The precision pin fit gauge shown in Figure 2-91 reads in what increments?

52. What is maximum allowable valve seat runout when doing a valve job?

53. What will be the results if new piston rings are installed in excessively worn grooves?

54. Where does the greatest amount of top ring groove wear usually occur?

55. How would you correct for top ring groove wear ?

56. What are two important uses for the engine prelubricator?

57. When using a torque wrench type valve spring tester, multiply the reading by _____ .
58. Valve springs should test within _____ of tension specifications.
59. Explain lifter leakdown.
60. Should new lifters be primed with engine oil?
61. List six methods of locating cracks.

CHAPTER 3

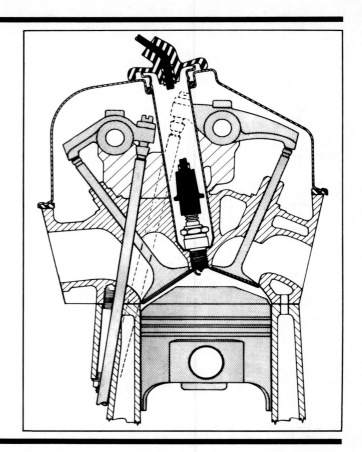

BASIC REVIEW OF ENGINE OPERATION

The automobile runs by overcoming road surface resistance. For this purpose it is necessary that power be supplied to its wheels by a *prime mover*, more generally known as an engine.

When the heat energy for conversion into motive power is produced outside the engine (as in a steam engine or steam turbine), the engine is an *external combustion* type. When the fuel is burned inside the engine to produce the heat energy (as in a gasoline or diesel engine), it is an *internal combustion* one.

OPERATING PRINCIPLES

The process of converting heat energy into power in the gasoline engine is illustrated in Figure 3-1. The fuel mixture (air and gasoline) is drawn into the cylinder, compressed by raising the piston, and ignited and burned to produce heat energy. High pressure and high temperature results. The piston, which moves up and down in the cylinder, is subjected to this high pressure and is pushed down. A connecting rod and a

crankshaft are provided to change the piston's motion to rotary motion. Then, the piston will move the connecting rod, which in turn will rotate the crankshaft connected to it.

It is necessary to exhaust the burnt fuel mixture and supply a fresh charge at regular intervals in order for the piston to continue to work. The action in the cylinder from the intake of the fuel mixture to its compression, combustion, and to the exhausting of the burnt fuel mixture is called the *engine cycle* (Figure 3-2). In most automobile engines, four piston strokes are taken to complete one cycle. This requires two turns of the crankshaft.

Four-Stroke Cycle Gasoline Engine

The piston moves up and down in the cylinder in reciprocating motion. The highest point reached by the piston is called *top dead center (TDC)*, and the lowest point *bottom dead center (BDC)*. The distance or the movement from TDC to BDC is called the *stroke* (Figure 3-3).

The operation of the four-stroke cycle gasoline engine will be explained (see Figure 3-4).

70

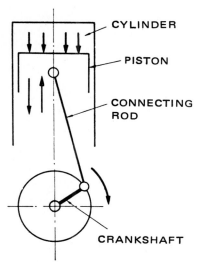

FIGURE 3-1
Piston and cranking mechanism.

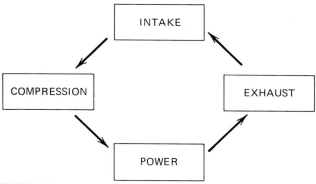

FIGURE 3-2
Engine cycle produces power in four actions.

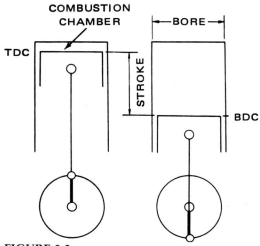

FIGURE 3-3
Piston stroke.

1. **Intake stroke**   In the intake stroke, the fuel mixture is drawn into the cylinder.

   For a moment, consider this example. The needle is removed from the end of a hypodermic syringe and the plunger is withdrawn slightly while holding the open end closed with a finger. When the finger is released, air will rush into the syringe with a popping noise. This phenomenon occurs because the pressure inside the syringe has become lower than atmospheric pressure (a partial vacuum has been created).

   The same action occurs in the engine. The piston in its downward motion from TDC to BDC creates a partial vacuum in the cylinder and the fuel mixture rushes in. During this time the intake valve will remain open and the exhaust valve closed.

2. **Compression stroke**   In this stroke, the fuel mixture drawn in the cylinder is compressed

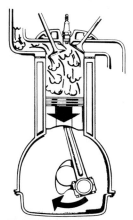
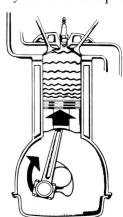

INTAKE STROKE   **COMPRESSION STROKE**

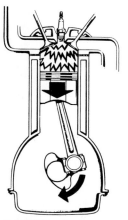
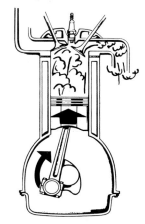

POWER STROKE   **EXHAUST STROKE**

FIGURE 3-4

Four stroke-cycle gasoline engine (Courtesy of Ford Parts and Service Division).

by the piston as it moves upward from BDC to TDC. As both the intake and exhaust valves will be closed during this time, the pressure and temperature of the fuel mixture will rise. The more the fuel mixture is compressed, the greater will be the explosive pressure and force to push down the piston. The piston has now made two strokes (one return trip), and the crankshaft one revolution.

3. **Power (expansion) stroke** In this stroke, the fuel mixture drawn in is ignited to cause combustion. The resulting pressure pushes down the piston to create the motive power. During this time, both the intake and exhaust valves remain closed. The piston has now made three strokes and the crankshaft one and one-half revolutions.

4. **Exhaust stroke** In this stroke, the piston pushed down to BDC rises once more to TDC to drive the burnt gas out of the cylinder. During this time, the exhaust valve is open.

When the piston reaches TDC after performing this action, it has returned to the position to start the intake stroke. The engine has now completed the four strokes (intake, compression, power, exhaust). The crankshaft has rotated twice (720°), and one power producing cycle is finished.

COMMENT: In an actual engine, the opening and closing of valves do not occur at exact TDC or BDC. This will be taken up in detail later in Chapter 18 when we discuss valve timing.

VALVE ACTION

In a four-stroke cycle engine, each cylinder is equipped with an intake valve and an exhaust valve. The series of parts that open and close the valves is called the *valve train*. The valve train can be classified into side valve, overhead valve, and F-head systems.

Side Valve System

In the side valve system, there are no valves in the cylinder head. As shown in Figure 3-5, the valves are opened directly by valve lifters. Note the camshaft location at the side of the crankcase.

Compared to the overhead valve system, the side valve system is less complicated, has fewer parts, and creates less noise. Its disadvantages include poor combustion chamber shape and a limited valve area. Side valve engines are not used much today. *NOTE: A side valve engine is sometimes referred to as an L-head, flathead, or valve-in-block engine.*

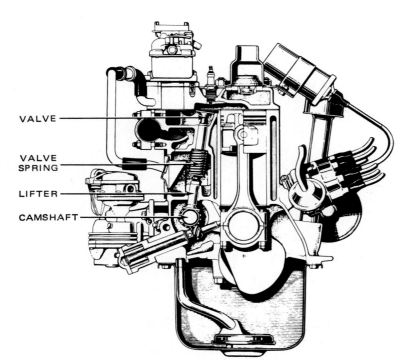

FIGURE 3-5
A six-cylinder L-head engine (Courtesy of American Motors Corporation).

VALVE

VALVE SPRING

LIFTER

CAMSHAFT

Overhead Valves (OHV)

As shown in Figures 3-6 and 3-7, the overhead valve system is composed mainly of the camshaft, valve lifters, push rods, and rocker arms. In operation, the camshaft lobe raises the valve lifter and forces the push rod up. When the push rod lifts up one end of the rocker arm, the other end pushes down on the valve and causes it to open. As the camshaft lobe turns further, the valve is closed by valve spring tension.

The OHV arrangement permits making the combustion chamber more compact, air flow is short and straight, and service is relatively easy. These advantages have resulted in wide adoption of this system. *NOTE: An OHV engine is sometimes referred to as an I-head engine.*

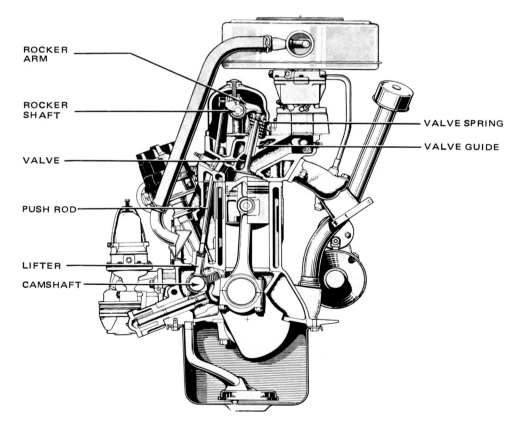

FIGURE 3-6

A six-cylinder OHV engine (Courtesy of American Motors Corporation).

FIGURE 3-7
Cutaway view of a six-cylinder Dodge OHV engine (Courtesy of Chrysler Corporation).

F-Head System

This system is a combination of the L-head and I-head arrangements (Figure 3-8). The intake valves are in the head and operated by push rods and rocker arms. The exhaust valves are in the block and operated directly by lifters on the camshaft.

Some of the four-cylinder engines produced by Jeep Corporation use an F-head system.

Overhead Camshafts (OHC)

In OHV engines, when the camshaft is brought to the top of the cylinder head (Figure 3-9) the camshaft can open the valves almost directly and eliminate inertia due to valve train weight.

The single overhead camshaft (SOHC) is the type that is in general use. For special applications, two overhead camshafts may be used (Figure 3-10). This is called a *double overhead camshaft (DOHC)* design.

Desmodramic Valve Train

Valves are opened and closed precisely by mechanical means (no valve springs are used). Two cams per valve are used, one for opening and the other for closing (Figure 3-11). The system (patented in 1910) was never put to practical use until Mercedes-Benz used it in their Grand Prix cars during the 1950s. The main purpose of the desmodramic system is to obtain extreme rpm without valve float. The Mercedes-Benz 300 SLR

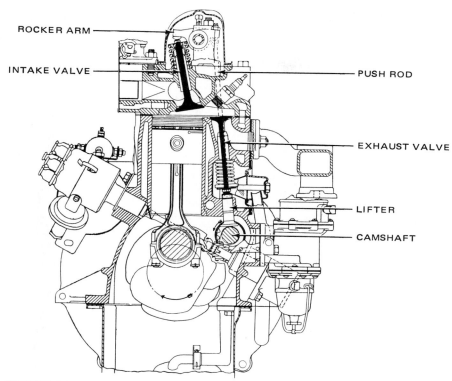

FIGURE 3-8
F-head engine (Courtesy of Jeep Corporation).

FIGURE 3-9
The overhead cam on this Pinto engine operates directly against the rocker arm (Courtesy of Ford Parts and Service Division).

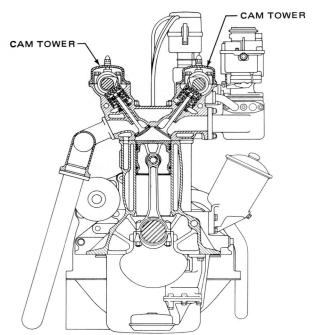

FIGURE 3-10
A chain drive DOHC six-cylinder engine.

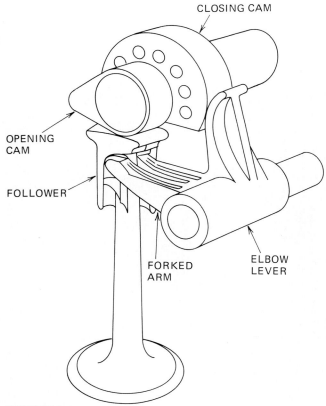

FIGURE 3-11
A desmodramic valve train assembly.

engines could be held at 8000 rpm for as long as five minutes at a time. Today, the only production vehicles using a desmodramic valve train are Ducati motorcycles.

PERFORMANCE FACTORS

Displacement

The volume that the piston displaces as it moves from BDC to TDC is called *stroke volume* or *displacement*. Therefore, the total displacement of the engine would be the stroke volume multiplied by the number of cylinders in the engine. Displacement is expressed in cubic inches, cubic centimeters, or liters. Generally, larger displacement means greater power output.

Compression Ratio

The compression ratio is the measurement of how much the fuel mixture is compressed in the compression stroke. It is the ratio between the volume when the piston is at BDC (cylinder volume and combustion chamber volume) to the volume remaining at the top of the cylinder when the piston is at TDC (combustion chamber volume). This is illustrated in Figure 3-12.

As the compression ratio rises, combustion pressure increases and greater engine power can be attained. However, raising it too high can cause fuel knock. Compression ratio is generally from 8:1 to 11:1 in gasoline engines and 16:1 to 20:1 in diesel engines.

Volumetric Efficiency

The amount of fuel mixture an engine takes in during the intake stroke compared to the stroke

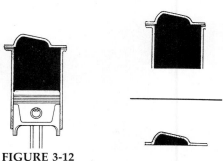

FIGURE 3-12
This is compression ratio.

volume is known as *volumetric efficiency*. On naturally aspirated engines, a 100% fill value is impossible due to intake system resistance and residual gas effects. The general fill value is from 65% to 85%. The curve in Figure 3-13 shows how the efficiency percentage varies with piston speed.

By using a supercharger or other means of forcing in the gas, you can increase the volumetric efficiency to 100%.

Torque

A force that turns or twists is called *torque*. For example, in tightening a bolt with a wrench, the force used to turn the bolt is torque (Figure 3-14). In this case, the torque turning the bolt is the product of the force and the distance from the bolt center to the point where the force is applied. If a greater torque is required, a longer wrench should be used, or more force applied.

If you compare the bolt example with that of an actual engine, the force applied on the wrench is the combustion pressure acting on the piston, and the distance is the length of the crank arm ($\frac{1}{2}$ the piston stroke).

Power

Before explaining what power is, it is necessary to define work. When an object is moved by applying force to it, work has been done. *Work is measured by a unit called the foot-pound.* A foot-pound is the amount of work done in raising one pound a distance of one foot (Figure 3-15).

It does not matter how long it takes to do work, as long as it gets done. In contrast, *power is the measurement of how much work can be done in a given time and measured in horsepower. One horsepower is required to do 33,000 ft-lb of work in one minute.*

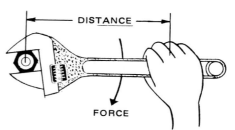

FIGURE 3-14
This is torque.

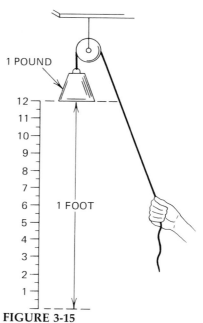

FIGURE 3-15
One foot-pound.

Originally, the term horsepower was derived from the fact that one average horse could raise 330 pounds 100 feet in one minute.

Valve Timing

The torque and horsepower produced by an engine is influenced greatly by valve timing.

Sometimes, in order to make a performance change, an offset crankshaft key or a special sprocket is used to advance or retard valve timing (Figures 3-16 and 3-17). Advancing the cam will increase low-end torque and midrange horsepower. Retarding the cam will kill low-end torque and horsepower, and increase top-end power.

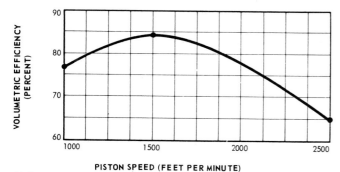

FIGURE 3-13
Volumetric efficiency curve.

FIGURE 3-16
An offset crankshaft woodruff key.

FIGURE 3-17
A multi-index crankshaft sprocket.

Thermal Efficiency

A certain amount of the heat energy produced inside the cylinder is used effectively, and the rest of it is lost. This is called *thermal efficiency*, and is shown in Figure 3-18.

The effective power that can be obtained from the crankshaft after these losses ranges from 23% to 28%.

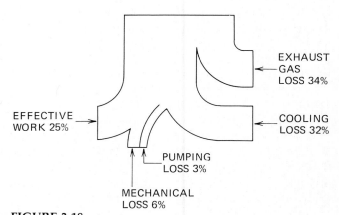

FIGURE 3-18
Diagram showing thermal efficiency.

TRUE CASE HISTORY

Vehicle	Chevrolet
Engine	8 cylinder, 327 CID
Mileage	Engine was bored and fitted with 350 CID Chevrolet pistons.
Problem	1. The vehicle owner said that the engine lacked power.
	2. A compression test showed each cylinder to have only 85–90 psi.
Cause	The wrong pistons were used. The pins in 350 pistons are located higher than in 327 pistons. This caused the pistons to be positioned lower in the bores (hence less compression).

CHAPTER 3 SELF-TEST

1. What is the difference between an internal combustion engine and an external combustion engine? Give an example of each.
2. Explain the operation of a four-stroke cycle gasoline engine.
3. How many revolutions does the crankshaft have to make in order for all the cylinders to fire in a four-stroke cycle V-8 engine?
4. What is TDC? BDC?
5. Describe a side valve system. What are the advantages and disadvantages?
6. Describe an overhead valve system. List three advantages.
7. What is the advantage of placing the camshaft overhead?
8. What is a DOHC system?
9. Describe an F-head engine.
10. What is unique about a desmodramic valve train?
11. How is the displacement of an engine expressed?
12. What is compression ratio?
13. What is volumetric efficiency?
14. What is the difference between torque and power?
15. Retarding the camshaft has what effect on engine performance?
16. Advancing the camshaft has what effect on engine performance?
17. What is thermal efficiency?

CHAPTER 4

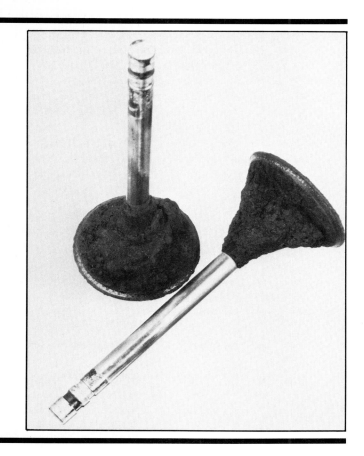

DIAGNOSING THE NEED FOR ENGINE TEARDOWN

Under normal operating conditions and with proper preventive maintenance, an internal combustion engine will provide satisfactory performance for thousands of miles. However, like any other mechanical device, parts wear with service. In time, performance drops to a point where it is no longer practical or economical to operate the engine.

The vehicle owner usually becomes aware of this when oil must be added frequently to the crankcase. This high oil consumption problem is then associated with the need for major engine work. Although this is true sometimes, there are many reasons for an engine to consume oil. To find and fix the real cause of excessive oil consumption, the mechanic should know these reasons and how to diagnose them.

An engine can develop many different noises. In some cases a complete engine rebuild may be necessary. In many cases only a partial teardown of the engine is required. To find the source of an abnormal engine noise, the mechanic must understand and follow step-by-step diagnosis procedures.

A number of engine cooling system problems can exist. Maybe coolant seepage into the oil has seized the crankshaft and a complete teardown is required. Perhaps an engine is overheating because of a blown head gasket.

The vehicle owner may feel a ring job is necessary because of poor compression. This could be true, but a loose timing chain could also be the cause.

Always ask these questions before you repair an engine problem.

1. For what reason(s) is the engine being disassembled? Is a complete teardown necessary?

2. Can you prove that the suspected parts are guilty prior to disassembly?

3. If parts are being replaced, have you corrected the cause behind the failure?

4. Is the cost justified in relation to the value of the vehicle? For instance, a $900 overhaul on a $200 car.

This chapter will provide answers to these questions.

81

VALVE COVER GASKET

CAMSHAFT
EXPANSION PLUG

OIL FILTER

REAR OIL SEAL

OIL PAN GASKET

FUEL PUMP GASKET

FRONT OIL SEAL

TIMING GEAR COVER

FIGURE 4-1
Some places where external oil leaks can occur (Courtesy of Dana Corporation—Service Parts Group).

CONDITIONS THAT CAN CAUSE ENGINE OIL CONSUMPTION

External Oil Leaks

Start by examining the engine thoroughly. Some points where external leaks may occur are oil lines, crankcase drain plug, oil pan gasket, valve cover gasket(s), fuel pump gasket, timing gear cover gasket, front oil seal, rear oil seal, oil filter, and camshaft expansion plug (Figure 4-1).

No possible source of leakage should be neglected, because even a very small leak will cause high oil consumption. Take for example, a vehicle driven at 40 mph down the highway. *A leak of one drop of oil every 20 feet will approximately equal a loss of one quart of oil every 100 miles.*

One quick check for external leaks is to raise the vehicle on a hoist and run the engine (it may be necessary to steam clean the engine first). Use a light or an oil dye (Figure 4-2) to find the actual leak source. Any area underneath an engine that is washed clean with oil indicates that there is an external leak somewhere in front of the washed area (Figure 4-3).

NOTE: A very high percentage of rear main oil leak complaints are leaks in the oil pan gasket, rocker covers, side covers, and so on, and are not leaking rear

FIGURE 4-2
The actual oil leak source can be pinpointed by using oil dye. The fluorescent dye is allowed to mix with the crankcase oil and a black light is directed on the suspected leak area.

FIGURE 4-3

This oil wash shows that there is an external leak somewhere in front of the pan (Courtesy of Dana Corporation—Service Parts Group).

main oil seals. Do not be misled just because you find oil on the bottom of the bell housing or on the rear end of the oil pan.

Front or Rear Main Seal Leaks

Worn or damaged front or rear main bearing seals will result in oil leakage. A grooved vibration damper hub will allow the front seal to leak (Figure 4-4). *NOTE:* *Several aftermarket seal manufacturers design their seals with a relocated lip. The chevron then will rub on a different location and a leak is prevented (Figure 4-5).*

Stuck Closed Positive Crankcase Ventilation (PCV) System

A PCV system closed with sludge and varnish deposits will result in plugged oil rings, oil consumption, rapid wear due to sludge buildup, ruptured gaskets and seals, oil thrown out around the filler cap or inside the air cleaner assembly, carburetor problems, and consequent rough idle.

You can test the positive crankcase ventilation system for clogging with one of the air flow testers on the market (Figure 4-6). If the test gauge indicates "repair," PCV system servicing is re-

FIGURE 4-4

A worn (grooved) vibration damper hub caused by the front cover seal lip.

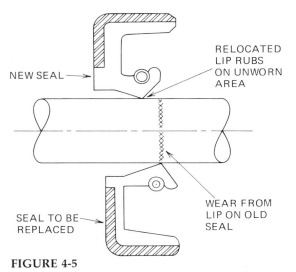

FIGURE 4-5

Some aftermarket seals are designed to rub on a new (unworn) area, to equal a factory fresh installation.

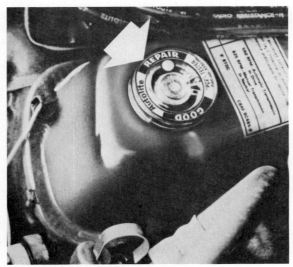

FIGURE 4-6
Testing a PCV system (Courtesy of Ford Parts and Service Division).

quired. *Make sure you check out the vacuum passage that goes through the spacer located under the carburetor on some V-8 engines.* Make sure the vacuum passage is not clogged. Some of today's engines contain an oil separator filter in the PCV system. This should be cleaned or replaced periodically to prevent clogging.

A PCV valve can become stuck open, or wear to a point where it is always operating at maximum flow. This then can allow oil mist to be drawn into the intake manifold. Oil consumption, fouled spark plugs, and rough idle result. *NOTE: Do not make the mistake of thinking that if a PCV valve rattles when you shake it, then it is okay. This is a common misconception. Shaking will not disclose seat and plunger wear inside the valve.*

A good, quick procedure for checking out a PCV valve is as follows.

1. Locate and disconnect the PCV valve, leaving the pilot (low pressure) hose attached.
2. Connect a tachometer to the engine.
3. Start the engine and read the idle rpm.
4. Place your finger over the end of the PCV valve opening (a strong vacuum should be felt).
5. The idle speed should normally drop no more than 80 rpm. If the rpm drops more than this, suspect a worn valve or the wrong valve.

Excess Oil in the Crankcase

Check to see if the engine is equipped with the correct dipstick. If possible, compare dipstick length and markings with several other identical engines. Compare dipstick tube length. Do not automatically assume that a satisfactory reading on the dipstick is a guarantee of correct crankcase oil level. Incorrect engine application or dipstick marking is not as rare as it may sound.

Excess oil in the crankcase will cause oil aeration (air bubbles in the oil). One immediate giveaway to aerated oil is that all hydraulic valve lifters are noisy. If oil level gets too high, the crankshaft end of the connecting rod will splash excessive quantities of oil on the cylinder walls. The extra oil will work its way up into the combustion chamber, resulting in excessive oil consumption and poor engine performance by fouling the spark plugs.

Oil Pressure Too High

While relatively rare in occurrence, an oil pump pressure relief valve can stick in a position causing the oil pressure to be too high. The result will be that the cylinder walls will be flooded with an abnormally large amount of oil in a manner similar to that which occurs with too much oil in the crankcase.

If high oil pressure is the suspected problem, install a master gauge at the engine oil pressure sensor port. Check the pressure against specifications.

Sometimes foreign material (pieces of valve stem oil seals or nylon particles from the camshaft sprocket) will get into the oil pump (Figures 4-7 and 4-8) and jam the pressure relief valve in a closed position. Pressure then becomes dangerously high. This can rupture the oil filter, pop out oil galley plugs, damage the oil pump, and erode bearings.

Clogged Oil Passages

Failure to change the oil frequently or to take proper care of the oil filter may cause the oil to become very dirty. This will promote clogging of the oil passages in the rings and pistons (Figure 4-9).

The oil passages carry the excess oil back to the crankcase. When the passages become clogged, the excess oil cannot be returned to the crankcase and some of it is burned.

Oil ring clogging is promoted by:

1. Use of a lubricating oil of too low an additive level.
2. Excessive combustion blow-by from worn rings and cylinders.
3. Dirty oil.
4. Low engine operating temperature (stop-and-go type of driving).

Severe carbon accumulation on a multiple-piece oil ring can cause the ring to become unitized,

that is, a condition where all parts are interlocked with carbon to form a "dead" ring. A unitized oil ring can be recognized easily when the piston and ring assembly is removed from the engine. The ring is tight in its groove. Its face remains flush with the piston skirt.

Cylinder head oil return passages have to be free of foreign material and sludge deposits. These drain back holes are located at each end of the cylinder head. If these holes are clogged, oil will flood the valve guides and result in high oil consumption (Figure 4-10).

If you suspect the engine is consuming oil because of clogged oil passages, try adding a chemical cleaning concentrate. When you pour the

FIGURE 4-7
Foreign material can get into many oil pumps through the screen bypass hole. The strap has been removed in order to see the hole. This screen is loaded with nylon particles from a cam sprocket.

FIGURE 4-9
A badly clogged multiple-piece oil ring that has become unitized (Courtesy of Dana Corporation—Service Parts Group).

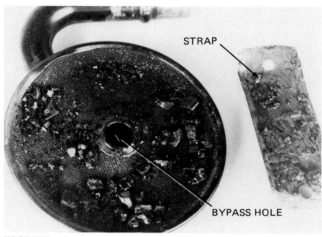

FIGURE 4-8
Umbrella valve stem oil seals become hard and brittle with age. Pieces sometimes break off and enter the oil pump.

FIGURE 4-10
Cylinder head oil return passage (Courtesy of Ford Parts and Service Division).

concentration into the crankcase, it mixes with the regular oil and quickly dissolves sludge and gums. *WARNING: Many cleaning concentrates contain acetone. When acetone cuts gum and sludge away, jellylike masses drop to the bottom of the crankcase. This can stop up the oil pump screen.* On a badly sludged engine the use of a nonacetone cleaner is recommended.

Piston Rings Stuck in Grooves

Any condition that causes the piston to operate at above normal temperature will aggravate ring sticking and often result in ring breakage.

Oil cannot be controlled by piston rings stuck in their grooves. Rings should never be installed without sufficient side clearance. Never spiral on compression rings. This can distort the ring and cause binding in the groove.

Worn or Broken Piston Rings

When piston rings are worn or broken, correct tension and clearances are not maintained (Figure 4-11). This will allow oil to be drawn into the combustion chamber on the intake stroke. Hot combustion gases will be blown down the cylinder past the piston on the power stroke. Both of these actions will result in burning and carboning of the oil on the cylinders, pistons, and rings.

The loose jagged end on broken rings can groove the cylinder wall. In addition, the sides of the piston grooves are sometimes cut into, causing breakage of the piston lands. Complete destruction of the piston assembly then results.

Cracked or Broken Ring Lands

Cracked or broken ring lands prevent the rings from seating completely on their sides and cause oil pumping (Figure 4-12).

Worn Ring Grooves

The sides of the ring grooves must be true and flat (not flared or shouldered) for the rings to form a good seal. The rings must have the correct side clearance in the grooves.

As the pistons move up and down, the rings must seat on the sides of the grooves to prevent leakage. *New rings in worn grooves will allow oil to be pumped into the combustion chamber.* Excessive side clearances also create pounding forces by the rings on the sides of the piston grooves (Fig-

ure 4-13). This promotes groove wear, and if the condition is not corrected, may cause breakage of rings or lands.

Ring groove wear, if detected before it has gone into the bottom of the groove, is sometimes corrected by the installation of GI (groove insert) spacers (Figure 4-14).

Piston Rings Fitted with Too Little End Clearance

When fitting new rings, sufficient end clearance is needed for expansion due to heat (Figure 4-15). During engine operation, the rings expand

FIGURE 4-11
Broken rings on a piston (Courtesy of Dana Corporation—Service Parts Group).

FIGURE 4-12
Broken ring lands.

FIGURE 4-13
Severe top ring groove wear (Courtesy of Dana Corporation—Service Parts Group).

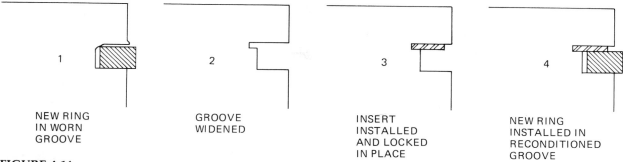

1 — NEW RING IN WORN GROOVE

2 — GROOVE WIDENED

3 — INSERT INSTALLED AND LOCKED IN PLACE

4 — NEW RING INSTALLED IN RECONDITIONED GROOVE

FIGURE 4-14
How ring groove wear correction can be made.

FIGURE 4-15
Checking end clearance (gap) on a new ring.

FIGURE 4-16
These standard rings were used in a .060" oversize bore. The excessive end clearance is what caused the darkened portion of the rings near the gaps (Courtesy of Sealed Power Corporation).

FIGURE 4-17
Several piston ring manufacturers make their oil ring expander spacers with an integral teflon insert at the ends. This prevents a mechanic from overlapping the ends during assembly and assures end abutment.

more than the cylinder. Provision for this expansion must be made by allowing a gap (end clearance) between the two ends of the ring. If there is not sufficient end clearance, the ends of the rings will butt while the engine operates. Butting will cause scuffing and scoring of rings and cylinders which, in turn, leads to oil consumption. Severe cases of butting may cause ring breakage.

Incorrect Diameter Piston Rings

Using incorrect diameter rings in a cylinder will lead to excessive oil consumption and severe combustion blow-by. Rings operated in a cylinder .020" or more larger than the ring size can be identified by carbon deposits on the ring faces near the gap (Figure 4-16).

Compression Rings Installed Upside Down

One compression ring installed upside down will cause the engine to use 30% to 40% more oil. Such incorrect installations will cause loss of oil control by scraping oil up into the combustion chamber instead of down into the crankcase.

Oil Control Ring Expander Spacers Installed Incorrectly

When you install a multiple piece oil control ring, care must be used to *make sure that the expander ends are butted together and not overlapped* (Figure 4-17). Sometimes a mechanic cuts off part of the expander with the idea of making installation easier. Engines assembled under these conditions will not give satisfactory oil control. This is because the reduced expander tension will not force the rails out against the cylinder wall with sufficient pressure.

Badly Worn Rings Due to Abrasion and Scuffing

Badly worn rings are ineffective in sealing combustion gases. Worn rings reduce the engine's horsepower and efficiency. Badly worn compression rings can be identified by their reduced radial wall width, low tension, excessive ring gap, and machine marks worn off the faces. When ring faces are covered with dull gray vertical scratches, the rings have been worn by abrasives (Figure 4-18).

Scuffing results when the surface temperature of two rubbing metal surfaces reaches the melting point of the material. *Scuffing is always the result of an excessive amount of heat at the rubbing surface.* Excessive heat discolors metal. You may, therefore, identify scuffing by looking for discolored areas on the surface of rings, pistons, or cylinders. The coloring may vary from a light straw through deep blue. Under a magnifying glass, scuffed metal is always burnished and

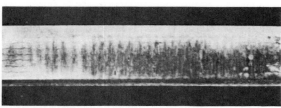

FIGURE 4-18
Dull gray vertical scratches on ring faces are caused by abrasives that have entered the engine (Courtesy of Dana Corporation—Service Parts Group).

FIGURE 4-19
This scored piston was caused by improper pin fit. Notice how the metal on the skirt has melted and smeared in the direction of motion.

smeared in the direction of motion of the part. Severe scuffing is called *scoring* (Figure 4-19).

Vacuum Accessory Leaks

Most of today's passenger vehicles are equipped with many vacuum accessories (Figure 4-20). There are power brake units, door locks, automatic transmission modulators, heater controls, speed control units, trunk openers, windshield washer controls, and so on. These units generally work by atmospheric pressure acting against a diaphragm. Some are directly vented to the atmosphere, while others have small fibrous filters. In either case, if the diaphragm becomes cracked or porous, there is a direct vacuum leak to the outside. Abrasives can now enter the engine. In some cases, these accessories are located where there is a tremendous amount of road dust and dirt.

Some vehicles have vacuum junction blocks with a number of large and small-diameter hose feeds. When one of these hoses splits due to age, weathering, or vibration, there is a direct leak to the outside air. This can lead to accelerated engine wear.

NOTE: In some cases a badly smoking engine can be caused by a faulty automatic transmission modulator (Figure 4-21). The diaphragm becomes porous and allows transmission fluid to be drawn into the engine. This can lead to transmission failure as the fluid level drops in the transmission. Additionally, the high detergent low viscosity automatic transmission fluid (ATF) will "wash" the cylinder walls of engine oil. This can lead to ring scoring and cylinder wall damage.

Cylinder and Piston Corrosion

A piston severely corroded by coolant leakage is shown in Figure 4-22. Coolant (ethelyne glycol) leakage will cause dilution of the oil and wash the cylinder walls. This will lead to wear of bearings, cylinders, and rings.

Connecting Rod Misalignment

When connecting rods are misaligned, the rings do not have proper face contact with the cylinder wall (Figure 4-23). Pistons wear unevenly, the engine is more prone to scuff and score, and oil consumption becomes excessive.

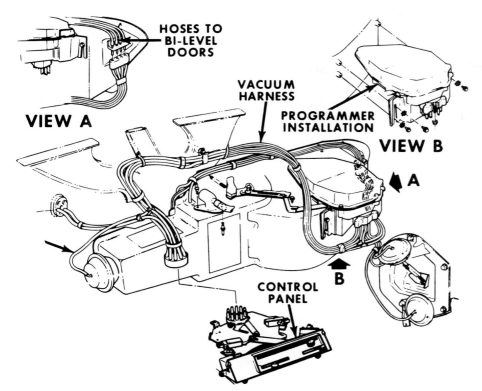

HOSES TO
BI-LEVEL
DOORS

VIEW A

VACUUM
HARNESS

PROGRAMMER
INSTALLATION

VIEW B

A

B

CONTROL
PANEL

FIGURE 4-20

A typical vacuum controlled heater/ air-conditioner used in passenger cars is illustrated. Abrasives can enter the engine if there is a vacuum leak (Courtesy of Sealed Power Corporation).

FIGURE 4-21

An automatic transmission vacuum modulator can cause an engine to smoke badly and to use oil.

FIGURE 4-22

The mottled pitted appearance on this piston skirt was caused by coolant leakage into the cylinder (Courtesy of Dana Corporation—Service Parts Group).

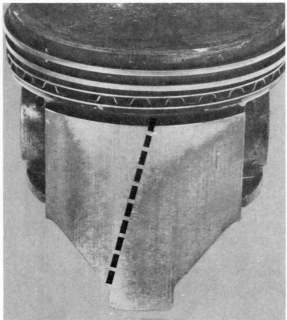

FIGURE 4-23

The diagonal wear pattern on this piston skirt was caused by a bent connecting rod.

Improperly Fit Wrist Pins

Wrist pins that are fit too tightly in the pin bosses prevent the pistons from expanding and contracting freely. This causes piston distortion that results in scuffing and scoring. This leads to blow-by and high oil consumption.

A precision pin fit is critical in today's engines. The pin bosses have to be free to move inward and outward on the wrist pin while maintaining a specific amount of oil clearance (Figure 4-24).

Worn Timing Gears or Chain

Worn timing gears or chain (Figure 4-25) cause the valves to be out of time with the crankshaft and will result in late valve timing. Late closing intake valves increase vacuum in the cylinders. Higher vacuum has a greater tendency to draw oil past the pistons, rings, and valve guides, and into the upper part of the cylinder where it will be burned.

Defective Fuel Pump

Occasionally a single-action fuel pump will develop a diaphragm crack where the pull rod is attached (Figure 4-26). This allows gasoline to

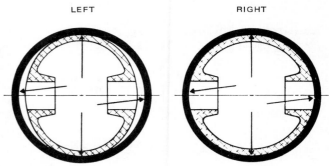

LEFT   RIGHT

FIGURE 4-24

The left diagram shows a cam-ground piston in a cold engine. The piston has cylinder wall contact in the direction of thrust, and clearance along the axis of the pin bosses.

The right diagram shows a cam-ground piston at engine operating temperature. The piston has expanded until it has even clearance in the cylinder. Note the pin bosses are further apart than when the piston was cold (Courtesy of Sunnen Products Company).

FIGURE 4-25

This timing chain is worn out. The extreme sag is caused by wear on the link pins.

PULL ROD   DIAPHRAGM

FIGURE 4-26

A fuel pump diaphragm crack or tear near the pull rod attachment point sometimes occurs. This will allow gasoline to pass into the crankcase.

pass into the crankcase, diluting the oil. The rings, cylinders, and bearings will then be quickly damaged.

Defective Intake Manifold Gasket

Some V-8 engines are designed with the manifold serving as the tappet chamber cover. A very narrow section of gasket separates the intake

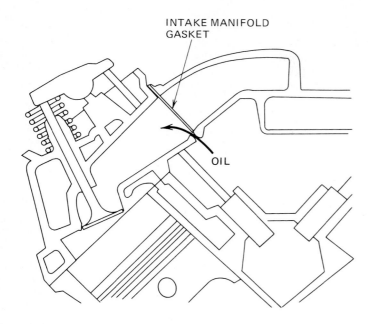

INTAKE MANIFOLD
GASKET

OIL

FIGURE 4-27
On some engines, a defective intake manifold gasket can cause oil to be pulled into the combustion chamber.

manifold ports from the oil mist in the tappet chamber and/or valve cover. An internal gasket leak at the parting face between the manifold and the cylinder head allows oil to be drawn into the intake ports (Figure 4-27). Oil consumption, oil-fouled spark plug(s) on one cylinder bank, and a rough idle result. Poor sealing at the end portions of the intake manifold gasket may cause an external oil leak.

Before removing the intake manifold, you can verify a suspected internal vacuum leak with this test.

1. Hook up a vacuum gauge to the intake manifold.

2. Fully close the carburetor throttle plates by turning out the curb idle speed screw.

3. Plug up the PCV valve opening.

4. Disconnect and plug up any vacuum hoses attached directly to the intake manifold.

5. Disconnect the battery feed wire to the ignition coil to prevent the engine from starting.

6. Crank the engine and observe the vacuum gauge. From 3″ to 7″ of vacuum (at a minimum of 250 cranking rpm) is normal. A defective intake manifold gasket will give, in most cases, a zero vacuum reading.

CAUTION: *If you are using a remote starter button to crank the engine on a 1961 or later General Motors* *vehicle, turn on the key. Failure to do so can possibly burn out the starter solenoid contacts, the key switch contacts, and the resistor bypass wire.*

Installing the Rocker Arm Shaft Upside Down

Occasionally, a mechanic will remove V-8 cylinder heads and find that excessive oil consumption is confined to only one bank (side). This situation may be caused by the rocker arm shaft on that bank being installed upside down. The rocker arm shaft oil passage holes (with several exceptions) should normally face the bottom side of the rocker arm (Figure 4-28). In this po-

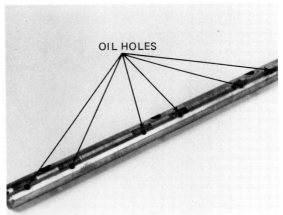

OIL HOLES

FIGURE 4-28
Rocker arm shaft oil passage holes normally face down.

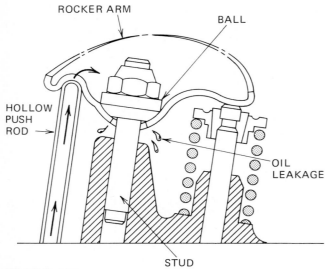

FIGURE 4-29

Oil leakage at the stud hole should occur on canoe type rockers.

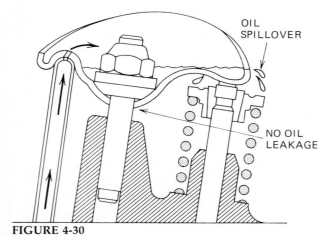

FIGURE 4-30

Excessive oil consumption can result if the ball seats into the rocker arm.

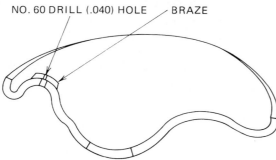

FIGURE 4-31

Modifying the rocker arm as shown will reduce oil supply.

sition, the upward force on the rocker arm from the valve spring tends to restrict oil flow from the rocker arm shaft. If the shaft is installed upside down, too much oil can flow over the rocker arms and enter the combustion chamber through the valve guides.

Worn Rocker Arm Pivot Balls

The rocker arms on many engines are lubricated by oil through hollow push rods (Figure 4-29). When these engines are relatively new, there is enough leakage at the bottom of the rocker arm to prevent excess oil building up and spilling over the rocker arm lip when the valve opens. As the rocker arm ball seats in, it forms a near perfect seal between the ball and the rocker arm. This allows excessive oil to puddle in the canoe shaped rocker arm. When the valve opens, the oil is spilled over the rocker arm lip (Figure 4-30). This excess oil is difficult for the valve stem seal to control. As a result some of the oil is then drawn into the combustion chamber through the valve guides.

To prevent oil buildup in the canoe rocker arms, different methods are used.

1. Braze shut the oil outlet hole in the push rod end of the rocker arm. Redrill with a number 60 (.040") drill to reduce the oil supply to the rocker arms (Figure 4-31).

2. File a groove approximately 1/16" deep and 1/16" wide, across the bearing face of the ball. This will allow trapped oil to partially drain out of the rocker arms (Figure 4-32).

3. Drill a .060" bleed hole into the side of the push rod (Figure 4-33). This will result in less oil reaching the rocker arms. ***NOTE:*** *There are some production engine remanufacturing facilities that do this as a matter of routine when rebuilding small block Chevrolets. It is their opinion that these engines were originally designed with too much oil flow to the rocker arms.*

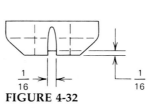 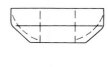

FIGURE 4-32

Grooving the rocker arm ball will allow oil drainage, yet will allow sufficient oil to be retained for proper lubrication.

Sometimes, excessively squirting push rods can be the cause of oil consumption. When the rocker arm push rod socket becomes very worn, the oil hole may be blocked during valve lift, and open when the valve closes (Figure 4-34). As a result, pressure is built up and released in surges. This sudden release of oil can be difficult for the valve stem seal to control. *COMMENT: Wrong valve lifters may cause excessive squirting of push rods.*

Defective Rocker Arm Shaft Support Bolt Holes

On certain Ford V-8 engines, a crack or pinhole can develop in the bottom of the rocker arm shaft support bolt holes (Figure 4-35). This can lead to oil burning. These bolt holes are directly adjacent to the cylinder head runners. If the bolt holes intersect into the runners, oil can be drawn into the combustion chamber. To correct this condition, install studs.

Oil Loss Past the Head Gasket

In some engines, the block has a drilled oil gallery that extends to the deck surface to carry oil to the rocker arms. If the head gasket does not properly seal this area, oil can escape into a cylinder or into the engine coolant (Figure 4-36). Engine oil found in the engine coolant, or con-

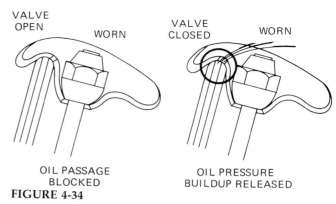

FIGURE 4-34

Worn rocker arm push rod sockets can cause excessively squirting push rods.

FIGURE 4-35

Oil burning can be caused by cracks in rocker arm shaft support bolt holes (Courtesy of Ford Parts and Service Division).

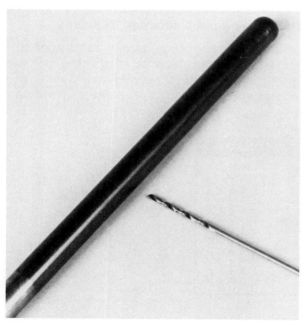

FIGURE 4-33

A .060" drill is used to make a bleed hole in one side of the push rod.

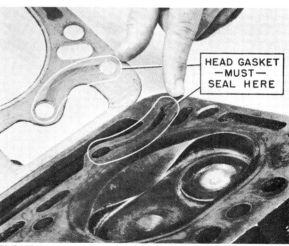

FIGURE 4-36

On this engine, the head gasket must seal in the area indicated. Otherwise, oil can escape into the cylinder or engine coolant (Courtesy of Dana Corporation—Service Parts Group).

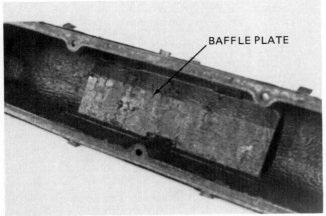

FIGURE 4-37
This valve cover baffle prevents the PCV valve from drawing oil vapor out of the rocker arm chamber.

tinued oil fouling of the spark plug next to this block oil gallery indicates an internal oil leak problem. The head gasket needs to be replaced.

Missing Baffle Plate or Vapor Separator

Some engines have a metal baffle plate or vapor separator installed in the valve cover(s) to prevent oil vapors from being drawn through the PCV valve into the combustion chamber (Figure 4-37).

Worn or Damaged Main Bearings

Worn or damaged main bearings allow an excessive amount of oil to be thrown off the crankshaft up onto the cylinder walls. This floods the cylinders with more oil than the pistons and rings can control. Burning of oil and carboning of pistons and rings results.

In some cases, a large loss of oil at the main bearings may reduce oil pressure to such an extent (especially at low speeds) that insufficient oil may be thrown on the cylinder walls. This will cause pistons and rings to wear quickly.

The amount of oil thrown off increases rapidly when bearing wear increases. For instance, if the bearing is designed to have .0015" clearance for proper lubrication and cooling, the oil thrown off will be normal as long as this clearance is maintained and the bearing is not damaged in any way. However, when the bearing clearance increases to .003", the amount thrown off will be five times normal. If the clearance is increased to .006", the throw off amount will be 25 times normal (Figure 4-38).

Worn or Damaged Connecting Rod Bearings

Clearances on connecting rod bearings affect oil throw off in the same proportions as mentioned for main bearings. Additionally, the oil is thrown more directly on the cylinders. This large volume of oil overloads the pistons and rings, and oil escapes to the combustion chamber (Figure 4-39).

Worn or Damaged Camshaft Bearings

Camshaft bearings are generally lubricated under pressure. If cam bearing clearances become too great, large quantities of oil will drain off into the crankcase. This excess oil is then thrown on

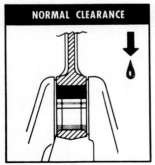 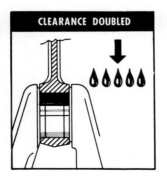 

FIGURE 4-38

Normal clearance in automotive bearings is from .001" to .003". When normal clearance is *doubled* the oil throw off at higher speeds is multiplied approximately *five* times. When normal clearance is *quadrupled*, the oil throw off is multiplied approximately 25 times (Courtesy of Imperial Clevite Inc., Engine Parts Division).

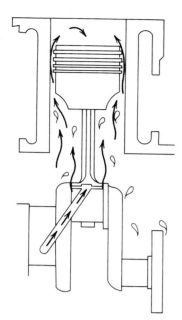

FIGURE 4-39

Worn bearings create oil control problems for the piston rings (Courtesy of McQuay-Norris Manufacturing Company).

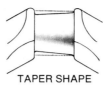

TAPER SHAPE

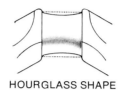

HOURGLASS SHAPE

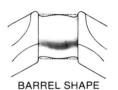

BARREL SHAPE

FIGURE 4-40

Out-of-shape crankshaft journals will not provide even oil clearance (Courtesy of Imperial Clevite Inc., Engine Parts Division).

the cylinders by the rotating crank throws, causing more oil on the cylinders than can be controlled by the piston rings. *Worn main, rod, or cam bearings will cause a drop in oil pressure.* This can then produce scuffing and scoring of parts.

Worn Crankshaft Journals

Worn crankshaft journals have the same effect on oil consumption as worn bearings. When they are worn, they will not give uniform oil clearance (Figure 4-40).

Distorted Cylinders

In a cylinder that is true, the oil can be controlled by the pistons and rings. However, with taper and out-of-roundness, oil control becomes more difficult. This is due to several factors. The increased piston clearances permit the pistons to rock in the worn cylinders. While tilted momentarily, an abnormally large volume of oil is permitted to enter on one side of the piston. The rings, also tilted in the cylinder, permit oil to enter on one side. Upon reversal of the piston

on each stroke, some oil is passed into the combustion chamber.

Cylinders that are distorted from causes other than wear, such as unequal heat distribution or boring a block without the main bearing caps torqued in place, present a surface on which the rings may not be able to follow completely. As a result, in some areas the rings will not remove all the excess oil. When combustion takes place, this oil will be burned, causing high oil consumption.

Unequal Tightening of Cylinder Head Bolts

The strains developed by unequal tightening of cylinder head bolts may cause serious cylinder distortion and result in oil pumping. When reinstalling a cylinder head, a torque wrench always should be used on the head bolts. The engine manufacturer's instructions should be followed for the torque readings and for the sequence in which the bolts are tightened (Figure 4-41).

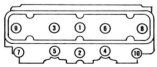 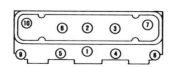 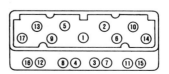 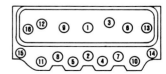

FIGURE 4-41

Cylinder head bolt tightening sequence diagrams are available for all engines.

Improper Carburetor Adjustment

Improper carburetor adjustment may lead to excessive oil consumption. Too lean a mixture may cause overheating and ring sticking and ring breaking. Too rich a mixture dilutes the oil with fuel, which can cause wear of bearings, cylinders and rings. In either case, the final result is high oil consumption. See Figure 4-42.

Dirty Cooling Systems

Rust, scale, sediment, or other debris in the water jackets and radiator will prevent a cooling system from performing its job efficiently (Figure 4-43).

FIGURE 4-42

Do you recognize these carburetor parts? If defective, each can cause too rich a mixture and oil dilution.

FIGURE 4-43

All of this junk was found in a water jacket cavity of an engine during a rebuild. When replacing freeze plugs, do not leave the old ones in the block. This can cause localized "hot spots" and result in scored parts.

This can cause cylinder, piston, and valve distortion.

Worn Valve Stems and Guides

On overhead valve (OHV) engines the manufacturers have found it necessary to maintain a generous supply of oil in the rocker arm area. This is done to reduce noise and help prevent wear of the rocker arms, valve stems, and guides.

A number of operating forces want to force this oil down through the stem-to-guide clearance (Figure 4-44). When the engine is running there is a reduced pressure (partial vacuum) existing in the intake valve port. In the rocker arm area above the valves there is approximately 15 psi. This pressure differential acts as the major force in pushing oil down past the intake valve guides and stems. As guide and stem wear increases, it is easier for oil to be drawn through and burned in the combustion chamber.

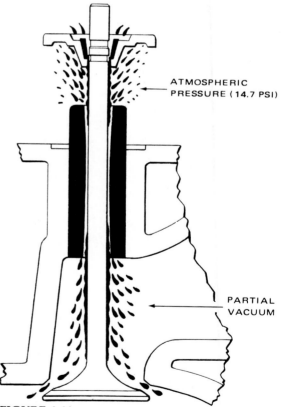

ATMOSPHERIC PRESSURE (14.7 PSI)

PARTIAL VACUUM

FIGURE 4-44

Gravity, inertia, and pressure differences force oil through intake valve guides (Courtesy of Dana Corporation—Service Parts Group).

At one time it was thought that oil loss through exhaust valve guides was not possible. However, *significant oil loss can occur through exhaust valves.* When the engine is operating there is an atomizer effect at the lower end of the exhaust valve guide (Figure 4-45). Exhaust gases flowing past the end of the guide reduce pressure at this point. Atmospheric pressure then pushes oil down the valve stem where it is carried out the exhaust pipe in the form of blue-gray smoke.

Oil passing through exhaust valve guides will cause the gradual buildup of hard sulfur on the valve stems. These deposits can cause valve sticking and prevent proper seating (Figure 4-46). Oil passing through intake valve guides will form hard black colored "coked" deposits under the valve head (Figure 4-47).

A smoke test is sometimes made to check for oil loss through the valve guides. Warm up the engine and idle for five minutes. Rapidly accelerate the engine and watch the exhaust for blue-gray smoke (Figure 4-48). Verify by driving the car at 40 mph, then decelerating with the throttle fully closed. Watch for large puffs of blue-gray smoke.

FIGURE 4-46
Sulfur deposits on the stem of an exhaust valve.

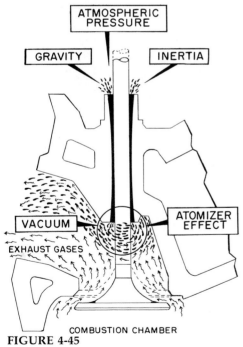

FIGURE 4-45
Oil can escape through exhaust valve guides (Courtesy of Dana Corporation—Service Parts Group).

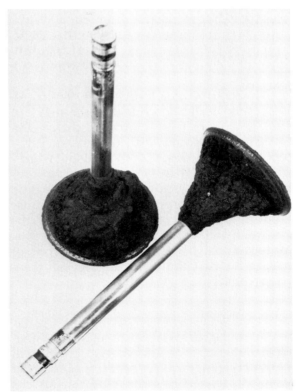

FIGURE 4-47
"Coked" oil deposits on an intake valve.

BLUE—GRAY = OIL BURNING
BLACK = RICH FUEL MIXTURE
WHITE = PRESENCE OF WATER

FIGURE 4-48
Exhaust smoke color can indicate several conditions (Courtesy of Ford Parts and Service Division).

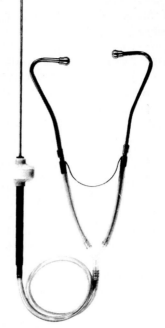

FIGURE 4-49
This tool allows you to trace piston noise, valve train click, water pump squeak, and other mechanical failures.

DIAGNOSING ENGINE NOISES

One of the most difficult problems facing the mechanic is locating abnormal noises coming from the interior of the engine. Abnormal engine noises vary in sound intensity and frequency. When diagnosing engine noises, make sure that an accessory or body part is not the cause. Also, noises can fool you by echoing the sound to some other location.

A very useful tool for locating the source of engine noise is a stethoscope (Figure 4-49). A length of 3/8" rubber hose can be used as a substitute to carry the sound directly to your ear.

It is very difficult to describe noises using words. The following descriptions are intended to serve as general guidelines. Experience has to be built up over the years. When you determine the cause of an abnormal noise condition, remember the situation for future reference. Service bulletins, trade publications, and training school courses can provide the mechanic with additional troubleshooting skills.

Crankshaft Noises

Excessive main bearing clearance noises are generally *dull, heavy metallic knocks.* The knock will increase in frequency as the speed and load on the engine is increased. Most main bearing knocks (due to excessive clearance) are not noticeable at an even idle speed. Main bearing knock is usually apparent when the engine is pulling hard,

FIGURE 4-50
The center of this automatic transmission flex-plate had completely broken out. The noise produced while the engine was running, sounded just like a rod knock.

when an engine is started, during acceleration, or at speeds above 35 mph.

Excessive crankshaft endplay will cause a sharp rap to occur at irregularly spaced intervals. The noise is noticeable at idling speeds. By the alternate release and engagement of the clutch, the noise becomes pronounced.

A loose flywheel or broken flex-plate (Figure 4-50) can usually be detected by this procedure.

1. Advance engine idle to 2000 rpm.
2. Turn *off* the ignition switch.

FIGURE 4-51
This separated vibration damper produced a rattling noise heard only at idle and off-idle speed.

FIGURE 4-52
This broken crankshaft piece came out of an engine that had a history of severe dieseling. The engine ran with this broken crank. A very audible heavy knock sound could be heard at all engine speeds.

3. When the engine has almost stopped, turn the switch *on* again.

4. Repeat this procedure several times.

5. If the flywheel is loose or the flex-plate is broken, a distinct knock will be heard every time the ignition switch is turned back on.

The crankshaft vibration damper assembly can become separated (Figure 4-51). This will generally produce a *heavy rattling noise* that can be heard at low speed.

Sometimes a crankshaft will break off at the

FIGURE 4-53
The skirt on this piston had broken off and was found in the bottom of the oil pan.

snout or at one of the front bearing journals (Figure 4-52). A *heavy knocking rattling sound* is then heard while the engine is running. Occasionally a crankshaft keyway will shear off causing the vibration damper to rattle.

Connecting Rod Noises

Excessive connecting rod bearing clearance noises are usually a *light rap or clatter* much less in intensity than main bearing knocks and the loudest when the engine is "floating" or running with a light load at from 25 to 35 mph. The noise will become louder as engine speed is increased. By grounding out each of the spark plugs, one at a time, you can determine from which cylinder the noise is coming. The noise may not be eliminated entirely by grounding, but it will be reduced considerably in intensity.

Piston Noises

The most common piston noise is known as *slap*. This is a *hollow, muffled, bell-like sound* due to rocking of the piston from side to side in the cylinder. This piston rocking condition is brought about when cylinder wall clearance becomes excessive or when a piston skirt collapses or breaks (Figure 4-53).

Piston slap is loudest when the engine is cold, and lessens or disappears after the engine is warm. When driving the car (at from 25 to 30 mph) the noise will increase in intensity as the throttle is opened and additional load applied.

To detect piston slap, try the following procedure.

1. Pour several ounces of 40 weight engine oil into the suspected cylinder(s).
2. Crank the engine for several revolutions with the ignition turned *off*. This will allow the oil to work itself down past the piston rings and act as a cushion.
3. Install the spark plug(s).
4. Start the engine.
5. If the noise is eliminated, the engine has a piston slap condition.

Piston Pin Noises

Piston pin noise is usually the result of excessive piston pin clearance. This will cause a *sharp, metallic, double-knock sound* most noticeable when the engine is idling. Sometimes the noise is more audible at car speeds of from 25 to 35 mph.

Occasionally a misaligned rod will cause "boss knock." This is a condition where the eye (small end) of the rod comes into contact with one of the pin bosses. The resulting sound is sometimes wrongly diagnosed as piston slap (Figure 4-54). To test for excessive piston pin clearance noise, use this procedure.

1. Run the engine at idle speed.
2. Retard the spark to reduce the intensity of the knock.
3. Return the spark timing to the normal setting.
4. Short out each spark plug, one at a time. The double-knock sound will become more audible at the cylinder with the loose pin.

Valve Train Noises

Valve train mechanism noise has a characteristic *"clicking" noise* or a sound like a sewing machine in operation. The frequency of valve train noise is at one-half crankshaft speed. A very common valve train noise is lifter clicking. If the engine is equipped with solid lifters, an adjustment is probably needed. If the engine is equipped with hydraulic lifters, many things could be wrong. For a detailed analysis, see Chapter 18.

Detonation and Preignition Noise

Detonation and preignition are forms of abnormal combustion in the engine.

During normal operation of the engine, the burning of the air-fuel mixture produces a steady, smooth push on the pistons. At the instant of spark plug firing, the combustion chamber flame moves rapidly outward like the waves when a stone is dropped into a pool of water (Figure 4-55). Certain conditions may allow pressures to start a second flame burning. A violent explosion will result when the normal flame front runs into the secondary flame front. This produces a *knocking, "pinging" sound* that is called detona-

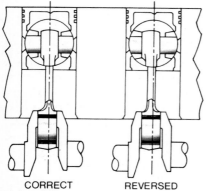

CORRECT   REVERSED

FIGURE 4-54

Assembling a connecting rod with the offset to the wrong side can result in interference noise as shown on the right.

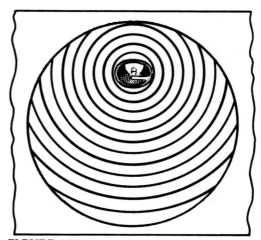

FIGURE 4-55

Normal combustion flame propagation.

tion. Detonation will cause piston and ring damage, top ring groove wear, scoring, sticking rings, burned-out head gaskets, and even possible complete engine failure (Figure 4-56). Detonation can be caused by the following:

1. A lean fuel mixture.
2. Gasoline with a too low octane rating.
3. Improperly set ignition timing.
4. Lugging the engine.
5. Excessive carbon deposits.
6. Excessive milling of the heads or block.

Preignition is (as the term suggests) the ignition of the air-fuel mixture before regular plug firing occurs. When the regular spark occurs shortly after the preignition, the two colliding flame fronts will cause a "pinging" noise. Preignition causes loss of engine power and can cause severe damage to pistons, rings, and valves (Figure 4-57). Preignition and detonation are very closely related, and it is very difficult to distinguish one from the other by sound. Each can lead to the other. Preignition can have the following causes:

1. Incandescent carbon particles.
2. Spark plugs that have too hot a heat range (Figure 4-58).

3. Spark plugs not firmly tightened.
4. Detonation.
5. Sharp edges in the combustion chamber.
6. Valves operating at higher than normal temperature.
7. A head gasket edge protruding into the cylinder bore area.
8. Overheating.
9. Ignition crossfire.

FIGURE 4-57
This piston was damaged by preignition.

FIGURE 4-56
The severe detonation damage on this truck piston was caused by continued lugging.

FIGURE 4-58
When antifoulers are used in an attempt to stop engine oil consumption, the spark plug heat range is made much hotter. This often burns up the plugs (as shown), and causes preignition.

Other Under the Hood Noises

Water pump bearing failure noise can be located easily by using a stethoscope (you can feel for bearing looseness by rocking the tips of the fan blade). If still in doubt, remove the drive belt for a short operating period. If the noise remains, it is not in the water pump. See Figure 4-59.

Belts frequently develop an idle or low-speed *"squish" sound*. Test by squirting water on the suspected belt(s) to try eliminating the noise temporarily. Correct by cleaning the belt and pulley groove. Then readjust the belt. Belt "squeal" is generally caused by glazed or burned sidewalls. Tighten or replace as necessary.

Alternator bearing noise can vary from a *harsh rumble to a siren sound*. It is always present at any speed. The noise can be made to come and go by loosening the adjustment bolt(s) and pulling the alternator against the drive belt at different tensions.

Listed below are some additional under the hood noises you may encounter.

1. Loose timing chain hitting the inside of the front cover (small block Chevy and Ford) (Figure 4-60).

2. Rocker arm "squeak" from worn push rod tips shutting off the oil delivery (on engines with drilled push rods).

3. Manifold heat control valve rattle (Figure 4-61).

4. Mechanical damage to the piston (Figures 4-62 and 4-63).

5. Connecting rod bolts hitting the inside of a dented oil pan.

6. Top of connecting rod bolts turned around and striking the camshaft.

7. Fuel pump attachment bolts too long and rubbing on the timing chain.

8. New piston rings hitting the ridge at the top of the cylinder bore.

9. Piston striking the edge of the head gasket.

10. Piston pin rubbing against the cylinder wall.

11. Bad mounts (Figure 4-64).

12. Defective fluid coupling on the fan.

13. Fuel pump rocker arm return spring broken.

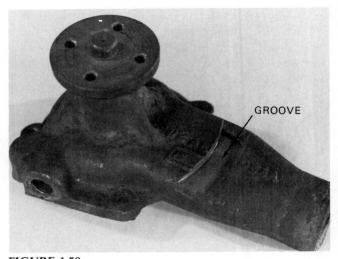

FIGURE 4-59

Can you believe this? The water pump bearing became so worn that the pulley machined a groove in the casting.

FIGURE 4-60

The grooves cut into this small block Chevrolet front cover were caused by a loose timing chain. The noise sounded just like a rod knock.

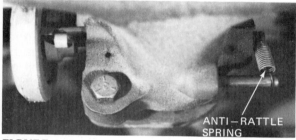

FIGURE 4-61

A missing or disconnected anti-rattle spring can cause manifold heat control valve noise.

FIGURE 4-62

A mechanic had removed the carburetor from an engine to make some adjustments but failed to cover the intake manifold opening. One of the accelerator pump ball checks fell into an intake port. The picture tells the rest of the story.

FIGURE 4-63

This piston damage was caused by a carburetor base nut falling into the intake manifold. *Always* cover the manifold opening when a carburetor is removed for service.

FIGURE 4-64

Separated engine mounts can cause clunking noise during acceleration.

COOLING SYSTEM PROBLEMS

Engine overheating, coolant loss, electrolysis, and slow warm up are the most common cooling system troubles. According to a recent nationwide fleet study, 53% of all premature engine failures are caused by neglecting to maintain the cooling system. From a maintenance standpoint, a cooling system requires attention in these areas.

1. The radiator, pressure cap, fan, belts, pulleys, water pump, oil cooler, thermostat, hoses, gaskets, and heater.
2. The heat transfer or "wet side" of the cooling system (where the coolant meets the metal). This area of the cooling system is seldom considered until the damage is done.

Overheating

Coolant loss (internal and external leaks), the accumulation of contaminants in the water jackets, and the passing of exhaust gases into the coolant are the main causes of overheating.

On some cars with automatic transmissions, a leak in the transmission intercooler may develop. Automatic transmission fluid, a poor conductor of heat, can be mixed with the radiator coolant. As a result, pistons, rings, valves, and valve seats will operate at a higher temperature than normal, and engine and automatic transmission damage can result.

Rust and Scale Corrosion

Oxygen from the air is the most common factor in promoting rust (Figure 4-65). No automobile cooling system is entirely free of air. Aeration (mixing air with water) of the coolant will increase rust corrosion. *Aeration can speed up the corrosion rate by as much as 30 times.* There are several principle causes for aeration.

1. Turbulence in the upper radiator tank mixes entrapped air and coolant.

2. Small leaks at the water pump or hoses can accelerate corrosion by allowing air to be drawn in (Figure 4-66). This is especially true of the lower radiator hose (suction side hose).

FIGURE 4-65
Rust and scale can cause an engine to overheat.

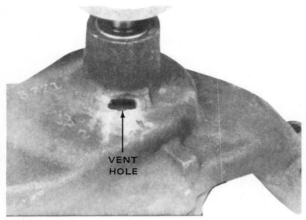

FIGURE 4-66
Coolant leakage from the water pump vent hole indicates a defective seal and will allow air to be drawn in (Courtesy of Dana Corporation—Service Parts Group).

Scaling is a condition caused by minerals (magnesium and calcium) present in water. As soon as engine temperatures reach 140°F, these minerals deposit on the water jacket walls, forming a white crusty coating. This coating is most effective as a heat insulator. *A 1/16" layer of scale deposited on a one-inch thick section of cast iron is equivalent to $4\frac{1}{4}"$ of cast iron as far as heat transfer is concerned.*

Electrolysis

Whenever two different metals are immersed in an electrical conducting liquid, and grounded to each other, an electric current will flow. This condition exists in today's engines. Here is an interesting test you can make using a voltmeter.

1. Ground one voltmeter prod on the radiator.
2. Insert the other prod in the coolant (do not touch the inside of the radiator).
3. Read the meter. You may find anywhere from .2 volt to 3 or 4 volts.

Any value up to .4 volt can be considered normal. However, as voltage increases, so does electrolysis, which can begin to "eat" up core plugs, front covers, water pump impellers, and so on, and cause excessive cooling system corrosion (Figure 4-67).

If you should find a higher than normal voltage reading, perform cooling system service—that is, drain, flush, and refill the radiator.

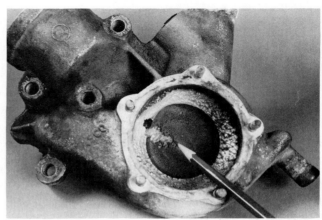

FIGURE 4-67
Electrolysis has started to "eat" into this Toyota water pump cavity.

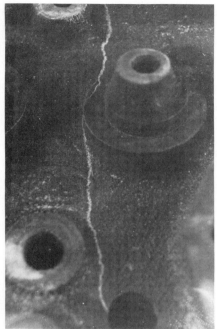

FIGURE 4-68
The severe crack in this six-cylinder Ford head allowed water to mix with oil. The inside of the valve cover was coated with a frothy, yellowish film.

Loss of Coolant

External loss of coolant can often be checked visually by the rust stains that form around the leak area. Remember, rust in the coolant is damage that has already occurred. Rust represents engine metal loss and abrasive particles in the cooling system. Another way to check a cooling system for external leaks is to use a pressure tester. The system is pressurized by a hand pump, and then a gauge is checked to see if pressure holds.

Loss of coolant through an internal crack or leak usually is detected by noting water bubbles on the dipstick or oil filler cap. Some of the more common internal coolant leaks are the result of

1. A hairline crack at either side of the engine block on the lifter chamber wall.
2. An intake manifold gasket that is either improperly installed or defective.
3. A cracked cylinder head (Figure 4-68).
4. A front timing gear cover that is "pinholed" from electrolysis (Figure 4-69).
5. A head gasket that is either improperly installed or defective.

To locate the exact cause for an internal leak is not always easy; it may require partial disassembly of the engine. Observe for "dirty mayon-

FIGURE 4-69
This Pontiac aluminum front cover is "pinholed" from electrolysis. Coolant from the water pump cavity leaked into the timing chain area and into the crankcase.

naise"-colored deposits (usually more evident at the crack or leak area). A method that is sometimes used to pinpoint the exact location of an internal crack prior to engine component disassembly is:

1. Drain and flush the cooling system until it is clean.
2. Fill the system with water.
3. Add two fluid ounces of red food coloring.
4. Allow the engine to reach operating temperature, and then continue to run for ten minutes.
5. The water-soluble red food coloring will show up in the crack.

Combustion Gas Leakage

A defective head gasket or cracked head or block can allow hot exhaust gas to pass into the coolant. If a head gasket combustion leak is allowed to continue, the gasket will deteriorate and allow coolant to enter the crankcase. Rapid oil contam-

ination, corrosion of engine parts, and formation of sludge will occur. Gummy deposits (from ethylene glycol type antifreeze) can cause engine seizure. There are several ways to test for internal combustion gas leakage into the coolant.

1. Start the engine and watch a pressure tester gauge after operating temperature is reached. If the gauge needle fluctuates up and down as the engine is revved up, the head gasket is probably faulty (Figure 4-70).
2. Use a combustion leak tester (Figure 4-71*a* and *b*).
3. Use an infrared tester to "sniff" for the presence of combustion gas at the radiator filler opening.

To simply look into the radiator filler neck to watch for bubbles when the engine is running is not the best test for a combustion gas leak. Bubbles can also be caused by air trapped in the system, or by natural turbulence of the coolant as it returns to the upper radiator tank.

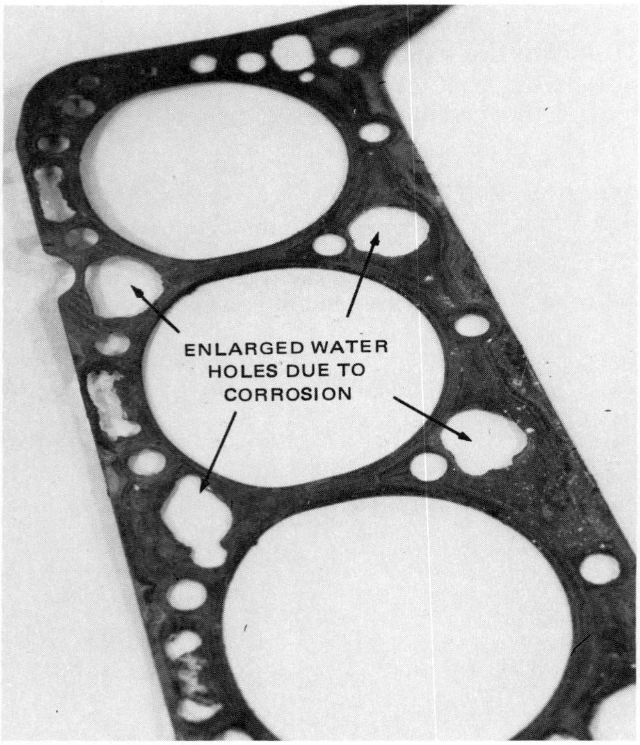

ENLARGED WATER HOLES DUE TO CORROSION

FIGURE 4-70

The water holes on this head gasket are enlarged because of corrosion. With less gasket material separating the coolant from the cylinders, combustion gas entered the cooling system.

P&G
BLOC-CHEK
COMBUSTION LEAK TESTER
U.S. PATENT 2.869.231
Other Patents Pending
TEST FLUID
16 FLUID OUNCES
NOT FOR HUMAN CONSUMPTION
P & G MANUFACTURING CO.
PORTLAND, OREGON

(a)   *(b)*

FIGURE 4-71

(a) This test fluid will change color in the presence of combustion gas. (b) The tube is filled with test fluid and placed in the radiator filler opening. Air above the radiator coolant is drawn into the test fluid while the engine is running.

FIGURE 4-72

A head gasket blown-out in two locations.

FIGURE 4-73

A head gasket with a leak to the outside.

ENGINE COMPRESSION PROBLEMS

A properly performed compression test will tell a mechanic which cylinder (or cylinders) is leaking. Engine cylinder pressures normally should not vary more than 25% from the highest to the lowest one. For instance, if the highest cylinder in an engine reads 200 psi, the minimum pressure allowed in any of the remaining cylinders would be 150 psi. If cylinder pressures are not correct, the engine cannot be tuned properly. Variations beyond specifications will cause uneven idling and loss of power. Engine compression problems are illustrated in Figures 4-72, 4-73, and 4-74.

A properly performed compression test is conducted in the following way.

1. The engine must be at *operating temperature.*
2. *Remove all the spark plugs.* If necessary, wear a pair of garden gloves to prevent burning your hands.
3. Open the choke plate and carburetor throttle valve(s) to a *wide open position.* A closed throttle valve will pull raw gasoline from the idle-discharge port in the carburetor, washing the oil film off the cylinder walls. This can cause excessive friction, slow cranking speed, and compressed gases to leak.
4. Disconnect the battery feed wire to the coil. When this wire is completely removed, the ignition system turns off. Grounding the coil high-tension wire will eliminate high voltage, but still allows the primary (low voltage) circuit to function. If you simply remove the

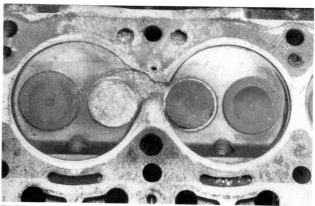

FIGURE 4-74

This is what can happen if a vehicle continues to be driven with a blown head gasket. In this case, the cylinder head was ruined.

FIGURE 4-75
This explosion was caused when the coil operated on open circuit during a compression test.

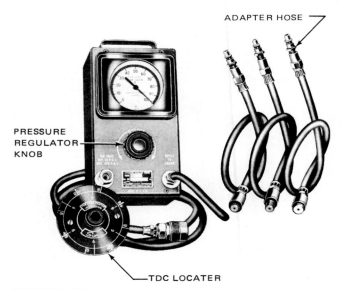

FIGURE 4-76
A popular cylinder leakage tester (Courtesy of Sun® Electric Corporation).

coil high-tension wire, the coil will operate on open circuit and possibly explode (Figure 4-75).

5. Crank the engine over so that each cylinder goes through *five* compression strokes.

6. Read the compression gauge. *NOTE: The battery should be fully charged during the test. An undercharged battery can cause low cranking rpm, and the cylinders will have lower pressure due to the increased time allowed for the compression stroke.*

A compression test will identify only what cylinder is at fault. To more accurately determine a cylinder leakage point use a cylinder leakage tester (Figure 4-76). The *cylinder leak test* (sometimes called an air test) will point out intake and exhaust valve leaks, leaks between cylinders or into the water jacket, bad rings, or defective pistons. This tester applies air to the cylinder at controlled volume and pressure, and then measures the percentage of cylinder leakage.

DIAGNOSING A WORN TIMING CHAIN AND SPROCKET ASSEMBLY

The owner of a Pontiac with a 400 CID engine complains of rough idling, hard starting, poor performance, and "popping" noise coming out of the carburetor top. According to the owner, the car was running fine until last weekend. Now, every day, it seems to get worse. How would you diagnose this problem?

The following criteria are suggested to help verify a suspected worn (loose) timing chain.

1. What is the size and make of engine? Certain model engines seem to have a higher frequency of timing chain problems than others. This author has seen several big block Chevrolet chains so loose after 30,000 miles that the chain was hitting the inside of the timing gear cavity (sounded like a rod knock).

2. How many miles are on the engine? Most vehicles need timing chain and sprocket replacements after 80,000 miles.

3. Does the engine have the symptom of idling

rough? If so, advance spark timing by turning the distributor until the engine smooths out. Then check the timing with a timing light. If the chain is bad you will see the timing set at from 25° to 30° advance according to the marks—yet the engine starts easily! The reason for this is simple. A loose timing chain will retard valve timing. Therefore, when the spark is advanced, you tend to compensate for the chain retard lag.

4. As you check the timing, are the "marks jumping back and forth"? A stretched or loose chain will prevent timing marks from holding steady across from each other.

5. Check the respective movement between the distributor rotor and the crankshaft. Turn the engine over slowly by hand, rocking the crankshaft back and forth. Carefully watch the movement between the rotor and the crankshaft to determine the amount of timing chain play. A 4° movement of the crankshaft, without any rotor movement, indicates excessive chain play.

6. Is an unusual amount of noise coming from beneath the timing gear cover? As the nylon protective coating on the cam sprocket (gear) wears and flakes off, soft aluminum is exposed. These teeth now are meshed with the much harder steel chain (Figure 4-77). A *grinding, gnashing type sound* results (similar to dragging a long tow chain on a concrete sidewalk).

7. Do all cylinders have low compression? A loose timing chain will cause cylinders to fill improperly. A lower than normal compression reading will result. A slipped timing chain will cause all of the cylinders to read near zero psi. To quickly diagnose a slipped chain, place the palm of your hand over the tailpipe opening and crank the engine. If your hand is drawn in, the chain has slipped.

QUICK CHECK
FOR A BURNED VALVE

Consider this example. A vehicle (with 78,000 miles on the odometer) is driven into the shop. The engine idle is very rough (the radio antenna is shaking badly). The tailpipe exhaust beat is irregular with a *"motorboat" type sound*. How would you diagnose the problem?

FIGURE 4-77
All of the nylon coating has been worn off this Buick cam gear (sprocket) teeth.

The following procedure enables you to quickly and accurately determine if the engine has a burned valve.

1. Note the symptoms during engine idle. Rough idle, and an irregular "motorboating" exhaust sound will occur when a valve is burned.

2. Raise the engine speed to approximately 2500 rpm. With ignition malfunction, the engine will still be shaking and running rough, but with a burned valve, the engine will smooth out.

3. Listen to the cranking rhythm. Disconnect the battery feed lead to the coil and crank the engine over. A burned valve (which means that compression is low in one of the cylinders) will cause the cranking speed to be uneven.

QUICK CHECK FOR
EXCESSIVELY WORN VALVE GUIDES

The owner of a Chevrolet pickup truck complains of excessive oil consumption. Large puffs of blue-gray exhaust smoke are evident during engine deceleration. If this problem is being caused by worn valve guides, how could you quickly diagnose it?

Use this procedure to determine if engine valve guides are worn excessively.

1. Note the symptoms. Excessively worn valve guides will almost always cause engine oil consumption problems. During deceleration watch for a dense cloud of blue-gray smoke.

2. Adjust the carburetor for the best idle smoothness.

3. Raise the engine speed to approximately 3000 rpm (transmission in neutral). Hold this rpm for three seconds.

4. Suddenly release the throttle and allow the engine to return to curb idle.

5. The instant the engine returns to curb idle speed, observe if the engine momentarily shakes. If the engine runs rough for just a moment, it indicates that the valve guides are worn.

Some mechanics prefer to check for valve guide wear by first removing the valve covers, oper-

FIGURE 4-78

Soaking the valve stem area with oil to test for valve stem-to-guide noise (Courtesy of Ford Parts and Service Division).

ating the engine at idle speed and listening for valve stem-to-guide noise. The valve stem seal area is then saturated with oil to see if the noise can be dampened (Figure 4-78). *NOTE: Apply side thrust with a hammer handle against the spring retainer when the valve has just started to open.* You will be able to see the relative movement between the stem and guide.

TRUE CASE HISTORY

Vehicle	Pontiac
Engine	8 cylinder, 389 CID
Mileage	82,000
Problem	1. No oil pressure.
	2. A loud rod clatter sound while running the engine.
	3. Rough idle.
Cause	A nylon timing gear particle entered the oil pump through the inlet bypass valve in the screen. This piece of nylon wedged between the oil pump gears (Figure 4-79). As a result, the oil pump driveshaft twisted out of position and broke (Figure 4-80). One of the connecting rods then started hitting this bent up driveshaft, resulting in the clatter noise.

FIGURE 4-79
A piece of nylon (from the cam gear) jammed this oil pump.

FIGURE 4-80
Twisted oil pump driveshaft caused by the pump jamming.

TRUE CASE HISTORY

Vehicle	Ford Van
Engine	8 cylinder, 302 CID
Mileage	68,000
Problem	1. Engine would not start.
	2. ''Spitting'' from the top of the carburetor during engine cranking.
	3. 10–20 psi compression in each cylinder.
	4. Palm of hand is drawn into tailpipe opening during cranking.
Cause	A timing chain that slipped because the nylon was worn off the cam gear teeth. See Figure 4-81.

FIGURE 4-81
Badly worn timing chain assembly. The nylon has started to flake off the cam gear.

TRUE CASE HISTORY

Vehicle	Ski Boat
Engine	8 cylinder, 289 CID
Mileage	Unknown
Problem	1. Engine suddenly quit during water skiing.
	2. Sea water started pouring out of the top of the carburetor.
Cause	One of the cylinder head water plugs had rusted through and caused an internal leak (Figure 4-82). This allowed the coolant (sea water, in this case) from the water jackets to flood into the engine and fill it up. The defective plug is shown in Figure 4-83.

FIGURE 4-82
Some cylinder heads have threaded-in water plugs.

FIGURE 4-83
This water plug rusted through and filled the engine with water.

CHAPTER 4 SELF-TEST

1. List eight points on a typical engine where an oil leak could occur.
2. What is indicated by an area underneath an engine that is washed clean with oil?
3. A grooved vibration damper hub will allow what to leak?
4. What is the procedure for checking wear on a PCV valve?
5. What is aerated oil?
6. Name one immediate giveaway to aerated oil.
7. What happens to engine oil pressure when the oil pump pressure relief valve jams in a closed position?
8. List three conditions that can promote oil ring clogging.
9. How can clogged cylinder head oil return passages cause high oil consumption?
10. What precaution must be observed when using a chemical cleaning concentrate in a badly sludged engine?
11. Give three reasons why a piston ring might stick in its groove.
12. How can excessive ring groove side clearance cause oil leakage into the combustion chamber?
13. What are GI spacers?
14. Why is piston ring end clearance needed?
15. When disassembling an engine, you notice carbon deposits on the ring faces near the gap. What does this indicate?
16. Why will installing one compression ring upside down cause a loss of oil control?
17. What precaution must be observed when installing an oil control ring expander?
18. What do dull gray vertical scratches on ring faces indicate?
19. What is meant by the term scuffing? Scoring?
20. How can a vacuum leak lead to accelerated engine wear?
21. How can a badly smoking engine be caused by a faulty automatic transmission modulator?
22. How can a misaligned connecting rod cause excessive engine oil consumption?
23. How can improperly fit wrist pins lead to blow-by and high oil consumption?
24. How will a loose or worn timing chain affect valve timing?
25. How can a defective fuel pump cause cylinder and bearing damage?
26. How can a torn intake manifold gasket allow oil to be drawn into the engine and burned?
27. What is the test procedure for verifying a suspected internal vacuum leak?
28. Why is it important to turn on the key when using a remote starter button to crank a 1961 or later General Motors vehicle?
29. Rocker arm shaft oil passage holes normally face what direction?

30. How can worn rocker arm pivot balls cause excessive engine oil consumption?
31. Describe two methods to prevent oil buildup in canoe type rocker arms.
32. Explain how defective rocker arm shaft support bolt holes on certain Ford engines can allow oil to be drawn into the combustion chamber.
33. What is the purpose of the metal baffle plate installed in some valve covers?
34. Worn engine bearings will increase oil throw off. How can this cause oil burning?
35. Why should a block never be bored without the main bearing caps torqued in place?
36. How can an excessively rich carburetor mixture lead to excessive engine oil consumption?
37. How can oil loss occur through worn intake valve guides on an OHV engine? Through worn exhaust valve guides?
38. What is a smoke test? How is it performed?
39. What are two useful tools to locate engine noises?
40. How can you test for a broken flex-plate?
41. When are most main bearing knocks noticeable?
42. How can you verify a connecting rod knock?
43. When is piston slap the loudest?
44. What is a good test for piston slap?
45. Piston pin noise usually causes what type of sound?
46. What condition is sometimes wrongly diagnosed as piston slap?
47. What characteristic is noted about the frequency of valve train noise?
48. What is the difference between detonation and preignition?
49. List five engine conditions that can cause detonation.
50. List five engine conditions that can cause preignition.
51. How can you verify a noisy drive belt?
52. What three engine conditions are the main causes of overheating?
53. How can a defective automatic transmission intercooler cause engine damage?
54. What are the principal causes of cooling system aeration?
55. What is scaling?
56. What is electrolysis?
57. How do you use a voltmeter to test for electrolysis in an engine?
58. Name two ways to check for external loss of coolant.
59. List four possible causes of an internal coolant leak in an engine.
60. How can red food coloring be used to locate an internal crack?
61. Why will exhaust gases, leaking into the water jackets, cause engine overheating?
62. What are three good tests for exhaust gases leaking into the coolant?
63. Why must the choke plate and carburetor throttle valve(s) be wide open when you make a compression test?

64. Engine cylinder pressures normally should not vary more than what amount from the highest to lowest?

65. What effect does low cranking rpm have on a compression test?

66. How can you accurately determine the source of compression leakage?

67. How can a loose timing chain produce a rod knock sound?

68. List five conditions that will help to diagnose a loose timing chain.

69. What is a quick way to diagnose a slipped timing chain?

70. What quick check procedure can be used to determine if an engine has a burned valve?

71. What quick check procedure can be used to determine if an engine has excessively worn valve guides?

CHAPTER 5

ENGINE IDENTIFICATION AND REFERENCE MATERIAL

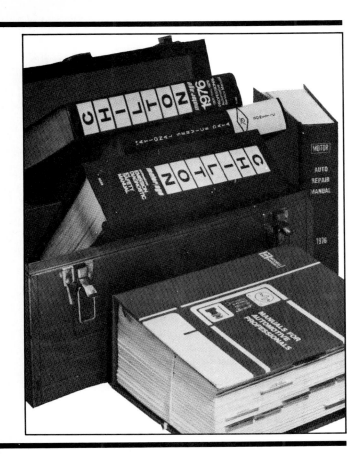

Not knowing the exact year or type of engine you are working on may lead to unnecessary headaches. For instance, you can unknowingly order the wrong part, improperly install an engine component, set incorrect clearance, or make a faulty adjustment. Such problems can be virtually eliminated by becoming familiar with certain details and available references.

VEHICLE IDENTIFICATION PLATE

All automobile manufacturers permanently stamp to the vehicle a serial number or attach a vehicle identification number (VIN) plate. This number defines exactly the specifications and engine option that was included when the vehicle was built. If the engine is still in the original production body and has not been changed, it can be identified by the engine code. The engine code is a certain digit of the vehicle identification number. Refer to Figure 5-1 for an example of a VIN system. Complete decoding information is always included in the applicable shop (factory) manual.

Sometimes decoding information is in the owner's manual. The code system varies with the different vehicle manufacturers. On most current automobiles the VIN plate is located on the top of the dash in the left corner where it can be seen through the windshield (Figure 5-2). On some automobiles the plate will be found on the left front door jamb or on the firewall.

ENGINE IDENTIFICATION

On some vehicles the engine code that is shown on the VIN plate can be verified by an alphabetical letter that is located on the engine block (Figure 5-3). This allows for positive identification of the engine. As a point of interest, General Motors was recently faced with several lawsuits because customers discovered that their new Oldsmobiles had Chevrolet engines.

Sometimes a tag mounted under the coil or at the front of the engine identifies engine cubic inch displacement and year (Figure 5-4a and b).

Table 5-1 is a reference table to determine which

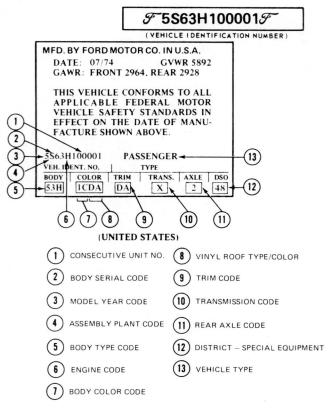

1. CONSECUTIVE UNIT NO.
2. BODY SERIAL CODE
3. MODEL YEAR CODE
4. ASSEMBLY PLANT CODE
5. BODY TYPE CODE
6. ENGINE CODE
7. BODY COLOR CODE
8. VINYL ROOF TYPE/COLOR
9. TRIM CODE
10. TRANSMISSION CODE
11. REAR AXLE CODE
12. DISTRICT – SPECIAL EQUIPMENT
13. VEHICLE TYPE

FIGURE 5-1

A VIN system and code (Courtesy of Ford Parts and Service Division).

FIGURE 5-2

The VIN plate is usually riveted to the cowl bar in the lower left-hand corner of the windshield (Courtesy of Cadillac Motor Division of General Motors Corporation).

FIGURE 5-3

This engine code number is located on a machined pad adjacent to the distributor. The letter in this code number denotes the cubic inch displacement of the engine (Courtesy of American Motors Corporation).

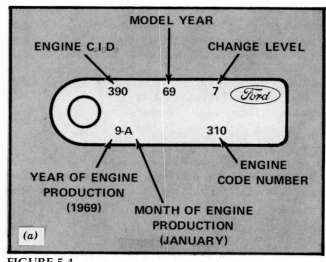

FIGURE 5-4

(a) Ford engine tag (Courtesy of Ford Parts and Service Division). (b) Engine tag used on American Motors (Courtesy of American Motors Corporation).

TABLE 5-1
Reference table for vehicle identification

AMERICAN MOTORS CORPORATION

Code (7th digit)	Engine Size (Cylinders and Cubic Inches)
A	6-258 with 1 bore carburetor
C	6-258 with 2 bore carburetor
E	6-232
H	8-304
N	8-360 with 2 bore carburetor
P	8-360 with 4 bore carburetor
Z	8-401

CHRYSLER CORPORATION

Code (5th digit)	Engine Size (Cylinders and Cubic Inches)
C	6-225 with 1 bore carburetor
D	6-225 with 2 bore carburetor
E	Special 6 cylinder
G	8-318
J	8-360 with 4 bore carburetor
K	8-360 with 2 bore carburetor
L	8-360 high performance
N	8-400 with 4 bore carburetor
P	8-400 high performance
T	8-440
U	8-440 high performance
Z	Special 8 cylinder

FORD MOTOR COMPANY

Code (5th digit)	Engine Size (Cylinders and Cubic Inches)
A	8-289 with 4 bore carburetor
	8-460 except police
B	8-406 high performance
	6-240 police
C	8-289 with 2 bore carburetor
	8-429 Cobra Jet
	8-460 police
D	8-302 police and taxi
E	6-240 taxi
F	8-260
	8-302 with 2 bore carburetor
G	8-406 high performance with 6 bore carburetor
	8-302 boss
H	8-390 with 2 bore carburetor
	8-351 with 2 bore carburetor
J	8-430
	8-302 with 4 bore carburetor
	8-429 high performance ram air
K	8-289 high performance
	8-429 with 2 bore carburetor
L	6-250
M	8-390 high performance with 6 bore carburetor
	8-410
	8-351 with 4 bore carburetor
N	8-429 with 4 bore carburetor
P	8-390 police
	8-428 police
	8-429 police
S	6-144
	8-390 GT
	8-400
T	6-200
U	6-170
V	6-223
	6-240

(continued)

(continued)

	FORD MOTOR COMPANY
Code (5th digit)	Engine Size (Cylinders and Cubic Inches)
W	8-292
	8-427
	4-98 (1600 cc)
X	8-352 with 2 bore carburetor
	4-122 (2000 cc)
Y	8-352 with 4 bore carburetor
	4-140 (2300 cc)
Z	6-171 (2800 cc)
	8-390 with 4 bore carburetor

	GENERAL MOTORS
Code (5th digit)	Engine Size (Cylinders and Cubic Inches)
A	4-140 Chevrolet with 1 bore carburetor
B	4-140 Chevrolet with 2 bore carburetor
C	6-231 Buick
D	6-250 Chevrolet
E	4-98 Chevrolet
F	8-260 Oldsmobile
H	8-350 Buick with 2 bore carburetor
I	4-85 Chevrolet
J	8-350 Buick with 4 bore carburetor
K	8-403 Oldsmobile
L	8-350 Chevrolet
P	8-350 Pontiac
R	8-350 Oldsmobile
R	8-350 Cadillac with electronic fuel injection
S	8-425 Cadillac with 4 bore carburetor
T	8-425 Cadillac with electronic fuel injection
U	8-305 Chevrolet
V	4-151 Pontiac
Y	8-301 Pontiac
Z	8-400 Pontiac

engine is being used in a vehicle. Just read the VIN plate and compare with the codes shown.

Casting/Forging Numbers

Often, the only available means of engine identification is the block casting number. Examples are shown in Figures 5-5 and 5-6. Numbers or special identification markings may be visible on the head, camshaft, or crankshaft (Figures 5-7 and 5-8). *NOTE: Refer to the Appendix for a list of block, cylinder head, and crankshaft casting/forging numbers.*

Port Configuration

On certain engines, the exhaust port configuration can serve as a means of identification (Figure 5-9).

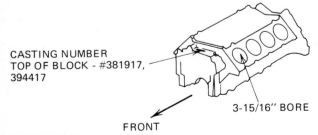

CASTING NUMBER
TOP OF BLOCK - #381917, 394417

3-15/16" BORE

FRONT

FIGURE 5-5
This block casting number is for a 330 CID Oldsmobile.

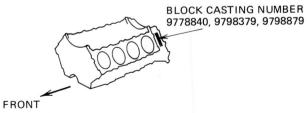

BLOCK CASTING NUMBER
9778840, 9798379, 9798879

FRONT

FIGURE 5-6
This block casting number is for a 326 CID Pontiac.

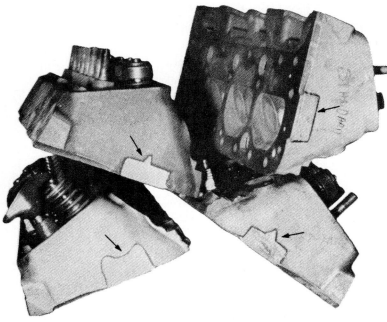

FIGURE 5-7

Identification of various Chevrolet heads can be established by casting marks at the end (Courtesy of *Popular Hot Rodding* magazine).

FIGURE 5-8

Camshaft casting numbers give the type and year of cam.

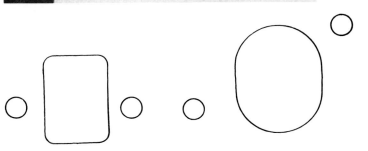

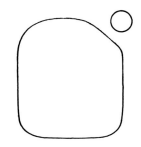

FIGURE 5-9

Small block Ford exhaust port configuration.

260-302
WINDSOR

302-351
2V CLEVELAND

302-351
4V CLEVELAND

REFERENCE MATERIALS

There are many sources of engine specifications and repair procedures.

Repair Manuals

Independent companies publish yearly manuals that include engine specifications, adjustments, and repair procedures. Three popular general repair manuals are:

1. *Motor's Auto Repair Manual* (Figure 5-10).
2. *Chilton's Auto Repair Manual* (Figure 5-11).
3. *Mitchell Manuals—National Service Data* (Figure 5-12).

Factory Manuals

The original complete specifications and repair data for all vehicles are printed by the car manufacturer in a shop or factory manual (Figure 5-13). To obtain factory manuals, write to the car manufacturer; the dealership parts department can provide the company's address.

Parts Company Literature

Engine parts companies have specification tables, installation booklets, service manuals, bulletins, and parts catalogs with interchange listings. The items listed below are extremely useful to the engine rebuilder.

1. *Federal-Mogul Engine Bearing and Shop Specification Manual.*
2. *McQuay-Norris Engine Shaft and Bearing Specification Book.*
3. *Mechanics Engine Bearing Reference Manual* by Clevite.
4. *Perfect Circle Doctor of Motors Service Manual.*
5. *Engine Bearing Service Manual* by Federal-Mogul.
6. *NAPA Engine Parts Catalog.*
7. *A Practical Guide to Understanding and Using Piston Rings* by Muskegon Piston Ring.
8. *Engine Manual* by TRW.
9. *Service Tips for the Automotive Mechanic* by Hastings.
10. *Sealed Power Replacement Engineering Service Bulletins.*
11. *TRW Tech Talks.*
12. *Michigan Engine Bearings Master Catalog* and *Shop Manual* by D.A.B. Industries, Inc.
13. *Engine Parts Catalog for Imported Vehicles* by Repco, Inc.
14. *McQuay-Norris Engine Bearing Installation Tips Book.*

Tool and Equipment Manufacturer Publications

The publications listed below provide valuable technical information.

1. *The Principles of Valve and Valve Seat Reconditioning* by Sioux Tools, Inc.
2. *Valve Seat Stone and Pilot Handbook* by The Black and Decker Manufacturing Company.
3. *Just What Is A Pin Fit?* And *Why Recondition Rods?* by Sunnen Products Company.
4. *The Basics of Cylinder Head Reconditioning* by Kwik-Way.
5. *Crack Repair Manuals* by Irontite Products Company, Inc.
6. *Connecting Rod Aligner Instructions* by K.O. Lee Company.
7. *Instruction Manuals, Blockmaster and Headmaster Surfacing Machines* by Storm-Vulcan, Inc.
8. *Operating Instructions for Sunnen CK-10 Automatic Cylinder Resizing Machine* by Sunnen Products Company.
9. *Boring Machine Operating Instructions* by Rottler Manufacturing Company.
10. *Pin Fitting Machine Instruction Manual* by Tobin Arp Manufacturing Company.
11. *Line Boring Machine Instruction Manual* by Tobin Arp Manufacturing Company.
12. *VN/IDL Valve Service Shop Instructions* by Van Norman Machine Company.

Magazines

There are two very outstanding magazine publications for the engine rebuilder.

1. *The Complete Book of Engines* by Peterson Publishing Company. New model factory data, technical information, and high performance modifications are included in this annual

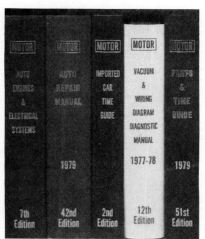

FIGURE 5-10
Motor Manuals.

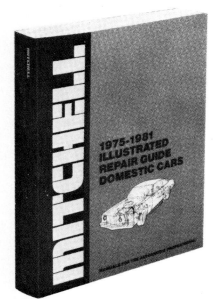

FIGURE 5-12
A Mitchell Manual.

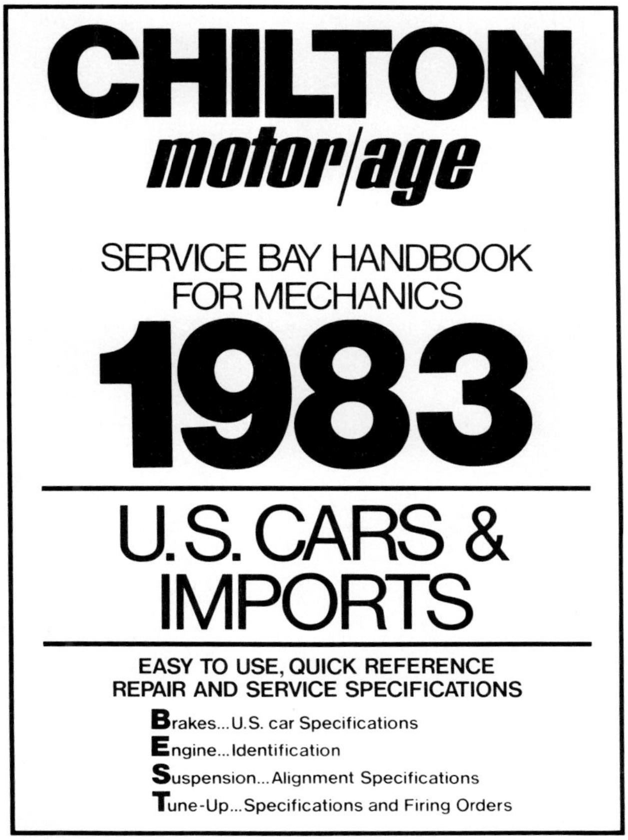

FIGURE 5-11
A Chilton reference manual.

FIGURE 5-13
Manufacturer's shop manual (Courtesy of Ford Parts and Service Division).

publication. Particularly outstanding are the specification charts. Much hard-to-find information has been included, such as bore spacing, piston and rod weights, amount of pin offset, connecting rod center-to-center length, rocker arm ratios, crankshaft material, and ring groove depths.

2. *Automotive Rebuilder* by Babcox Publications, Inc. This monthly magazine is considered the voice of engine remanufacturing.

Trade Organization Material

There are several very worthwhile engine rebuilding organizations. The Automotive Engine Rebuilders Association (AERA) provides core identification guides, engine specifications, and periodic technical service bulletins (Figure 5-14). Each year a national convention is held. A proceedings summary of the technical and administrative topics is sent to each member. Many valuable references, "shop tips," and so on are contained in these proceedings.

The Production Engine Remanufacturers Association (PERA) is an organization for passenger

```
CAUTION ON FLYWHEEL, CRANKSHAFT & DAMPER ON SOME

1981 FORD, LINCOLN-MERCURY 5.0 L ENGINES

Ford Motor Co. announced that some 1981 Ford and Lincoln-Mercury engines
are manufactured with a 1980 crankshaft, flywheel and damper.  These en-
gines may be identified by a decal having a red "S" on a white background,
located on the left hand rocker cover.

The new 1981 flywheel, crankshaft and damper cannot be used on the 1980
engines unless all three components are replaced as a unit.  Otherwise,
engine vibration may result.

1981 engines may be identified by a daub of green paint on the flywheel,
crankshaft and damper.
```

FIGURE 5-14
Service bulletin information (Courtesy of the Automotive Engine Rebuilders Association).

car and truck engine remanufacturers. Every year they compile an outstanding core identification and common numbering parts catalog.

NOTE: Refer to Appendix D for the addresses of the companies and organizations mentioned in this chapter.

TRUE CASE HISTORY

Vehicle	Ford Mustang
Engine	8 cylinder, 289 CID
Mileage	Zero mileage. A used 289 CID engine was purchased and rebuilt because the old engine was damaged beyond repair.
Problem	When installing the rebuilt engine, it was discovered that the bell housing casting on the block had six bolt holes (Figure 5-15a). The original engine had a five-bolt hole pattern (Figure 5-15b).

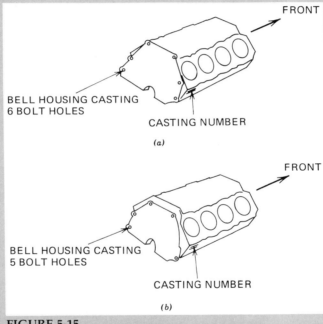

FIGURE 5-15

(a) **Late model 289 CID block used by Ford.** *(b)* **Early model 289 CID block used by Ford.**

Cause	Early model 289 CID blocks have five bolt holes. The block casting number is C40E-C or C40E-F. Late model 289 CID blocks have six bolt holes. The block casting number is C5AE-E.

CHAPTER 5 SELF-TEST

1. What is a VIN plate?
2. Where is the VIN plate located on most current model vehicles?
3. Complete VIN coding information always is found where?
4. Besides the VIN plate, how else can you identify an engine?
5. Name three general repair manuals.
6. What is a factory manual?
7. Name five parts companies that provide technical literature.
8. Name three tool and equipment manufacturers that distribute technical publications.
9. Name two magazine publications for the engine rebuilder.
10. Name two engine rebuilding organizations.

CHAPTER 6

REMOVAL OF ENGINE FROM VEHICLE

Although some engine work can be performed with the engine in the chassis, it is recommended that it be removed for a major overhaul. The time spent removing and replacing the engine is well worthwhile, and will result in a cleaner, more accurate job.

When removing the engine from a vehicle, certain definite procedures need to be followed. This chapter will describe the typical steps for engine removal from the chassis. There will be some minor variances depending on the make and model of vehicle, type and size of engine, type of transmission, and so on. If available, you should refer to an appropriate manual.

UNDER THE HOOD PREPARATION FOR ENGINE REMOVAL

1. Place fender covers on fenders.
2. Mask the edges of the hood, fenders, and cowl if there is any possibility of scratching the paint.
3. Scribe a line to mark the position of the hood hinge attachment brackets (Figure 6-

1). This will enable you to quickly align the hood when installing it.

4. Remove the hood by taking out the bracket bolts. Obtain assistance to help hold the hood when you unbolt and remove it. Be careful not to damage the fenders or break the windshield!

Disassembly organization is extremely important from this point on. Valuable assembly time will be saved by keeping close track of nuts and bolts, wire terminals, and other removed items. If necessary, place small parts in marked cans. Place masking tape around disconnected wires and vacuum hoses. Use a permanent ink felt pen to mark the location on the tape. A photograph of the engine compartment can be very useful.

SHOP TIP

Adhesive number tags used by electricians can be used to mark disconnected wires and hoses. These tags can be purchased at most electrical supply stores.

127

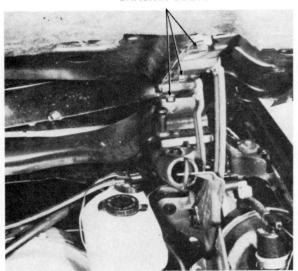

BRACKET BOLTS

FIGURE 6-1
Scribe the hood hinge location on underside of hood panel. This aids in repositioning when reinstalling the bracket bolts (Courtesy of Oldsmobile Motor Division of General Motors Corporation).

FIGURE 6-2
Radiator drain petcock.

PETCOCK

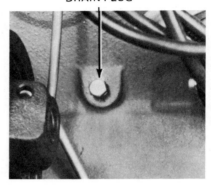

DRAIN PLUG

FIGURE 6-3
Cylinder block drain plug.

FIGURE 6-4
Special tool for removing radiator hoses (Courtesy of Mac Tools, Inc., Washington Court House, Ohio 43160).

5. Remove the battery. On some vehicles it will be necessary to disconnect the ground cable at the cylinder head before the engine can be lifted.

6. Drain the cooling system at the radiator and the cylinder block (Figures 6-2 and 6-3).

7. Remove the air cleaner and intake duct assembly.

8. Disconnect the upper and lower radiator hoses. It may be necessary to use a screwdriver or special tool (Figure 6-4) to pry the hoses loose. Disconnect the heater hoses at the block fitting and water pump.

9. On a vehicle with an automatic transmission, disconnect the transmission oil cooler lines from the radiator. Always use a flare nut wrench when removing tubing fittings (Figure 6-5) to prevent rounding off the corners of the fitting.

10. Remove the bolts attaching the fan shroud to the radiator, if so equipped.

11. Remove the radiator. Remove the fan, spacer, belt pulley, and shroud.

12. On an air-conditioned vehicle, remove the compressor from the mounting bracket, and position it out of the way, *leaving the refrigerant lines attached.* If this is not possible,

FIGURE 6-5
Flare nut wrenches (sometimes called tubing or line wrenches) prevent rounding off fitting corners.

isolate and remove the compressor. Refer to the air conditioning safety section in Chapter 1.

13. On a vehicle with power steering, disconnect the power steering pump. Remove the drive belt. Position the power steering pump out of the way and in a position that will prevent the fluid from draining out.

14. On a vehicle with power brakes, disconnect the brake vacuum line from the intake manifold. Disconnect any other lines or hoses attached between the engine and chassis.

15. Remove the alternator and mounting brackets. Move them out of the way, leaving the wires connected.

16. Disconnect the fuel inlet line at the fuel pump. *Plug or cap the line to prevent fuel loss.*

17. Disconnect the accelerator cable or linkage at the carburetor and intake manifold. Disconnect transmission downshift linkage, if so equipped.

18. On some vehicles it will be necessary to remove the transmission fluid filler tube attaching bolt from the cylinder head.

19. Disconnect the engine wire harness at the ignition coil, water temperature sending unit, and oil pressure sending unit. Remove the wire loom from the hold-down clips.

20. Remove the carburetor from the engine to prevent damage from the lift sling as the engine is being removed. *Use two wrenches to avoid twisting the fuel line.*

21. Remove the distributor, distributor cap, and spark plug wires from the engine. This will prevent firewall or lift sling interference damage as the engine is tilted during removal.

22. Make one last check around the engine compartment for any additional items to be removed.

UNDER THE VEHICLE PREPARATION FOR ENGINE REMOVAL

1. Raise the vehicle.
2. Drain the engine oil.
3. Disconnect the exhaust pipe(s). See Figure 6-6.
4. Remove the nuts or bolts attaching the front

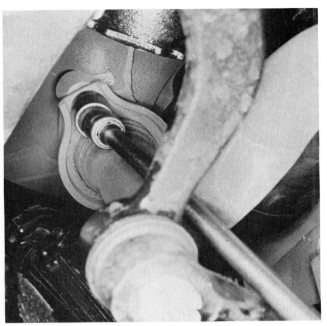

FIGURE 6-6
Disconnecting the exhaust pipe from the exhaust manifold is best done from underneath the vehicle.

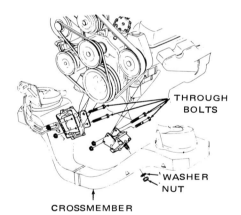

FIGURE 6-7
Typical engine mounts (Courtesy of Cadillac Motor Division of General Motors Corporation).

motor mount rubber insulators to the crossmember (Figure 6-7).

5. Remove the starter (this is only necessary on some vehicles).
6. On a vehicle with a manual shift transmission, disconnect the clutch linkage.
7. Remove the clutch or converter housing bolt access cover.

8. On a vehicle with a manual shift transmission, remove the transmission. It is not necessary to remove the clutch assembly at this time. It will come out with the engine.

9. If the vehicle has an automatic transmission, disconnect the converter from the flex-plate (Figure 6-8). Rotate the assembly in the normal direction of engine rotation as necessary. Mark the converter and flex-plate position in order to maintain correct balance.

10. Remove the clutch or converter housing upper and lower attaching bolts. *The upper bolts can be removed by lowering the rear of the transmission/engine assembly from 5° to 6° and using a long socket extension.*

11. Lower the vehicle. Support the transmission and clutch or converter housing with a jack or support bar.

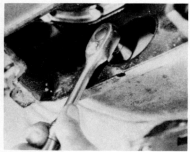

FIGURE 6-8
Removing the torque converter mounting bolts through the service hole at rear of engine.

FIGURE 6-9
Engine lifting eye bolts and sling (Courtesy of Ford Parts and Service Division).

Lifting Out the Engine

1. Attach an engine lifting sling (Figure 6-9). If a piece of chain is used, place flat washers under the attaching bolt heads. *CAUTION: The sling attachment bolts must thread into the block a minimum of 1½ times the bolt diameter. Engines are quite heavy, and this will prevent the bolts from pulling out during lifting.*

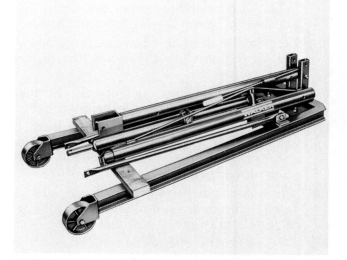

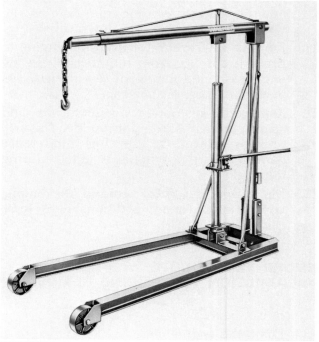

FIGURE 6-10
This "cherry picker" disassembles for storage and transportation (Courtesy of Lincoln St. Louis, Division of McNeil Corporation).

2. Select a chain hoist or a portable hydraulic crane (commonly called a "cherry picker" in the trade). See Figures 6-10 and 6-11. Attach the engine sling to the lift hook.

3. Raise the engine slightly and carefully pull it forward from the transmission to disengage the converter (Figure 6-12). *Never remove the engine with the converter; make sure that the converter remains attached to the transmission.*

FIGURE 6-11

A ½ ton chain hoist.

TRANSMISSION   SCREWDRIVER
SUPPORT BAR

FIGURE 6-12

Using a screwdriver to disengage the converter from the engine after the converter mounting bolts have been removed.

Carefully tilt the engine (it may be necessary to reposition the sling) and lift it out of the engine compartment. Be careful not to bend or damage any components.

4. Support the engine on the floor, or install it on a stand. *NOTE: After using the "cherry picker," be sure the lift ram is always fully retracted.* If this is not done, rust will form on the polished ram surface and can damage the seals and cause a leak.

WARNING: Do not let the converter hang unsupported in the transmission while the engine is out being worked on. This will ruin the front seal. Either install a support or remove the converter.

REMOVING THE ENGINE AND THE TRANSMISSION

In some cases, the manual's suggested procedure for engine removal calls for unnecessary steps. When taking out an engine (particularly on late model Fords) many shops pull transmission and all in order to save time.

On transverse mounted engines with front wheel drive, it is recommended to remove the engine and transaxle assembly as a unit (Figure 6-13).

Under the Vehicle Preparation for Removing the Engine and Transmission

1. Raise the vehicle; for best accessibility use a twin-post hoist.

2. Drain the engine oil.

3. Drain coolant from the engine block by removing the drain plugs or opening the petcocks.

4. Remove the nuts or bolts that attach the exhaust pipe(s) to the manifold(s).

5. Disconnect the speedometer cable where it attaches to the transmission (Figure 6-14).

6. On automatic transmissions, disconnect the oil cooler lines at the transmission (Figure 6-15).

7. Disconnect any electrical wires that attach to the transmission from under the dash.

8. Disconnect any transmission or clutch linkage that is not attached directly to the engine.

9. On front wheel drive vehicles, pull the drive

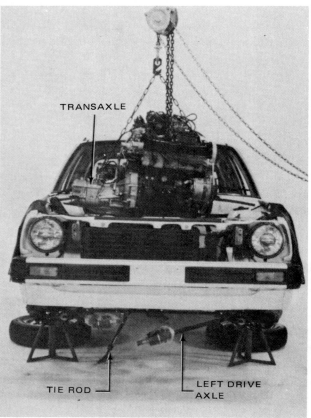

FIGURE 6-13

Raising and removing a transverse engine with the transaxle assembly attached (Courtesy of Honda Motor Co., Ltd. All rights reserved).

FIGURE 6-14

Loosen the serrated collar and disconnect the speedometer cable.

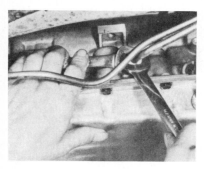

FIGURE 6-15

Disconnect oil cooler lines from transmission fittings.

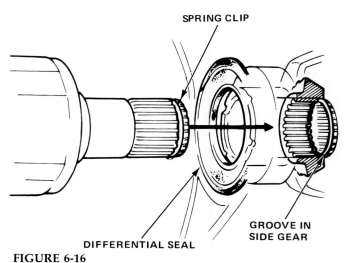

FIGURE 6-16

Front wheel drive axles are splined to the differential side gears (Courtesy of Honda Motor Co., Ltd. All rights reserved).

FIGURE 6-17

Marking driveshaft and companion flange with a center punch. This will ensure correct orientation of driveshaft during installation.

axles from the differential gear splines (Figure 6-16). *CAUTION:* *Do not damage the differential seals.*

10. On conventional rear wheel drive vehicles, scribe or chalk a reference mark between the driveshaft and rear end companion flange (drive pinion yoke). See Figure 6-17. This will allow the same balance to be retained upon reassembly.

Disconnect the rear universal joint and remove the driveshaft. *CAUTION:* *Do not let a driveshaft hang unsupported.* This can distort the yoke or bend the centering ball out of position, resulting in a severe vibration. To correct, a new driveshaft has to be installed.

Install a spare universal yoke or commercially made plug over the transmission output shaft to eliminate any oil spillage (Figure 6-18).

FIGURE 6-18

This set of transmission tailshaft plugs come in three different sizes (1½" OD, 1 11/16" OD, and 1⅞" OD). Using a plug eliminates the spilling of automatic transmission fluid during transmission removal.

11. Remove the nuts or bolts that fasten the transmission mount to the frame crossmember.

12. Disconnect any electrical wires that attach to the starter motor or solenoid.

13. Disconnect the fuel inlet line from the fuel pump.

14. Remove the nuts or bolts that attach the front motor mounts to the frame crossmember.

15. Make a final visual inspection for anything left connected between the engine/transmission assembly and the chassis.

16. Lower the vehicle.

Under the Hood Preparation for Removing the Engine and Transmission

1. Position fender covers on fenders. Use tape to mask the edges of any painted surfaces that might be scratched.

2. Remove the hood. (Refer to Figure 6-1.)

3. Remove the air cleaner and intake duct assembly.

4. Remove the battery.

5. Drain the radiator. Remove the radiator and fan shroud.

6. On vehicles with power-steering, disconnect the hoses attached to the pump.

7. On air-conditioned vehicles, remove the compressor and position it out of the way, leaving the refrigerant lines attached. If this is not possible, isolate and remove the compressor.

8. On vehicles with power brakes, disconnect

the brake vacuum line that attaches to the intake manifold. Disconnect any other lines or hoses attached between the engine and chassis.

9. Disconnect the accelerator pedal linkage from the carburetor and intake manifold.

10. Disconnect any electrical wires attached between the engine and the chassis.

11. Remove the carburetor, distributor, and distributor cap from the engine. This will prevent possible breakage when the engine and transmission are tilted during removal.

12. Make a final visual inspection for anything left connected between the engine and the chassis.

Lifting Out the Engine/Transmission Assembly

1. Attach an engine lifting sling. Position the sling so that the engine/transmission assembly will be sharply tilted during lifting. *COMMENT: Many transaxle units have a factory installed lifting bracket attached to the case (Figure 6-19).*

2. Attach a chain hoist or "cherry picker" to the lift sling.

3. Carefully raise and pull the engine forward, making sure the transmission clears the bottom of the firewall.

4. Support the engine/transmission assembly on the floor.

FIGURE 6-19

This Hondamatic transaxle unit has a factory installed lifting bracket.

AFTER THE ENGINE HAS BEEN REMOVED FROM THE VEHICLE

Once the engine has been removed, visually inspect all external surfaces for any damage or sources of leaks. Do this before steam cleaning. *NOTE: Have the radiator rodded out, not back flushed.* Back flushing will not remove scale deposits from the tubes. If the radiator core is damaged, replace it. In many cases defective radiators have enough capacity to cool the old engine, since it is loose and has very little internal friction. However, the rebuilt engine will run hotter than normal during the break-in period. The radiator will have to be in top condition to handle this extra heat.

An evaluation of the automatic transmission condition needs to be made. Explain to the vehicle owner that even though the automatic transmission was working properly behind the old engine, the rebuilt engine (which develops more power) will often "take out" a forward clutch assembly. This results in a no-drive condition and an unhappy customer. If any transmission defects are corrected now, it will be ready to perform with the new engine.

Separating an Automatic Transmission from the Engine

1. Place a drain pan under the transmission pan and drain the fluid. If the transmission does not have a drain plug, allow the fluid to drain by loosening the attaching bolts at the rear of the pan.

2. Remove the converter access cover from the lower end of the converter housing.

3. Remove the converter-to-flex plate attaching nuts or bolts. Mark the converter and flex-plate position in order to maintain correct balance. Place a wrench on the crankshaft pulley attaching bolt to turn the converter to gain access. *CAUTION: On belt-driven overhead camshaft engines, never turn the engine backward because the belt can skip and change the valve timing. This can result in valves hitting the pistons.*

4. If the converter has a drain plug (Figure 6-20), remove the plug and catch the fluid in a container.

5. Remove the bell housing bolts that attach the automatic transmission to the back of the block.

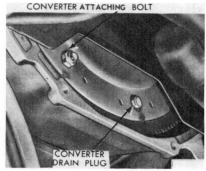

FIGURE 6-20
Converter drain plug location (Courtesy of Ford Parts and Service Division).

FIGURE 6-21
A converter holding strap. When removing or installing the transmission this device prevents the converter from falling out, regardless of how much the transmission is tilted.

FIGURE 6-22
This guide pin was made from a bolt.

6. At this point, the transmission can be pulled back away from the block. When the transmission starts to separate from the block, hold the converter in the bell housing (Figure 6-21). This will prevent the converter from pulling out of the front pump and falling.

Separating a Standard Transmission and Clutch Assembly from the Engine

1. Remove the bolts that attach the transmission to the flywheel housing (bell housing). *Slide the transmission straight back, using a jack or guide pins to support the weight* (Figure 6-22). If the transmission is allowed to rest un-

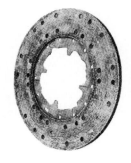 

CRACKED
SEGMENTS

FIGURE 6-23

The segments on this clutch disc were cracked during transmission installation. Once in service the complete hub broke out.

supported in the clutch disc, the hub can be bent out of position and ruined (Figure 6-23). *COMMENT: Some mechanics will remove and replace a standard transmission by "shaking" the extension housing up and down, or back and forth. This is a poor shop practice that will ruin the clutch disc.*

2. Remove the bolts that attach the bell housing to the cylinder block. On some engines it may be necessary to remove a metal cover plate on the front bottom edge of the bell housing.

3. Move the housing back just far enough to clear the pressure plate assembly, and remove the housing. Be careful not to damage the clutch linkage.

4. Loosen the pressure plate cover, attaching bolts evenly to release the spring tension without distorting the cover. If the same pressure plate and cover assembly are to be reinstalled, mark the cover and flywheel so

FIGURE 6-24

Marking clutch and flywheel (Courtesy of Chrysler Corporation).

the pressure plate can be reinstalled in the same position to maintain correct balance (Figure 6-24).

TRUE CASE HISTORY

Vehicle	Dodge Van
Engine	6 cylinder, 225 CID
Mileage	Zero mileage. The old engine was removed and an exchange engine was installed.
Problem	1. A loud clatter noise was heard the moment the new engine started. 2. The noise sounded similar to a rod knock. 3. The noise would speed up or slow down in relation to crankshaft rpm.
Cause	When the clutch housing bottom cover was reinstalled, different bolts were used. They were too long and several were hitting the flywheel ring gear as it turned.
Comment	This problem could have been avoided if the mechanic had kept bolts better organized during engine removal.

CHAPTER 6 SELF-TEST

1. Prior to removing a hood, what should be done to enable you to maintain alignment when reinstalling?
2. How can the location of disconnected wires and vacuum hoses be easily marked?
3. What type of wrench should be used to remove tubing fittings?
4. Why should two wrenches be used to disconnect the fuel line from the carburetor?
5. Why should the distributor be removed before lifting the engine from the vehicle?
6. On a vehicle with a manual shift transmission, is it necessary to remove the clutch assembly before lifting out the engine?
7. Before disconnecting the converter from the flex-plate, why should the position be marked?
8. What procedure helps when you remove the converter housing upper bolts?
9. Sling attachment bolts must thread into the block a minimum of how much?
10. Why isn't it a good idea to lift an engine with the converter attached?
11. After using a "cherry picker", why should the lift ram be fully retracted?
12. Why shouldn't the converter be allowed to hang unsupported in the transmission while the engine is being worked on?
13. On transverse mounted engines, is it recommended to remove the engine and transaxle assembly as a unit?
14. Why shouldn't a driveshaft be allowed to hang unsupported?
15. Why should you chalk a reference mark when disconnecting the drive pinion yoke?
16. What can be done to eliminate oil spillage from the transmission output shaft during removal?
17. In transaxle units, where is the factory installed lifting bracket often attached?
18. When a rebuilt engine is being installed in a vehicle, why should the radiator be rodded out, instead of back flushed?
19. Before installing a rebuilt engine, why is it wise to evaluate the automatic transmission?
20. Why shouldn't a belt-driven OHC engine be turned backward?
21. What are guide pins?
22. When replacing a standard transmission, why is it poor shop practice to "shake" the extension housing up and down?
23. If the same pressure plate and cover assembly is to be reinstalled in an engine, what should be done to maintain balance?

CHAPTER 7

DISASSEMBLING THE ENGINE

Certain definite procedures need to be followed in the disassembly of all automobile engines. This work requires correct diassembly procedure and the ability to visually recognize wear conditions and/or problems. As the engine is disassembled, the reason for the teardown needs to be verified. For instance, if the engine is being overhauled because of a compression problem, the mechanic should be able to point out the exact cause.

The following engine disassembly and visual analysis procedure can be considered typical. The steps may vary a slight degree depending on the make and model of the engine.

TOP END DISASSEMBLY

1. If possible, attach the engine on a stand (Figure 7-1a and b).
2. Drain the engine coolant if not previously done.
3. Drain the engine oil if not previously done, and inspect for evidence of foreign sub-

stance. This would include metal, plastic, rubber, dirt, sand, or coolant.

"Smell" the oil for the presence of gasoline. More than 4% to 5% dilution is ex-

FIGURE 7-1a
Toyota 8R-C OHC engine mounted on a rebuilding stand.

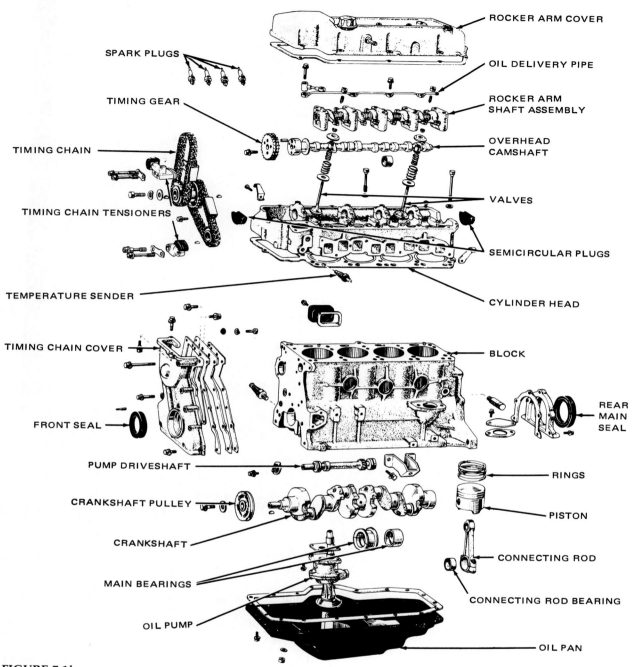

FIGURE 7-1b

Exploded view of the Toyota 8R-C engine shown in Figure 7-1a (Courtesy of Toyota Motor Sales U.S.A., Inc.)

cessive and leads to rapid engine wear. It takes only eight ounces of fuel added to a five-quart crankcase to change an SAE 30 oil to SAE 20 viscosity. Cold-engine stop-

start driving contributes to rapid dilution. So does excessive idling, overchoking and poor compression.

"Feel" the oil for appreciable viscosity in-

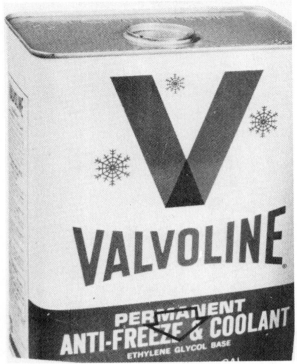

FIGURE 7-2
Most antifreeze solutions used today are ethylene glycol based.

FIGURE 7-3
Oil fouled plugs are usually caused by oil leaking through worn piston rings or valve guides.

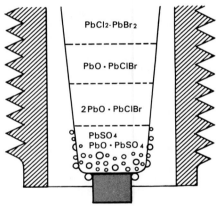

FIGURE 7-4
Spark plugs frequently show powdery or granular deposits around the insulator tip. This coating is due to lead fouling. Such coatings remain nonconductive when cold, but become conductive at driving temperatures and can cause misfiring.

crease. Thickening indicates higher than normal solid contaminants in the oil. The amount of solids in used oil is a good indicator of engine condition. The longer an oil is used, the more solids it contains. Tars, gums, lead compounds, soot, and carbon are the common solid contaminants that form in engine oil. Oil viscosity increase can also be a result of high oil temperature, or by permanent antifreeze (ethylene glycol) seeping into the engine (Figure 7-2).

4. Remove the distributor cap and wires if not previously done.

5. Remove the distributor and hold-down if not previously done.

6. Remove the spark plugs. Analyze the plug insulator and shell condition. Properly interpreted plugs can reveal operating conditions within an engine (Figures 7-3 and 7-4).

7. Remove the oil filter assembly, dipstick, any side cover(s), and fuel pump and lines.

8. Remove the carburetor.

9. Drain fuel (carburetor, fuel pump, and lines).

10. Remove the alternator, thermostat, water pump, belts, hoses, and so on, that are not part of the basic block assembly. *NOTE: If you plan on reusing the water pump, store it submerged in water while the engine is being rebuilt. Failure to do so can cause seals to dry out and the pump will leak when put back into service.*

11. Remove the intake manifold and the lifter chamber cover (if one is used). See Figure 7-5. If the manifold is stuck, jar the ends by

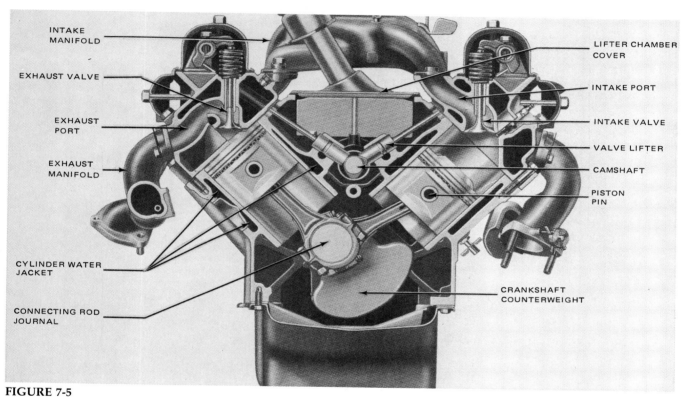

FIGURE 7-5

Cross-section of an engine that uses a separate metal cover over the valve lifter chamber (Courtesy of Buick Motor Division of General Motors Corporation).

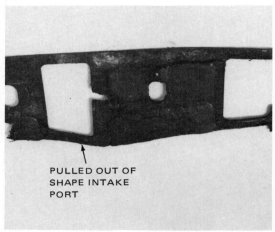

FIGURE 7-6

This defective intake manifold gasket came off a small block Ford. Notice how the intake port has been pulled out of shape. This caused an internal vacuum leak which resulted in hard starting, rough idle, and excessive oil consumption.

FIGURE 7-7

The wet, oily condition on this intake valve head was caused by improper intake manifold installation.

FIGURE 7-8

Excessive amounts of carbon on intake valve stems is usually caused by worn valve guides.

hitting upward with a block of wood. *If you use a screwdriver to pry between the manifold and the block you can ruin the gasket surfaces.*

Carefully check the gasket(s) for tear damage, and for evidence of water, vacuum, or oil leakage (Figure 7-6). Shine a light down the valve ports and observe for signs of oil being drawn toward the intake valves from the lifter valley (Figure 7-7). These signs often occur on engines using a manifold that encloses the lifter valley. Warpage, loose bolts, or improper gasket installation can allow oil to be drawn into the intake ports and be burned in the cylinder.

If possible, inspect the underside of valve heads (especially intakes) for hard carbon buildup (Figure 7-8). This indicates that oil is passing down between the valve stem and guide.

Inspect the intake manifold port passages by probing with a finger. If oil wetness shows up here and in the cylinder head intake ports, oil is being introduced from an outside source. A defective automatic transmission vacuum modulator, malfunctioning PCV system, or a faulty fuel pump are common culprits. Occasionally an oil bath type air cleaner (on older vehicles) gets overfilled or has a light viscosity oil added. This will result in oil being pulled into the engine by the incoming path of air.

12. Some engines have a sheet metal oil deflector riveted over the outside of the intake manifold exhaust crossover passage (Figure 7-9). If so equipped, use a cold chisel and lift out the attaching rivets. The deflector can later be reattached by pop riveting, or by tapping the rivet holes and using machine screws.

Remove any carbon buildup and visually check for cracks. Carbon is an excellent heat

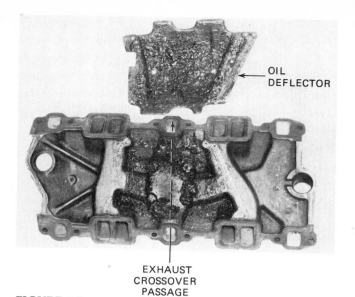

FIGURE 7-9

When disassembling an engine, remove the oil deflector from the intake manifold and check the crossover passage for cracks.

FIGURE 7-10

This exhaust manifold gasket was leaking at the break.

FIGURE 7-11

The deposits inside this valve cover are a "dirty mayonnaise" color. They resulted when a cracked cylinder head allowed water to mix with engine oil.

insulator. Occasionally, carbon forms under this deflector to such an extent that the crossover passage splits open.

13. Remove the exhaust manifold(s). Inspect for broken gaskets and for any casting cracks (Figure 7-10).

14. Remove the valve cover(s). Observe for sludge or any unusual deposits (Figure 7-11).

An engine mechanic should know the basic causes of sludge deposits and be able to point out remedies to prevent sludge troubles.

The prime cause of engine sludge deposits is

oil contamination. If an engine is operated with low cylinder wall temperatures, condensation and washing of combustion chamber products past the rings will cause oil contamination. If crankcase temperatures are also undesirably low, and if the engine suffers from faulty ventilation, moisture and fuel contamination cannot be purged from the oil. The blow-by contaminants in the oil then accumulate and increase in quantity until a critical point is reached where they begin to coagulate and separate out as sludges. In this first phase of formation, the sludge is soft and pasty and can easily be carried to engine parts

FIGURE 7-12
Thermostats are used to warm up the engine quickly and hold the coolant to the most efficient temperature. They are available in different temperature ratings. Note that these thermostats use a "jiggle" pin to help remove trapped air.

FIGURE 7-13
Removing union bolts and oil delivery pipes from a Toyota 8R-C OHC engine.

**ROCKER SHAFT
REMOVAL ORDER**

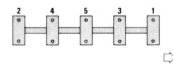

FIGURE 7-14
Removing rocker shaft support (stand) bolts. Always remove a rocker shaft assembly in correct sequence in two or three steps.

where oil flow is slow or restricted. This accounts for the often found deposits in valve lifter galleries, timing chain covers, oil pan sumps, oil pump screens, etc. To prevent these conditions, always install a new thermostat of the correct temperature when overhauling an engine (Figure 7-12). *Never operate an engine without a thermostat.* Once sludge has formed, it may gradually dry out and become converted to grainy "coffee-ground" or hard deposits.

15. Remove any rocker arm shaft lubrication delivery pipes (Figure 7-13).

16. On shaft mounted rocker arms, remove in sequence in two or three steps (Figure 7-14). This will help prevent broken rocker arms and cracked stands.

17. Remove the push rods, and check for visible defects. *On big block Ford V-8 engines keep the push rods in order, because of possible undersize or oversize* (Figure 7-15). Other engines may use different lengths for the intakes and exhausts.

    If the engine has individual mounted rockers, back off the stud nuts just enough to remove the push rods. Leave the rocker arms attached to the cylinder head. They can be hot-tank cleaned with the heads.

18. On OHC engines, remove the camshaft timing gear, bearing caps (if equipped), camshaft, and cam bearings. See Figures 7-16 and 7-17. *Keep the bearing caps attached to the head to prevent losing them.*

19. On V-8 engines, mark each cylinder head as right or left before removal. Remove the cylinder head attaching bolts. It is ex-

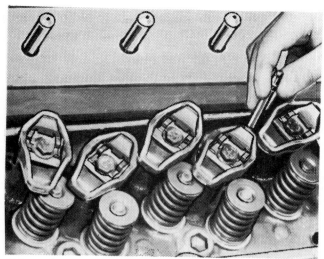

FIGURE 7-15
Removing push rods from a Ford 460 CID engine (Courtesy of Ford Parts and Service Division).

FIGURE 7-16
Removing the camshaft gear from an OHC engine.

FIGURE 7-17
The camshaft bearing caps on this OHC engine are stamped with location numbers. The prominent ends must face toward the front.

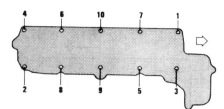

CYLINDER HEAD REMOVAL ORDER

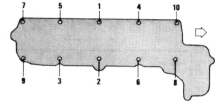

CYLINDER HEAD TIGHTENING ORDER

FIGURE 7-18
Always remove a cylinder head by loosening the attaching bolts in correct order.

FIGURE 7-19
This mechanic is removing this aluminum head *wrong*. The head is still warm and he is starting in the middle.

tremely important that the head bolts be removed in correct sequence (Figure 7-18). Otherwise, head warpage or cracking can result. *CAUTION: Do not remove an aluminum head unless it has completely cooled down; the head can become warped if this procedure is not followed (Figure 7-19).*

20. Lift the cylinder head(s) from the block. Because dowel pins are installed on many blocks, never attempt to slide or push the head off the block.

SHOP TIP

Sometimes a cylinder head becomes stuck to the block and removal can be difficult. Here is a procedure that can be used if the engine is still in the vehicle. First, check to make sure that every head bolt is removed. Next, reinstall all the spark plugs and crank the engine over with the starter motor. The compression pressure will then break the bond between the block and head.

---

21. Remove the valve lifters (Figure 7-20). If the lifters stick and are difficult to remove, leave them in the engine for the moment and raise each one far enough to clear the camshaft lobes. Then after the camshaft is removed, each lifter can be tapped out toward the bottom of its bore.

Examining the Top of the Engine

After removal of the cylinder head(s), visually inspect the combustion chambers for cracks, burn throughs, electrolysis damage, or other irregularities (Figure 7-21). Check combustion chamber and valve head coloring. Refer to Figures 7-22 and 7-23. Disassemble the heads. Check the valves and seats for irregularities (Figures 7-24 and 7-25).

Check the tops of the pistons. An oily, caked deposit on the top of a piston indicates too much oil is getting into the combustion chamber. If the piston head shows "washing" on the edge, excessive oil flow past the rings is indicated (Figure 7-26).

Look at the cylinder walls above each piston.

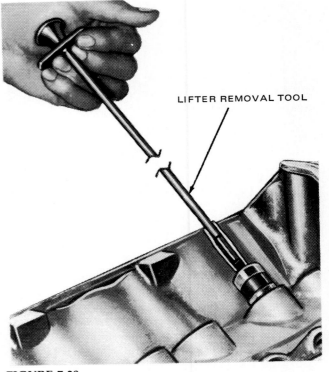

LIFTER REMOVAL TOOL

FIGURE 7-20
Removing a valve lifter or tappet (Courtesy of Ford Parts and Service Division).

Inspect for vertical scratches (usually caused by a broken ring) and cracks. "Feel" the ring ridge depth to provide a quick determination of whether the cylinders have significant taper wear.

Examine the block deck surface for cracks, corrosion around water jacket holes, or other defects (Figure 7-27). Look closely at the head gasket(s)

FIGURE 7-21
This cylinder head had a leak between combustion chambers. Do you see where?

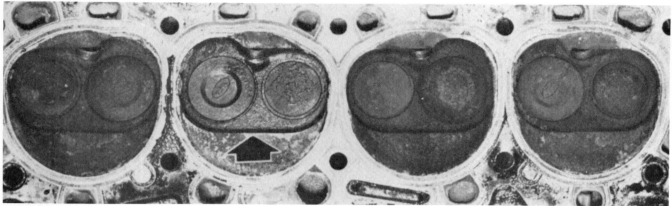

FIGURE 7-22
The dark, wet deposits in this combustion chamber were caused by oil leakage past the rings. The rest of the chambers indicate a rich air-fuel mixture.

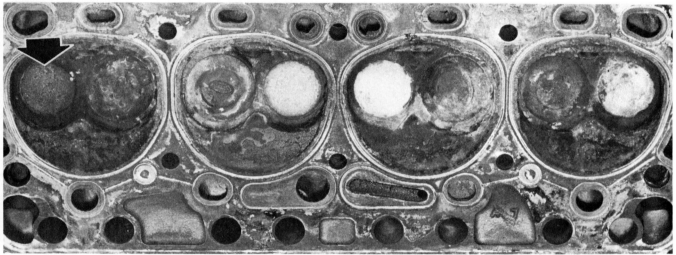

FIGURE 7-23
The sooty, black appearance of the combustion chamber at the left indicates an excessively rich air-fuel mixture. The rest of the chambers show normal color.

for problem signs. Check for burn-through between cylinders, seepage around water holes, and incorrect application (Figure 7-28).

ENGINE FRONT DISASSEMBLY

1. Remove the crankshaft snout bolt and washer. Using the proper puller and step plate, remove the crankshaft pulley or vibration damper (harmonic balancer). See Figure 7-29. *CAUTION: Never pull on the outside of a damper.*

   Sometimes, it becomes necessary to prevent the engine from turning over while breaking the snout bolt loose. To do this, insert a bolt or screwdriver to hold the flywheel (Figure 7-30).

SHOP TIP

Frequently, a mechanic has to break the snout bolt loose while the engine is still mounted in the chassis. It can be a problem to prevent the engine from turning over. Here is a trick that will help. Place the required size socket on the snout bolt (use ½" drive size). Put a long breaker bar on the socket and place the handle end against the frame. Crank the engine over using the key switch. The bolt will be broken loose as the engine is cranked against the wedged breaker bar handle.

FIGURE 7-24

This valve is an example of what can happen when an air-pump equipped engine becomes overheated.

FIGURE 7-25

The seat ring in this head became loose and wedged under the valve head as shown.

FIGURE 7-26

The lack of carbon around this piston head edge indicates that a large quantity of oil was passing the rings.

WASHED AREA

FIGURE 7-27

Corrosion on this block deck surface prevented proper head gasket seating and allowed coolant to seep into the cylinder.

FIGURE 7-28

Head gasket burn-through caused by a warped head.

Inspect the vibration damper hub for groove wear caused by the timing cover seal. Harmonic balancer repair sleeves are available for a number of engines (Figure 7-31). This can eliminate the leakage problem that often occurs to the seal when you replace a timing cover seal without replacing the harmonic balancer.

Check the condition of the rubber dampening ring. If the rubber shows no ''life'' (spring) when jabbed with a pencil, is loose or not completely intact, has pieces missing,

FIGURE 7-29

Removing a cast crankshaft pulley from an engine.

FIGURE 7-30

Using a screwdriver to hold the flywheel from turning.

FIGURE 7-32

An excessively loose timing chain cut through this front cover.

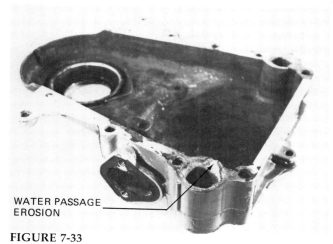

WATER PASSAGE EROSION

FIGURE 7-33

Electrolysis caused this water passage erosion and let coolant into the crankcase.

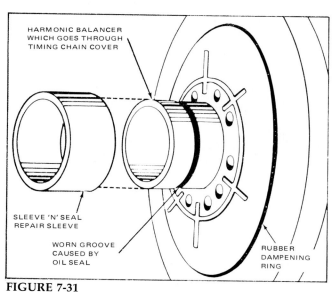

HARMONIC BALANCER WHICH GOES THROUGH TIMING CHAIN COVER

SLEEVE 'N' SEAL REPAIR SLEEVE

WORN GROOVE CAUSED BY OIL SEAL

RUBBER DAMPENING RING

FIGURE 7-31

This repair sleeve is installed with a .002″ press-fit over the worn harmonic balancer hub (Courtesy of Fel-Pro, Inc.).

or is severely cracked, purchase a new vibration damper. *Crankshaft failures due to breakage are often caused by a defective damper.* Remove the key from the crankshaft snout

and check its fit in the damper hub. If any keyway wear exists, replace the vibration damper. If the crankshaft keyway shows wear, the snout will require reconditioning.

2. Remove the front timing cover. **NOTE:** *On some engines, the oil pan will have to be removed first.* Inspect the front cover for rubbing wear (Figure 7-32). Occasionally a loose timing chain will hit the inside cover wall, producing a rod knocking sound. On engines that use aluminum front covers, electrolysis is often a problem. Carefully inspect for pinholing and erosion damage in the water pump cavity and around coolant passage holes (Figure 7-33).

   Check for excessive timing chain slack. In most cases, the chain and sprockets (gears) will be replaced as routine procedure during an engine rebuild.

3. Remove the crankshaft oil slinger and note

FUEL PUMP
ECCENTRIC

ALIGNMENT
PIN

OIL SLINGER

FIGURE 7-34

Details of the crankshaft oil slinger and fuel pump eccentric on one popular model OHV engine (Courtesy of Ford Parts and Service Division).

CHAIN
TENSIONER
PLUNGER

FIGURE 7-35

This chain tensioner is actuated by engine oil pressure.

the direction that it faces (Figure 7-34). Check for any deformation.

4.   Remove any timing chain tensioners (Figure 7-35).

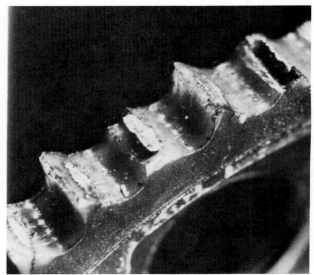

FIGURE 7-36

The nylon coating on this timing gear had worn off, causing the timing chain to slip.

FIGURE 7-37

Typical wear pattern on a crankshaft gear.

FIGURE 7-38

The fuel pump and the distributor are driven by a separate shaft on some engines.

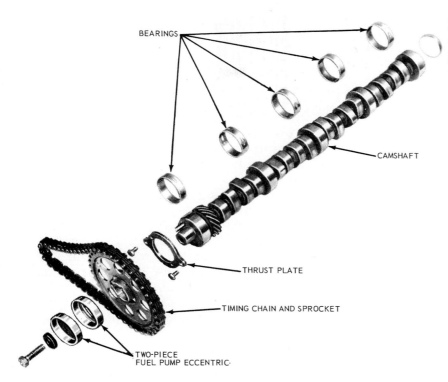

FIGURE 7-39

Camshaft and related parts. Note the thrust plate (Courtesy of Ford Parts and Service Division).

BEARINGS

CAMSHAFT

THRUST PLATE

TIMING CHAIN AND SPROCKET

TWO-PIECE FUEL PUMP ECCENTRIC·

5. Remove the timing chain or belt, if so equipped. It may be necessary to first remove the timing gear from the camshaft. Examine the gear for wear, cracks, and the flaking off of nylon from the teeth (Figure 7-36). Hold up the timing chain and check for excessive bowing.

6. If possible at this time, remove the crankshaft timing gear with an appropriate puller. Examine the gear teeth for wear and cracking (Figure 7-37).

7. If the engine is equipped with a separate fuel pump eccentric, push rod, or driveshaft, remove it now (Figure 7-38).

8. Remove the camshaft (Figure 7-39). *In some cases, you may find it easier to remove the camshaft after the rods and crankshaft have been removed and the engine is upside down.*

 NOTE: On some engines, it will be necessary to unbolt a thrust plate first. Exercise care when removing the camshaft by pulling gently. Installing a long bolt or a homemade support tool (Figure 7-40) in one of the holes in the front end of the camshaft will help reduce possible damage to lobes and journals. Visually examine the camshaft for any obvious defects (rounded lobes, edge wear, galling, and so on).

9. Remove the valve lifters if not previously done. *CAUTION: The lifters must be kept in order if they are going to be used again with the old camshaft.*

 If either the camshaft or valve lifters are worn at all, both should be replaced. *Do not install worn valve lifters with a new camshaft. Do not install new valve lifters on a worn cam-*

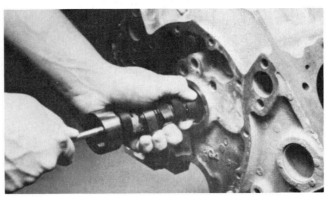

FIGURE 7-40

Using a bolt to help remove the camshaft.

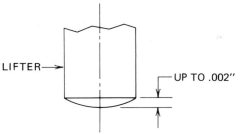

LIFTER → ⊢ UP TO .002″

FIGURE 7-41
Lifters for most engines have up to a .002″
dome on the bottom when made.

shaft. If the lifter bottom is not domed, the
lifter is considered worn (Figure 7-41).

BOTTOM END DISASSEMBLY

1. Remove the oil pan if not previously done.
   Observe the settlings in the sump and on
   the screen (Figure 7-42). Identification can
   provide insight to wear areas in the engine.

   Foreign material found in the oil pan can
   include ferrous metal (iron or steel), aluminum, bronze or brass, grinding wheel
   particles, sand, glass beads or lead shot,
   rubber from valve stem seals, and plastic
   from the timing gear.

The most common contaminant is iron or
steel (easily identified by its magnetic at-
traction). Iron or steel in sliver or flake form
generally means adverse wear of camshaft
lobes, lifters, timing gears, oil pump gears,
cylinder walls, piston rings, push rods, or
rocker arms. Fine iron or steel particles with
a uniform grain size are generally from machining operations on parts that were not
thoroughly cleaned. Sources of such particles can be from boring and honing of cylinders, reboring main bearing bores, reaming or knurling of valve guides, regrinding
the crankshaft, and reconditioning connecting rod big-end bores.

Nonferrous particles found in the oil pan
can come from pistons, pin bushings, main
bearing thrust washers, oil pump and distributor drive bushings, rocker arm bushings and valve guides (when bronze is used).
The metal machinings left from pin fitting
and rocker arm bushing sizing can be a
source of bronze, copper, or aluminum in
an engine.

Grinding wheel and honing stone particles can damage the engine. They will keep
on cutting metal as long as they circulate.
These particles are white or colorless like
ordinary sand. A distinguishing feature is
the shape of the particle when viewed under magnification. Grinding wheel and

FIGURE 7-42
These timing gear particles were inside this oil pump
pickup when the engine was disassembled.

honing stone particles are always irregularly shaped with sharp corners. Sand always has rounded corners.

Sand can find its way into the engine through a number of sources—using contaminated cleaning solvent, reusing a dirty air filter, or working in an open windy area are ways in which sand can be put into an engine. Glass beads or lead shot (.005" to .050" in diameter) are sometimes used to clean and stress relieve engine parts. A special machine uses compressed air to blast the bead or shot against the metal part. Unfortunately, glass or lead becomes stuck in crevices, corners, holes, and in casting irregularities. Cleaning can be very difficult and time consuming. *If you are going to glass bead clean or lead shot blast, do it only on parts that can be easily solvent cleaned and hand scrubbed afterward.*

2. Remove the oil pump assembly (Figure 7-43). This would include the pump, screen, and driveshaft.

3. Check all connecting rod and main bearing caps for correct position and numbering prior to removal (Figure 7-44). Consult a repair manual if you have any doubt. If the numbers cannot be seen or are missing, establish markings with a center punch or number stamp. *Never file or scribe marks on connecting rods. This creates a stress point that could cause the rod to fail.*

4. Remove the cylinder wall ring ridge (Figure 7-45).

5. Unbolt one rod cap at a time and push the connecting rod and piston assembly out of the cylinder from the oil pan side. Lightly tap or push with a hammer handle or wooden dowel. Avoid nicking the crankshaft journals by covering the rod bolts with rubber hose or with special protective boots. *The crank throw should be positioned at BDC when removing or installing pistons. This places the connecting rod at the straightest angle and provides for the least interference during removal.*

As the rods are removed, keep the caps and bearings in correct order. Mark each bearing as to location and position. Scribe markings on the bearing back such as 1U (meaning front upper), 1L, 2U, 2L, and so on. Clean and wipe the bearings dry. Lay them out on a flat surface in order of po-

FIGURE 7-43
The oil pan has been removed on this four-cylinder import engine. Observe the lower end parts.

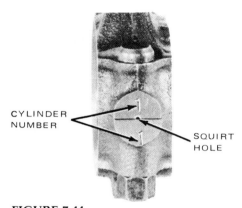

FIGURE 7-44
Connecting rod numbering and squirt hole location on one in-line engine model.

FIGURE 7-45
A good ridge removal job blends the cut smoothly into the cylinder wall surface (Courtesy of Dana Corporation—Service Parts Group).

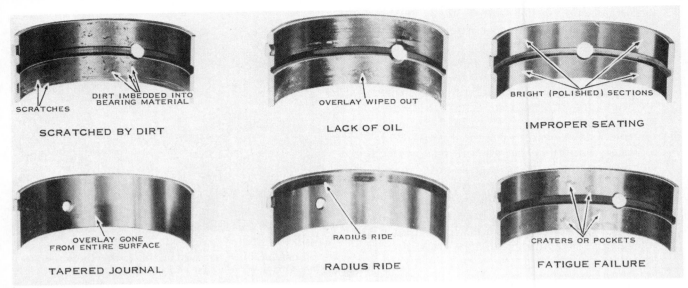

FIGURE 7-46
Some typical bearing distress patterns (Courtesy of Ford Parts and Service Division).

sition. Closely examine the operating surface and note any distress areas (Figure 7-46). Examine the bearing backs for correct fit by noting the transfer pattern on each half. Look at the backs for proper locking lip fit, and for possible dirt particle interference (Figure 7-47). Usually a general conclusion can be reached at this point. For example, the bearings had overheated, were improperly fitted, or had just worn out. Re-

fer to Chapter 22 for more detailed bearing analysis information.

6. Wipe the pistons dry and examine them carefully for cracks, scuffing or scoring, cracked rings, and bent or broken lands (Figure 7-48). Look at the skirt wear pattern

FIGURE 7-47
Dirt particle interference usually shows on both sides of the bearing.

FIGURE 7-48
Engine overheating caused the cracks in this piston head.

FIGURE 7-49
Removing a pressed-in piston pin with a press and special fixtures.

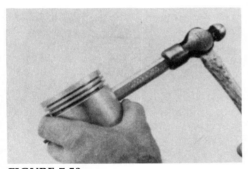

FIGURE 7-50
Removing a full floating piston pin by tapping it out after heating under a hot water faucet.

for any indication of connecting rod misalignment.

7. Remove the piston pins (Figures 7-49 and 7-50). Keep the pistons and rods in correct order and position.

8. Wipe the cylinder walls dry. Shine a light into each cylinder from the bottom and inspect the walls. Check for pitting, cracks, and scuffing or scoring. Shortly after the block is hot-tank cleaned, each cylinder needs to be measured for taper and out-of-round wear. These measurements will help determine if it is necessary to rebore.

9. The flywheel (or drive flex-plate) can be removed at this time. Check flex-plates closely for cracking. There have been numerous engines unnecessarily overhauled (broken piston or rod knock noise) due to this problem. Flex-plate noise can be detected easily with a stethoscope.

Sometimes, the mechanic will not remove the flywheel at this point. When the crankshaft is removed with the flywheel attached, the flywheel becomes a convenient upright stand for the crankshaft.

10. Remove the main bearing caps and lower bearings. Carefully lift out the crankshaft. *Store the crankshaft on end, making sure it cannot fall over.* A crankshaft should never be stored laying down unsupported. It can sag and become warped.

11. Remove the upper main bearings. *Bolt the main bearing caps back on the block so they do not get lost.* The factory precision fits all main bearing caps to the block by machining. Losing a cap will result in lots of extra time and expense.

12. Mark the location and position of each main bearing. Place them in order on a bench top for visual inspection analysis.

13. Wipe the crankshaft main and rod bearing journals dry and inspect for obvious defects (scores, nicks, ridges, and so on). The "copper penny test" can be used to check the surface finish of a crankshaft. Rub the coin several times across the bearing surfaces. If the surface picks up copper from the coin, the shaft is too rough and must be reconditioned.

14. Remove the rear main bearing seal from the block.

15. Visually inspect the entire block casting for any defects that would prevent reuse.

16. Remove the dipstick tube if it sticks up higher than the block deck surface. This will prevent accidental bending or breaking. The tube is normally pressed into place, and will have to be hit out from the bottom.

17. Remove all freeze (core) plugs. Cup type freeze plugs can be easily removed by driving them in, and then pulling them out with

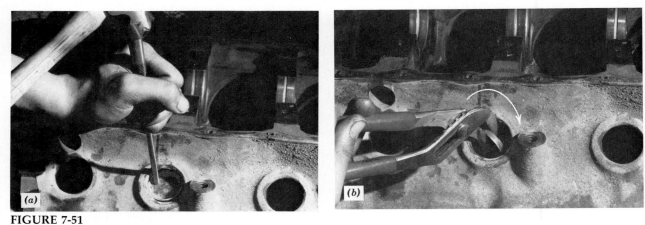

FIGURE 7-51

(*a*) Pushing in a cup type core plug prior to removal. (*b*) Extracting the core plug with channel lock pliers.

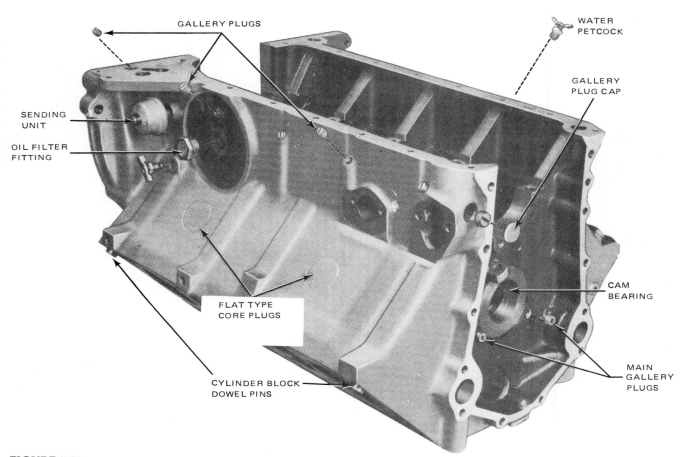

FIGURE 7-52

This block uses flat type core plugs. Do not drive them in to remove. Take note of the other attached items that must be removed before hot-tank immersion (Courtesy of Ford Parts and Service Division).

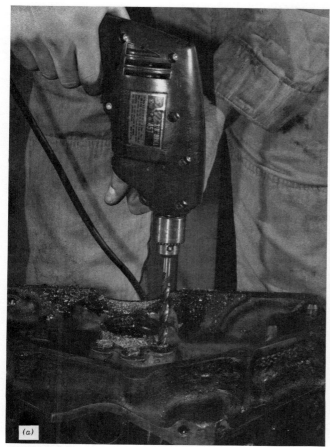 

FIGURE 7-53

(*a*) Drilling out a rear oil gallery plug in order to remove it from the block. (*b*) After drilling out the plug, drive in an "easy-out" (screw extractor) and turn out with a crescent wrench.

a pair of channel lock pliers (Figure 7-51*a* and *b*). *CAUTION: Do not attempt to drive in flat type freeze plugs (Figure 7-52), as this can break off the support ledge inside the hole.* In order to remove this type of plug, first drill a hole near the center, then insert a slide hammer dent puller and pull the plug out.

18. Remove all front and rear oil gallery plugs. Sometimes threaded plugs can be difficult to remove. Using a drill and "easy-out" works quite well (Figure 7-53*a* and *b*).

SHOP TIP

Threaded oil gallery plugs can be removed by using heat. Heat the plugs with a welding torch. Then, while the plugs are still hot, turn out with a ½" drive air impact wrench.

The cup type oil gallery plugs in some engines can be removed easily by using a dent pulling tool, or by hitting them out from the back side with a long rod.

19. Remove the cam bearings (Figure 7-54). This is important before cleaning the block in the hot-tank. Otherwise, the hot-tank solution will be diluted as the bearings are "eaten up," and all metal in the hot-tank will be coated with bearing material.

20. Remove any other nonferrous pieces from the block before hot-tank cleaning. This would include distributor drive bushings, water petcocks, sending units, and rubber mountings.

21. The engine block should now be completely disassembled. Cleaning parts will be discussed in the next chapter.

FIGURE 7-54
Driving out a cam bearing.

TRUE CASE HISTORY

Vehicle	Ford Truck
Engine	8 cylinder, 390 CID
Mileage	88,000 miles
Problem	1. Hard starting after engine was warm.
	2. Rough idle.
	3. Oil would suddenly get black after being changed.
Cause	A split intake manifold crossover passage (Figure 7-55). This split extended into an intake runner, causing an internal vacuum leak. Additionally, exhaust gas was passing into the engine lifter valley during warm-up. This caused the oil to be quickly contaminated.

FIGURE 7-55
Split on the bottom side of the intake manifold crossover passage.

TRUE CASE HISTORY

Vehicle	Toyota Crown
Engine	6 cylinder, 2563 cc (cubic centimeters)
Mileage	58,000
Problem	1. Difficult to start engine when cold because of slow cranking speed. 2. After warming up, the engine would start okay.
Cause	Permanent type antifreeze had seeped into the lubrication system because of a head gasket failure. This resulted in thermoplastic coolant-oil deposits throughout the engine.
Cure	The head gasket was replaced and butyl cellosolve (ethylene glycol mono-butyl ether) was used to flush out the engine. The procedure used was as follows. After draining the old engine oil, the crankcase was filled to proper level with a mixture of 2 parts butyl cellosolve and 1 part SAE 10 motor oil. The engine was run at fast idle for one hour. Then the crankcase was drained, and new oil and a filter installed.
Comment	Butyl cellosolve can be purchased through sales offices of the Union Carbide Chemicals and Plastics Division of Union Carbide Corporation.

CHAPTER 7 SELF-TEST

1. What are some contributing causes of fuel dilution in the crankcase?
2. What does oil thickening indicate?
3. If you plan on reusing a water pump, how should it be stored while the engine is being rebuilt?
4. How should a stuck intake manifold be removed?
5. On engines that use an intake manifold to enclose the lifter valley, what conditions can allow oil to be drawn into the cylinder?
6. What does a hard carbon buildup on the valve head underside indicate?
7. If the intake manifold port passages are wet with oil, what is indicated?
8. When rebuilding an engine, what is the reason for removing the sheet metal oil deflector on the intake manifold bottom? How can this deflector be reattached?
9. What is the prime cause of engine sludge deposits?
10. Why should an engine never be operated without a thermostat?
11. What engines use undersize or oversize push rod lengths?
12. Why should head bolts be removed in a certain sequence?
13. Why should an aluminum head be removed only when cold?
14. Why should you never attempt to slide or push a head off the block?
15. What is indicated if a piston head shows "washing" on the edge?
16. Why should you never pull on the outside of a vibration damper?
17. What trick can be used to loosen the crankshaft snout bolt when the engine is mounted in the chassis?
18. What is a harmonic balancer repair sleeve?
19. When checking the condition of a harmonic balancer dampening ring, what should you look for?
20. When inspecting an aluminum front cover, what should you look for?
21. Should you install worn valve lifters on a new camshaft?
22. Should you install new valve lifters on a worn camshaft?
23. If fine steel particles of a uniform grain size are found in an oil pan sump, what is indicated?
24. Steel sliver and flake settlings in an oil pan generally means what?
25. Bronze particles found in an oil pan could have come from where?
26. How can you differentiate between grinding particles and sand?
27. What precaution must be observed when glass bead cleaning parts?
28. Why shouldn't you ever file identification marks on connecting rods?
29. Where should the crank throw be positioned when removing or installing pistons?
30. What special measure should be taken to avoid nicking the crankshaft rod journals when removing or installing pistons?
31. Flex-plate noise can be easily detected with the use of what tool?
32. Why should a crankshaft never be stored laying down unsupported?

33. What is the "copper penny test"?

34. Why shouldn't you attempt to remove flat type freeze plugs by driving them in?

35. How can cup type oil gallery plugs be easily removed?

36. How can threaded type oil gallery plugs be easily removed?

CHAPTER 8

CLEANING PARTS

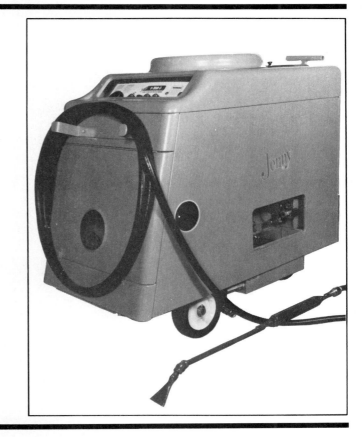

After the engine block has been disassembled and the attached pieces removed, the parts need to be properly cleaned. Thorough inspection cannot be made if grease, oil, dirt, rust, or scale are masking defects.

How a part is cleaned depends on the item being cleaned and the type of equipment available. The wrong cleaning method or cleaning agent often can be as harmful as no cleaning at all. Bearing bores, polished surfaces, and gear teeth exposed to moisture, acids, or caustic solutions during the cleaning process can quickly water spot, stain, rust, or corrode. When returned to service such parts will wear rapidly and premature failure will follow.

Here are the eight methods used by shops to clean engine parts.

1. Hand cleaning.
2. High-pressure spraying.
3. Cold soaking.
4. Hot-tank immersion.
5. Steam cleaning.
6. Glass bead cleaning.
7. Airless shot blasting.
8. Pyrolytic oven cleaning.

HAND CLEANING

Many engine parts will require hand cleaning. This is especially true when the engine is being overhauled in the chassis. Carbon, varnish, sludge, paint, sealer, and gasket material deposits are often removed by scraping or by brushing (Figure 8-1a and 8-1b).

Solvent Tank Brushing

Some parts and assemblies are best cleaned by brush scrubbing in a solvent tank (Figure 8-2). Special nylon brushes that will not flare out or get flabby are used. Most modern solvent tanks are equipped with a pump that delivers filtered solvent through a pistol grip nozzle or flexible metal hose. Solvent tanks are required to have a fusible safety link on the lid (Figure 8-3). If a fire

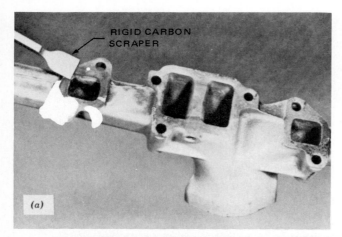

FIGURE 8-1

(*a*) Cleaning the exhaust manifold flange surface (Courtesy of Ford Parts and Service Division). (*b*) Scotch-Brite* surface conditioning discs are excellent for cleaning aluminum, cast iron, and even plastic. The discs can be used for cleaning oil pans, intake manifolds, heads, thermostat housings, and many other parts (Courtesy of Automotive Trades Division/3M).

should start in the tank this link will melt at 165° F. This will cause the lid to fall shut and suffocate the fire.

CAUTION: Prolonged immersion of your hands in solvent may produce a burning sensation. In some cases, a skin rash may develop. When cleaning parts in a solvent tank, direct the solvent stream away from your hands as much as possible or wear rubber gloves.

Wire Brushing Carbon

Carbon is an excellent insulator. If allowed to remain on parts, proper heat dissipation will not take place.

Carbon can be removed by using a twist type wire brush driven by an electric or air drill motor (Figure 8-4). Brushes come in different shapes and sizes. Straight and flare end wire brushes (Figure 8-5) are used to clean the cylinder head combustion chambers, top of the block, and the piston head. Deep-hole end brushes (Figure 8-6) are very effective for detailing cylinder heads

and cleaning valve parts. *Wire brushes must be used lightly on aluminum heads and pistons to avoid scoring. CAUTION: Always wear a face mask when cleaning parts by a power-driven wire brush. Wire* bristles are often thrown off from the brush. These wire pieces can be compared to minature straight pins. A face mask will provide for eye and skin protection.

SHOP TIP

Wire brushing carbon from cylinder head, valve ports and combustion chambers after degreasing is very often a stubborn time-consuming operation. Some shops use a unique method of cracking and chipping the carbon prior to using a wire brush. An air gun with a round-tipped peening tool is used. The peening tool is moved in a light circular motion against the carbon. This will crack and dislodge the carbon for easier wire brushing. *Avoid peening gasketed surfaces.*

---

*Registered trademark of the 3M Company, St. Paul, MN

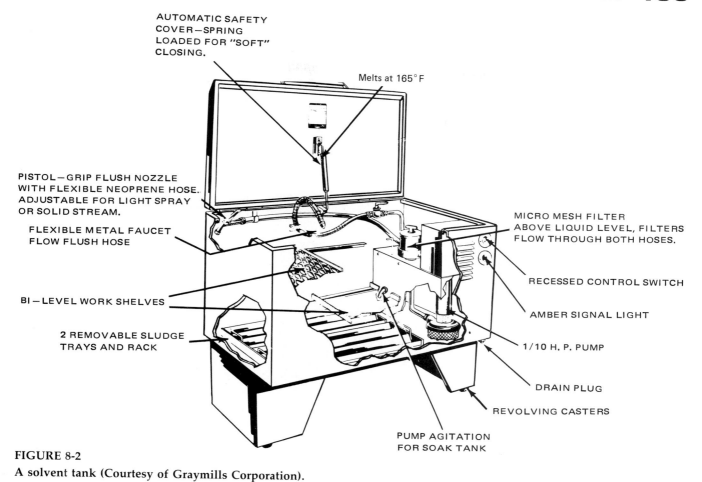

AUTOMATIC SAFETY
COVER—SPRING
LOADED FOR "SOFT"
CLOSING.

Melts at 165° F

PISTOL—GRIP FLUSH NOZZLE
WITH FLEXIBLE NEOPRENE HOSE.
ADJUSTABLE FOR LIGHT SPRAY
OR SOLID STREAM.

FLEXIBLE METAL FAUCET
FLOW FLUSH HOSE

MICRO MESH FILTER
ABOVE LIQUID LEVEL, FILTERS
FLOW THROUGH BOTH HOSES.

RECESSED CONTROL SWITCH

AMBER SIGNAL LIGHT

BI—LEVEL WORK SHELVES

1/10 H. P. PUMP

2 REMOVABLE SLUDGE
TRAYS AND RACK

DRAIN PLUG

REVOLVING CASTERS

PUMP AGITATION
FOR SOAK TANK

FIGURE 8-2
A solvent tank (Courtesy of Graymills Corporation).

Valve Guide Cleaning

Valve guides require very careful cleaning. Any carbon or gum deposits left in a guide will deflect valve grinding pilots, resulting in inaccurate work. Knurling tools often become jammed because of improper guide cleaning.

Valve guides are cleaned by using a drill motor

FIGURE 8-3
In case of a solvent tank fire, this link is designed to melt and come apart. This allows the lid to fall shut automatically and suffocate the fire.

FIGURE 8-4
Wire brush cleaning combustion chambers (Courtesy of Ford Parts and Service Division).

FLARED END   KNOT TYPE   HOLLOW CENTER

FIGURE 8-5

Wire end brushes. The hollow center type is especially designed for aluminum heads (Courtesy of Sioux Tools, Inc.).

FIGURE 8-6

Deep-hole end brushes. The cables are tipped with hard-facing and tungsten carbide.

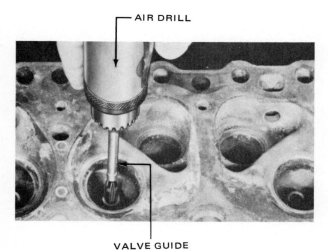

AIR DRILL

VALVE GUIDE SCRAPER

FIGURE 8-7

Removing carbon from valve guides with a spring-type scraper (Courtesy of Black and Decker Manufacturing Company).

and a correct size nylon brush or spring-type scraper (Figure 8-7). Lacquer thinner or carburetor cleaner will help dissolve the deposits. To check how clean the guides are, hold a light at one end of the guide, and look down the guide hole from the other end. A properly cleaned guide will look polished on the inside diameter.

Valves

After the valves are removed from the cylinder head, carbon and varnish deposits need to be removed.

Valve head and fillet area deposits can be removed by wire wheel brushing on a power grinder (Figure 8-8). *CAUTION: Wear a face mask.* Do not wire brush the valve guide contact area on the stem. This will produce fine scratch marks that can score and ruin the guide. Instead, clean the stems by using fine grit crocus cloth.

Another way to remove carbon is by soaking the valves in carburetor cleaner. Generally this will soften the carbon enough where it can be peeled off with a scraper or putty knife.

Many shops clean valves by using a glass bead blaster. Hard carbon from the head and varnish from the stem can be removed very quickly this way.

Valve Springs

Clean the valve springs, keepers, and retainers by brushing them in a solvent tank or by soaking them in carburetor cleaner. *If the valve springs are factory painted, do not use carburetor cleaner.* Some valve springs are painted to prevent etching from acid buildup in the oil. Carburetor cleaner will remove this special acid resist paint.

FIGURE 8-8

A wire wheel, driven by a grinder motor, is sometimes used to clean carbon from valves (Courtesy of Black and Decker Manufacturing Company).

Rocker Arm Assembly

Rocker arm assembly parts can be cleaned by hand scrubbing the parts in a solvent tank. Sometimes the parts are cleaned best by soaking in a carburetor cleaner or by soaking in a hot-tank. *CAUTION: Do not put aluminum or magnesium rocker stands in the hot-tank solution. Do not put bronze bushed rocker arms in the hot-tank solution.* The hot-tank chemical solution will react on these metals and erode them away.

Pistons

Clean pistons by soaking them in carburetor cleaner or by solvent tank brushing (Figure 8-9). It may be necessary to hand scrape the ring grooves to remove hard carbon. A ring groove cleaning tool, or an old piston ring broken in half and used as a scraper will do a good job. *Do not gouge or undercut the sides of the ring grooves.* This surface must remain flat to help maintain oil control. Make sure that the oil ring drain back holes or slots are clean (Figure 8-10).

Many shops today clean pistons by glass bead blasting. This is a fast and thorough method. *However, be careful not to round off the ring groove edges. Always solvent clean and hand scrub parts after bead blasting.*

Oil Passages

All oil passage holes need to be hand brushed with solvent and blown out with compressed air. This would include drilled holes and sludge traps in the crankshaft (Figures 8-11 and 8-12), engine block main gallery holes (Figure 8-13), rocker arm and rocker shaft lubrication holes, drilled push rods, and connecting rod spit holes. Some OHC engines have passageways in the camshaft, and chain tensioner oil holes that must be thoroughly cleaned. A rifle brush cleaning kit or a length of welding rod with the end split (Figure 8-14) are useful tools for cleaning small oil passages.

HIGH-PRESSURE SPRAYING

High-pressure sprayers provide hot or cold cleaning solutions or clean hot tap water at pressures up to 1000 psi. These popular cleaning machines are widely used by engine and transmission rebuilders (Figure 8-15). This system of spraying uses jet streams of hot detergent similar to a dishwasher, and is capable of cleaning an engine block

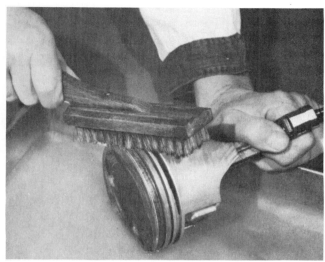

FIGURE 8-9

Cleaning pistons using solvent and a soft bristle brush. Never use a caustic cleaner or a wire brush (Courtesy of Ford Parts and Service Division).

FIGURE 8-10

Probe and open up all piston oil drain holes or slots (Courtesy of Ford Parts and Service Division).

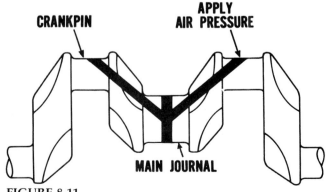

FIGURE 8-11

Crankshaft oil holes should be reverse flushed after solvent cleaning (Courtesy of Imperial Clevite, Inc., Engine Parts Division).

FIGURE 8-12

Some crankshafts have sludge trap plugs. They can be removed in order to thoroughly clean oil holes.

FIGURE 8-13

Using a small brush and compressed air to clean an oil gallery (Courtesy of Imperial Clevite, Inc., Engine Parts Division).

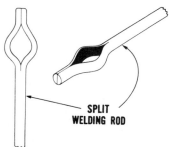

FIGURE 8-14

A split welding rod can be used for cleaning.

in ten minutes. Several metals (iron, steel, and aluminum) can be cleaned in the same solution.

High-pressure sprayers start cleaning almost immediately. They do not require several minutes to increase temperature levels like a steam cleaner.

SHOP TIP

In order to save energy, most of the chemical companies now market low-temperature cleaners. These chemicals will work with little heat. However, you will have to use a higher concentration. If you want to switch from high- to low-temperature chemicals, do it gradually. Start at the same temperature and concentration that you have been using, then lower the temperature 10° F. If cleaning efficiency is lost, increase the chemical concentration. Keep lowering the temperature in 10° F increments, adding the necessary amount of chemical. Continue this until you are satisfied with the results.

COLD SOAKING

Carburetor cleaner is the most commonly used cold soak chemical (Figure 8-16). It effectively removes gum, varnish, carbon, dirt, and sludge from parts. Carburetor cleaner is available in various size containers, with or without a dip basket. Use the cleaner by placing metal parts to be cleaned in the dip basket or suspend by wires. Figure 8-17 shows a carburetor dip tank. Thor-

FIGURE 8-15

A "jet clean" high-pressure spray tank. Parts to be cleaned are placed on a revolving turntable. Close the door and set the timer for desired cleaning cycle time (Courtesy of Storm-Vulcan, A Division of the Scranton Corporation).

FIGURE 8-16
Cold-dip nonflammable carburetor cleaner and recharge booster (Courtesy of McKay Manufacturing Company).

oughly mix the chemical by moving the dip basket or parts up and down four or five times before leaving parts standing in the solution. Water should be used to flush parts after dipping. Finish by drying with compressed air. *CAUTION: In case of eye or skin contact with carburetor cleaner, flush immediately with water and get medical attention. Do not leave magnesium parts in carburetor cleaner for more than one hour, or discoloration of the metal may occur.*

HOT-TANK IMMERSION

An immersion tank, using a hot caustic solution, is one of the most efficient and economical ways of cleaning engine parts (Figures 8-18 and 8-19). Hot-tanks are very effective for removing oil,

FIGURE 8-17
A commercially available two-compartment carburetor dip tank.

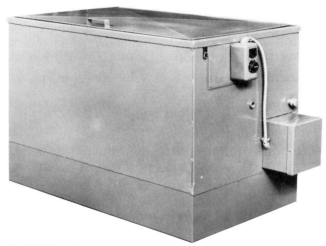

FIGURE 8-18
An electric heated hot-tank.

HYDRAULIC CRANE

HINGED LID

BASKET

SAFETY PILOT ASSEMBLY

THERMOSTATICALLY CONTROLLED

110 VOLTS

FIGURE 8-19

A natural gas heated hot-tank with a crane and basket for lowering parts into the cleaning solution (Courtesy of Storm-Vulcan, A Division of the Scranton Corporation).

grease, varnish, sludge, paint, gasket cements, and rust deposits on the inside of water passages. Parts are placed in the hot caustic solution kept at approximately 200° F. *CAUTION: Always wear gloves, suitable clothing, and a full face shield when handling parts in the hot-tank solution.*

Most hot-tanks can agitate the solution to make cleaning more effective. Three commonly used methods of agitation are

1. Mechanically moving the basket with a traverse motion. This creates a turbulence like an ocean tide that helps dissolve grease and dirt and wash it away (Figure 8-20).
2. Sending compressed air through the solution.
3. Pump or propeller agitation to give physical motion to the solution.

Hot-Tank Soap

Most hot-tank alkali cleaning soaps, regardless of the brand, have the same chemical properties. These compounds are a fairly equal ratio of sodium hydroxide (NaOH) and sodium carbonate (Na_2CO_3). The pH of the water solution of the material is between 11 and 12, which is very basic.

Sodium hydroxide (also known as caustic soda; sodium hydrate; lye; white caustic) can be a dangerous irritant to the eyes, skin, and mucous membranes. It is corrosive to tissue in the presence of moisture.

Sodium carbonate (also known as soda ash; sal soda; crystal; soda) is normally found as a monohydrate, which is $Na_2CO_3 \cdot H_2O$. The combined action of sodium hydroxide and sodium carbonate is as an oil saponification and reaction

AGITATION

ENGINE BLOCK

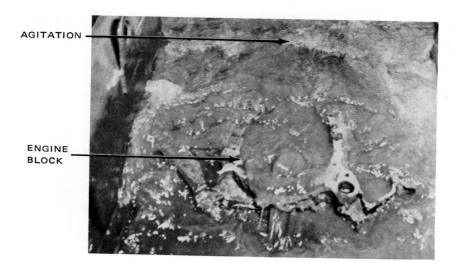

FIGURE 8-20

Actual photo showing hot-tank solution turbulence created by moving the basket (Courtesy of Storm-Vulcan, A Division of the Scranton Corporation).

agent. This makes oil, water soluble and causes dissolution from the metal surface of parts.

Adding cleaning compound to the hot-tank is commonly referred to in the trade as "charging the tank." The stronger the "charge," the greater the cleaning ability. Most engine rebuilding shops will initially "charge" their hot-tank with 1 lb. of soap for every gallon of water. This will give a pH of about 12.90 and provide a solution concentration adequate for heavy cleaning. Each week the tank should be "charged" with additional soap to retain proper strength. *CAUTION: Hot-tank soap should never be handled with your hands.* Use a small container as a scoop. Always slowly add the soap to the water in order to avoid spillover and to give an even distribution. Skim the solution occasionally to remove oil and floating debris on the surface.

Hot-tank caustic solutions are harmful to some metals. Immerse only engine parts made of ferrous metals (iron and steel). Remove bronze distributor drive bushings, oil and water temperature sending units, oil filter bypass adapters, cam bearings, and so on. Brass, bronze, aluminum, babbit, and magnesium will react with the chemical solution and be "eaten up." In addition, the solution becomes diluted and may require changing.

Removing Scale

Hot-tank soap concentrate will not effectively remove water jacket scale (calcium and lime deposits). Scale is an excellent insulator, and must be removed when rebuilding an engine. Many ring jobs have gone "sour" because the mechanic

let scale remain in the cooling system. The rings scored because proper heat transfer could not take place. See Figure 8-21.

Calcium and lime deposits require that an acid be used for cleaning. Descaling may be accomplished in one of several ways.

1. Circulation of an acid solution through the cooling system just before engine disassembly. There are some excellent radiator additives on the market that are designed to clean scale from water passages (Figure 8-22).

2. Immersion of parts after disassembly. Some engine rebuilding companies have acid immersion tanks next to the hot-tanks. After parts are degreased in the hot-tank, acid im-

CYLINDER WALL DISTORTION

CYLINDER WALL SECTION

PISTON AND RING ASSEMBLY

SCALE

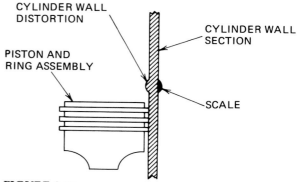

FIGURE 8-21

Scale deposits result from magnesium and calcium contaminants present in water. These deposits retard heat transfer and cause localized hot spots. In turn, this can cause the cylinders to bulge in the hot spot area. As the rings hop and skip across the bulged surface, scuffing and scoring result.

FIGURE 8-22
This chemical cleanser is designed to remove rust and scale from the engine cooling system.

mersion is performed. A final alkaline rinse is then required to neutralize any remaining acid.

If your shop does not have a tank for acid dipping parts, this improvised procedure will work.

1. Purchase one gallon of muriatic acid from a swimming pool or masonry supply company. Dilute the acid to 50% concentration with tap water, using a plastic or glass container in which to mix the solution. Pour the acid slowly into the water–*never add the water to the acid! CAUTION: Always wear a rubber apron, rubber gloves, and a full face shield when handling muriatic acid.*

2. Seal off all the outside water jacket holes in the engine block and cylinder head(s). This would include freeze plug holes, petcock drain holes, the water temperature sending unit hole, water transfer holes on the block and head mating surfaces, and coolant holes on the cylinder head-to-intake manifold surface. Duct tape used for household air conditioning installations can be used to seal off these holes. All surfaces must be completely dry and free of any oil film before applying the tape. Press firmly down on the tape.

3. Position the block and head(s) so as to allow complete filling of the water jackets.

4. Pour in the acid solution. *CAUTION: Do not inhale the acid fumes or spill any acid on yourself.* If the acid contacts your skin, flush the area immediately with cold water and consult a doctor.

Allow the acid to remain in the water jackets for one-half hour. Dump the solution and then neutralize the parts by placing them back into the hot-tank.

STEAM CLEANING

One of the most common types of machines used for removing baked-on soil, grime, and grease is the steam cleaner (Figure 8-23).

COMMENT: Technically speaking, the name "steam cleaner" is a misnomer. The spray actually is not steam, but a boiling hot vapor. Steam itself does not have either the ability to carry compounds or the frictional power to penetrate into cracks and crevices. Yet, because the vapor coming from the nozzle looks like steam, the machines have come to be called steam cleaners, despite manufacturer's efforts to have them referred to as vapor spray cleaners.

Fundamentally, the steam cleaner combines four basic cleaning elements—water, heat, soap, and friction. Water that contains soap is pumped through a water coil. Heat is applied to the coil until the water temperature reaches approximately 300° F. A restrictor in the discharge line increases pressure in the coil, preventing the water from boiling and turning to steam. When this high temperature water is released to the atmosphere at the cleaning gun nozzle, approximately 10% of the water turns to steam, literally exploding hot water and soap against the object being cleaned.

The usual capacity of a steam cleaner is from 45 to 150 gallons of water per hour, delivered at around 100 psi. The nozzle temperature is 212° F. Vapor temperatures drop to between 140° and 160° F at the surface being cleaned. Detergents can be added in the vapor stream in any desired concentration. Units are available with either gasoline driven engines or electric motors.

Operating the Steam Cleaner

To prevent damage to the steam cleaning machine, it is important to follow certain steps during lighting and shut down. The following lighting procedure is generally used.

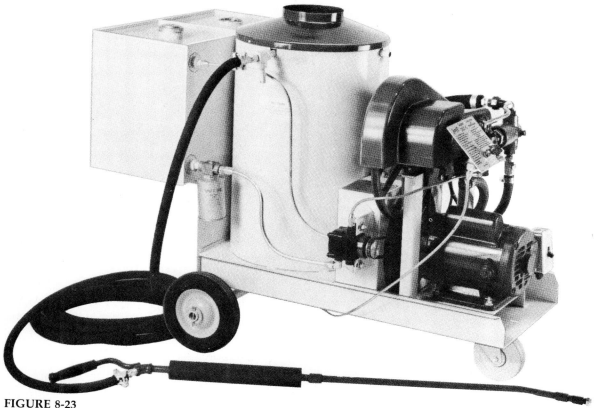

FIGURE 8-23

A kerosene fired, electric motor driven, portable steam cleaner (Courtesy of Jenny Division, Homestead Industries, Inc.).

1. Turn on the water.
2. Turn on the pump motor and wait until a solid stream of water is flowing from the steam gun nozzle.
3. Open the fuel valve 1/4 turn. This will light the burner and allow water pressure to build up. Ideal steam cleaner operating pressure is from 90 psi to 100 psi. This pressure can be regulated by turning the fuel valve from the 1/4 turn setting.
4. Wait until hot water is flowing from the gun nozzle, then open the soap solution valve.

When shutting down the steam cleaner, it is important to follow this procedure.

1. Turn off the fuel completely.
2. Close the soap solution valve.
3. Wait until cool water is flowing from the steam gun nozzle.

4. Turn off the water and the pump motor.

NOTE: The burner flame and the pump motor should never be turned off at the same time. Otherwise, the water coils will quickly boil off the remaining solution and become excessively hot. This can cause the coils to warp or crack.

As you steam clean, always be aware of the sound of the vapor spray leaving the nozzle. If a spitting action occurs, and the hose and nozzle start to jerk, reduce pressure by shutting off the fuel supply. Under normal conditions, approximately 10% of the nozzle discharge will be steam. Above-normal operating levels can increase the steam to 30 or 40% and cause hose damage.

In general, steam cleaners require more frequent inspection, service and understanding than other pieces of cleaning equipment. Improperly working steam cleaners are often simply the result of a poor preventive maintenance program. Table 8-1 lists some of the more common steam cleaner problems.

TABLE 8-1
Common problems in steam cleaning

Troubles	Possible Causes
Insufficient solution pressure	a. Insufficient fuel to burner b. Dirt in burner nozzle c. Restrictor not installed in cleaning gun d. Fuel pressure too low e. Solution discharge valve not fully open
No water or insufficient water at cleaning gun nozzle	a. Loose drive belt b. Water supply inadequate because of low water supply pressure *NOTE: Water supply must be able to keep float tank 1/3 full at all times while operating.* c. Loss of water through open or leaking valves d. Excessive scale inside heating coil e. Motor running slow
Sudden rise in solution pressure	a. Restricted hose, gun, or nozzle b. Discharge valve partly closed or restricted c. Water pump not operating properly because of low oil level in crankcase
Relief valve leaks during normal operation	a. Dirt or loose scale in relief valve or restricted heating coil
Insufficient soap at cleaning gun nozzle	a. Soap pump not primed b. Congealed or undissolved compound in solution tank c. Soap meter not adjusted properly
Low or no fuel pressure	a. Fuel filter clogged b. Fuel pressure not adjusted properly
Burner fails to ignite or stops during operation	a. Faulty ignition transformer b. Defective ignition cable
Sooty smoke from flue outlet *NOTE: This condition must be corrected immediately to prevent sooting of heating coil and burner.*	a. Fuel pressure not adjusted properly b. Improper air supply to burner c. Carboned, loose, or worn burner nozzle
Fluttering fire	a. Restricted or sooted flue pipe causing back pressure in combustion chamber b. Burner nozzle carboned over c. Improper air supply to burner
Motor fails to start or stops during operation	a. Blown fuse b. Loose electrical connections or defective motor c. Restricted heating coil causing the motor to overload and heat up, which trips a thermal breaker and shuts off the motor

Steam Cleaning Soap

The steam cleaning soap solution, liquid or powder, should have a normal concentration of about 10%. *Steam cleaners should not be operated without cleaning compound, or with cheap materials that contain resin, laundry soaps, or inert or insoluble ingredients.* Otherwise, quick buildup of deposits will start to form on the inside of the heating coil. This will cause higher fuel consumption, require a hotter flame, restrict water flow, and eventually ruin the coil.

COMMENT: Many shops that specialize in steam cleaner parts and service are firmly against using powdered soap, because it is difficult to dissolve and leads to premature coil scaling. They recommend using only liquid soap in a steam cleaner.

GLASS BEAD CLEANING

Glass bead cleaning (sometimes called glass bead blasting) is a process where compressed air is used to propel microscopic glass beads against surfaces to be cleaned. This process is ideally suited for removing hard carbon, paint, rust, and stains on aluminum. When parts are cleaned they have a nice new looking matte finish.

Oil retention is improved when parts are bead blasted. Glass bead cleaning in engine rebuilding applications is designed to be a secondary cleaning and/or finishing operation. It is usually done after a primary cleaning of parts (hot-tanking, pressure spraying, and so on) has been completed.

There are basically two different types of glass bead blasting machines. The most simple is the *siphon gun type (Figure 8-24).* Compressed air is passed through a venturi in the blasting gun. This creates a vacuum in the gun body, and glass beads are drawn from a storage hopper into the air stream.

The second type is the *direct pressure cabinet (Figure 8-25a and b).* The glass bead supply is held under constant air pressure in a storage pot. When the air valve is actuated, a metered quantity of glass beads are forced out of the blasting gun nozzle.

The biggest limitation of glass bead cleaning is the fact that it requires an operator 100% of the time. Another limitation is the difficulty in knowing if all glass beads have been removed from the cleaned parts.

AIRLESS SHOT BLASTING

The airless blasting machine is a fairly new concept in cylinder head cleaning (Figure 8-26). Machines are airless, and do not require an air compressor. They operate on a very simple principle. A large motor (from 7 to 15 horsepower) turns an impeller at approximately 2000 rpm. Fine steel shot (.020″ in diameter) is fed onto the impeller blades. The shot is propelled and scattered at a high rate of speed throughout the cabinet. The cylinder heads turn on a table, thus exposing all surfaces to the cleaning action. The normal cleaning cycle ranges from four to ten minutes depending on the model of machine. Although these machines have been designed to handle primarily cylinder heads, they can be used to clean valves, camshafts, and manifolds. It is safe to clean aluminum parts with the steel shot if a reduced cycle time is used and no machined surfaces are exposed. Sometimes crushed walnut hulls or cherry

FIGURE 8-24
A portable siphon fed type blasting machine. Connect air supply to the gun and push the wand into a bag or container of glass beads or sand (Courtesy of Davis Sandblasting Machines).

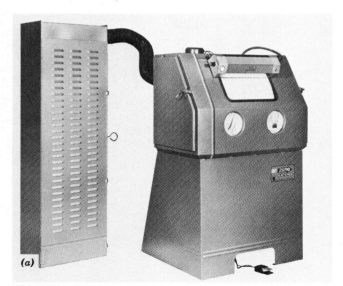

FIGURE 8-25*a*
A direct pressure glass bead machine with a reclaimer unit (Courtesy of Zero Manufacturing Company).

This is a sample listing of some media and typical applications.

Type Media	Sizes Normally Available	Use-Type of Result
Glass Beads	8 to 10 Sizes From 30 to 440 Mesh Also many special graduations	Decorative Blending, light deburring, peening, general cleaning, texturing. Non-contaminating. Peening media for non-metal removal cleaning (holds tolerances), retains critical part dimensions.
Aluminum Oxide	10 to 12 Sizes From 16 to 325 Mesh	Fast cutting, matte finish. Descaling and coarse, sharp textures.
Garnet	6 to 8 Sizes Wide Bank Screening From 16 to 325 Mesh	Non-critical cleaning, cutting, texturing. Non-contaminating for brazing steel & stainless steel.
Crushed Glass	5 Sizes Wide Band Screening From 30 to 400 Mesh	Fast cutting, low cost, short life, abrasive. Non-contaminating.
Silicon Carbide	36-220 Mesh	Extremely fast cutting.
Steel Shot	12 or more Sizes Close graduation From 8 to 200 Mesh	General purpose, rough cleaning foundry operations, etc. Peening.
Steel Grit	12 or more Sizes Close graduation From 10 to 325 Mesh	Rough cleaning, coarse textures, foundry welding applications, some texturing.
Cut Plastic	3 Sizes Fine, Medium, Coarse Definite size particles	Deflashing thermosetting plastics. Cleaning, light deburring.
Crushed Nut Shell	6 Sizes Wide Band Screening	Deflashing plastics, cleaning, very light deburring fragile parts.
Corn Cobs	Various Sizes	Deflashing plastics, cleaning & deburring where the surface must not be marred.

(b)

FIGURE 8-25*b*

A sample of some grit media and typical applications for glass bead machines (Courtesy of Zero Manufacturing Company).

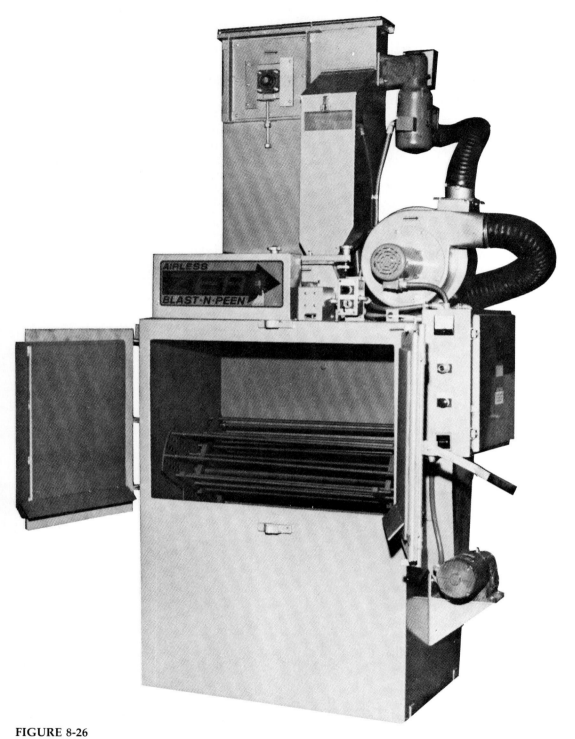

FIGURE 8-26

An airless shot blasting machine with a bottom impeller (Courtesy of Zero Manufacturing Company).

pits are used in conjunction with an airless blaster for cleaning aluminum parts.

After the cleaning cycle is completed, *make certain that no shot is left in parts.* Some machine models have features that allow the parts to tumble free or vibrate for a few minutes after cleaning in order to remove the shot.

COMMENT: For an airless blaster to be effective, parts must be thoroughly degreased and dry before being placed in the machine.

PYROLYTIC OVEN CLEANING

This is a dry cleaning technique (similar to a self-cleaning oven) used mainly by high-volume production rebuilders. A special large oven is used to bake the moisture from grease and oil, leaving only a dry ash deposit on the part that can be blown off with compressed air.

MARKING CLEANED ENGINE PARTS

The identification of cleaned parts can be a problem in the shop. A recommended method of metal marking (widely used in the steel industry) is to use a special self-contained white roll-on tube marker. After cleaning and drying, cylinder heads, blocks, crankshafts, and so on, can be easily identified by writing the desired information on the part or casting with this marker. The identification mark is very durable and is not removed by additional wash cycles in caustic or soak cleaning chemicals. See Appendix D for a source.

TRUE CASE HISTORY

Vehicle	Volkswagen Beetle
Engine	4 cylinder, 1600 cc
Mileage	250 miles after a rebuild
Problem	1. Engine seizure. 2. Fine sandlike grit throughout the engine.
Cause	The engine case halves were glass bead cleaned during the rebuilding process. Even though each case was solvent washed and water rinsed three times, all of the abrasive beads were not removed. This caused severe scoring of internal parts and the resulting seizure.
Comment	If you are going to glass bead or lead shot blast, do it only on parts that can be easily solvent cleaned and hand scrubbed afterward.

TRUE CASE HISTORY

Vehicle	Cadillac
Engine	8 cylinder, 472 CID
Mileage	Zero mileage on a newly installed short block. The old engine failed when the timing chain broke, damaging several pistons and valves.
Problem	1. As soon as the new engine assembly started, a loud internal noise was heard. 2. The engine started to run extremely rough. It was immediately shut off. 3. Coolant could be heard running into the oil pan.
Cause	The intake manifold was not checked for the presence of foreign material. Some piston metal was lodged inside a runner when the old engine failed. This metal was drawn into the new engine, cracking a piston and knocking a hole in the cylinder wall.
Comment	Always check and clean the inside of the intake manifold prior to installation.

CHAPTER 8 SELF-TEST

1. Why is it generally necessary to clean parts before inspection?
2. List eight ways to clean engine parts.
3. What safety device are solvent tanks required to have on the lid?
4. What precautions are necessary when cleaning parts with a power-driven wire brush?
5. Wire brushing hard carbon from combustion chambers can be made easier by doing what first?
6. How are valve guides generally cleaned?
7. What solvents are especially effective for removing gum and varnish from valve guides?
8. A properly cleaned valve guide will look _____ on the inside diameter.
9. Why shouldn't valve stems be cleaned by power wire brushing?
10. How can carbon deposits be removed from under the valve head?
11. Why should you avoid using carburetor cleaner to clean factory painted valve springs?
12. List three methods to clean pistons.
13. How can an old piston ring be used to clean the carbon from the piston ring grooves?
14. Do not put aluminum rocker arm stands in the hot-tank solution. Why?
15. What precautions are necessary when cleaning pistons by glass bead blasting?
16. What are two useful tools for cleaning small oil passages?
17. What advantages are offered by high-pressure spray cleaning?
18. After dipping parts in carburetor cleaner, flush with _____ .
19. What can happen if magnesium parts are soaked overnight in carburetor cleaner?
20. What are the three common methods of hot-tank agitation?
21. The hot-tank solution temperature should be kept at approximately what degrees Fahrenheit?
22. The hot-tank solution pH should be kept at about _____ .
23. What safety precautions are important when handling hot-tank soap?
24. What metals will be harmed by the hot-tank caustic solution?
25. What is scale? How can it be removed from water jackets?
26. What safety precautions are important when handling acid solution?
27. Why is a final alkaline rinse needed after descaling?
28. What is the typical procedure for turning on a steam cleaner?
29. What precaution is necessary before turning off the steam cleaner pump motor?
30. The steam cleaning soap solution should have a normal concentration of about _____ %.
31. Why are some shops against using powdered soap in a steam cleaner?

32. Name and describe the two basic types of glass bead blasting machines.
33. What are the disadvantages of glass bead cleaning?
34. Describe the operating principle of an airless shot blasting machine.
35. How can aluminum parts be safely cleaned with an airless shot blasting machine?
36. What is pyrolytic cleaning?
37. What simple method of marking cleaned engine parts can be used in the shop?
38. Why is it important to always check and clean the inside of the intake manifold prior to installation?

CHAPTER 9

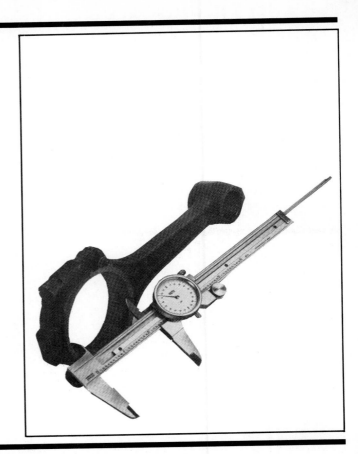

MEASUREMENT INSPECTION

Accurate measurements must be taken after cleaning to determine which parts need machining. Certain parts will have to be measured for size, while wear and warpage will be measured on others. Measurements are made with inside and outside micrometers, calipers, dial gauges, straightedges, and feeler gauges. Parts should be at room temperature and should be completely free of any foreign material when measurements are made. Most important, the mechanic has to have the correct "sense of feel" when using measurement tools in order to obtain correct readings.

CYLINDER WALL WEAR

A used cylinder always has ring ridge, at the unworn area at the top of the bore (Figure 9-1). In today's engines this is a band about ¼" wide all around the cylinder. A pocket, or area of maximum wear is directly below the ring ridge. This worn area is due to several factors.

1. Lack of lubrication at this portion of the cylinder wall.

2. The diluting effect of the raw gas on the engine oil.

3. Built-up pressure behind the rings at their uppermost position during the compression and power strokes.

Cylinder wear diminishes toward the bottom of the ring travel. The cylinder surface below the ring travel shows practically no wear since it is well lubricated and has relatively light wall pressure on it by the piston skirt.

Measuring Cylinder Taper

Cylinder taper is one of the main measurements used to determine the condition of an engine. The amount of taper is the difference in diameter between the top of the cylinder ring travel (in the direction of piston thrust) and the bottom of the cylinder (Figure 9-2). Subtracting the smaller diameter from the larger gives the amount of

FIGURE 9-1
Ridge at top of cylinder (Courtesy of Ford Parts and Service Division).

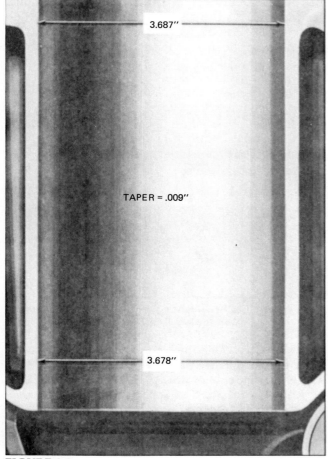

3.687''

TAPER = .009''

3.678''

FIGURE 9-2
Determining cylinder bore taper (Courtesy of Ford Parts and Service Division).

cylinder taper. For example, if the diameter at the top is 3.687" and the bottom diameter is 3.678", the taper would be .009". A cylinder bore dial gauge, an inside micrometer, or a telescoping gauge and an outside micrometer can be used for this measurement. *NOTE: Failure to measure the cylinder diameter directly underneath the ring ridge will result in a false wear figure.*

There is an alternate method sometimes used to quick check taper. A piston compression ring of cylinder size and a feeler gauge set is all that is required.

1. Place the ring in the cylinder, directly under the ridge. Square it up in the cylinder by using the head of the piston as shown (Figure 9-3). Measure the end gap with a feeler gauge.

2. Check the ring gap at the bottom of the cylinder in the same manner (Figure 9-4).

3. Subtract the bottom gap measurement from the top gap measurement. *Divide by three.* This figure equals the approximate amount of taper. See Figure 9-5 as an example.

FIGURE 9-3
Pushing a compression ring into the bore using the head of the piston. This squares the ring and locates it at proper depth for an accurate end gap measurement (Courtesy of Ford Parts and Service Division).

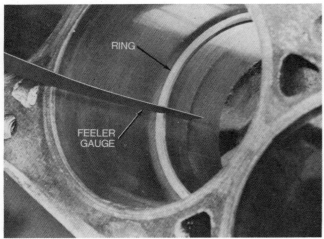

FIGURE 9-4

Measuring ring end gap (Courtesy of Ford Parts and Service Division).

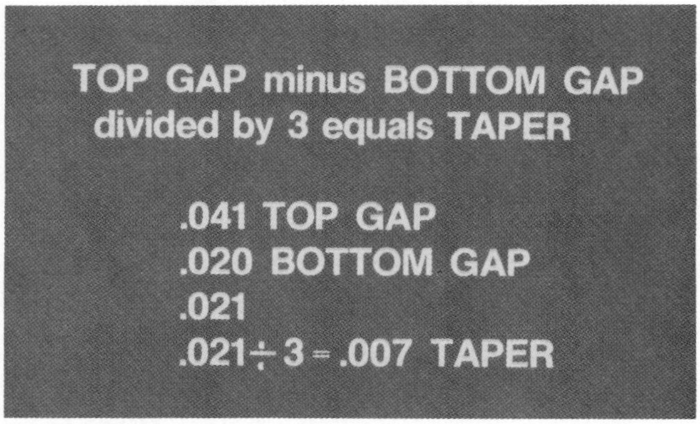

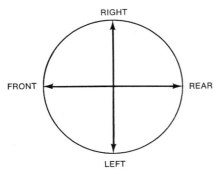

FIGURE 9-5

An example of calculating cylinder bore taper using the ring end gap method (Courtesy of Ford Parts and Service Division).

DIFFERENCE IN DIAMETERS = OUT-OF-ROUND
FIGURE 9-6

Measuring cylinder out-of-round (Courtesy of Ford Parts and Service Division).

Measuring Cylinder Out-of-Round

Cylinder out-of-roundness is the difference between right-to-left and front-to-back bore diameter (Figure 9-6). Take the measurements directly underneath the ring ridge.

Maximum Allowable Taper and Out-of-Round Before Reboring

Maximum allowable taper without reboring depends on a number of things. The general condition of the engine, type of engine, cost involved, and type of service used are factors to be considered. Always try to follow the manufacturer's specified limit. If no specifications are available, the following guidelines can be used.

1. *If maximum cylinder taper is .012" or over, boring is strongly recommended.*

2. If the cylinders are tapered between .003" and .012", the pistons should be expanded by knurling and a ring set engineered for re-ring applications installed (Figure 9-7a and b). The cylinder walls should be trued up by using a rigid hone. ***COMMENT:*** *Several ring manufacturers make claims that their rings will provide proper oil control in cylinders tapered up to .018". In most cases this is true. However, in order to maintain ring contact with the tapered walls, heavy pressure expanders are used. This will increase internal engine friction.*

FIGURE 9-7a

Piston skirt diameter can be increased by knurling. The piston head in this picture has been cutaway in order to show the knurling rollers in operation.

FIGURE 9-7*b*

A typical ring set engineered for re-ring application. Note the expander and rail used with the second compression ring.

If cylinder out-of-roundness is greater than .005", the engine should be rebored. It is much more difficult for piston rings to conform to out-of-roundness than to taper.

CAUTION: There are some current General Motors engines that cannot safely be rebored beyond .020".

The Oldsmobile 403 CID engine is one such example. If this block is rebored more than .020", the head gasket may fall into the cylinder bore chamber and fail. Also, the walls between the cylinders are very narrow on this block. This results in a minimum load area for head gasket sealing.

Selecting the Correct Oversize Ring Set

When measuring the unworn portion of the cylinder, it is important to determine if this dimension is standard size or an oversize size. It may be necessary to refer to specification tables in a manual. Sometimes a rebored cylinder size can be verified by checking the piston heads or rings for a stamped oversize number. Cylinder oversizes with recommended ring set oversizes are specified in Table 9-1.

TABLE 9-1

Cylinder and ring set oversizes

Cylinder Oversizes	Ring Set Oversizes
Standard and .010"	Standard
.020"	.020"
.030"	.030"
.040" and .050"	.040"
.060" and .070"	.060"
.080" and .090"	.080"

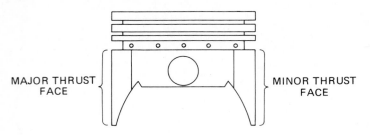

FIGURE 9-8
Piston thrust faces.

PISTON CONFIGURATION

Every piston has a major and a minor thrust face (Figure 9-8). The side of the piston which presses against the cylinder wall on the power stroke is the major thrust face. The minor thrust face is the opposite side of the piston.

All pistons have a major and minor axis (Figure 9-9). The minor axis runs across the piston, parallel to the pin axis. An imaginary centerline drawn from the center of the major and minor thrust faces is the major axis. The piston manufacturers make the minor axis smaller than the major axis. This results in a piston skirt contour that is cam shaped (elliptical).

Piston skirts are cam-ground so as to compensate for thermal expansion (the ring area of the piston is not cam-ground, but is circular). A perfectly round piston would not expand equally in all directions because of metal thickness differences. There is more metal in and around the pin boss area, while skirt faces are generally thin. The thin areas do not expand as much as the thick areas. In order to allow for this, the piston minor axis is designed smaller than the piston major axis (Figure 9-10). When the piston reaches

operating temperature and has expanded, it will be round.

Piston skirts are slightly tapered (Figure 9-11). The piston manufacturers make the top portion smaller in diameter than the bottom (about .0015") to allow for thermal expansion. The top of the skirt will expand more than the bottom. At operating temperature the piston expands and fits correctly. Most OEM (original equipment manufacturer) and aftermarket piston skirts are cut with a straight taper.

Piston Measurement

The piston diameter can be checked by using an outside micrometer. *The diameter is generally checked at the centerline of the piston pin bore and at 90° to the pin* (Figure 9-12). However, some manufac-

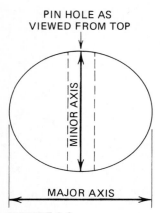

FIGURE 9-9
Piston axes.

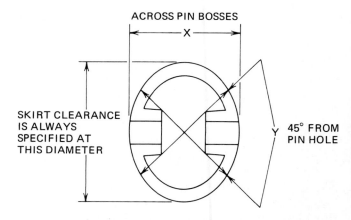

Cam Type	Diameter Reduced at X	Diameter Reduced at Y
A	.005 to .007"	.001 to .002"
B	.006 to .008"	.003 to .004"
C	.009 to .011"	.008 to .010"
D	.012 to .014"	.010 to .012"
E	.013 to .015"	.0065 to .0085"

FIGURE 9-10
Piston cam contours.

turers specify that piston measurement be taken at a point different than at pin height (Figure 9-13). In other cases, the measurement is taken just below the oil ring groove, or the bottom of the skirt may be the specified point.

Measure all pistons and compare to specifications. Check for collapsed skirts by measuring with a micrometer. *Most pistons should measure between .001" to .003" smaller (across the widest part of the skirt) than the cylinder bore size.* A collapsed piston skirt will permit the piston to rock back and forth in the bore. Piston noise, increased blow-by, spark plug oil fouling, and excessive oil consumption will result. If any piston is collapsed, replace it.

Measuring Ring Groove Wear

Piston ring grooves are normally machined .0015" wider than the ring. If there is excessive clearance at this point due to wear (Figure 9-14), oil pump-

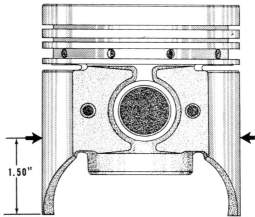

FIGURE 9-13
In this case, the engine manufacturer specifies that piston diameter be measured at a point different than at pin height.

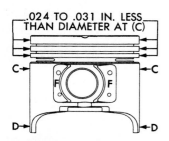

DIAMETERS AT (C) AND (D)
CAN BE EQUAL OR DIAMETER
AT (D) CAN BE .0015 IN.
GREATER THAN (C)
FIGURE 9-11
Piston head and skirt profile.

FIGURE 9-12
Using a micrometer to check piston outer diameter (Courtesy of Ford Parts and Service Division).

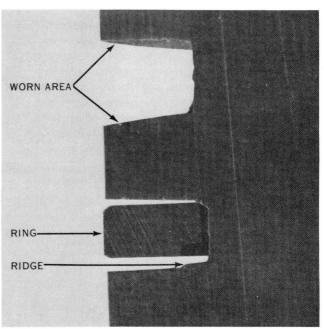

FIGURE 9-14
Cross-section of a piston with badly worn ring grooves (Couresty of Dana Corporation—Service Parts Group).

FIGURE 9-15
Checking for top ring groove wear with a feeler gauge and a new ring (Courtesy of Dana Corporation—Service Parts Group).

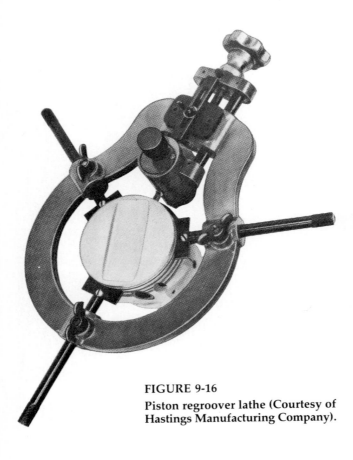

FIGURE 9-16
Piston regroover lathe (Courtesy of Hastings Manufacturing Company).

ing will occur. Each piston should be checked for top ring groove wear (Figure 9-15). If you can insert a .006″ feeler gauge leaf between a new ring and the land, the piston must be replaced, or the groove reconditioned (Figure 9-16). This check can also be made with a ring wear gauge. See Chapter 2.

CRANKSHAFT MEASUREMENT

Each rod journal (often called a crankpin or throw) and main journal (often called simply a journal) needs to be carefully measured for taper, out-of-round, and undersize. A micrometer is generally used to make these measurements (Figure 9-17). Figure 9-18 shows the points to measure. Crankpins will develop the most wear on the underside. In the trade, the term "a flat crank" is often used. This refers to a crankpin that is out-of-

FIGURE 9-17
Measuring a crankshaft throw. Note how the micrometer is being held (Courtesy of Imperial Clevite, Inc., Engine Parts Division).

round because of underside wear. *An out-of-round or a tapered condition of .001" or more on any crankpin or journal can justify regrinding.*

Sometimes the bearing journals will show hourglass or barrel-shaped wear. *The maximum limit is .001".*

Most current engines use the combination of a grooved upper main bearing and a plain (non-grooved) lower one. This can subject the journal to ridging (Figure 9-19). *If this ridge exceeds .0003", the journals should be reground to a standard underside.* Failure to remove ridging can distress and fatigue a new set of main bearings.

Crankshaft alignment (straightness) needs to be checked. The recommended method requires using vee blocks and a dial indicator. *Alignment from journal to journal should be within .001". Overall tir must be within .002".* The crankshaft will require straightening if these limits are exceeded. Refer to Chapter 2's discussion on vee blocks for additional information.

```
A VS. B = VERTICAL TAPER
C VS. D = HORIZONTAL TAPER
A VS. C AND B VS. D = OUT-OF-ROUND
CHECK FOR OUT-OF-ROUND
AT THE END OF EACH JOURNAL
```

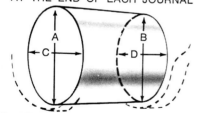

FIGURE 9-18

Crankshaft taper and out-of-round measurements (Courtesy of Ford Parts and Service Division).

FIGURE 9-19

Badly worn and ridged journals (Courtesy of Imperial Clevite, Inc., Engine Parts Division).

Check runout on the crankshaft snout, and face runout on the flywheel flange. *Maximum allowable runout is .001".*

When checking for taper and out-of-round, compare micrometer readings with the standard specification size of the crankshaft to determine if the shaft has been ground undersize. The usual undersizes to which bearing journals are ground is .010", .020", or .030". Bearings are commonly stocked in these undersizes to fit reground shafts.

Crankshaft Grinding Limits

In most instances, automotive crankshaft rod or main journals should not be ground more than .030" undersize. Engine remanufacturers will usually not accept crankshaft cores that will not clean to .030". Removing metal beyond this limit can lead to crankshaft breakage.

COMMENT: There are some heavy-duty and industrial engines that have crankshafts stiff enough to permit grinding to .060" undersize. As a guide, check on the availability of undersized bearings below .030". When miking for wear, allow a minimum of .008" clearance at the lowest point for grinding purposes. Remember to inspect the crankshaft flanges, as "end thrust" wear may result in a junk core.

CONNECTING ROD AND PISTON PIN MEASUREMENT

The first step in checking the connecting rods is to install the bearing caps. Tighten the nuts to specification with a torque wrench.

Measure the big-end bore for out-of-roundness (stretch). Stretch generally takes place in a vertical direction up to an angle of about 35° (Figure 9-20). It almost never occurs across the parting line. *If any bore is elongated by more than .001", the rod should be replaced or reconditioned.* Installing new inserts in a stretched or worn big-end bore is foolish shop practice.

Measure the diameter of the connecting rod pin bore (often called the little end or "eye" of the rod). If it is greater than specification, an oversize pin may have to be installed. Piston pins are available for most engines in .0015" and .003" oversizes. If an oversize full-floating pin is used, it will have to be slip fitted to the rod bushing ID, and to the pin boss holes. If an oversize press-fit pin is used, it will have to be slip fitted to the pin boss holes, and the rod "eye" must be machined larger. This is necessary in order to maintain the correct press fit (usually − .0008″ to − .0012″).

All connecting rods must be checked for alignment. This is done on a special fixture (Figure 9-21). *Big-end and little-end bores must be parallel within .001″ per six inches of rod length. The twist between these bores must not exceed .001″ per six inches of rod length (Figure 9-22).* Misaligned connecting rods will cause the pistons to be out-of-square with the cylinder walls. This can result in noise, blow-

by, oil pumping, bearing wear, and ring damage (Figure 9-23).

To measure the amount of wear on a piston pin, check the diameter at a number of points across the pin (use a measuring tool that can discriminate to .0001″). Wear is shown as the

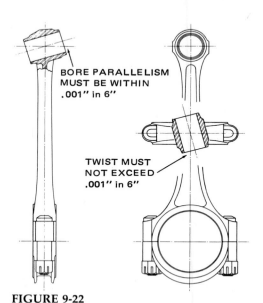

FIGURE 9-22
Exaggerated drawings of recommended limits for connecting rod alignment (Courtesy of Federal-Mogul Corporation).

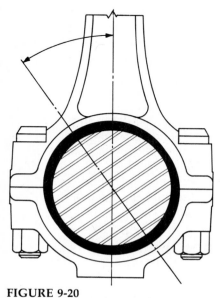

FIGURE 9-20
Typical rod bore elongation (Courtesy of Sunnen Products Company).

FIGURE 9-21
Checking connecting rod alignment (Courtesy of Dana Corporation—Service Parts Groups).

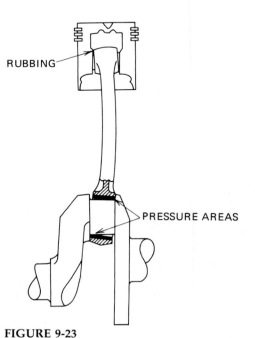

FIGURE 9-23
A bent connecting rod will create pressure areas and can cause premature failure of parts.

amount of difference in diameter readings. Any difference whatsoever justifies pin replacement. This is because pin clearance is the most critical engine measurement in terms of closeness of fit and tolerance. For example, an aluminum piston with a floating pin normally requires a .0001" to .0003" boss fit.

VALVE GUIDE AND VALVE STEM MEASUREMENT

Valve stems and guides need to be checked for wear. Too much stem-to-guide clearance will allow air and oil to be drawn into the combustion chamber. This can cause a rough idle, increase oil consumption, foul spark plugs, allow carbon buildup, and upset carburetion.

Valve guides will typically show bell mouth wear at the extreme top and bottom (Figure 9-24). The direction of greatest wear is in line with rocker arm movement. Valve stem wear will show the greatest at the extreme top and bottom of the guide rub area (Figure 9-25). There are several methods for measuring valve guide wear.

1. A telescoping or split ball gauge is used to size the guide. Remove the gauge and transfer the size to an outside micrometer (Figure 9-26).

2. Use a special valve guide dial bore gauge (Figure 9-27). The gauge is "zeroed" to the actual valve stem size using either an outside micrometer or setting fixture. Insert the dial gauge into the guide and read wear at any point.

3. Insert a new valve into the guide. Position a dial indicator at right angles to the valve stem (Figure 9-28). Rock the valve stem back and forth against the indicator plunger. If the clearance is more than specified, wear is indicated.

4. Valve grinding pilots are often used as go and no-go plug gauges for checking wear in guides. However, with this method it is dif-

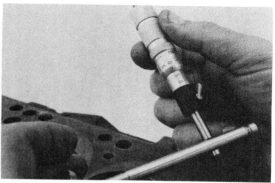

FIGURE 9-25
Checking for valve stem wear at the top of the guide rub area.

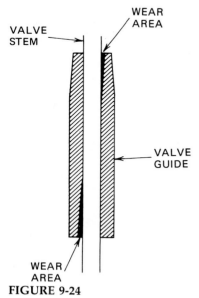

FIGURE 9-24
Valve guides generally wear bell mouthed.

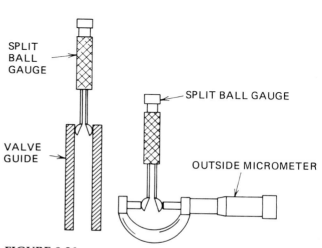

FIGURE 9-26
Measuring valve guide wear.

ficult to measure egg-shaped bore wear and bell mouthing. This is because most valve pilots center in the middle of the least worn section of the guide.

The valve stem is best measured for wear by using a micrometer. *Replace any valve that has more than .001" stem wear. When the clearance between the stem and guide together exceed the original clearance by .002", make the necessary correction to the valve and/or guide as may be required.*

COMMENT: Some valves are made with tapered stems (smaller at head end than at tip). This does not indicate the valve was improperly manufactured. Valves designed with a taper (generally .001" taper) have been a requirement on numerous original equipment exhaust valves for some time. As a result, when supplying aftermarket replacement valves the taper is also employed. The taper is added to compensate for thermal expansion differences between the stem and guide. The exhaust valve runs hotter than the guide, and some alloy steels also have a higher coefficient of thermal expansion than the cast iron guide. Also, the valve reaches operating temperatures sooner than the guide. Therefore, the stem taper has been incorporated to reduce wear and scuffing at the port end of the guide.

CAMSHAFT MEASUREMENT

Very close inspection must be given to the camshaft. Using an outside micrometer, or vee blocks and a dial indicator, check the wear on each cam lobe (Figure 9-29). Check each intake lobe against the other intakes. Check each exhaust lobe against the other exhausts. *Maximum lobe height variation is .006".*

Visually inspect the oil pump—distributor drive gear and the fuel pump eccentric. Replace the

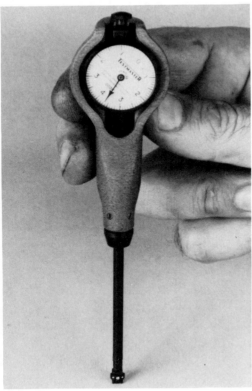

FIGURE 9-27
Special dial bore gauge for measuring valve guide wear.

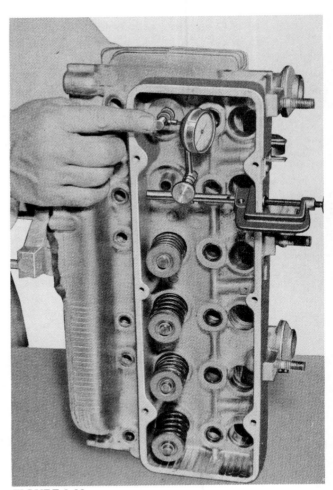

FIGURE 9-28
Determining of valve stem clearance with a dial indicator. (Courtesy of Chevrolet Division of General Motors Corporation)

FIGURE 9-29
Using a micrometer to check camshaft lobes for wear.

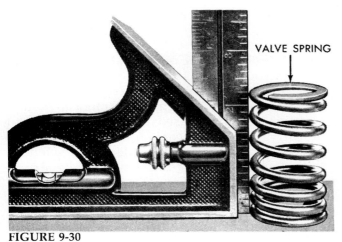

FIGURE 9-30
Inspecting valve spring squareness (Courtesy of Chrysler Corporation).

FIGURE 9-31
Checking saddle bore alignment with a precision ground arbor (Courtesy of Federal-Mogul Corporation).

camshaft if there is any obvious wear or damage to these surfaces.

Measure the size of each cam journal and compare to specification. *Maximum wear limit on any journal is .001".*

Check the camshaft for straightness by using vee blocks and a dial indicator positioned on the bearing journals. *Maximum runout is .002".*

VALVE SPRING MEASUREMENT

The valve springs function to ensure positive closing of the valves. After much mileage, the springs become weak and lose tension. Lifters may then bounce and cause the springs to close at the wrong rate. The result is an engine power loss.

The valve springs can be measured for tension by using a spring tester. *All springs should read within 90% of specifications.* Stand all springs on a flat surface and check free length height. *All springs should be within 1/16" of specifications.*

Check valve springs for squareness. Stand each spring up next to a square on a flat surface and turn it slowly one full turn (Figure 9-30). Note the distance the top coil moves away from the square. *The maximum allowable out-of-squareness limit is generally recognized as being 5/64".* This specification will apply to the majority of automotive valve springs. However, this 5/64" limit is based on a 2" (plus or minus ¼") free length. If springs are out of this range, it is advised to check manufacturer's specifications.

MAIN BEARING BORE MEASUREMENT

The engine block main bearing bores need to be checked for alignment and for excessive out-of-roundness. *The main caps have to be located correctly and torqued to specifications before making these measurements.*

Main bearing bore alignment (often called saddle bore alignment) can be checked best by using a precision ground arbor (Figure 9-31). The arbor should be sized .001" less in diameter than the original low limit diameter of the saddle bores. Place the arbor into the saddles and install the main caps. Try to turn the arbor with a 12" long handle. If you can turn the arbor, the saddle bores are considered to be in alignment. If the

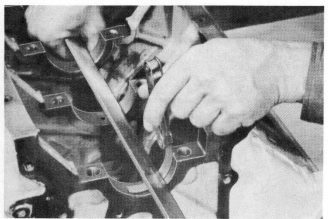

FIGURE 9-32
Checking saddle bore alignment with a precision straightedge and feeler gauge (Courtesy of Federal-Mogul Corporation).

arbor will not turn, the crankcase is warped or an out-of-round bore condition exists.

If a ground arbor is not available, saddle bore alignment can be checked by using a precision straightedge and a feeler gauge (Figure 9-32). Try to insert a .0015" feeler gauge under the straightedge at various parallel positions in the bores. If this can be done, the crankcase is out of alignment and should be corrected.

Saddle bore out-of-roundness can also be checked by using a dial bore gauge, inside micrometer, or telescoping gauge. The vertical bore dimension should not be larger than the horizontal. A .001" maximum out-of-round is permitted if the horizontal dimension is larger than the vertical. For opposed engines with vertical split cases (Volkswagen, Corvair, and so on) this would be vice-versa.

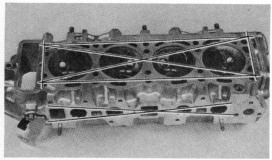

FIGURE 9-33
Check cylinder head surfaces for flatness in several directions.

CYLINDER HEAD WARPAGE MEASUREMENT

Check the cylinder head gasket surface for warpage by using a precision straightedge and a feeler gauge. The surface should be checked in both diagonal directions, across the head, and lengthwise (Figure 9-33). *Maximum allowable warpage is .004" at any point.*

Check flatness of the intake and exhaust manifold mounting surface on the head. *Maximum deformation allowed is .004".*

INTAKE AND EXHAUST MANIFOLD WARPAGE MEASUREMENT

Mounting surfaces *must be straight within .010"* and free of pits and erosion.

BASIC ENGINE REBUILDING MATH

The United States is slowly converting to the metric system. At present, the American system of measurement is used in most of American industry. In a few years, however, the metric system should predominate. Mechanics will need to be familar with both systems and be able to convert from one to the other. Help yourself by working out in your own mind the values of the metric system as compared to the American. Memorize the conversion values given below.

1 millimeter (mm) = .0394"
Inches × 25.40 = millimeters
Cubic inches (cu. in.) × 16.39 = cubic centimeters
Cubic centimeters (cc) × .06102 = cubic inches
Liter × 1000 = cubic centimeters
Cubic inches × .01639 = liters
Liters × 61.02 = cubic inches

Example   1.6 liters × 61.02 = 97.6 cubic inches

Problem   An engine has a 3½" bore. What will be the bore size in millimeters?

Displacement

The volume of a cylinder is given by the area of the bore in square inches times the stroke in inches. The result is cubic inches. The area of a cylinder is easy to calculate. Multiply the bore by itself and then again by .7854 or 3.1416/4.

The formula for figuring displacement would, therefore, be

Bore^2 × Stroke × .7854 × Number of cylinders

Problem   A six-cylinder engine has a bore of 3.625″. The stroke is 3.562. Calculate the displacement.

How Much Displacement Is Gained by Going .060″ over Bore?

You have an eight-cylinder engine with a 4.00″ bore and 3.5″ stroke. Going to .060″ oversize on the cylinders will give you 4.060″ bore. The total piston area can be obtained the hard way by multiplying 4.060 × 4.060 × .7854 × 8 (cylinders) = 103.57 square inches. The new displacement is 103.57 × 3.5 = 362.49 cubic inches. However, there is an easier way to get a close approximation. You can find out how much you add to the total piston area by using a 12.66 factor, times the bore, times the overbore.

For example, a 4.00″ bore is increased to 4.060″. The extra piston area will be 12.66 × 4.00 × .060 = 3.01 square inches. To figure out how much displacement is gained, multiply this answer by the stroke. You have gained 3.01 × 3.5 = 10.5 cubic inches. The answer is accurate within half a cubic inch.

Calculating Compression Ratio (CR)

The compression ratio of an engine is the comparison between combustion chamber volume and the combined volume of the combustion chamber and the displacement. In other words

$$CR = \frac{V + CV}{CV}$$

V = displacement in one cylinder
CV = compression volume

NOTE: *Both values must be in consistent units. In other words, cc's and cubic inches can't be mixed.*

Example   The displacement volume is 46 cu. in. and the compression volume is 5 cu. in.

$$CR = \frac{46 + 5}{5} = 10.2$$

Calculating Compression Volume (CV)

The above formula can be reworked to read

$$CV = \frac{V}{CR - 1}$$

The displacement in a particular engine is 54 cubic inches per cylinder. You want a compression ratio of 10:1. The compression volume should be

$$\frac{54}{10 - 1} = 6 \text{ cubic inches}$$

Compression Ratio Versus Piston Displacement

Generally speaking, compression ratio will change as a direct function of percentage of increase (or decrease) in piston displacement. In other words, a change in compression ratio is usually brought about when the piston bore size is increased. This change is overlooked too often by the engine builder, with the consequences being possible engine damage.

Example   We have a popular 302 CID engine with a compression ratio of 10.4:1. Cylinder bore size is increased by .060″. This will change the displacement to 311 cubic inches. This represents a percentage of piston displacement change of + 3.02%. Therefore, the amount of compression ratio increase would be 3.02%, or an increase from 10.4:1 to 10.7:1.

COMMENT: *There are some piston manufacturers that compensate for CR increase when using oversize pistons by changing the pin position.*

Decimal Equivalents

Frequently a mechanic needs to change a fraction into an equivalent decimal. Most shops have a decimal equivalent chart posted in the working area. However, there may be times when you do not have access to such a chart, and you need the information immediately.

RULE: *To convert a fraction into a decimal, divide the numerator by the denominator.*

Example   Change 7/8 to a decimal. Since a fraction is an indicated division problem, we divide 7 by 8. Since 8 does not go into 7, we place a decimal point after the 7, add zeros, and carry out the division. Hence, 7/8 = .875.

Problem The valve clearance specification on an engine is .016″. The mechanic misplaced his feeler gauge. Could a piece of 1/64″ thick shim stock be used instead?

Sometimes it may be necessary to mathematically change a decimal to a fraction. For instance, you need to reduce .750 to a fraction.

Solution The number after the decimal point is the numerator of the fraction, while the denominator is a 1 with 3 zeros (because there are three places to the right of the decimal point). Thus, .750 = 750/1000. This fraction is then reduced to lowest terms. First divide both the numerator and denominator by 25 and get 30/40. Then di-

viding each term of 30/40 by 10 gives 3/4. Hence, .750 = 3/4.

ENGINE TEARDOWN AND ASSEMBLY GUIDE

The following checklist guide is designed to serve as a job aid when rebuilding an engine. The sequence and steps listed will apply to most engines. Included are spaces for recording key measurements. If used in a teaching situation, the student has to stop at certain points and obtain an instructor sign-off.

ENGINE TEARDOWN AND ASSEMBLY GUIDE

Make of engine_____

Cubic inches_____

Number of cylinders _____

Firing order _____

Bore and stroke _____ × _____

Engine on stand   Yes ☐ No ☐

Prior to teardown, record any compression data in the notes column

Drain engine oil ☐

Remove distributor cap and wires ☐

Remove distributor and hold-down ☐

Remove spark plugs ☐

Remove oil filter assembly, any side cover(s), and fuel pump and lines ☐

Remove the carburetor assembly (Figure 9-34) ☐

Drain fuel (carburetor, fuel pump, and lines) ☐

Remove alternator, waterpump, belts, hoses, and so on that are not part of the main block assembly ☐

Remove manifolds   Intake ☐
                          Exhaust ☐

Check manifold surfaces for warpage (Figure 9-35) ☐

How much? _____

Machine   Yes ☐ No ☐

Notes:

FIGURE 9-34

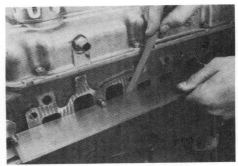
FIGURE 9-35

Remove cylinder head cover(s)
(Figure 9-36) ☐

Remove any rocker arm shaft
lubrication delivery pipes ☐

Remove the rocker arm assembly ☐

Diagram below and show the sequence
used in disassembly and in tightening ☐

Remove push rods and lifters on
those engines equipped ☐

On OHC engines, remove camshaft
timing gear, bearing caps, bearings,
and camshaft (Figure 9-37) ☐

Camshaft bearing caps marked ☐

How? _____

Remove cylinder head(s) ☐

Diagram and show the disassembly
and tightening sequence

Remove valves from head(s) ☐

Remove carbon from block and
head surfaces ☐

Check block and head surface
for warpage (Figure 9-38) ☐

How much? _____

Machine   Yes ☐ No ☐

Check cylinder head casting
for cracks ☐

What detection method was
used? _____

Remove the crankshaft pulley ☐

Invert the engine and remove the
oil pan (Figure 9-39) ☐

Remove core and gallery plugs ☐

Remove oil pump assembly ☐

Remove the timing chain cover ☐

Remove any chain tensioners and
dampers ☐

Remove timing gears, or sprocket
and chain assembly ☐

Notes:

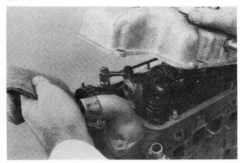

FIGURE 9-36

FIGURE 9-37

FIGURE 9-38

FIGURE 9-39

Mark parts to keep in correct
relationship ☐

Visually inspect parts for wear
and damage ☐

Cylinder ridge cut
(Figure 9-40) Yes ☐ No ☐

Connecting rod caps marked
before removal ☐
How?

Remove rod and piston assemblies ☐

Note 1. Use covers on rod bolts (Figure 9-41).
2. Position crank throw at BDC when
   installing or removing pistons.

Check connecting rods for bend
and twist ☐

Were any bad? Yes ☐ No ☐

Remove the piston rings and
piston pins ☐

Remove the connecting rod bushings
if so equipped (Figure 9-42) ☐

OEM (Original Equipment Mfg.)
cylinder bore _____

Oversize Yes ☐ No ☐

#1 top (thrust direction) _____
    bottom (unworn portion) _____

#2 top _____   bottom _____
#3 top _____   bottom _____
#4 top _____   bottom _____
#5 top _____   bottom _____
#6 top _____   bottom _____
#7 top _____   bottom _____
#8 top _____   bottom _____

Amount of cylinder taper (Figure 9-43)

#1 _____   #2 _____   #3 _____
#4 _____   #5 _____   #6 _____
#7 _____   #8 _____

Amount of cylinder out-of-round

#1 _____   #2 _____   #3 _____
#4 _____   #5 _____   #6 _____
#7 _____   #8 _____

_____
Signature

Notes:

FIGURE 9-40

FIGURE 9-41

BUSHING
DRIVER
FIGURE 9-42

FIGURE 9-43

Remove flywheel   Yes ☐ No ☐

Main bearing caps marked before
removal (Figure 9-44)   ☐
How?

Remove crankshaft and main bearings
from engine   ☐

Or leave crankshaft installed and
use a roll-out pin to remove
main bearings   ☐

Remove camshaft and/or oil pump
driveshaft from those engines
having placement in the block   ☐

Engine unit should now be
completely disassembled   ☐

_____
Signature

Main bearing journals OEM size _____

Undersize   Yes ☐ No ☐

#1 (taper)_____

   (out-of-round)_____

#2 _____   #3 _____

   _____   _____

#4 _____   #5 _____

   _____   _____

#6 _____   #7 _____

   _____   _____

Maximum allowed main bearing
journal taper _____

Maximum allowed main bearing journal
out-of-round _____

Connecting rod journals OEM size _____

Undersize   Yes ☐ No ☐

#1 (taper) _____

   (out-of-round) _____

#2 _____   #3 _____

   _____   _____

#4 _____   #5 _____

   _____   _____

Notes:

FIGURE 9-44

#6 _____   #7 _____

_____   _____

#8 _____

_____

Maximum allowed rod journal
taper _____

Maximum allowed rod journal
out-of-round _____

_____

Signature

OEM camshaft journal measurement
#1 _____   #2 _____
#3 _____   #4 _____
#5 _____

Present camshaft journal measurement
#1 _____   #2 _____
#3 _____   #4 _____
#5 _____

Camshaft lobe height measurement

#1 _____   #2 _____   #3 _____
#4 _____   #5 _____   #6 _____
#7 _____   #8 _____   #9 _____
#10 _____   #11 _____   #12 _____
#13 _____   #14 _____   #15 _____
#16 _____

Camshaft height limits
Intake _____
Exhaust _____
Camshaft lobes worn   Yes ☐ No ☐

Camshaft runout measurement
(Figure 9-45) _____

Crankshaft runout measurement
(Figure 9-46) _____

Camshaft gear runout measurement
checked   Yes ☐ No ☐

How much? _____

Crankshaft gear runout measurement
checked   Yes ☐ No ☐

How much? _____

Oil pump driveshaft runout
measurement checked   Yes ☐ No ☐

Flywheel runout checked   Yes ☐ No ☐

Notes:

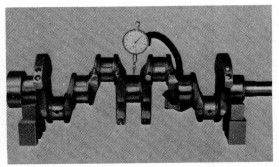

FIGURE 9-45

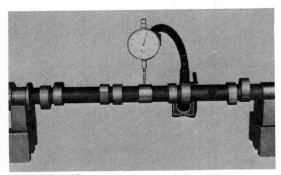

FIGURE 9-46

How much? _____

Piston measurement of skirts
Perpendicular to pin (Figure 9-47)

#1 _____   #2 _____   #3 _____
#4 _____   #5 _____   #6 _____
#7 _____   #8 _____

Piston-to-cylinder wall clearance

#1 _____   #2 _____   #3 _____
#4 _____   #5 _____   #6 _____
#7 _____   #8 _____

Were any pistons collapsed?   Yes ☐ No ☐
Were any pistons knurled?   Yes ☐ No ☐
How much? _____

Piston pin fit (check if satisfactory)

#1 _____   #2 _____   #3 _____
#4 _____   #5 _____   #6 _____
#7 _____   #8 _____

_____
Signature

Connecting rod bore out-of-round
measurements (without precision
bearings)

#1 _____   #2 _____   #3 _____
#4 _____   #5 _____   #6 _____
#7 _____   #8 _____

Cylinder block main bearing bore
out-of-round measurements

#1 _____   #2 _____   #3 _____
#4 _____   #5 _____   #6 _____
#7 _____   #8 _____

Rod bore out-of-round limits
are _____

Block main bearing bore out-of-round
limits are _____

Check main bearing bore saddle
alignment   Satisfactory ☐
                            Unsatisfactory ☐
How much out? _____

All threads chased   Yes ☐ No ☐

Cylinder walls deglazed
or honed   Yes ☐ No ☐

Walls washed with soap and hot
water three times   Yes ☐ No ☐

Notes:

FIGURE 9-47

Walls oiled to prevent rusting   Yes ☐ No ☐

New camshaft or oil pump
driveshaft bearings installed   Yes ☐ No ☐

Oil holes checked for alignment Yes ☐ No ☐

Core and oil gallery plugs
installed   Yes ☐ No ☐

Camshaft bearing clearance
measurement (on OHC engines perform
after cylinder head has been installed)

#1 _____ #2 _____ #3 _____
#4 _____ #5 _____

Measurement limits are _____

_____
                    Signature

Camshaft installed (except OHC engines)   ☐

Camshaft end play measurement
checked   Yes ☐ No ☐

How much? _____

Oil pump driveshaft end play
measurement checked   Yes ☐ No ☐

How much? _____

Camshaft and main bearings
installed   ☐

OEM main bearing clearance   ☐

Main bearing clearance on this engine
#1 _____ #2 _____ #3 _____
#4 _____ #5 _____ #6 _____
#7 _____

Crankshaft end play measurement
(Figure 9-48) _____

Crankshaft rear main seal assembly
installed   ☐

Crankshaft to oil seal drag: Light ☐
                                   Medium ☐
                                    Heavy ☐

Main bearing bolt torque
_____

Inspect each of the timing gears
for wear and damage   ☐

Notes:

FIGURE 9-48

Visually inspect chain(s) for wear
and damage and for smooth
connections between the links ☐

Measure chain stretch with a
vernier caliper Yes ☐ No ☐

Determine timing gear clearance
or chain deflection Yes ☐ No ☐

How much? _____

What is the OEM specification? _____

Method used

Inspect the chain tensioners
for visual wear
and damage (Figure 9-49) Yes ☐ No ☐

Are the tensioners in need of
replacement? Yes ☐ No ☐

Inspect the chain vibration dampers
(if equipped) ☐

How is the camshaft timed to the
crankshaft? Make a diagram and
show markings

_____

Signature

Fuel pump eccentric, crankshaft front
oil slinger, oil seal, front
cover and crankshaft pulley (and/or
damper assembly) installed ☐

Hang rods on pistons, observing
proper relationship (Figure 9-50) ☐

How marked? _____

Check new pinlocks (if used) for
proper seating and the open end
facing down ☐

Check piston ring to land clearance
(all pistons) ☐

Top groove
All satisfactory ☐

Unsatisfactory ☐

Which pistons are not to specs? _____

Notes:

FIGURE 9-49

FIGURE 9-50
In this example, the front of the piston
is marked with an "indent."

Second groove
All satisfactory ☐
Unsatisfactory ☐
Which pistons are not to specs? _____

Third groove
All satisfactory ☐
Unsatisfactory ☐
Which pistons are not to specs? _____

Check piston ring butt or end
clearance (all cylinders) ☐

Satisfactory ☐
Unsatisfactory ☐
Check ring depth clearance ☐

Establish proper ring position,
and install on pistons (Figure 9-51) ☐

Piston ring end gaps staggered
correctly (Figure 9-52) ☐

Piston and rod bearing assemblies
installed (use covers on rod bolts) ☐

Connecting rod bearing clearance
(Figure 9-53)
#1 _____ #2 _____ #3 _____
#4 _____ #5 _____ #6 _____
#7 _____ #8 _____

Connecting rod end play checked ☐
How much? _____

Connecting rod bolt or nut
torque _____

New nuts used   Yes ☐ No ☐
Double check torque ☐

Check piston deck height
relationship (optional)
#1 _____ #2 _____ #3 _____
#4 _____ #5 _____ #6 _____
#7 _____ #8 _____

_____

Signature

Engine valves, related parts, and
guides cleaned ☐
Valve guides worn   Yes ☐ No ☐
Valve stems worn   Yes ☐ No ☐

Notes:

FIGURE 9-51

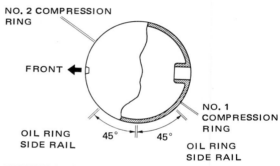

NO. 2 COMPRESSION
RING

FRONT ←

NO. 1
COMPRESSION
RING

OIL RING
SIDE RAIL   45°   45°

OIL RING
SIDE RAIL

FIGURE 9-52
Position piston rings on piston.

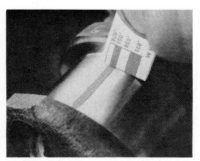

FIGURE 9-53

What is the installed spring height specification? _____

Valve guides knurled or replaced

Yes ☐ No ☐

Protruded height (Figure 9-54) _____

Valve keepers worn   Yes ☐ No ☐

Valves refaced   Yes ☐ No ☐

What angle? Intake _____
Exhaust _____

Valves tipped and chamfered
(Figure 9-55)   Yes ☐ No ☐

Valve seats refaced (Figure 9-56)

Yes ☐ No ☐

What angle? Intake _____
Exhaust _____

Seat runout held within _____

Interference angle used   Yes ☐ No ☐

Valve margin (Figure 9-57)

All satisfactory   ☐

Unsatisfactory   ☐

Which valves are not to specs? _____

Valve seat width and location

All satisfactory   ☐

Unsatisfactory   ☐

Which seats are not to specs? _____

Valve springs checked against
specifications   Yes ☐ No ☐

Check for

Free length ☐ Tension ☐

Installed height   ☐

Squareness (Figure 9-58)   ☐

Coil bind and retainer interference
(check visually with head
assembled on engine
while turning by
hand)   ☐

Valves assembled
with new seals   Yes ☐ No ☐

Cylinder head(s) and gasket(s)
installed   ☐

Cylinder head bolt (or nut)
torque _____

Push rods and lifters visually
checked (if equipped)   Yes ☐ No ☐

Check hydraulic lifter leakdown
rate   Yes ☐ No ☐

Notes:

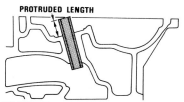

FIGURE 9-54

FIGURE 9-55
Resurfacing valve tip.

FIGURE 9-56

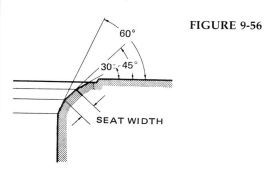

FIGURE 9-57
Check margin.

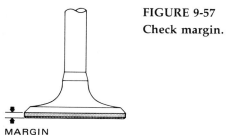

FIGURE 9-58

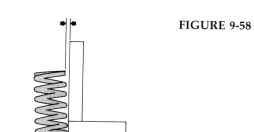

Lifters replaced   Yes ☐ No ☐
Check lifter bore clearance   Yes ☐ No ☐
How much? _____

_____

Signature

Rocker arms and shafts
visually worn   Yes ☐ No ☐
Rocker arms refaced   Yes ☐ No ☐
Replaced   ☐
Rocker arm shafts and/or
bushings replaced   Yes ☐ No ☐
Check rocker arm to shaft
clearance (Figure 9-59)   Yes ☐ No ☐
How much? _____
Install rocker arm assemblies   Yes ☐ No ☐
Locate exact TDC
and establish factory
marking accuracy (optional)   ☐

FIGURE 9-59

Oil pump clearances checked   Yes ☐ No ☐
What type of pump? _____
Measurements should be as follows

Oil pump bushing in block
worn   Yes ☐ No ☐
Has none   ☐
Oil pump and screen, pan gasket,
and oil pan installed   ☐
Oil pan bolt torque
specifications _____
Oil pan plug tight   ☐

_____

Signature

On OHC engines, assemble the camshaft onto the cylinder head   Yes ☐ No ☐

Torque camshaft bearing caps ☐

Check thrust clearance on camshaft   Yes ☐ No ☐

Install valve rocker support assembly and any oil delivery pipes (Figures 9-60 and 9-61)   Yes ☐ No ☐

Install timing gears (Figure 9-62)   Yes ☐ No ☐

Adjust the valve clearance (statically) in two revolutions of the crankshaft   Yes ☐ No ☐

Explain the procedure below

Install the valve cover(s) ☐

Install the manifold assemblies (make sure passages are clean) ☐

Carburetor, fuel pump and lines installed ☐

Distributor spark timed to engine ☐

Oil filter replaced   Yes ☐ No ☐

Oil in engine ☐

Number of quarts _____

Engine lubrication system primed before starting   Yes ☐ No ☐

Fuel pump checked   Yes ☐ No ☐

Check and reinstall or replace any remaining parts ☐

Engine is now ready to start   Yes ☐ No ☐

_____
Signature

Oil pressure at 1000 rpm

Valves adjusted (hot)   Yes ☐ No ☐

Notes:

FIGURE 9-60

FIGURE 9-61

FIGURE 9-62

Oil to rocker arms   Yes ☐ No ☐

Cylinder head bolt (or nut)
torque rechecked   Yes ☐ No ☐

Check for oil, water, fuel,
vacuum, or exhaust leaks   ☐

REMARKS *Method One* The recommended
method of running in a completely rebuilt engine
is to install it on a test stand. **Operate it at a fast
idle speed,** increasing the speed intermittently
between minor adjustments. While it is running,
check the oil pressure on a pressure gauge to be
sure that it is normal.

Method Two An alternate to test stand running is
to install the engine in the vehicle. Start it and
test it out on the road for about an hour, set to
a fast idle, but drive it up to 50 mph in high gear,
with alternate deceleration at least 10 times. Perform minor adjustments as required.

Engine operation:

Satisfactory   ☐

Unsatisfactory   ☐

_____

Signature

_____

Date

Notes:

TRUE CASE HISTORY

Vehicle   Dodge Truck

Engine   8 cylinder, 361 CID

Mileage   Zero mileage on engine after installing a new camshaft and hydraulic lifters.

Problem
1.   At idle the engine operated normally with no particular noise.
2.   When engine rpm was increased, severe lifter noise could be heard.

Cause   Standard size hydraulic lifters had been installed in .008″ oversize lifter bores.

Comment   Some Chrysler-Dodge-Plymouth V-8 engines were manufactured with nonstandard size components. A diamond mark on the right front of these blocks (Figure 9-63) means the block has .008″ oversize lifters. A Maltese cross indicates the engine is equipped with a crankshaft that has one or more undersize rod and/or main bearing journals.

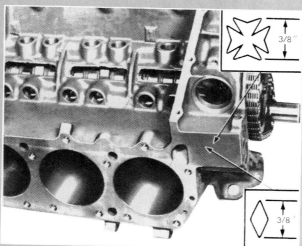

FIGURE 9-63
Courtesy of Chrysler Corporation.

CHAPTER 9 SELF-TEST

1. What is the ring ridge?
2. Why does the greatest amount of cylinder wall wear occur directly below the ring ridge?
3. Define cylinder taper.
4. What instruments can be used to measure cylinder taper?
5. How can a piston ring and a feeler gauge be used to determine cylinder taper?
6. Define cylinder out-of-roundness.
7. If maximum cylinder taper is _____ " or over, boring is strongly recommended.
8. If cylinder out-of-roundness is greater than _____ ", the engine should be rebored.
9. Why can't certain model General Motors engines safely be rebored beyond .020"?
10. Which side of a piston is the major thrust face?
11. What is the piston major axis?
12. Why are piston skirts cam-ground?
13. Why are most piston skirts slightly tapered?
14. Piston diameter is generally measured at what point?
15. What is a collapsed skirt?
16. Piston ring grooves are normally machined _____ " wider than the ring.
17. How are pistons checked for top ring groove wear?
18. What is meant by a "flat" crank?
19. Any crankpin or journal that is out-of-round or tapered more than _____ " should be reground.
20. What is ridging?
21. Overall crankshaft tir must be within _____ ".
22. What is maximum allowable flywheel flange runout?
23. What are the common crankshaft bearing journal undersizes?
24. Generally, automotive rod or main journals should not be ground more than what amount?
25. Connecting rod bore stretch generally takes place in what direction?
26. If a connecting rod big-end bore is elongated by more than _____ ", the rod should be replaced or reconditioned.
27. Piston pins are available for most engines in what oversizes?
28. Connecting rod twist must not exceed what limits?
29. Too much valve stem-to-guide clearance can cause what?
30. Valve guides typically show wear where?
31. What are three methods for measuring valve guide wear?
32. Replace any valve that has more than _____ " stem wear.
33. What is maximum camshaft lobe height variation?

34. What is maximum camshaft runout?
35. Valve spring tension should be within _____ " of specification.
36. Valve spring free length height should be within _____ " of specification.
37. How are valve springs checked for squareness? What is maximum allowable out-of-squareness?
38. What two methods can be used to check main bearing bore alignment?
39. How is saddle bore out-of-roundness checked? What are the limits?
40. What is maximum allowable cylinder head warpage?
41. Intake and exhaust manifold mounting surfaces must be straight within what amount?
42. A five-liter engine has how many cubic inches?
43. What is the formula for figuring displacement?
44. What is the formula for calculating compression ratio?
45. How do some piston manufacturers compensate for compression ratio increase when using oversize pistons?
46. How is a fraction changed into a decimal?
47. How is a decimal changed into a fraction?

CHAPTER 10

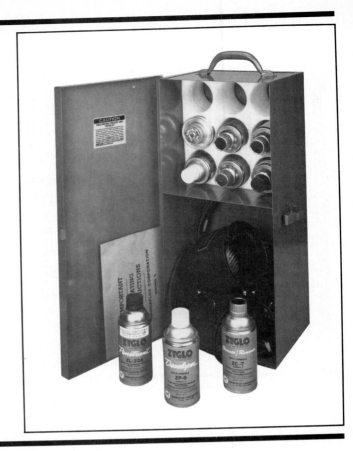

LOCATING CRACKS AND CASTING REPAIR

A crack in an engine part indicates a failure process is occurring. If the part is a moving one, sudden destruction may result. Crack failures often start with the formation of small cracks that slowly develop into larger flaws and eventual damage.

Cracks are generally caused by one of the following reasons.

1. Stress in the metal due to improper alloy distribution when castings are "poured." The mix of metal must be evenly dispersed, with no internal material structure defects, inclusions, or voids (Figure 10-1).

2. Overheating due to defects in the cooling system. Heat cracks usually occur in regions of the greatest water space and around exhaust valve seats.

3. Chilling of the metal by a sudden rush of cold air or water over the surface, after it has become extremely hot (Figure 10-2). Trucks equipped with vacuum brakes often show chill cracks in every intake valve port after the driver has suddenly closed the throttle

and applied the brakes on an overheated engine.

4. Strain put on parts by excessive tightening, or from misalignment (Figure 10-3). On rare occasions, freezing coolant will cause enough strain from expansion to crack a head or block.

5. Fatigue (the phenomenon that leads to fracture under repeated or fluctuating stress cycles). Fatigue fractures are progressive, beginning as minute cracks that grow under the action of the cycles (Figures 10-4 and 10-5).

6. Flexing of the metal due to lack of rigidity. The early model Toyota Corolla with the 2T engine is a prime example. The head would often crack between the no. 2 and no. 3 cylinders on the rocker arm side (Figure 10-6). Japanese engineers realized that they had a problem, and improved the head design. Strength improvement was made by adding reinforcement ribs to the manifold stud bosses (Figure 10-7) and by increasing metal thickness on the rocker arm side of the head.

212

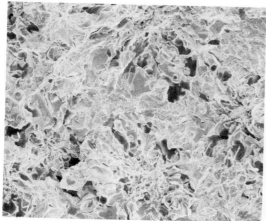

FIGURE 10-1

This 100× electron microscope scan of gray cast iron shows a normal mix of iron (Fe) and carbon (C) evenly dispersed. There are no internal material structure defects, and no inclusions or voids.

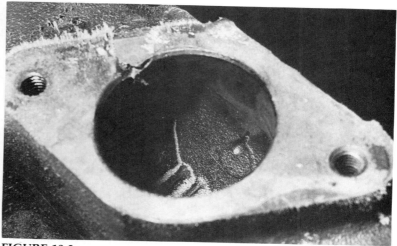

FIGURE 10-2

The inside of this intake manifold shows a crack. Sudden metal chilling is often the cause.

FIGURE 10-3

Incorrect mounting bolt installation torque caused this exhaust manifold to crack.

FIGURE 10-4

This lifter fracture was due to a tension failure in the cross-section minor area, with the oil hole being a stress point.

FIGURE 10-5

Poor lubrication occurred in the crankpin radius (fillet) area and caused this failure.

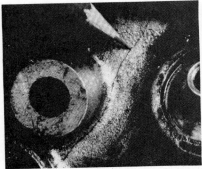

FIGURE 10-6

This is a common head crack on early model Toyota 2T engines. Flexing of the aluminum, due to lack of rigidity, is the cause.

FIGURE 10-7

Comparing the early and late model Toyota 2T cylinder heads. Note that the bottom (late) head has ribs added to the manifold stud bosses for reinforcement.

LOCATING CRACKS

There are several trade accepted methods (besides visual) of locating cracks. *To avoid wasted time and money, parts should always be checked for cracks before any machine or repair service.* Regardless of the crack detection method used, metal must be cleaned thoroughly beforehand.

Pressure Testing

Pressure testing is a very popular method used by engine rebuilders to locate cracks in heads and blocks. This method will find cracks in all areas of the casting. Magnetic inspection, however, will only detect external cracks; it will not detect internal cracks, or leaks around rocker arm studs, valve guides, cylinder sleeves, and injector tubes.

The head or block pressure testing operation typically consists of the following basic steps.

1. Mount the casting on a special test fixture (Figure 10-8).
2. Close the water ports on the casting for pressurizing, using various size and shape closure pads.
3. Pressurize the head or block by applying from 40 psi to 50 psi to the air pressure regulator.

4. Spray soap or shampoo solution on the surfaces of the casting. Bubbles will appear where the head is leaking.
5. Chalk mark any cracks that are discovered. *Be sure to chalk the entire length of the crack.*
6. After the casting has been repaired, again close off all the water ports and pressurize to make certain that the crack has been completely repaired.

Some shops prefer using hot water to pressurize, rather than air. They feel hot water will cause expansion and make the test more sensitive. This method, however, requires using a circulator tank and pump (Figure 10-9).

NOTE: An area in which very few people realize that leaks occur is around rocker arm studs. This will allow coolant and oil to mix.

Cracks in the rocker arm oil delivery line in Ford 330 and 390 CID engines are among the most common. These cracks are located near the bottom of the casting and difficult to locate. Therefore, a pressure test is always recommended on these blocks. If there is a crack, it can be repaired by drilling out the oil line and installing a piece of tubing (Figure 10-10).

← AIR INLET LINE

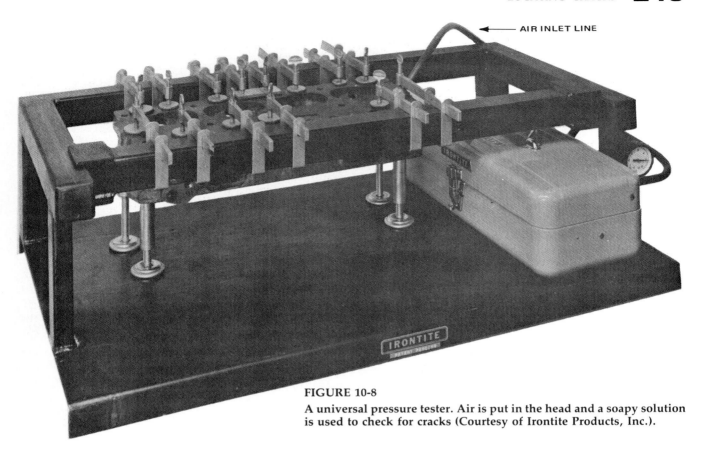

FIGURE 10-8

A universal pressure tester. Air is put in the head and a soapy solution is used to check for cracks (Courtesy of Irontite Products, Inc.).

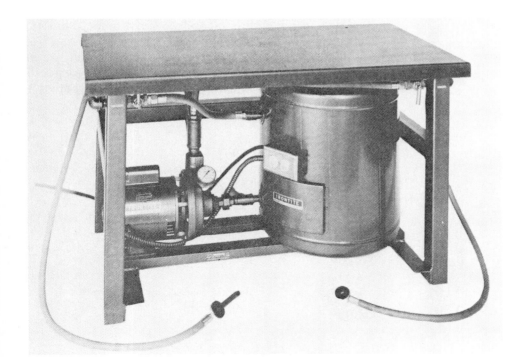

FIGURE 10-9

A circulator can be used to pressurize the head with hot water when checking for cracks (Courtesy of Irontite Products, Inc.).

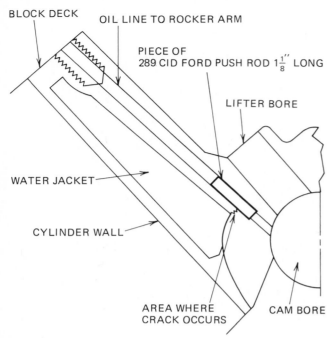

FIGURE 10-10

Oil line crack repair on Ford 330 and 390 CID engines. The oil line is drilled out to a depth of 5½" with a letter *N* drill and reamed to .312". Next, install a tube 1⅛" long with a diameter of .314". The tube can be cut from a push rod used in a 289 CID Ford engine (Courtesy of Automotive Engine Rebuilders Association).

Magnetic Inspection

Magnafluxing is the general term used in the trade to refer to the magnetic inspection of metal. Actually, Magnaflux is the name of a company that manufactures crack detection equipment. There are many other companies that make similar equipment (Figure 10-11).

Magnetic crack detection works on the broken magnet principle—opposite and attracting poles are established across each break (Figure 10-12). When the part being inspected is magnetized, local poles will be set up at any crack. Fine iron powder dusted on the surface will be attracted. A definite gray-colored outline of the crack can then be seen (Figure 10-13). *NOTE: Yellow and white colored powder is also available.*

Magnafluxing is applicable only on materials that can be magnetized. This method will not work on stellite valve seats, aluminum, and bronze.

Helpful Hints

Always apply the magnet before dusting on the powder. Hold the atomizer at least 6" from the area being checked to allow the powder to drift into space between the magnetic poles. Place the magnet in two positions at right angles to each other when inspecting an area. Do this because a straight crack running parallel to the poles of the magnet may not show. The magnetic field must cross a break in order to excite the powder.

Engine blocks are best inspected while they are suspended in the air (Figure 10-14), since they can then be easily rotated. Be certain to check main bearing webs, lifter bores, and cylinder wall areas.

Often cylinder head cracks are found that have not leaked. The mechanic may feel that it is, therefore, not necessary to have the head repaired. This is a mistake—these cracks may open up when the head is installed and retorqued. Also, when a crack is indicated, it almost always goes entirely through the casting.

FIGURE 10-11

Permanent magnet crack inspection equipment (Courtesy of Irontite Products, Inc.).

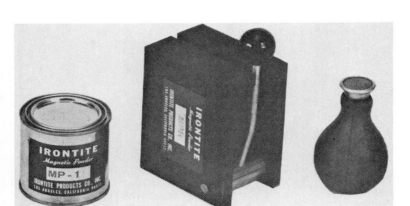

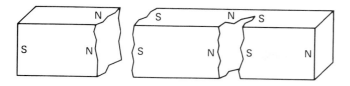

FIGURE 10-12
Opposite and attracting poles will be established across casting breaks when the part is magnetized. Iron powder dusted on the casting surface will then be drawn to the crack.

FIGURE 10-13
The gray line on the outside of this block indicates a crack (Courtesy of Seal-Lock International).

FIGURE 10-14
The cables surrounding this block electromagnetize it. A special air gun is used then to blow iron particles over suspected areas.

Magnetic Fluorescent Inspection

Sometimes the iron powder is not applied dry. Instead, it is applied in suspension as a liquid and viewed under black light.

Magnaglo is one very popular such method (Figure 10-15). A special fluorescent paste containing iron particles is mixed with oil. A magnetic field is induced in the part to be inspected and the fluorescent solution is then sprayed on. Under black light, any cracks will show as vivid white streaks (Figure 10-16).

Dye Penetrant Inspection

There are several methods of locating cracks using penetrants. This method of inspection can be used on magnetic and nonmagnetic materials.

One very simple old-time method is to use a 50/50 mixture of ether (starting fluid can be used) and kerosene. Brush the solution on the area to be checked. Allow it to soak for about ten seconds, and then blow dry the surface with light air pressure. Any solution that has penetrated

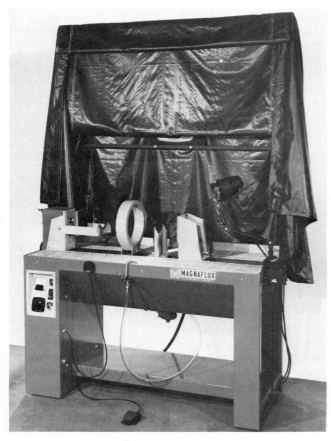

into a crack will rise to the surface to evaporate, making it easy to see the crack.

Zyglo (a trademark of the Magnaflux Corporation) is a fluorescent penetrant that is viewed under black light. Any crack will show up as a brilliant fluorescent trace (Figure 10-17). This method works as follows.

1. Clean the area to be checked by using the special cleaner supplied with the Zyglo kit.
2. Spray on the fluorescent oil base penetrant.
3. Allow several minutes for the penetrant to work, then dry the area by blotting with a paper towel.
4. Spray on a thin coat of developer, allowing it to dry to a white color.
5. Inspect with black light.

Met-L-Check is a three-step spray system that uses a red stain as a penetrant (Figure 10-18). Defects show up as bright red traces (Figure 10-19).

FIGURE 10-15
A magnetic fluorescent inspection booth (Courtesy of Magnaflux Corporation).

FIGURE 10-16
The white streaks indicate cracking in the fillet area of this crankshaft (Courtesy of Magnaflux Corporation).

FIGURE 10-17
A cracked piston as viewed under black light using the Zyglo method (Courtesy of Magnaflux Corporation).

REPAIRING CRACKS

Cracks are normally repaired by cast iron pinning ("cold" welding is the term often used) or by oxyacetylene welding. Minor cracks and sand holes on the outside of the block are sometimes repaired with epoxy.

Before attempting to repair any casting crack, the following points need to first be considered.

1. Can the extremities of the crack be completely contained?

2. Is the length of time required to make the repair reasonable? In other words, is the labor price worth the risk of failure and lower than the cost of purchasing an uncracked head or block?

3. Is the crack located in a stress area that will cause the crack to flex, thereby opening it up again? This is why combustion chamber cracks, and cracks subject to valve spring pressures, are not normally attempted.

4. If using cast iron pins, will the pins go to water and cool themselves? Sometimes pins

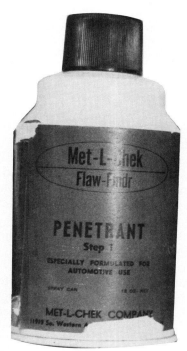 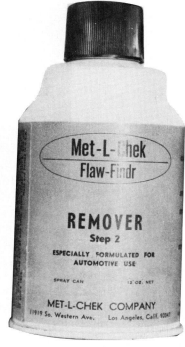 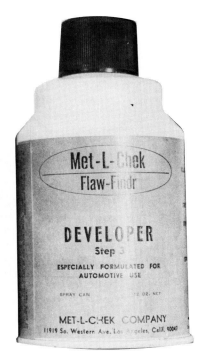

FIGURE 10-18
Met-L-Chek dye penetrant inspection kit.

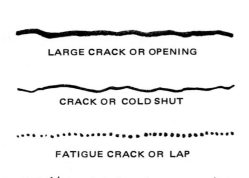

LARGE CRACK OR OPENING

CRACK OR COLD SHUT

FATIGUE CRACK OR LAP

POROSITY OR PITS

FIGURE 10-19

The four sketches above show how various imperfections show up in different patterns when treated by the Met-L-Chek crack detection system.

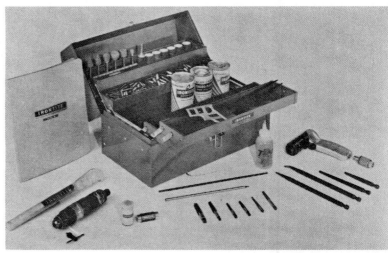

FIGURE 10-20

This special kit enables you to salvage heads and blocks by pinning (Courtesy of Irontite Products, Inc.).

FIGURE 10-21

These tapered cast iron plugs are used to repair casting cracks. They are designed for easy peening without chipping or breaking off (Courtesy of Irontite Products, Inc.).

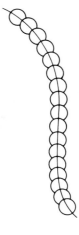

FIGURE 10-22

When pinning internal cracks, make certain all of the plugs are locked together. Drilling the holes with care will help ensure success.

will overheat and cause the crack to open up.

5. After repairing a crack by pinning, can the repaired area be reached for a peening operation with an air hammer?

Basically, there are two kinds of cracks—*internal* (as in the combustion chamber area) and *external* (where the crack can be seen from the outside).

Cast Iron Pinning

Pinning a cracked casting is a proven repair process that has been around for many years. Several companies make specially designed kits for pinning cracked components (Figure 10-20).

After locating the crack, tools are used to drill, ream, and tap holes along the crack. Threaded, special taper plugs are then screwed into the hole (Figure 10-21). The protruding pin ends are sawed off, peened, and ground smooth.

Internal cracks should be interlaced with a solid line of pins completely containing the crack (Figure 10-22). On external cracks this is not always required. It is a common practice to allow ¼" to ⅜" spacing between pins on outside cracks (Figure 10-23). The total area is peened then to close up the intermittent cracks and to eliminate stress and strain.

One of the most common problems in pinning cracks is failure to completely capture the cracks

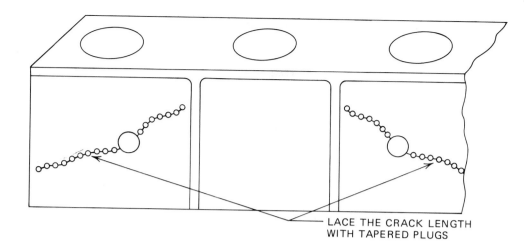

LACE THE CRACK LENGTH WITH TAPERED PLUGS

FIGURE 10-24
A common crack on the 230 CID Chevrolet head. Repair should be made by welding (Courtesy of Automotive Engine Rebuilders Association).

at the ends. Always magnetically inspect the drilled holes at the extreme ends of the crack to determine whether or not the crack is completely contained.

Some Typical Crack Examples

Let us analyze some examples of cracks you might encounter, and see whether a satisfactory pin repair can be made.

Figure 10-24 shows a 230 CID Chevrolet head. A crack has formed between two seat ports. This is a dangerous crack to try and repair. Drilling will weaken the area to a point where the pins will not hold. This type of crack is best repaired by preheating and torch welding.

Figure 10-25 shows an older Chevrolet 6 head that is cracked on the side walls of the combus-

FIGURE 10-25
This is a common crack on older model Chevrolet cylinder heads that can be successfully pinned.

FIGURE 10-26
International V-8 heads often crack in this area. This crack should be welded, as pins tend to not hold (Courtesy of Automotive Engine Rebuilders Association).

tion chamber. This somewhat common crack can be repaired by pinning.

On International V-8 engines, cracking between the spark plug hole and the intake valve port may occur (Figure 10-26). This innocent little crack looks like a simple repair. However, field experience shows that pinning will not hold. The pins will overheat because they do not extend into water. This will cause the crack to open up and drop the seat ring out. This crack should be welded.

FIGURE 10-27
A setup for heat welding a cylinder head (Courtesy of Automotive Engine Rebuilders Association).

Welding Method of Crack Repair

Figure 10-27 shows a cylinder head being prepared for heat welding. Note the setup using firebrick and a fire nozzle to heat up the work before welding. Heat or furnace welding is definitely the best way to weld cast iron. Preheating the entire casting eliminates the problem of stress cracks forming during the cooling-off period. *As a general rule, heat welding is not for the beginner.* This method not only requires proficient welding skills and a good heat source, but it also requires certain pieces of machinery to clean up the weld afterward. If you attempt this method of repair, take the following precautions.

1. Slowly heat the casting to approximately 1400° F. When welding, try to stay up in the range of from 1200° to 1300° F. The casting will be a dull, cherry red color when it is ready to be welded. After the welding is completed, bring the temperature back up to 1400° F for a period of 30 minutes. Then shut the furnace down and allow the part to cool at not over 100° per hour. This procedure allows for proper stress relieving and normalizing.

2. When heating the casting, cover it with a sheet of asbestos to help hold the heat in. Then make an access hole in the asbestos over the area to be welded.

3. Weld using an acetylene torch adjusted to a neutral flame and a pure cast iron rod. Use a high-quality glass flux. Puddle the base metal and then melt the rod into the puddle.

Epoxy Repair

Minor external cracks and casting sand holes are sometimes repaired with epoxy adhesive (Figure 10-28). Repair with this material in the following way.

1. Thoroughly clean the surface to be repaired. Start by rotary filing, and follow by using lacquer thinner.

2. Mix the epoxy according to the directions on the container.

3. Use a putty knife to press on the epoxy mixture.

4. Allow the mixture to cure. Twelve to 14 hours of air drying is normally required for proper hardening.

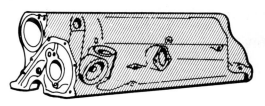

FRONT AND LEFT SIDE

TYPICAL FOR 6-CYLINDER ENGINE

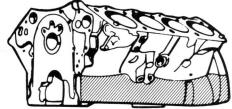

FRONT AND LEFT SIDE

TYPICAL FOR V-8 ENGINE

REAR AND RIGHT SIDE

REAR AND RIGHT SIDE

FIGURE 10-28
The shaded areas may be repaired with epoxy (Courtesy of Ford Parts and Service Division).

REPAIRING DAMAGED THREADS USING A HELI-COIL

The engine rebuilder will encounter threads that have been damaged from corrosion, wear, and overtorquing. Fortunately there is an effective and reliable repair available. This repair is often accomplished through the use of the Heli-Coil thread repair insert (Figure 10-29). This insert is a precision formed thread liner made of coiled stainless steel wire.

When installing a Heli-Coil, begin by drilling out the damaged threads using the drill size specified on the package. After drilling, use a special tap to cut threads for the Heli-Coil. *COMMENT: Spark plug hole inserts require no drilling; a Heli-Coil reamer tap is used.*

A special inserting tool is needed to install. Make sure the tool is engaged with the tang at the bottom of the Heli-Coil. Screw the Heli-Coil in by turning the installation tool clockwise. *CAUTION: Apply pressure on the insert, not on the tool.* To complete the repair, remove the tang by carefully bending it up and down with a pair of needlenose pliers.

In applications where aluminum is used, the use of Heli-Coil inserts is a lifesaver. A case in point is the General Motors 140 CID engine (used in Chevrolet Monza and Vega, Pontiac Astre and Sunbird, and the Oldsmobile Starfire). These en-

gines experience frequent main bearing thread stripping and head bolt breakage. Corrosion, high engine temperatures, overtorque, and the use of impact wrenches during bolt removal are all contributing factors. Since a Heli-Coil insert is stainless steel, it won't corrode easily. It is stronger and will hold up better to repeated disassembly.

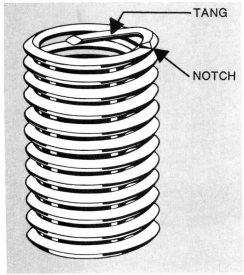

FIGURE 10-29
Heli-Coil insert.

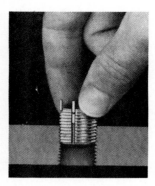

FIGURE 10-30

Thread in the insert with fingers. A prewinder tool is not required (Courtesy of Rexnord Tridair Industries).

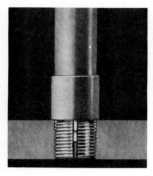

FIGURE 10-31

Using special installation tool, drive "kees" down with light taps of a hammer (Courtesy of Rexnord Tridair Industries).

REPAIRING DAMAGED THREADS USING KEENSERTS INSERTS

This type of thread repair does not require a special tap or prewinder tool. When installing a Keenserts* insert, begin by drilling out the old threads with a standard tap and thread in the insert (Figure 10-30). The "kees" act as a depth stop. Next, drive the "kees" down (Figure 10-31) to mechanically lock the insert in place.

REMOVING BROKEN BOLTS

Bolts will sometimes break, leaving the threaded portion stuck in the hole. The break might be

above, below, or flush with the top of the hole. Some common causes of breakage are listed below.

1. Overtightening.
2. Using the wrong type of bolt.
3. Bottoming out.
4. Rust or corrosion on the threads.
5. Physical damage.

Always begin the removal process by saturating the threads with a good penetrating oil and tapping lightly on the broken end. If the remaining portion of the bolt is above the surface, several methods can then be tried.

1. Grip and remove the bolt with a pipe wrench, stud puller, vise grips, or a drill chuck.
2. File the sides of the bolt flat, and use a crescent wrench for removal.
3. Cut a slot in the top, and use a screwdriver for removal.
4. Arc weld a nut on the end of the bolt. *COMMENT: The heat from the welding will often loosen a broken piece.* Use an impact wrench for removal. Start out on low pressure and gradually increase the pressure.

When the break is flush or below the surface, removing the bolt can be more difficult. One of these methods will usually work.

1. Position a sharp center punch on the bolt end. Hammer the punch at an angle trying to unscrew the broken piece.
2. Use a screw extractor (commonly called an "easy-out") (see Figure 10-32). A hole of the correct size is drilled in the center of the broken piece. The extractor is then installed in the drilled hole and turned. *COMMENT: When people have difficulty using an*

FIGURE 10-32

Screw extractors can be used to remove broken screws, studs, bolts, pipe, and fittings (Courtesy of Proto Tool Division, Ingersoll Rand Company).

---

*Keenserts is a registered trademark of the Rexnord Specialty Fastener Division.

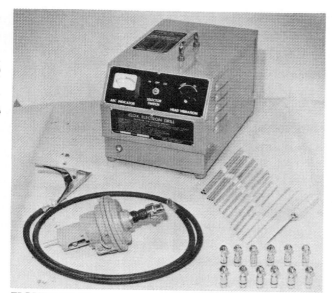

FIGURE 10-33

A portable EDM machine. The head is mounted in a drill press. Water coolant may be used with the machine (Courtesy of John Wiley and Sons, Inc.).

"easy-out," it's usually because too small a hole has been drilled, or the hole has not been completely drilled through the broken piece.

3. Using electrical discharge machining (EDM) (see Figure 10-33).

This is a process employed by the tool and die making industry. A carbon rod that discharges a high frequency electrical current is used to disintegrate metal. EDM is very effective for removing broken bolts being held by severe corrosion, taps or drills, hardened bolts, and broken off valve guide knurling tools. Your telephone directory should list companies that do EDM work.

TRUE CASE HISTORY

Vehicle	Toyota Corolla
Engine	4 cylinder, 2T-C model (push rod type)
Mileage	38,000
Problem	1. External coolant leakage.
	2. A crack was detected above the no. 4 core plug (Figure 10-34).

FIGURE 10-34
Toyota 2T-C engine block crack, above no.
4 core plug.

Cause Stress cracking in this area of the block was a problem on early production models. On later models, ribs were incorporated on the outer, left rear portion of the block. This added reinforcement and corrected the problem.

TRUE CASE HISTORY

Vehicle	Chevrolet Vega
Engine	4 cylinder, 140 CID (OHC)
Mileage	47,000
Problem	1. Engine overheating.
	2. Engine oil mixing with coolant inside the radiator.
	3. Excessive blue-gray smoke from the tailpipe.
Cause	An internal oil artery runs lengthwise along this cylinder head. If the engine is overheated, a crack often occurs and lets oil enter the adjacent water jacket. The head must be replaced as the crack is in an inaccessible area.

CHAPTER 10 SELF-TEST

1. List six possible reasons for crack formation in an engine part.
2. Magnetic inspection will not detect what type of crack?
3. How are cracks located when pressure testing a head or block?
4. What is the procedure for repairing a crack in the rocker arm oil delivery line on a Ford 390 CID engine?
5. What is Magnafluxing?
6. Magnafluxing isn't applicable on what materials?
7. How should the magnet be positioned when Magnafluxing an area?
8. Describe the Magnaglo inspection method.
9. How can ether and kerosene be used to locate a crack?
10. Describe the Zyglo inspection method.
11. What is the Met-L-Check inspection system?
12. What is "cold" welding?
13. Why is the repair of a combustion chamber crack not normally attempted?
14. How are the repair pins generally spaced on an outside crack?
15. How should a crack between two seat ports on a Chevrolet 230 CID head be repaired?
16. Where do older Chevrolet six-cylinder heads often crack?
17. What method of crack repair should be used on an International head that has cracked between the spark plug hole and the intake valve port?
18. When repairing a cast iron head by heat welding, why is it important to preheat the entire casting?
19. When heat welding a cast iron head, try to keep the casting in what temperature range?
20. How should a cast iron head be stress relieved and normalized after being repaired by heat welding?
21. Adjust the acetylene torch to a _____ flame when heat welding.
22. What kind of casting defects are sometimes repaired with epoxy adhesive?
23. What is a Heli-Coil?
24. What precautions are given when installing a Heli-Coil?
25. What is used to lock a Keenserts insert in place?
26. What are several advantages of using Keenserts inserts?
27. List five common causes of bolt breakage.
28. What methods can be used to remove a bolt that has broken above the surface? Below the surface?
29. What often causes problems when using an "easy-out"?
30. Describe EDM.

CHAPTER 11

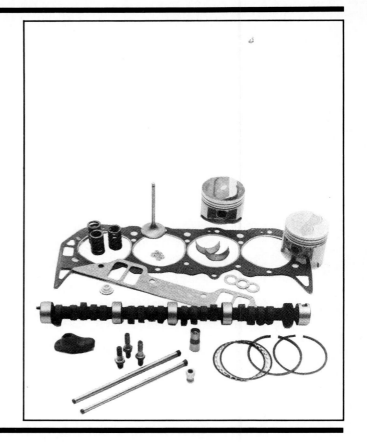

ORDERING ENGINE PARTS

With the many vehicle models in current use, it becomes important to be prepared with information to assure that you will receive the correct part. The following list outlines details often necessary to obtain correct engines or parts.

1. Year of vehicle.
2. Make.
3. Cylinder number.
4. Cubic inch displacement.
5. Bore size, spacing (Figure 11-1).
6. Engine code letter from VIN (for instance, a 1980 Pontiac 350 CID engine has the letter R, X, or N).
7. Passenger or truck.
8. Transmission—automatic, standard, number of speeds (three or four).
9. Hydraulic or solid lifters.
10. Equipped with or without air pump.
11. Rocker cover holes.
12. Carburetor—single, two, or four bore.

13. Filter—full flow or bypass.
14. Dip stick mounting—pan or block (Figure 11-2).
15. Oil drain plug—front, rear or side.
16. Timing cover—tin, aluminum, cast iron.
17. Pan sump—front, center, rear, single dip, double-dip (Figure 11-3).
18. Block casting number.
19. Head casting number.
20. Camshaft casting number.
21. Special factory markings (Figure 11-4).
22. Crankshaft forging number.
23. Crankshaft snout thread size.
24. Flywheel bolt size.
25. Clutch pedal linkage mountings (Figure 11-5).
26. Starter mounting holes on block—staggered or in-line.
27. Alternator mounting holes on block—location.

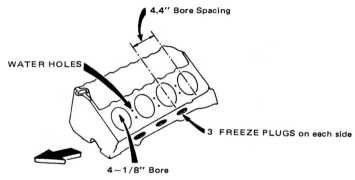

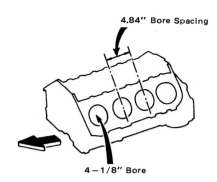

FIGURE 11-1

Chevrolet 400 CID blocks have two different bore spacings.

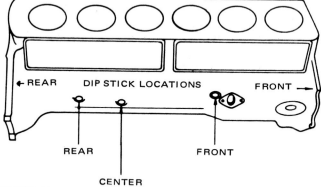

FIGURE 11-2

Chevrolet six-cylinder 194 and 230 CID engines.

28. Number of bell housing bolt holes.
29. Equipped with or without crankshaft pilot bushing (Figure 11-6).
30. Water holes in block: number and location (Figure 11-7).
31. Motor mount holes on block: number and spacing.

The parts person will not expect to get all this information. However, always be ready to furnish pertinent details to the item you are ordering. This can save a great deal of time.

Refer to Appendix E for a listing of casting and forging numbers.

ENGINE FAMILIES

Understanding engine "families" can assist the engine rebuilder. An engine family (Chart 11-1) is a group of engines with different cubic inch displacement ratings that evolved from one basic design. For example, one of Ford's most famous,

the *FE* family, was started with the 332-352 cubic inch engines. Later, improved versions included the 360, 361, 390, 406, 410, 428, and the high performance 427 cubic inch engines.

CHART 11-1

Ford V-8 engine family chart (Courtesy of Ford Parts and Service Division).

BASIC ENGINE DESIGN	THESE ENGINES EVOLVED FROM BASIC DESIGN
MEL	383-430-462 CID
Y	239-272-292-312 CID
90° V	221-260-289-289HP-302-302 BOSS-351 Windsor
335	351 Cleveland-351 BOSS-400 CID
FE	332-352-360-361 Edsel-390-406-410-427-428 CID
385	429-429 BOSS-460 CID

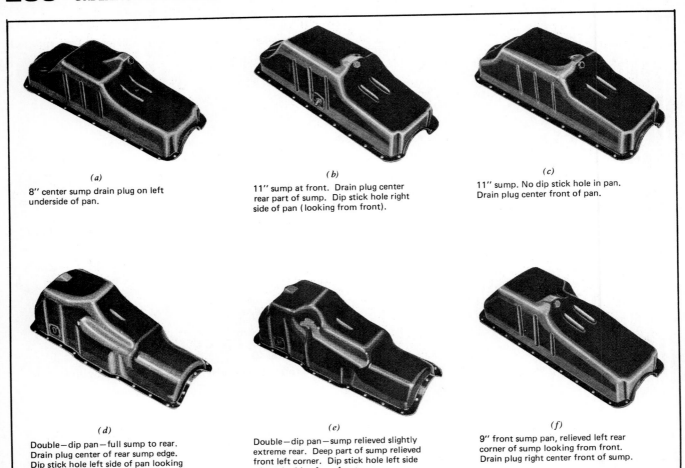

(a)
8" center sump drain plug on left underside of pan.

(b)
11" sump at front. Drain plug center rear part of sump. Dip stick hole right side of pan (looking from front).

(c)
11" sump. No dip stick hole in pan. Drain plug center front of pan.

(d)
Double—dip pan—full sump to rear. Drain plug center of rear sump edge. Dip stick hole left side of pan looking from front.

(e)
Double—dip pan—sump relieved slightly extreme rear. Deep part of sump relieved front left corner. Dip stick hole left side of pan looking from front.

(f)
9" front sump pan, relieved left rear corner of sump looking from front. Drain plug right center front of sump.

FIGURE 11-3
All these Ford six-cylinder truck pans are from the same year. On all engines, it is important to note sump size, drain plug location, and where the dip stick hole (if any) is located.

RAISED
LETTER
"O"

FIGURE 11-4
The raised letter "O" on this Chevrolet rocker arm indicates it is a high-performance type.

FIGURE 11-5
Some different clutch pedal linkage mountings used by Chevrolet.

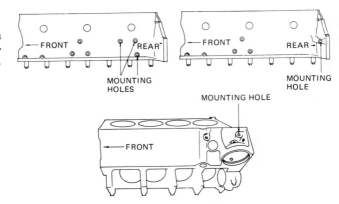

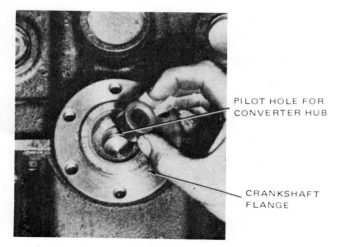

FIGURE 11-6
This engine is equipped with a Powerglide automatic transmission and uses no pilot bushing.

PILOT HOLE FOR
CONVERTER HUB

CRANKSHAFT
FLANGE

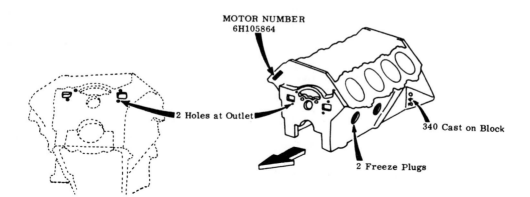

MOTOR NUMBER
6H105864

2 Holes at Outlet

340 Cast on Block

2 Freeze Plugs

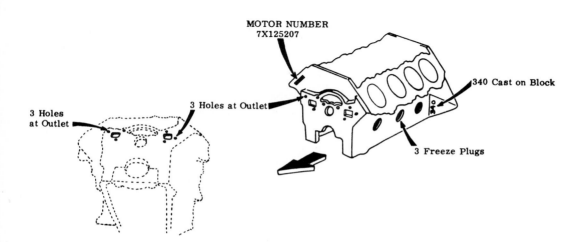

MOTOR NUMBER
7X125207

3 Holes
at Outlet

3 Holes at Outlet

340 Cast on Block

3 Freeze Plugs

FIGURE 11-7

Freeze plug and water outlet holes can be different on the same model block. This illustration shows the Buick 340 CID engine.

A Bore
B Stroke
C Bore Spacing
D Crankshaft Main Journal Dia.
E Crankshaft Rod Journal Dia.
F Cam Journal Dia.
G ₵ of Crank to Top of Block (Head Face)

H Block Deck Height Clearance
I Piston Compression Height
J Con Rod center-to-center length
K Intake and Exhaust Valve Head Dia.
L Valve Spring Installed Height

FIGURE 11-8

Key engine dimensions (Courtesy of Ford Parts and Service Division).

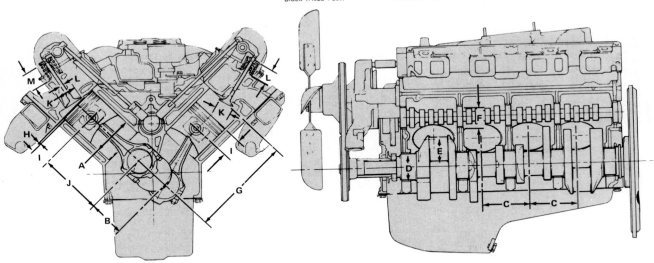

Component design may differ within a family, as may certain dimensions such as bore and stroke. However, many parts and dimensions are identical, making parts interchangeable. Maintaining certain key manufacturing dimensions makes this possible (Figure 11-8). Generally, an engine family has the same cylinder bore spacing (which affects crankshaft interchange), the same head bolt pattern (which affects head interchange), and the same dimension from crankshaft centerline to top of block head face (which may mean common connecting rod and possibly interchangeable pistons, and various valve train parts). See Chart 11-2 as an example.

REMANUFACTURED ENGINES AND KITS

Engine remanufacturing companies offer several kinds of kits. Partially assembled or complete engines are available.

Crankshaft Bearing Kit

A basic lower end engine kit (Figure 11-9) consists of a reground and microfinished crankshaft with main and connecting rod bearings in either .010″, .020″, or .030″ undersize.

Engine Parts Kit

An engine parts kit (Figure 11-10) normally consists of a crankshaft bearing kit plus pistons in .030″, .040″, or .060″ oversize, pins and rings, rod bushings where used, cam and crank drive gears or timing chain and sprockets.

Master Rebuild Kit

This kit would contain all the engine kit parts, plus a reground camshaft, cam bearings, timing chain gears, valve lifters, reconditioned oil pump (where pump is part of front cover, gears are only furnished), complete rebuild gasket and seal set, and water tube (where used).

Short Short Block

This basic unit (Figure 11-11) consists of a cleaned and machined block, rebored cylinders, reground camshaft, reground crankshaft, reconditioned rods, all new oversize pistons, pins, rings, cam bearings, connecting rod bearings, main bearings, timing gears or timing sprockets and chain. Also installed are all water plugs, oil gallery plugs, and fuel pump eccentric (if applicable).

CHART 11-2

Ford FE engine cylinder head interchange chart (Courtesy of Ford Parts and Service Division).

The chart shows FE service heads. Some combustion chamber volumes and valve head diameters are different than OEM, but the heads backfit the blocks as shown. Engine blocks are listed down the left column and heads across the top. Since head bolt patterns for all FE engines are the same, all heads physically bolt on. However, some problems exist per the following key: *Yes* indicates no major problems; *Yes* with a footnote indicates a problem easily corrected with parts and/or machining that is worthwhile because of benefits gained from head; *No* indicates valve clearance problem. The *Special Instructions* explain the footnotes.

Head Interchange

ENGINE	HEADS					
	352	390-410 428	406-428CJ 427LR	427MR	427HR	427TP
332-352 (4.00″)	Stock	Yes	Yes[1]	No[2]	No[2]	No[2]
390-410 (4.05″)	Yes	Stock	Yes	No[2]	No[2]	No[2]
406-428 (4.13″)	Yes	Stock	Stock	Yes[1]	Yes[1,3]	No[2]
427 (4.23″)	Yes	Yes	Stock	Stock	Stock[3]	Stock

Special Instructions

1 Requires notching of cylinder bores for valve clearance.

2 Physically fits, but notching breaks cyl. walls.

3 High riser heads requires special rocker stands and manifold.

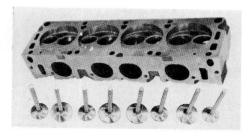

ROUND TUNNEL PORT HEADS
REQUIRE ROUND TUNNEL
PORT INTAKE MANIFOLD.

NOTE: DIFFERENCES IN THIS DISTANCE DISTINGUISH
BETWEEN HIGH, MEDIUM, AND LOW RISER DESIGNS.
INTAKE PORTS OF HEADS AND MANIFOLDS
MUST BE COMPATIBLE.

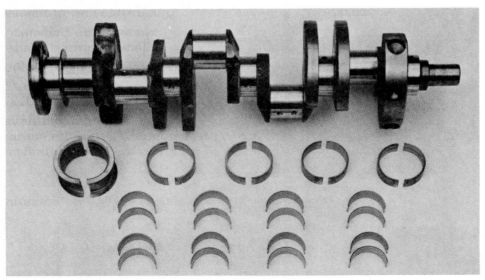

FIGURE 11-9 (above)
Crankshaft bearing kit.

FIGURE 11-10 (below)
Engine parts kit.

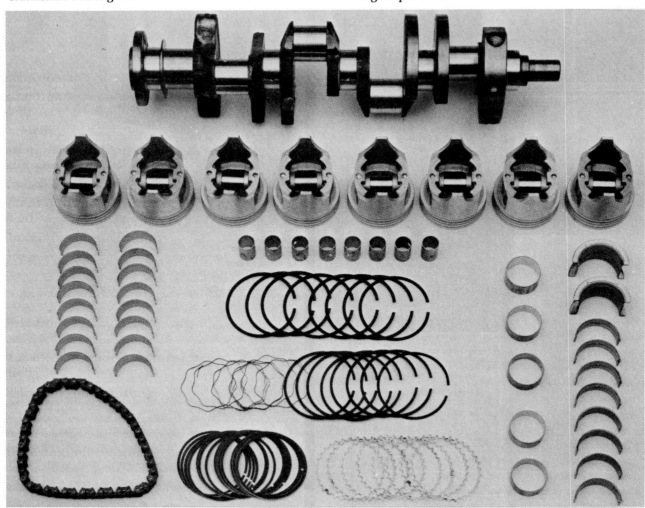

FIGURE 11-11 (above)
This is considered a short short block assembly.

FIGURE 11-12 (below)
The basic short block assembly.

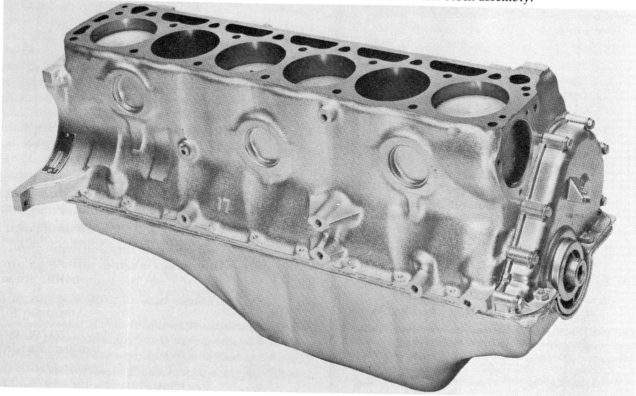

FIGURE 11-13
A complete remanufactured engine assembly. Note the parts that still have to be installed.

Short Block

A basic short block (Figure 11-12) would include all of the items in the short short block assembly plus an oil pump, oil pump drive (if applicable), oil pickup assembly, oil pan, oil pan gasket, timing cover, timing cover gasket, timing cover seal, mechanical lifters (if applicable), and all necessary gaskets to complete the engine installation.

Complete Engine

A complete engine assembly (Figure 11-13) would include all of the parts included in the short block plus remanufactured head assemblies, remanufactured rocker arms or assemblies, push rods, new hydraulic lifters, rocker arm covers, and rocker arm cover gaskets.

CORE CREDITS

Many engine remanufacturing companies have a full core credit guarantee that applies to non-disassembled engine cores only and guarantees that full credit will be issued regardless of the internal condition of the core. Cores with visible cracks or holes in the block, obvious overbores, or broken cranks are not covered. Disassembled engines are subject to company inspection. A salvage value may be allowed on a junk core, depending on the amount of usable parts. Individual parts submitted for core credit are subject to individual inspection. Rebuildable cylinder head cores with three or less combustion chamber cracks are generally acceptable for full core credit.

Because of the shortage of certain model cores, companies require that all cores be exchanged on a type-for-type basis.

PRICING

Companies generally have three different price schedule sheets.

1. List price.

2. Net price (sometimes called dealer or garage net).

3. Jobber price.

The list figure is the most expensive price, followed by the net, then the jobber price. These different price sheets have a color code recognized by the parts industry. A yellow colored price sheet often is used for determining net price. A blue colored price sheet often is used when determining jobber price.

When purchasing parts, ask what price schedule you qualify for. Be sure the parts person uses the correct price sheet.

IMPORTED VEHICLE PARTS

A reason why some repair shops avoid imported cars is that parts are difficult to obtain. A few years ago this was very true. Today, however, import parts are substantially easier to find—you just have to know where to look.

Sources of Imported Engine Parts

Several companies have emerged as national suppliers of a full line of import parts. Also, many domestic aftermarket companies now manufacture and warehouse distribute imported parts to jobbers. (*See Appendix D.*)

CORE SUPPLIERS

When rebuilding or repairing an engine, a major component like a cylinder head, block, or crankshaft is sometimes beyond repair. Obtaining duplicates of these is not easy. If the part is a limited production or a special high-performance item it may be very difficult to obtain. Certain companies specialize in supplying engine cores and the component parts. (*See Appendix D.*)

TRUE CASE HISTORY

Vehicle	Ford
Engine	8 cylinder, 351 Windsor
Mileage	Zero mileage after a complete engine overhaul.
Problem	1. Difficult to start engine.
	2. Once the engine started, it ran extremely rough.
Cause	A 302 CID camshaft (C90Z-6250-C) was mistakingly installed.
Cure	A 289/302 cam can be used in a 351 Windsor engine, provided the firing order is changed. 351W engines use a different firing order (1-3-7-2-6-5-4-8) than 289/302 engines (1-5-4-2-6-3-7-8). Once the secondary wiring in the distributor cap was changed to the 289/302 firing order, the engine ran normal.

TRUE CASE HISTORY

Vehicle	Ford Mustang
Engine	8 cylinder, 351 Windsor
Mileage	20 miles after a major tune-up. A new distributor was installed during the tune-up.
Problem	1. Vehicle suddenly quit running and had to be towed back into the shop.
	2. The camshaft was broken in half.
	3. The distributor gear teeth were badly damaged.
Cause	A 351 Cleveland distributor was installed by mistake. The distributor gear diameter for a 351C engine measures 1⁷⁄₁₆″. The distributor gear diameter for a 351W engine measures 1¼″. The Cleveland distributor, even though it could be installed, did not mate correctly with the Windsor camshaft. Severe gear tooth bind and a broken camshaft were the result.
Comment	You can quickly determine if an engine is a Cleveland or a Windsor. A 351C engine has 8 bolts holding each valve cover and requires the use of a ⁵⁄₈″ spark plug socket. A 351W engine has 6 bolts holding each valve cover and requires the use of a ¹³⁄₁₆″ spark plug socket.

CHAPTER 11 SELF-TEST

1. When ordering engine parts what details should you be prepared to give regarding the following items?
   a. Transmission
   b. Lifters
   c. Carburetor
   d. Oil filter
   e. Dip stick mounting
   f. Oil drain plug
   g. Timing cover
   h. Pan sump
   i. Clutch pedal linkage
   j. Starter mounting
   k. Alternator mounting
   l. Bell housing
   m. Pilot bushing
   n. Block water holes
   o. Block motor mount holes
   p. Crankshaft

2. What is an engine family? State an example.

3. A basic lower end engine kit would consist of what items?

4. An engine parts kit normally consists of what items?

5. What items would be included in a master rebuild kit?

6. What is the difference between a short short block and a short block?

7. A full engine core credit guarantee will generally not be issued under what conditions?

8. Rebuildable cylinder head cores with _____ or less combustion chamber cracks are generally acceptable for full core credit.

9. What are the three different price schedule sheets? Which sheet lists the most expensive price? The least expensive price?

CHAPTER 12

BLOCK RECONDITIONING

In this chapter we will discuss the machining and preparation operations frequently done to get the block ready for assembly. The order of the steps is important in determining accuracy and speed.

1. "Chase" (clean up) the threads in all tapped holes. Chamfer all bolt holes where the threads run up to the surface.

2. Remove all burrs and casting slag from the block interior.

3. Tap the main galleries at the front of the block so that pipe plugs can be installed.

4. Align bore or align hone.

5. Deck the block.

6. Hone the lifter bores.

7. Bore the cylinders.

8. Prepare the cylinder wall surface.

9. Scrub the block with soap and water.

10. Paint the block.

PREPARING THREADS

All threaded holes should be "chased" with the correct size bottoming tap (Figure 12-1) to eliminate any burrs or dirt that might throw off a torque wrench reading.

The areas around bolt holes should be checked carefully for "thread pull." See Figure 12-2 for an example of this condition. "Thread pull" can be corrected by smoothing with a file, chamfering, or counterboring (Figure 12-3). This often overlooked step can mean the difference between long gasket life and early gasket failure.

DEBURRING

Burrs and casting slag on the inside of the block (Figure 12-4) should be removed with a high-speed grinder. This procedure helps to get rid of bits of sand or casting that might jar loose later and damage the engine. Figure 12-5 shows various types of rotary files that can be used.

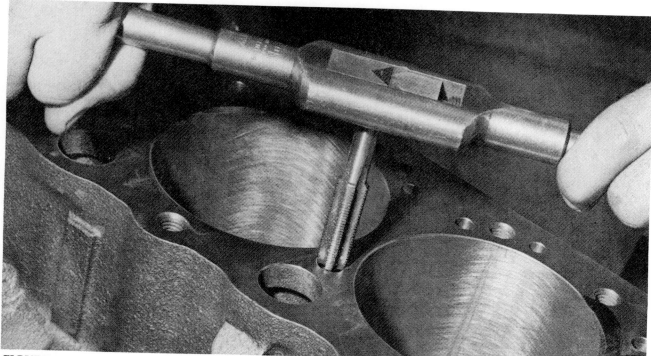

FIGURE 12-1

"Chase" (clean up) all block bolt holes. Dirty threads can distort torque readings.

OIL GALLERY PLUGS

Some engine builders like to tap the front oil galleries and install threaded pipe plugs (Figure 12-6). *NOTE: Tapping these holes with a pipe tap will often only give several threads. It is necessary then to epoxy the plugs in place when installing.*

ALIGN BORING

If the main bearing bores are not aligned, the condition can be corrected by boring the saddles into alignment, or by align boring semifinished bearings installed in the crankcase.

Let us review some of the causes of main bearing bore distortion. First, heating and cooling cycles over thousands of miles create stress in the engine. This stress causes the block to warp and distort. The result is misalignment of the main bearing tunnel (Figure 12-7). Since warpage takes place slowly, the original main bearings and crankshaft will compensate for this by wear-

ing unevenly. However, when a reground crankshaft and new bearings are installed in the warped block, the result is binding and rapid wear.

Another serious problem is main bearing cap stretch (Figure 12-8) that results when high loads

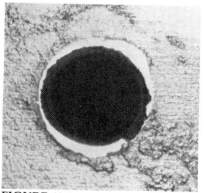

FIGURE 12-2

Running a mill smooth file over bolt holes will show if threads are pulled; a shiny circle of metal will be seen.

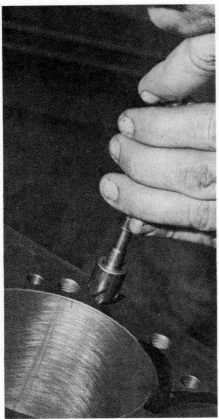

FIGURE 12-3

If threads run up to the surface, chamfer the hole. Inspect all bolt holes carefully (especially if the surfaces have been refaced). Failure to do this can cause premature gasket failure.

CASTING FLASH

FIGURE 12-4

Blocks often show rough metal, sharp edges, ridges, and bits of casting slag in the camshaft tunnel area. A well-deburred block is one that you can run your hand around without getting cut or nicked by any surface, top to bottom.

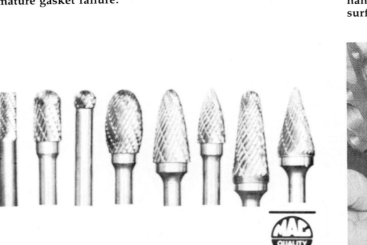

FIGURE 12-5

Rotary files (often called burrs) are usually used with a high-speed grinder when deburring a block (Courtesy of Mac Tools, Inc., Washington Court House, Ohio 43160).

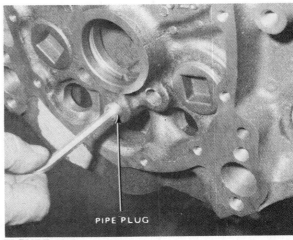

PIPE PLUG

FIGURE 12-6

The main oil gallery soft plugs on this Chevrolet engine are being replaced with 1/4" pipe plugs.

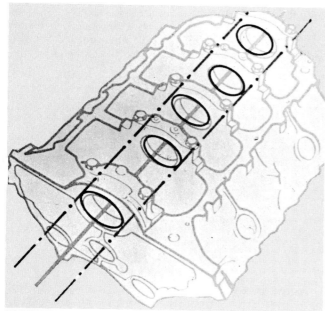

FIGURE 12-7
Block showing misalignment caused by warpage (Courtesy of Sunnen Products Company).

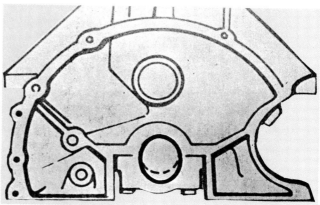

FIGURE 12-8
Block showing cap stretch (Courtesy of Sunnen Products Company).

are imposed on an engine. Again, the original main bearings and crankshaft will compensate for this. *COMMENT: A spun main bearing is often the result of an excessively stretched cap.*

The preferred method for correcting crankcase warpage is by boring the saddles (Figure 12-9). This procedure is basically as follows.

1. Place the main bearing caps in position. Check the cap-to-block recess fit. Torque the caps to specifications.

2. Carefully measure each bore for size, stretch, and misalignment. Refer to Chapter 9 if necessary. Record these measurements.

3. Reduce the diameter of the main bearing bores by removing stock at the parting face of the caps. Use a precision cap and rod grinder (Figure 12-10). This method will ensure that the parting face is kept square with the bore.

   Refer to your measurements taken in Step 2 above to determine how much material should be ground off the parting line. Grind off a sufficient amount to overcome the distortion. As an example, if you find maximum bore stretch is .004″, grind off .005″.

4. Install the bearing caps and torque them to specifications.

FIGURE 12-9
Saddle boring a block (Courtesy of Federal-Mogul Corporation).

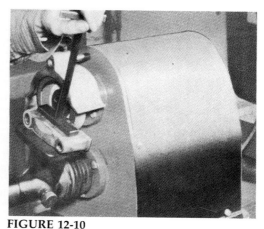

FIGURE 12-10
Precision cap and rod grinder (Courtesy of Federal-Mogul Corporation).

FIGURE 12-11
When boring a crankcase into alignment, camshaft to crankshaft center distance must not be changed more than .005" (Courtesy of Federal-Mogul Corporation).

5. Set the boring bar into position. Adjust the tool bit to cut to the original bore diameter size. With the bar centered, check the amount of stock to be removed from each bore. Compensate as necessary so that a minimum amount of stock will be removed on the block of the bore. *CAUTION: A .005" cut into the block side is considered maximum.* If it is necessary to cut into the block side, chain/sprocket sets are available for the closer camshaft to crankshaft center distance (Figure 12-11).

The other way to correct crankcase warpage is to align bore the mains using semifinished bearings (Figure 12-12). These bearings are undersize with extra wall thickness to allow for reboring.

FIGURE 12-12
Reboring the crankcase is the preferred way to correct crankcase warpage. The other way is to bore the mains using semifinished bearings as shown in this illustration (Courtesy of Federal-Mogul Corporation).

Sometimes, the cam bearing tunnel requires aligning. This is usually accomplished by installing semifinished cam bearings and boring to size.

ALIGN HONING

Many shops use a precision hone to correct main bearing tunnel misalignment (Figure 12-13). A drive-and-feed mechanism operates a mandrel with separate stones. There is a pair of guide shoes for each stone, which give a long, rigid alignment average.

There are several advantages to align honing. Less stock has to be generally removed from the cap parting lines than with the boring method. With align honing you can accurately remove bore high spots. The operation is fast.

Some shops do not like to use a hone on main bearing bores that are discolored from overheating. They claim the metal has become work hardened, and it is faster to use a boring machine.

SURFACE FINISH

Before proceeding with this chapter, let us first discuss finishes. The standard of surface finish measurement is the *microinch*. Literally, one microinch means one-millionth of an inch (.000001"). Stating this in a more meaningful way, we could say that .0001" is the same as 100 microinches.

FIGURE 12-13
A horizontal main bearing bore precision hone (Courtesy of Sunnen Products Company).

No surface is ever perfectly smooth. Machined surfaces have thousands of minute grooves of various depths. A *profilometer* is used to measure surface finishes (Figure 12-14). This instrument traverses a stylus back and forth over the area to be checked and automatically computes the average depth of the grooves. (This is the figure that is used whenever expressing microinches.) There are two methods of averaging used in the United States—the root-mean-square or *RMS method*, and the arithmetic average or *AA method*. The difference in readings between these two methods is so small that it can be ignored for our purposes. However, as a rule-of-thumb, the microinch reading of a surface is approximately ⅓ of the maximum peak-to-valley depth (Figure 12-15). The higher the RMS or AA number, the coarser the surface finish. As an example, the mirrorlike glaze often found in the cylinder ring travel zone will measure about 5 to 7 RMS. Below the ring travel the finish generally measures around 20 RMS.

A profilometer is an ideal instrument to measure surface finish, but it is expensive. Another way to measure surface finish is to use a specimen block kit. The kit contains a set of blocks that has various microinch roughness surfaces for comparison purposes.

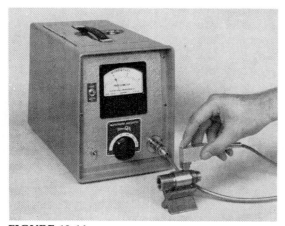

FIGURE 12-14
Surface roughness can be measured by using a profilometer (Courtesy of Federal-Mogul Corporation).

FIGURE 12-15
Cross-section drawing of a surface cross-section. The solid line indicates the theoretical surface. The dotted line indicates approximately ⅓ of the maximum peak to valley depth. In this case, if the distance from the lowest valley to the highest peak was 75 millionths, the surface analyzing equipment would register approximately a 25 microinch surface.

DECKING THE BLOCK

Sometimes, the cylinder block deck surface develops irregularities that can cause compression and water leaks. The flatness of the block deck can be easily checked by using a straightedge and a feeler gauge. *If maximum deformation at any point is greater than .004, the block should be decked.*

The term *decking* is used in the trade for the machine process that trues the head gasket surface of the block. Most of the time, the objective of decking is just to end up with a flat surface cut that is parallel with the centerline of the crankshaft and at the correct angle to the cylinder bores (generally perpendicular). However, a simple surface cut may not be what the customer requests. A performance specialist building a V-type engine would also want the surface of each deck the same height (distance) from the crank centerline.

Decking the block is usually done either in a machine that wet grinds the surface (Figure 12-16) or in a milling-type machine with a rotating cutter (Figure 12-17). Incidentally, a dead smooth finish on the deck is not desirable. The deck should be rough enough to "capture copper" when a penny is dragged across the surface. This helps head gasket seating.

Be sure to replace any dowel pins removed prior to decking.

Importance of Correct Deck Clearance

A V-type block with a different deck height on each bank will cause a varying deck clearance. *Deck clearance is the distance from the top of the piston to the top of the block surface.* Sometimes, this is a negative figure—this means the piston is below the block surface at TDC (Figure 12-18). A positive deck clearance is when the pistons are above the block surface at TDC. When building a performance engine, all piston-to-deck clearances must be the same. If not, an unbalanced running condition may result. Some cylinders will require different fuel mixtures and spark advance than others.

Besides deck height, there are other conditions that affect deck clearance. These include piston profile height (pin centerline to piston top distance), connecting rod center-to-center length, position of crankshaft in block, and crankpin indexing (the stroke at each rod journal).

FIGURE 12-16
Wet grinding a block deck surface.

The Effects of Decking

When metal is removed from the block or heads on a V-type OHV engine, the heads will be positioned closer to the crankshaft. This downward movement will cause the intake manifold to fit differently between the heads. As a result, ports may be mismatched, and manifold bolt holes may not line up. In order to return the intake manifold to its original alignment, corrective machining is required. Also, corrective action may have to be taken regarding rocker arm geometry, hydraulic valve lifter plunger position, piston-to-valve clearance, and emission levels.

If more than .010" of stock is removed from the block, the head, or the block and head combined, the intake side of the head should be machined. On some engines material may also have to be removed from the bottom of the intake manifold. Refer to Chapter 13 for more detailed information.

HONING THE LIFTER BORES

Clean the wall surfaces of the lifter bores with a wheel cylinder hone (Figure 12-19). Remove all rust, glaze, burrs, and high spots to help prevent scoring. *CAUTION: Do not increase lifter bore diameter by more than .0005", or valve train noise and wear problems may result.*

FIGURE 12-17
This setup shows the deck on a GMC truck block being milled (Courtesy of Storm-Vulcan, Inc., A Division of the Scranton Corporation).

CYLINDER BORING

The cylinder boring operation is accomplished by using a machine called a boring bar (Figure 12-20). A tool bit attached to a rotating cutter head is used to enlarge the cylinder diameter. Most boring bar machines attach to and reference off the block deck surface. If you are using this type of bar, *the deck must be parallel with the crank centerline prior to boring. If the deck is crooked, then the bores will be off.* A shop that specializes in high-performance machine work would use a machine that centers off the main bearing saddles. A deck plate (sometimes called a "stress" plate) would also be used. This is a two- or three-inch thick plate that is torqued onto the block during boring and honing (Figure 12-21). The deck plate pre-stresses the block in the same manner as the head does when it is installed, and results in a truer bore.

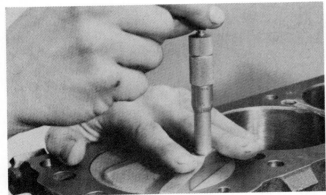

FIGURE 12-18
Checking deck height (Courtesy of Ford Parts and Service Division).

FIGURE 12-19
A wheel cylinder hone can be used to clean the lifter bores (Courtesy of *Popular Hot Rodding* Magazine).

FIGURE 12-20
A multiple (gang) boring bar setup.

FIGURE 12-21
The cylinders in this block are being honed with a deck plate attached.

With the advent of the thin wall block casting and its inherent flexibility, many shops are using deck plates as standard procedure on all boring and honing jobs.

When boring is done, the main bearing caps should be torqued into place. This will stress the lower end of the block just as it will be once everything is together. *If the caps are installed after*

boring, the bottom of the cylinder bores can become distorted and out-of-round.

Boring Bar Fixtures

A few blocks are made with an angled deck surface—in other words, the deck is not perpendicular to the cylinder bores. When using a block mounted boring bar on these blocks, an angle fixture must be used (Figures 12-22 and 12-23).

Boring cylinder barrels can be difficult because of the small support surfaces. Several companies manufacture fixtures that make the boring of barrels and cases an easy task (Fig. 12-24).

FIGURE 12-22
Boring a block with an angled deck surface.

FIGURE 12-23
This support platform is used for mounting a boring bar on a block which has the deck surface machined at an angle. It uses the axis of the existing bore as a locating index. Begin by installing the proper length stop bolt (A) and align clearence arch (B) over cylinder to be bored. Tighten wedge bolt (C) and adjust anchor bolt (D) to your boring bar (Courtesy of Storm-Vulcan, Inc., A Division of the Scranton Corporation).

FIGURE 12-24
A special clamping fixture for boring cylinder barrels and cases. Note the special long travel boring bar (Courtesy of Kwik-Way Manufacturing Company).

Cylinder Wall Thickness

Occasionally, one encounters a block that has had a core shift during manufacture (one side of the cylinder wall is much thinner than the other side). If a block with a core shift is bored, the machinist might be lucky and cut into the water jacket. Installing a sleeve will then fix the problem. However, there are times when a block with a core shift is bored and looks okay, but as soon as the engine is started, the cylinder wall will crack. Because of this, some shops now use ultrasonic test equipment to measure cylinder wall thickness on certain blocks. The tester looks like a small electrical testmeter with a probe. The probe is set into position on the cylinder wall with cup grease. Accuracy of the metal thickness reading is within .010".

Centering the Boring Bar

Should the boring bar be centered in the worn part of the bore, in the middle of the bore, or in the unworn portion of the bore? This question is often asked. To answer the question, let's look at a typical situation. A customer brings a block with .015" taper into the shop and wants it rebored to .020". This is possible if the bar is accurately centered in the most worn part of the bore. However, if the bar is centered in either the middle or unworn part, experience has proven that all cylinders will usually not clean up. This is due to the fact that *a cylinder does not wear equally on each side of its vertical centerline.*

Because of the preceding, many automotive machine shops will center the boring bar in the worn part of the bore. Then, the bore can be cleaned to a given size with the minimum amount of stock removal.

A few shops will only center in the unworn portion. They do not like the idea of the cylinder bore being shifted over with respect to the crankshaft and connecting rods. Some shops prefer to line up the boring machine halfway down in the bore. This provides an average location between the top and bottom.

Chamfering the Cylinder Bore

After boring, a chamfer should be cut into the top edge of the cylinder (Figure 12-25). This is done primarily as an aid when installing the piston assembly into the cylinder. The chamfer should be 1/16" wide, and cut at a 45° angle. If there is not enough chamfer, the edge of the compres-

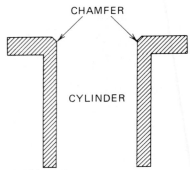

FIGURE 12-25
Correct cylinder chamfer.

sion rings will not enter the bore correctly and can be fractured. Too wide a chamfer can allow the oil ring rails to stick out and become bent after they leave the ring compressor.

Chamfering can be done one of several different ways. Most machinists put a chamfering tool bit in the boring bar, or simply extend the regular tool bit past the cylinder edge and use the top face (normally ground around 45°). The cutter head is then lowered by hand.

Sometimes the chamfer is put in freehand using a chamfering cone (Figure 12-26).

Piston Sizing

New pistons should always be purchased prior to boring. This is necessary for three reasons.

1. Manufacturers differ in the method of sizing pistons. *Most manufacturers build the clearance into the piston.* In other words, a piston marked

FIGURE 12-26
This chamfering cone consists of a tapered hard rubber cone covered by a replaceable abrasive sleeve. It can be used with a 1/4" drill motor (Courtesy of Sunnen Products Company).

+.060 is slightly smaller than .060" because the designed operating clearance has been removed during manufacturing. Thus, the automotive machinist does not have to figure in the operating clearance when boring and honing. The machinist simply finishes the cylinder to the size marked on the piston.

2. There are still some manufacturers who finish pistons to the exact bore size marked on them. The operating clearance must, therefore, be figured in the boring and honing.

3. Due to poor quality control the piston diameters in a set may vary (sometimes as much as .003").

For these reasons, always mike each piston before boring, and machine accordingly.

Pistons are normally available in standard, +.030", +.040", and +.060" sizes. Some vehicle manufacturers have standard size pistons in high-limit sizes. This is done to accommodate small variations during production. For instance, Ford has four different standard size pistons available for the 302-2V engine.

1. Coded red (3.9984–3.9990")
2. Coded blue (3.9996–4.0002")
3. Coded yellow (4.0020–4.0026")
4. .003 oversize (4.0008–4.0014")

Import manufacturers may further complicate matters. Toyota, for instance, manufactures all pistons in a set with an allowable variance of ±.005 mm (.0002"). Yet, one standard set of pistons could vary .004" in size from another standard set. For example, the following piston sets are sold by Toyota for the 8R engine.

Standard Size 85.97 to 86.02 mm (3.3846 to 3.3890")
.25 O/S 86.21 to 86.26 mm (3.3941 to 3.3961")
.50 O/S 86.46 to 86.51 mm (3.4039 to 3.4059")
.75 O/S 86.71 to 86.76 mm (3.4138 to 3.4158")
1.00 O/S 86.96 to 87.01 mm (3.4236 to 3.4256")

Sleeving

If a cylinder wall has a defect (crack, hole, deep score, severe pitting, or is excessively oversize), a sleeve can be installed. This operation consists of boring the damaged cylinder oversize and pressing or driving a cast iron sleeve into place

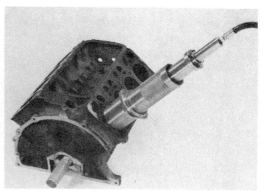

FIGURE 12-27
Portable cylinder sleeve installing press (Courtesy of Rottler Boring Bar Company).

(Figure 12-27). After being installed, the sleeve is bored to size. Some engines are designed with replaceable "wet" sleeves that can be removed and replaced by hand (Figure 12-28).

Automotive cast iron sleeves are generally stocked in regular ⅛" and special ³⁄₃₂" wall thickness. The OD of a ⅛" wall thickness sleeve is made .253" larger than the standard bore size of the engine. As you can see, an extra .003" stock is built into the sleeve. This provides for a good press fit in the block. The OD of ³⁄₃₂" wall thickness sleeve is made .1905" larger than the standard cylinder bore. Sleeves, when purchased, are longer than the cylinder and have to be cut to length. This can be done before or after installation. *Always purchase the sleeve before boring.* Do this because both sizes may not be available for the engine you are working on. Once you have the sleeve, installation can begin. The recommended procedure is given below.

1. Bore the damaged cylinder as smooth as possible and straight for the sleeve. Leave a small shoulder (about ⅛" wide) at the bottom of the bore to prevent the sleeve from "pulling" through. If installing a ⅛" wall thickness sleeve, bore the block .250" larger in diameter than the standard bore size. For example, an engine with a 3¾" standard cylinder bore should have the block bored to 4.000" (3.750" + .250"). If installing a ³⁄₃₂" sleeve, bore .1875" larger than the standard bore size.

2. Cut the sleeve ⅛" longer than the actual required length. A lathe or hacksaw can be used. Cut off the end that does not have the chamfer. Make sure the cut end is square.

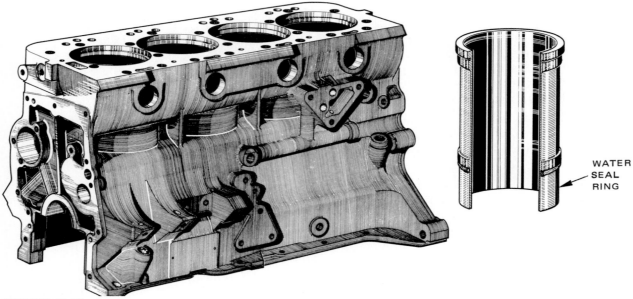

WATER
SEAL
RING

FIGURE 12-28
Some engines are built with wet-type replaceable sleeves (Courtesy of Peugeot).

3. Coat the entire outside surface of the sleeve with antiseize, a press fit lubricant, or graphite. This helps prevent galling when the sleeve is driven in and makes for a good seal.

SHOP TIP

Ordinary waterless hand cleaner will do an excellent job if none of the above is available.

---

4. Before driving in the sleeve, it helps to warm up the block. This can be done by moving a propane torch flame over the wall surface of the cylinder where the sleeve is going. Be sure to apply the heat uniformly throughout the entire cylinder. Some shops use a steam cleaner to warm up the block.

   A process of shrinking the sleeve, prior to installation, is occasionally used. Pour kerosene into a bucket, until it is three-fourths full, and stand the sleeve up in the bucket. Place dry ice around the sleeve (inside and outside). The refrigerated solution will chill the sleeve and shrink it without forming ice. This reduces the amount of press fit and makes installation of the sleeve easier.

5. Start the chamfered end of the sleeve in the block. Keep the sleeve square to prevent cocking. Drive the sleeve in place by using a large hammer and a suitable driver (Figure 12-29). Stop driving when the sleeve strikes the shoulder at the bottom of the bore. The sleeve should extend slightly above the top of the bore.

6. Recenter the boring bar and cut the top of the sleeve flush with the block. This is generally done with a special flat cutter that is fed by hand.

7. Bore the cylinder to the desired finish size.

8. *Carefully measure the cylinder(s) adjacent to the sleeved cylinder for trueness.* This is a very important step, because driving in a sleeve often causes other cylinders to become distorted. Correct as necessary by rigid honing.

CYLINDER WALL PREPARATION

Some type of cylinder wall preparation is necessary on all engine rebuilding jobs. If new rings are being installed in an engine with little cylinder wall wear, glaze breaking may be all that is necessary.

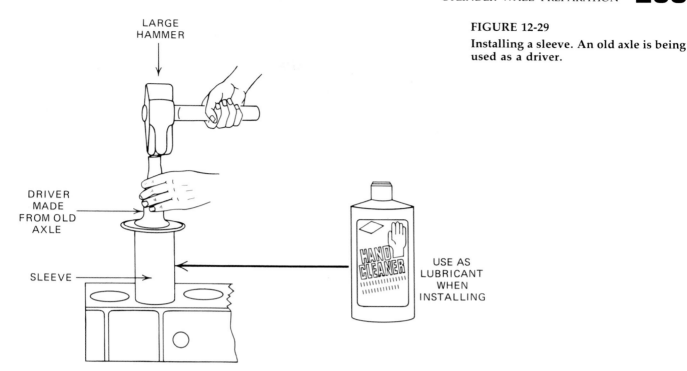

If the cylinders are bored, the wall surfaces must be properly finished using a rigid hone. Sometimes boring cannot be justified, yet the cylinder walls have excessive taper and out-of-roundness. The worn cylinders will have to be trued-up (reconditioned) according to a certain procedure using a rigid hone.

Cylinder Wall Finish

A cylinder wall surface finish must meet certain criteria to form a quality bearing surface for the rings. The requirements are as follows:

1. A plateaued area of from 60% to 80% of the surface. Figure 12-30 shows the profile of a

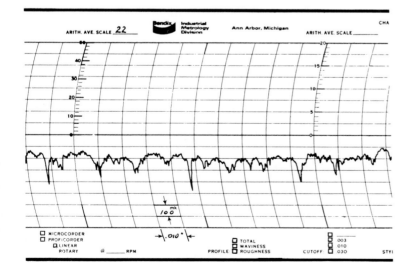

FIGURE 12-30

A Bendix Proficorder with a diamond stylus took this cylinder wall roughness reading. The distance between vertical lines is .010″ with 16 lines across (Courtesy of Brush Research Manufacturing Company, Inc.).

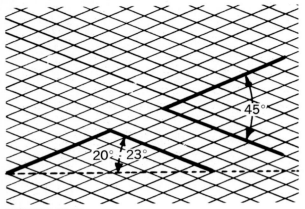

FIGURE 12-31

Cylinder wall cross hatch angle (Courtesy of Brush Research Manufacturing Company, Inc.).

properly prepared cast iron cylinder taken on a Bendix Proficorder. The roughness reading is 25 RMS (about right for the average engine). Note the percentage of plateaued area and the percentage of valleys. The plateaus form the ring seating surface and the valleys provide for oil retention.

2. A cross hatch angle with uniform cuts in both directions. Factory finish for most current engines calls for a relatively shallow *cross hatch of 22° to 32°* from the horizontal (Figure 12-31). The cross hatch angle produced is a combined result of the up and down actuation of the hone and drive motor rpm.

3. A surface free of voids, glaze, foreign material, tears, and boring tool marks.

4. A surface finish roughness to fit the ring material being used. For example, a chrome ring needs a coarser finish than a moly ring. Most aftermarket piston ring manufacturers recommend a 25 to 30 RMS surface finish for plain cast iron rings. For chrome-plated rings a 20 to 25 RMS finish is best. Moly rings require an ultrafine finish of 10 to 15 RMS or less. *CAUTION: Using moly rings on a surface prepared for chrome or cast iron rings will result in severe oil consumption.*

A hone with a given grit size will produce a specified cylinder surface finish so long as normal cutting pressure is applied to the stone. The same stone, however, will produce a finer finish if the cutting pressure is reduced, since the abrasive grains will not penetrate as deep. On the other hand, if the cutting pressure is increased, the

same stone will produce a rougher finish. When used properly, *a 280 grit stone will produce an approximate RMS finish of 20. A 320 grit stone will produce an approximate 15 RMS finish.* **NOTE:** *Catalog order numbers do not always correspond to grit size. For example, a Sunnen brand No. 500 stone is actually 280 grit, a No. 200 is 150 grit, and a No. 100 is 70 grit.*

See Figure 12-32 for some bore finish problems.

Glaze Breaking

In order for a ring job to be a success, all cylinder wall glaze (shine caused by friction and oil varnish) should be removed. Glazed cylinder walls will result in poor ring seating because the wall cannot hold oil. High oil consumption, blow-by, and scuffed pistons are conditions that then develop.

Glaze should be removed with a flexible-type hone or specially designed glaze breaker (Figure 12-33). Using sandpaper or emery cloth to "rough" up a cylinder by hand is not acceptable; yet there are shops still doing this! Glaze breaking tools are designed to hold the abrasive against the cylinder wall with light pressure. Almost no stock removal takes place. Several dozen or so strokes of the hone usually are sufficient to produce the proper finish for ring seating.

Drill speed should be between 200 and 450 rpm. Most authorities recommend using a lightweight oil when glaze breaking. This helps to lubricate and increase the life of the hone, to absorb heat, to prevent work hardening, and to catch and hold the cast-off abrasives. Do not use solvent, kerosene, or diesel oil. They will allow contaminants to be trapped in the valleys of the finish.

COMMENT: Brush Research Manufacturing Company makes an extremely popular glaze breaking hone. See Figure 12-34. Their Flex-Hone® is easy to use, is self-centering, has long life, provides a consistent finish, and makes it possible to carry the cross hatch into the pocket directly underneath the ring ridge. Grits are available from 20 to 800. For address information, see Appendix D.

Preparing the Cylinder Wall Surface After Boring

A boring bar cut can be compared to a plowed field. The field will have fractures all around each furrow made by the plow. Likewise, the boring

ITEM	PHOTOGRAPH	EFFECT ON ENGINE PERFORMANCE
1. CROSS HATCH GROOVES IRREGULARLY SPACED	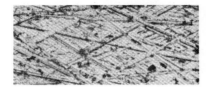	POOR OIL DISTRIBUTION, ERRATIC BREAK-IN OIL ECONOMY.
2. WIDE DEEP CROSS HATCH GROOVES		CAUSES ABNORMAL WEAR, EXCESSIVE OIL CONSUMPTION, POOR AND VARIABLE BREAK-IN PERIOD
3. ONE DIRECTIONAL CUT		CAUSES RAPID WEAR, POOR SEATING, RING ROTATION, LOWERS OIL CONTROL
4. LOW CROSS HATCH ANGLE		HIGH IMPACT FORCES CAUSING EXCESSIVE WEAR, POOR OIL DISTRIBUTION, SLOWER BREAK-IN OF RINGS

FIGURE 12-32

Common deviations from a good bore finish (Courtesy of Muskegon Piston Ring Company).

tool will fracture the metal on the sides as well as below the depth of cut. Because of this, *it is imperative to leave a minimum of .002" stock in the cylinder for finish honing.* The heavy pressure of a rigid hone is required (Figure 12-35).

If you are *hand honing* a cylinder that has just been bored, follow this procedure.

1. Use an air or electric slow-speed drill (200 to 450 rpm) with right-hand rotation. Stroke the full length of the cylinder at a rate of about 30 complete cycles per minute.

2. Always begin the honing operation by truing the stones to the radius of the cylinder. Lightly expand the hone and stroke it up and down in the cylinder a few times. Repeat this process until the entire cutting surface of each stone is making contact with the cylinder wall.

3. Rough hone with 180 grit stones to within .0005". If used dry, these stones will cut a lot faster. **NOTE:** *Honing stones which have been used for wet honing must never be used for dry honing as the stones will load up quickly and stop cutting.*

4. Hone to finish size with 280 or 320 grit stones. Apply a continuous flow of honing oil to the stones. If honing oil is not immediately available, you can use vegetable shortening (such as Crisco).

If you are using an *automatic hone,* such as the Sunnen CK-10, the same basic procedures apply (Figure 12-36).

See Chart 12-1 (Page 258) for references to Sunnen stone numbers.

FIGURE 12-34
A Flex-Hone® in operation (Courtesy of Brush Research Manufacturing Company, Inc.).

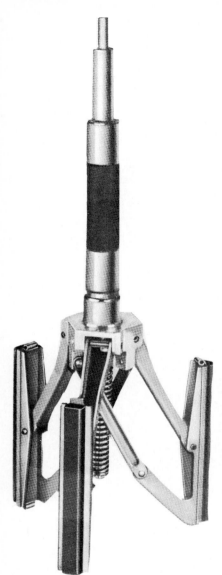

FIGURE 12-33
Spring loaded cylinder hones are designed for glaze breaking, not stock removal.

FIGURE 12-35
Rigid-type cylinder hone (Courtesy of Sunnen Products Company).

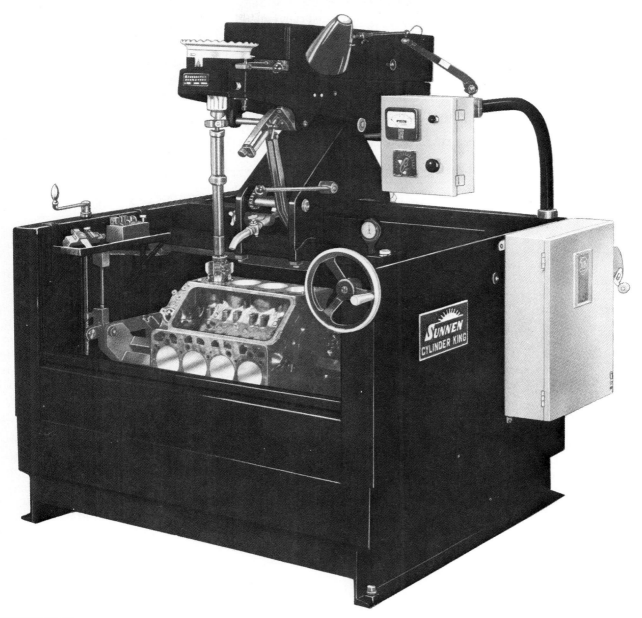

FIGURE 12-36

Automatic cylinder hone resizing machine. Stock removal of more than .008" per minute is possible (Courtesy of Sunnen Products Company).

Reconditioning Worn Cylinders by Rigid Honing

As previously mentioned, if the cylinders are tapered up to .012", the bore can be honed oversize and the old pistons expanded by knurling.

This procedure requires the hone operator to follow certain specific steps. Otherwise, the cyl-

inder can be easily ruined and boring will be required.

1. Start stroking at the bottom, or least worn section of the cylinder. Use short strokes in order to concentrate honing in the smallest section of the cylinder. Allow the stones to

CHART 12-1

Reference to Sunnen stone numbers (Courtesy of Sunnen Products Company).

APPROX. MICROFINISH	GRIT SIZE	CK-10 AUTOMATIC STONE SET NUMBER	HAND OPERATED STONE SET NUMBER
85-105	70	EHU-133	
135-170	70		AN-101
25- 40	150		AN-201
25- 35	220	EHU-525	
20- 25	220		AN-301
14- 23	280	JHU-625	
15- 20	280		AN-501
8- 13	400	JHU-820	
5- 10	400		N-37-J85
4- 8	600	C-30-CO3-81	
3- 5	600		NN40-CO5

FIGURE 12-37

A shop-made cabinet for rigid honing cylinders. A motor driven pump circulates honing oil. The weight of the drill motor is counterbalanced by a two garage door springs. Down travel can be adjusted by two stop collars (Courtesy of Automotive Engine Rebuilders Association).

pass through the lower end of the bore approximately 1".

2. Keep the stones revolving in the bottom of the cylinder and gradually lengthen the stroke as metal is removed. In this way, *the unworn section of the bore is used as a guide to keep the hone straight.* This is very important in preventing the cylinder from becoming bell mouthed.

SHOP TIP

Listen to the speed of the drill motor as the hone is worked in the cylinder. A reduction in drill speed indicates a smaller diameter. Localize stroking at such sections until drill speed is constant over at least 75% of the cylinder length; then stroke the full length of the cylinder. See Figure 12-37.

Frequent Problems When Rigid Honing

1. Stones do not cut.
   a. When honing dry, traces of grease, oil, or carbon in the cylinders will cause the stone to load up and stop cutting.
   b. Stones may be glazed. Dress stones with an abrasive stick, and increase stone pressure.
   c. Check to be sure you are using the proper grit.
2. The hone chatters.
   a. If honing wet, the wrong type of coolant

or not flowing the coolant directly on the stones. As the stones "break down" during the honing process, the coolant must flush the loose abrasive and metal particles from the stones and cylinder wall. This keeps a continuous supply of sharp cutting edges exposed.

b. If honing dry, failure to keep the stones clean.

c. Stone guides may be too tight. To check, expand stones and guides out against the cylinder wall. You should be able to wiggle the guides with your fingers. If you cannot, file or grind off about 1/32" from the wearing surface of each guide to relieve the pressure.

3. The stones and guides become tapered.

a. The stroking procedure is incorrect.

b. Not enough pressure is being applied to the stones.

CLEANING THE BLOCK

After all of the machining operations have been completed, the entire block must be thoroughly cleaned. This includes oil galleries over their full length, cylinder bores, lifter bores, and every hole and cavity.

Use laundry detergent, hot water, rags, and a scrub brush for cleaning. This combination will do a superior job. It is the only sure way to remove the grit that remains after honing. *Do not use cleaning solvent,* as this will not float out dirt and abrasive particles. Continue scrubbing with fresh solution until the soap suds no longer turn gray. Dry the block with air pressure.

PAINTING THE BLOCK

Some high-performance engine builders like to paint the inside of the block. They feel this serves two purposes.

1. The paint's smooth surface speeds up oil "drainback" to the pan.

2. The paint helps retain any casting grit which might later work loose.

The inside areas usually painted are the lifter galley chamber, the crankcase areas between the

FIGURE 12-38
Glyptal enamel paint.

cylinder bores, and the front of the block which is covered by the timing chain.

An ideal paint to use on the inside is *Glyptal* (Figure 12-38), a special paint coating manufactured by General Electric. Generator and electric motor windings are often coated with this red-colored paint. Before applying the paint, warm the area to be covered with a heat lamp to set up the coating harder than by drying at room temperature. Let the Glyptal harden for at least 24 hours.

TRUE CASE HISTORY

Vehicle   Chevrolet

Engine   8 cylinder, 350 CID

Mileage   1600 miles on a just installed rebuilt engine

Problem
1. Blue smoke from tailpipe.
2. Several spark plugs were oil fouling.
3. The ring seating process did not appear to be completed even after 1600 miles.

Cause   Moly rings were installed in this engine. The cylinder wall surface finish was not honed smooth enough after boring, and the moly coating on the top compression rings was ruined. Moly rings require a cylinder wall finish of from 10 to 15 RMS or less.

Comment   The glaze found on cylinder walls is an excellent antiscuff break-in surface for moly rings. However, the cylinders must be free of waves and scuff marks.

TRUE CASE HISTORY

Vehicle   Toyota

Engine   6 cylinder, 2M model (OHC)

Mileage   40 miles on the engine after a complete rebuild by the vehicle owner.

Problem   Engine seizure.

Cause   The block was sent to a local machine shop for cylinder boring. Prior to assembling the engine, the owner failed to thoroughly clean the block. Some boring machinings remained (Figure 12-39). The crankshaft lubrication holes became plugged and a rod bearing burned out (Figure 12-40).

FIGURE 12-39

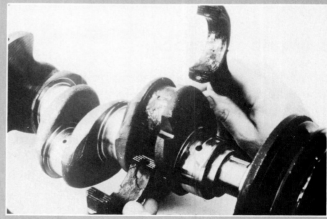

FIGURE 12-40

CHAPTER 12 SELF-TEST

1. What type of tap should be used when "chasing" threaded holes?
2. What is "thread pull" and how can it be corrected?
3. Why should burrs and casting slag be removed from the inside of a block?
4. When tapping the front oil galleries for threaded plugs, why should the plugs be installed with epoxy?
5. What two methods can be used to correct main bearing bore misalignment? Which method is preferred?
6. A spun main bearing is often the result of what condition?
7. What is considered to be the maximum cut into the block when saddle boring?
8. What are the advantages of align honing?
9. What is a microinch?
10. What is a profilometer?
11. Does the RMS number increase or decrease as the surface finish becomes smoother?
12. What is a specimen block kit?
13. How can flatness of a block deck be easily checked? What is the maximum allowable deformation?
14. What is decking?
15. Why is a dead smooth finish on the desk undesirable?
16. What is deck clearance?
17. List four conditions that can cause deck clearance to vary in an engine.
18. When assembling a V-type OHV engine, you discover that the intake manifold bolt holes do not line up on the heads. What could be the cause of this problem?
19. On a V-type OHV engine, if more than _____" of stock is removed from the block, the heads, or the block and heads combined, corrective machining is required.
20. Why should lifter bores be lightly honed?
21. What is a "stress" plate?
22. Most boring bar machines reference off what surface?
23. What is an angle fixture?
24. What is a core shift?
25. What is the advantage of centering the boring bar in the worn part of the bore?
26. Why should a chamfer be cut into the top edge of the cylinder after boring?
27. How is the cylinder chamfering operation performed?
28. How do most manufacturers size pistons?
29. What piston oversizes are generally available?
30. Why do some manufacturers have standard size pistons in high-limit sizes?

31. Automotive cast iron sleeves are generally stocked in what wall thickness sizes?

32. When boring a block for a sleeve, why is a small shoulder left at the bottom of the bore?

33. What should the outside of the sleeve be coated with, just before driving in? Why?

34. How can a sleeve be shrunk prior to installation?

35. When driving in a sleeve, what end of the sleeve should be started into the block?

36. Why should the cylinder(s) adjacent to a just installed press-fit sleeve be measured?

37. What is a "wet" sleeve?

38. Why must a cylinder wall surface finish consist of plateaus and valleys?

39. What is the cylinder wall cross-hatch angle for most current engines?

40. What is the recommended RMS surface finish for plain cast iron rings? For chrome rings? For moly rings?

41. What will happen if moly rings are used on a surface prepared for chrome or cast iron rings?

42. When used properly, a 280 grit stone will produce what approximate RMS finish? A 320 grit stone?

43. What should be the drill speed rpm when glaze breaking?

44. Why should a lightweight oil be used when glaze breaking?

45. After boring, how much stock should be left for finish honing? Why?

46. Why is it important to keep the hone stroking in the bottom of the bore when rigid honing?

47. Why must honing stones that have been used for wet honing never be used for dry honing?

48. What can cause the stones and guides to become tapered when rigid honing?

49. What can cause chatter when rigid honing?

50. How should cylinder walls be cleaned after honing?

51. Sometimes, certain inside areas of a block are painted. What are the reasons for doing this?

52. What is Glyptal?

CHAPTER 13

CYLINDER HEAD SURFACING

Every cylinder head needs to have the head gasket surface, the intake manifold mounting surface, and the exhaust manifold mounting surface inspected. Visually inspect for any coolant passage damage and gasket surface erosion (Figure 13-1a, b, and c). Use a straightedge and feeler gauge to measure for straightness. If surfaces are warped beyond limits, machining will be required.

If surface A on a V-type OHV engine is cut more than .010″, it is recommended to remove metal from surface B (Figure 13-2). Otherwise, the holes in the intake manifold and cylinder head might not line up when reassembled. In addition, port mismatch will occur. The amount of corrective machining varies with the angle on the intake manifold side of the head. A special stock removal gauge or protractor can be used to measure this angle (Figure 13-3).

TABLE OF CONSTANTS

Table 13-1 shows the amount of corrective machining required. The amount of stock removed at A, times the constant (multiplier), equals the amount of stock to be removed at B (refer to Figure 13-2).

Let's take a look at an example—a 327 CID Chevy has had .014″ surfaced off each head. Another .006″ has been surfaced from each deck. Thus, .020″ of metal has been removed at A. We now multiply .020″ by the 10° constant, which is 1.2. This gives .024″ as the amount to be removed from B.

If material has to be removed from B, the cut can be made on the *intake side of the head (Figure 13-4)*, or on the *intake manifold-to-head-mounting surface (Figure 13-5)*. However, the machinist will usually want to remove material from the head, for several reasons.

1. Cylinder heads are generally much easier to mount in the surfacing machine.

2. A replacement manifold will not have to be surfaced in order to fit.

263

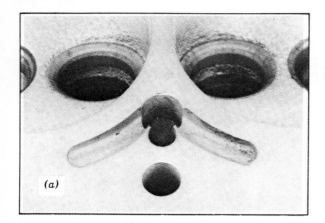

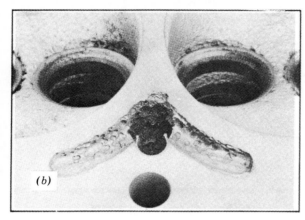

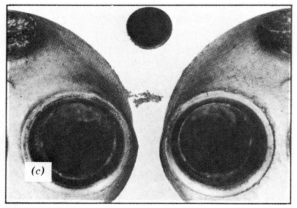

FIGURE 13-1

(*a*) Acceptable coolant passage. (*b*) The cylinder head should be repaired or replaced if erosion has widened the coolant passage as shown here. (*c*) Unacceptable head with erosion present between two combustion chambers. Attempt resurfacing before replacing a cylinder head for this reason.

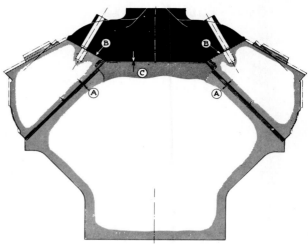

FIGURE 13-2

After machining the head at surface *A*, material should be machined from surface *B* for proper intake manifold alignment. Note how the bolt holes in the intake manifold do not line up after machining the head gasket surface *A* (Courtesy of Hastings Manufacturing Company).

TABLE 13-1

THE AMOUNT OF CORRECTIVE MACHINING REQUIRED VARIES WITH THE ANGLE AT WHICH THE HEAD IS MADE. HEAD ANGLE IS THE VARIANCE FROM 90°. FOR INSTANCE, A HEAD THAT HAS THE HEAD GASKET AND THE INTAKE MANIFOLD SURFACE MADE AT A RIGHT ANGLE (90°) TO EACH OTHER WOULD BE CLASSIFIED AS A 0° HEAD.

0°	Head angle (Big Block Ford) Amount removed at $A \times 1.0$ = Amount to be removed at B
5°	Head angle Amount removed at $A \times 1.1$ = Amount to be removed at B
10°	Head angle (Small Block Chevy and Ford) Amount removed at $A \times 1.2$ = Amount to be removed at B
15°	Head angle (Chrysler, Y Block Ford) Amount removed at $A \times 1.4$ = Amount to be removed at B
20°	Head angle (Oldsmobile) Amount removed at $A \times 1.7$ = Amount to be removed at B
25°	Head angle (Cadillac, Dodge, Plymouth) Amount removed at $A \times 2.0$ = Amount to be removed at B
30°	Head angle Amount removed at $A \times 3.0$ = Amount to be removed at B

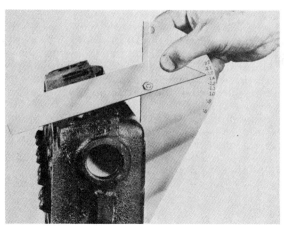

FIGURE 13-3

The stock removal gauge determines the correct amount of stock to remove from the manifold pad. Merely place one edge of the gauge on the head surface and the other on the intake manifold surface of the head. Multiply the amount removed from the head surface times the factor indicated on the gauge to determine stock removal (Courtesy of Storm-Vulcan, A Division of the Scranton Corporation).

FIGURE 13-4

A setup for milling the manifold pad surface on a V-8 OHV head (Courtesy of Storm-Vulcan, A Division of the Scranton Corporation).

SURFACING THE BOTTOM PAD OF THE INTAKE MANIFOLD

If the heads are surfaced on certain Ford and Chevrolet engines, metal removal from surface C will be required (refer to Figure 13-2). Stock can be removed from the manifold bottom, or

FIGURE 13-5

The above illustration shows one method of setting up an intake manifold for milling. Two long and two short jacks with a parallel bar are used to provide a rigid mounting (Courtesy of Storm-Vulcan, A Division of the Scranton Corporation).

from the top of the block. Figure 13-6 shows a setup for milling the intake manifold bottom. The amount of stock to be removed at C can be determined by using Table 13-2.

Let's look at the 327 Chevy from the previous example. A total of .020" had been removed at A. Therefore, multiplying .020" by 1.7 gives .034" as the amount to be removed at C.

COMMENT: *It may be necessary to install shims, equal in thickness to the material removed from the intake manifold bottom, under the distributor.*

TABLE 13-2

ON SOME ENGINES, IT IS NECESSARY TO REMOVE STOCK FROM THE INTAKE MANIFOLD BOTTOM IF THE HEADS HAVE BEEN SURFACED. THE AMOUNT REQUIRED VARIES WITH THE HEAD ANGLE.

0°	Head Angle (Big Block Ford) Amount removed at A × 1.4 = Amount to be removed at C
10°	Head Angle (Small Block Chevy and Ford) Amount removed at A × 1.7 = Amount to be removed at C

FIGURE 13-6
After Chevrolet V-8 head and intake manifold pads have been machined, it is necessary to remove stock from each end (*A* and *B*) of the manifold. The shield (*C*) does not permit machining all the way across. This makes it necessary to turn the manifold. The amount of stock to be removed is determined by multiplying the amount of stock removed from the head gasket surface by 1.71 (Courtesy of Storm-Vulcum, A Division of the Scranton Corporation).

METHODS OF SURFACING

Cylinder head surfacing most often is done by milling, broaching, or wet grinding. Which method you use makes little difference so long as the equipment operating instructions are followed.

Dry grinding of heads generally is not recommended. Heat distortion, burned spots, and loading of the grinding wheel have been a problem with this method.

Milling Machine Setup

Storm-Vulcan makes a widely used milling machine (Figure 13-7a and b). Basically, setup begins by leveling up the head until the cutters touch all four corners. Then the cutters are moved to the center of the head to find the low spot. How much of a cut you will have to make to "clean" the head is then easily determined.

Wet Grinder Setup

One of the most popular wet grinders is made by Peterson (Figure 13-8a). This piece of equipment sets up like a mill. First, level up the head. Then make a fast pass over the head to be sure the grinding wheel is "spotting" all four corners. If not, make the necessary screw jack adjust-

ments. Be sure the wheel is sharp, since a dull wheel will polish the surface and leave burned spots.

Another popular wet grinder is made by Kwik-Way (Figure 13-8b, c, and d).

Broach Setup

Winona Van Norman manufactures a machine that uses an underside rotary cutter (broach). A block, cylinder head, or intake manifold is held in an inverted position as the broach passes underneath (Figure 13-9). Machines of this style have a drawback because a mirror is required to see if the surface has "cleaned" up. Broaches and mills leave a factory type finish that the operator has no control over, whereas the wet grinder surface finish is controllable.

Leveling the Head

Surfacing machines are constructed with the support bed and the cutting wheel parallel to each other. If the machine sags (this happens frequently because of their heavy weight), it is possible to level a head with the earth and not be level with the cutter. The cut then will not be true. For this reason, if a bubble-type level (Figure 13-10) is used to level up a head, *make sure you level the head to the cutting wheel.*

(a)

(b)

FIGURE 13-7

(*a*) A head ready to be milled on the Storm-Vulcan Headmaster. (*b*) The head surface after milling.

FIGURE 13-8*a*

Peterson wet surface grinder.

(a)

FIGURE 13-8*b-d*

This Kwik-Way machine uses a surfacing belt that can wet grind heads, manifolds, oil pans, and so on (Courtesy of Kwik-Way Manufacturing Company).

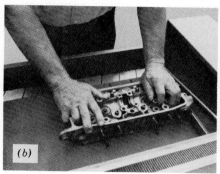

(b)

(c)

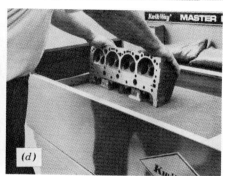

(d)

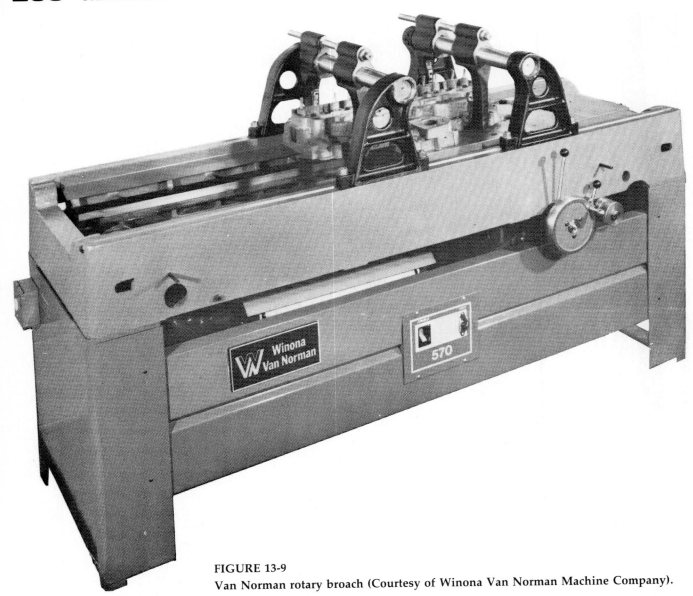

FIGURE 13-9
Van Norman rotary broach (Courtesy of Winona Van Norman Machine Company).

Always make it a point to periodically check level accuracy on a surface plate. Most shop levels can be adjusted by turning two hex nuts to bring the tube assembly to a level setting (Figure 13-11). To verify this setting, the level should be reversed in its positioning on the reference surface, making sure that the exact location is duplicated. If the reversal does not result in a repeat level reading, an error in the reference surface is indicated. A correction should be made in the leveling adjustment of the reference surface of one-half the error indicated, and the other half of the error taken up in the adjustment of the tube assembly. Repeat the reversal check. When the level can be reversed in position on the reference surface, and repeat readings obtained, it is correctly adjusted.

The most accurate method of head leveling is to use a dial indicator attached to the cutter (Figure 13-12).

Milling Aluminum

Sometimes aluminum can be a problem to mill—it does not machine right and wants to "chunk" out.

To help prevent "chunking" and to get a good finish, penetrating oil (such as WD-40) is often sprayed on the aluminum surface to be machined.

FIGURE 13-10
Leveling the head with a precision bubble level.

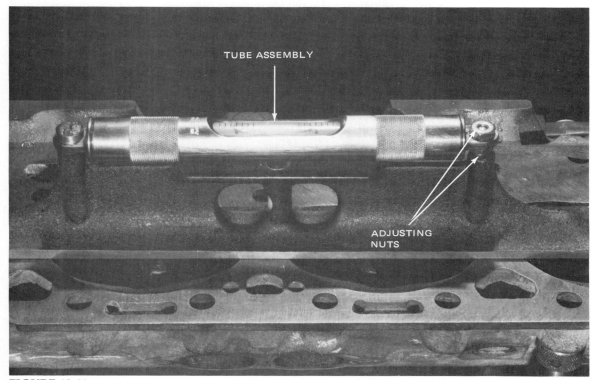

FIGURE 13-11
This shop level can be adjusted by turning the nuts at the end of the tube assembly.

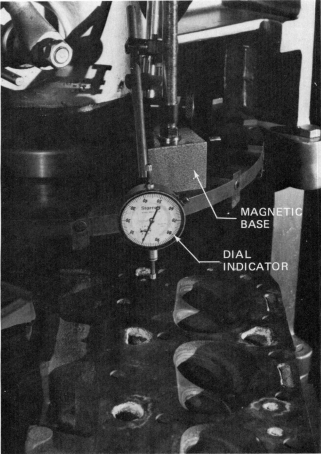

FIGURE 13-12
A dial indicator is sometimes used to level the head. This is done to ensure that the head is mounted level with the cutter.

SHOP TIP
Coating the head with liquid dishwashing soap will give excellent results if milling aluminum.

STOCK REMOVAL GUIDELINES

The amount of stock removed from the head gasket surface has to be limited for many reasons. Excessive surfacing can lead to these problems.

1. Abnormal combustion because of a higher compression ratio. *If .070" of metal is removed from an OHV cylinder head, the compression ratio will be increased approximately one step.*

2. Piston to valve interference. When the block

and/or head is surfaced, the piston-to-valve clearance during the overlap period becomes less. To prevent the valves from making contact with the piston, *a minimum of .070" piston-to-valve clearance must be maintained.*

3. Valve train problems. Surfacing will cause the valve tips, rocker arms, and push rods to be dimensioned closer to the camshaft. This will cause a change in rocker arm geometry (see Chapter 16), and can also cause hydraulic lifters to bottom out. When a large amount of stock is removed from the head, the valve train ''height'' dimension will require correction. There are several ways to correct this.

   a. Place shims equal in thickness to one-half the amount of stock removed, under the rocker stands.

   b. Shorten the push rods an amount equal to the amount removed from the cylinder head.

SURFACING OHC CYLINDER HEADS

After surfacing an OHC head, bind problems are often encountered. When this type of head warps, camshaft bearing alignment is affected (Figure 13-13). For example, let's say a head is warped .006" in the center. The head is milled and torqued down. You then discover the camshaft will not turn. What is the solution? Here are some approaches being used by shops.

1. Try to straighten the head before surfacing it. Apply heat in three places on the valve spring side of the head (Figure 13-14a). **CAUTION:** *This can be done successfully, but experiment on some junk heads.* This author knows of one shop that melted an aluminum BMW

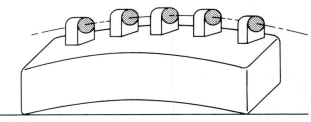

FIGURE 13-13
On OHC engines, the camshaft bearing tunnel will be affected when there is head warpage.

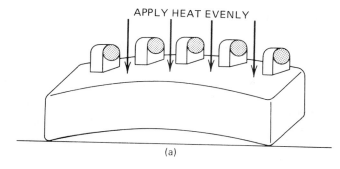
APPLY HEAT EVENLY

(a)

FIGURE 13-14*a*

Heat applied to the valve spring side of an OHC head can sometimes remove distortion.

FIGURE 13-14*b*

On OHC heads, the camshaft centerline can be realigned by boring the cam journals (Courtesy of Kwik-Way Manufacturing Company).

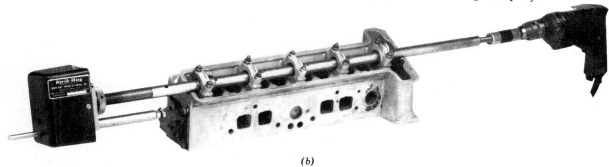

(b)

six-cylinder head. The new head (bare casting) cost $900.

2. When the head is being set up in the mill, pull the camshaft bores into alignment. Mill the head. When you take it off the machine, it will bow back. However, when torqued down on the block, it will be true again.

3. True the camshaft bearing tunnel by boring, or use semifinished bearings and machine to size (Figure 13-14*b*).

4. Certain OHC engines with aluminum heads have the camshaft bearing journals riding directly on the aluminum. In some cases, camshafts with oversize journals are available for these engines. The oversize journals allow for machining of the cam bearing bores and salvage of the head.

Datsun OHC Cylinder Heads

A special approach can be taken on Datsun OHC aluminum heads that are warped. They can be serviced instead of being thrown away as suggested in many manuals. The servicing procedure is as follows.

1. Remove the camshaft, camshaft bearing towers, rocker arms, rocker arm pivot nuts, and valve assemblies (Figure 13-15).

SHOP TIP

The rocker arm pivot nuts can be easily removed with an air impact wrench.

2. Surface an equal amount of material from both top and bottom of head.

3. Measure head thickness between top and bottom machined surfaces (Figure 13-16). *OEM thickness on these heads is 4.240" to 4.250".*

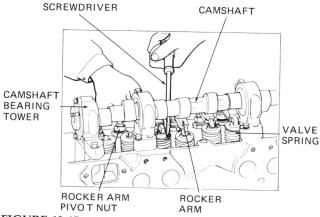

SCREWDRIVER · CAMSHAFT · CAMSHAFT BEARING TOWER · VALVE SPRING · ROCKER ARM PIVOT NUT · ROCKER ARM

FIGURE 13-15

Pressing down the valve spring in order to remove the rocker arm (Courtesy of Nissan Motor Company, Ltd).

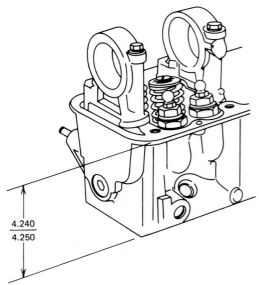

FIGURE 13-16
Datsun OHC head thickness when new.

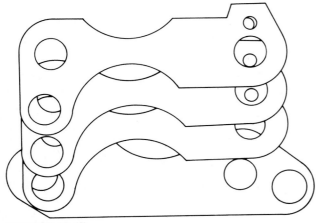

FIGURE 13-17
Datsun head-saver shim kits are available from the dealer.

ANGLE MILLING

This is a high-performance machining procedure used on small block or big block Chevrolet heads. With angle milling you can build a high-compression ratio Chevrolet engine and overcome the inherent problems of large piston domes, minimum piston-to-head clearance, small combustion chambers, poor flame travel, and spark plug shrouding.

To get around these problems, the heads are angle milled. As an example, consider a stock ZLI combustion chamber that holds 122 cc. If a final chamber volume of 104 cc is the blueprint specification, the heads will require a cut of .125". This would be marginal if the heads were milled in the conventional "level" manner (you would cut into the intake valve seat). However, with angle milling it is possible to obtain 104 cc without any problem.

Angle milling simply involves removing more material off the exhaust side of the head than the intake side (Figure 13-18). After determining the combustion chamber volume you want, position the head in the mill using a level and shims (Figure 13-19). Most heads will only require a 1° to 1½° cut to get the chambers where you want them. Usually, all the material is taken off the exhaust side of the head and nothing is taken off the intake side.

After angle milling, the use of a table of constants is not necessary. Just make sure the intake side of the head is at its original manufactured angle.

Before installing the head, the tops of the head bolt bosses will have to be cut parallel to the new head surface (Figure 13-20).

4. Replace material removed by surfacing with shims (Figure 13-17). Add shims until within .015" of OEM head thickness. Insert the shims between the top of the head and the cam towers.

5. Recenter the camshaft bearing towers and lightly dress the bearing bores with a hone.

SHOP TIP

After surfacing a cylinder head, a sharp edge is left around the outside of the combustion chamber. Lightly chamfer this edge by using a rotary file or a piece of emery cloth. This will help prevent preignition caused by the sharp edge acting as a "hot spot."

FIGURE 13-18
Angle milling a Chevrolet head (Courtesy of *Hot Rod* Magazine).

FIGURE 13-19
A level and feeler gauge leafs are used to position the head prior to cutting (Courtesy of *Hot Rod* Magazine).

FIGURE 13-20
Truing the tops of the head bolt bosses with a drill press and the correct size cutting tool (Courtesy of *Hot Rod* Magazine).

TRUE CASE HISTORY

Vehicle	International Truck
Engine	8 cylinder, 304 CID
Mileage	Unknown. The cylinder heads were removed, resurfaced, given a valve job, and reinstalled.
Problem	1. The engine would not crank. 2. Excessive starter draw. 3. Battery cables hot when touched. 4. The starter would make a "clunk" sound when attempting to crank the engine.
Cause	Interference condition between the valve heads and the piston tops had mechanically locked the engine.
Comment	Material in excess of .010" should not be removed from this model head. Overall head height can be measured to determine if the head has been resurfaced previously, and the amount of material removed. OEM height is shown in Figure 13-21. If the head measures below the low limit, it indicates that the head has been resurfaced.

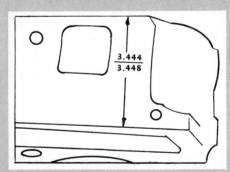

FIGURE 13-21

OEM cylinder head thickness for International V-266, 304, and 345 engines (Courtesy of Hastings Manufacturing Company).

TRUE CASE HISTORY

Vehicle	Chevrolet
Engine	8 cylinder, 327 CID
Mileage	Zero mileage. The cylinder heads were removed and replaced with an identical pair that had been extensively modified for drag racing.
Problem	The oil pump seized after a few minutes of engine operation.
Cause	The cylinder heads had been radically milled. As a result, the intake manifold was positioned lower on the block. When the distributor was installed, oil pump driveshaft end play was removed and gear-to-cover bind occurred.
Comment	Nylon shims are available from high-performance shops to eliminate this problem. Figure 13-22 shows a shim kit that contains one each .030″, .060″, and .100″ shim.

FIGURE 13-22
Nylon shim kit for correcting
distributor end play.

CHAPTER 13 SELF-TEST

1. On a small block Chevrolet, .016" of stock has been surfaced from each head. How much material should be machined off the intake manifold side of each head?

2. When corrective machining is required, why is material usually removed from the head rather than from the intake manifold?

3. When surfacing the bottom pad of the intake manifold on certain Ford and Chevrolet engines, why is it often necessary to install shims under the distributor?

4. List three methods of cylinder head surfacing.

5. Why is dry grinding of cylinder heads generally not recommended?

6. What simple method can be used to level a head on a Storm-Vulcan mill without using a bubble-type level?

7. What company manufactures a very popular wet grinder for surfacing heads?

8. What is required to see if the surface has "cleaned" up when using a Van Norman rotary broach?

9. How is it possible to level a head with the earth and not be level with the cutter?

10. How can the accuracy of a bubble-type level be checked?

11. What can be done to help prevent "chunk" out when milling aluminum?

12. How much metal has to be removed from an OHV cylinder head to raise the compression ratio approximately one step?

13. What is minimum piston-to-valve clearance?

14. List two ways to correct valve train "height" after removing a large amount of stock from the cylinder head.

15. What methods can be used to correct camshaft bearing tunnel alignment on an OHC head?

16. What special approach can be taken on Datsun OHC aluminum heads that are warped?

17. What is the purpose of angle milling?

CHAPTER 14

VALVE GUIDE AND ROCKER ARM STUD SERVICE

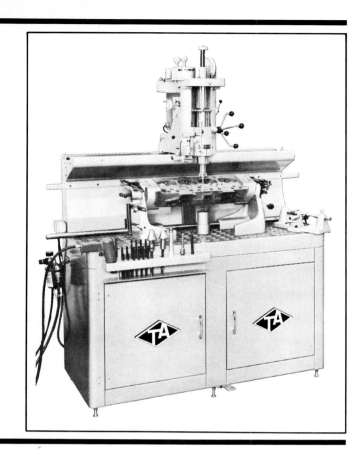

In this chapter we are concerned with valve guide reconditioning procedures for replaceable and integral type guides and rocker arm stud replacement. Valve guide oil control and the machining of guides for PC seals also will be covered.

DETERMINING VALVE GUIDE WEAR

The first step in guide reconditioning is to determine the amount of valve stem-to-guide clearance present (review Chapter 9 if necessary). This will help decide which reconditioning method best fits your particular situation.

If valve clearance exceeds specifications by 50%, replace or repair the guide. If specifications are not available, *use .003" maximum as the general rule-of-thumb for intake valve clearance, and use .004" maximum for the exhaust valves.* Let's now look at the common valve guide reconditioning methods.

REMOVING AND REPLACING REPLACEABLE GUIDES

A few engines are still equipped with replaceable guides (Figure 14-1). The old guides are pressed or driven out (by hand or with an air hammer), and new ones are installed. Here are some suggestions to follow when installing replaceable type valve guides.

1. If using a driver, do not allow the driver to bounce on the top of the guide. Otherwise, a cracked or broken guide can result.

2. Coat the guide with an antiseize lubricant before installing.

3. Whenever possible, remove and install the guide by driving or pressing from the valve spring side of the head. This is especially important on aluminum heads. Hard carbon often builds up in the valve seat port area around the guide. The guide will carry some of this carbon through its bore hole if removal takes place from the wrong side (Figure 14-2). This can score or broach out the

277

FIGURE 14-1
Set of eight replaceable valve guides.

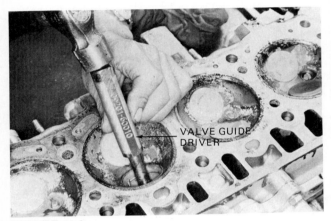

FIGURE 14-2
The guide in this aluminum Toyota head is being driven out *wrong*. It should be driven out from the valve spring side of the head.

FIGURE 14-4
Measuring guide installed height.

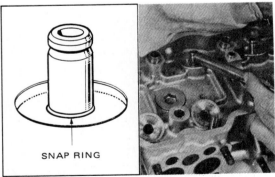

SNAP RING

FIGURE 14-3
On some engines, you must break the guide before removal.

ID of the bore hole in the head and the new guide will be loose. *CAUTION: 396 and 427 CID cast iron head Chevrolet engines have exhaust valve guides with a tapered OD. These guides are one exception and must be driven out from*

the valve seat side. Some Volkswagens use flanged type guide inserts. These also have to be removed by driving from the valve seat side.

4. On many Toyota engines, the guides are removed by first breaking off the top with a brass punch and hammer (Figure 14-3). The guide is then driven out with a hammer from the valve spring side of the head. The snap ring is only provided on the replacement guide, not on the originally installed one. The new guide is driven in until the snap ring contacts the head.

5. Measure the installed height of the guide before removing it (Figure 14-4). This is the distance the guide extends above the spring seat. Drive the new guide into the head, keeping this same protrusion. If the guide is not driven in far enough, the spring retainer can bottom out on top of the guide when the valve fully opens. If the guide is driven in

too far, accelerated valve stem wear will result and air flow in the port will be restricted.

6. When new replaceable guides are installed, distortion and high spots will occur in the bore ID. *Always make a pass with the correct size reamer, even though the valve may fit the guide.*

7. Dealership replacement guides for certain import engines are made in unusual oversizes. For instance, Datsun uses .008" oversize OD guides, while Fiat uses a .005" oversize OD. Rather than machine the bore hole in the head to accommodate the oversize dimension, it is easier to reduce the guide OD in a lathe.

Removing Bronze Guides

Some engines (usually imports) use replaceable bronze guides. Removing these guides is not always an easy task, especially on Volkswagen. Hitting on the top of the bronze may cause the guide OD to swell out. As the guide is driven out, the bore in the head will be enlarged. The new guide will then be loose. To prevent this from happening, the following method can be used.

1. Cut threads inside the guide (Figure 14-5*a*). It is best to do this on the spring side of the guide.

2. Screw a bolt into the guide (five or six turns). See Figure 14-5*b*.

3. Drive out the guide by hitting against the bottom of the bolt with a valve guide driver (Figure 14-5*c* and *d*).

4. Preheat the head with a propane torch or steam cleaner before installing the new bronze guides.

SHOP TIP

Some shops prefer to use a thermal method when removing bronze guides. The cylinder head is placed on an electric hot plate for a half-hour; then the guides are easily driven out.

VALVE GUIDE KNURLING

Knurling is a popular guide reconditioning operation normally used to recover clearances in

integral guides worn up to about .008". When wear conditions beyond this are encountered, most shops recommend core drilling the guide for installation of a thin-wall valve guide insert.

In the knurling operation, a special resizing arbor is run through the guide with an electric drill and speed reducer (Figures 14-6 and 14-7). This produces a spiral groove with raised plateaus that decrease the ID of the guide (Figure 14-8). In other words, metal is displaced out of the groove by pressure. After the knurling arbor forms the groove, a reamer is used to finish the guide to proper size. Because the knurling groove acts as an oil reservoir, tighter clearances are possible. *When knurling worn guides, the recommended clearance is one-half the minimum factory specification.*

Typical Operating Instructions for a Guide Knurler

1. Thoroughly clean the guide before starting the resizing operation.

2. Select the proper resizing arbor for the guide bore (the size is marked on the shank end).

3. Lubricate the inside of the valve guide with the special lubricant in the kit. Dip the starting end of the resizing arbor into the lubricant.

4. Start the electric drill, and hold the speed reducer with one hand to prevent turning. This will revolve the resizing arbor through the guide at the correct rpm. *NOTE: It is extremely important that the resizing arbor be held in visual alignment with the valve guide bore. It is easier to control arbor cocking if the resizing arbor is started from the seat side of the head.*

5. Thoroughly brush the guide bore. Flush with solvent and blow dry.

6. Select the proper reamer and ream the guides. Make sure the reamer is not cocked when starting the reaming operation. Again, clean the guide bore. Check the valve stem clearance with the valve to be used.

Special Knurling Tips

1. Hardened or chilled valve guides (such as found in 450 International engines) cannot be properly resized.

2. Should the resizing arbor stall in operation

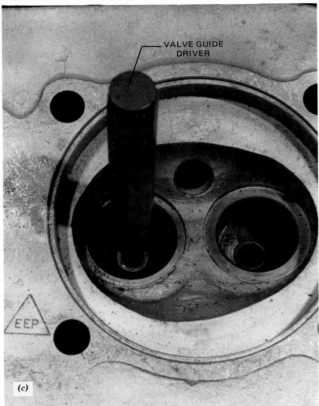

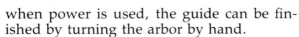

VALVE GUIDE DRIVER

FIGURE 14-5

(*a*) Removing VW bronze guides can be difficult. Easy removal can be accomplished by first tapping the guide. (*b*) Screw a bolt into the threaded guide. (*c*) Invert the head and drive against the bottom of the bolt. (*d*) The removed guide and attached bolt.

5. Brake fluid can be used as a knurling lubricant to prevent galling and binding.

6. When guides are knurled and the clearance is .001″ or less, it is not recommended to use positive guide seals.

SLEEVING INTEGRAL VALVE GUIDES

Sleeves should be installed when the guides are really worn. There are several systems available for this repair. Some require a special fixture in order to maintain the original centerline when boring out the worn guide (Figure 14-10). Other systems are self-centering as they feed. Regardless of the system used, the procedure consists

when power is used, the guide can be finished by turning the arbor by hand.

3. If the resizing arbor stalls on guides .343″ in size, run a .345″ reamer through the guide, then use the resizing arbor, and finish with a .343″ reamer.

4. A hand brace makes an excellent drive for valve guide reamers (Figure 14-9).

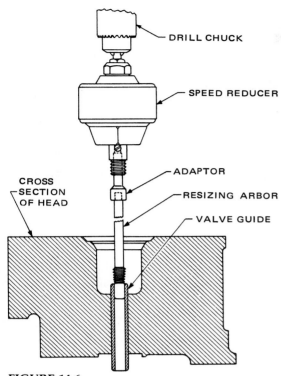

FIGURE 14-6
Setup for valve guide knurling (Courtesy of Hall-Toledo).

FIGURE 14-7
Closeup of the resizing arbor shown in Figure 14-6 (Courtesy of Hall-Toledo).

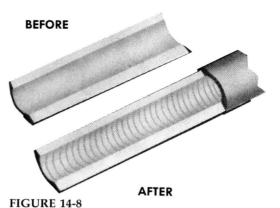

FIGURE 14-8
Knurling pattern (Courtesy of Hall-Toledo).

FIGURE 14-9
Using a hand brace to ream a knurled guide (Courtesy of M&L Motor Supply Company).

basically of boring the guide for a .001″ to .0015″ press fit, driving in the sleeve, and then reaming it to the correct size.

Repair sleeves come in several styles.

1. Thin-wall cast iron inserts.

2. Phospher bronze liners.
3. Silicon aluminum bronze bushings.

Bronze is strongly recommended for high-performance or heavy duty application. The guide wall surface is highly superior to the original cast

FIGURE 14-10

Boring out a worn integral guide. After boring, a sleeve will be installed. The special fixture maintains the original centerline when boring (Courtesy of Hall-Toledo).

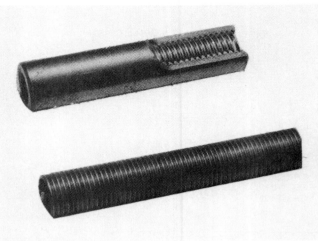

FIGURE 14-11

At the bottom is a bronze spiral bushing prior to installation. It is shown installed in the cutaway guide at the top (Courtesy of Winona Van Norman Machine Company).

iron. Also, for dry fuel application (liquid propane or butane) bronze will provide maximum resistance against galling, corrosion, and heat. *CAUTION: Silicon aluminum bronze valve guides cannot be run against a stainless steel stem.* The stem must be hard chromed. The granular structure of chrome allows oil to be trapped in the pores of the metal, thus ensuring lubrication at all times.

BRONZE WALLING

This reconditioning procedure uses a bronze spiral bushing that looks like a Heli-Coil (Figure 14-11). The process has several strong points.

1. It can handle badly worn integral guides.
2. Rapid heat dissipation is provided.
3. Bronze has good lubrication characteristics and low friction.
4. Valve stem-to-guide clearances of around .0005″ can be run.
5. Bronze has a wear resistance 25 times greater than cast iron.

The Winona Van Norman Machine Company makes a widely used bronze walling kit. The basic operation is as follows.

1. Make sure the valve guides are absolutely clean and dry. All varnish, carbon, and rust must be removed. Select a workbench or stand that will position the head at approximate knee level with the guides in a near vertical position.
2. Chuck the hex/square drive adaptor into the special drill motor provided with the kit. (Figure 14-12*a*).
3. Set the drill motor switch to the "tap" position. *CAUTION: Do not use any lubrication when tapping.* Otherwise, the guide will not be cleared of chips and a tap may break. Insert the tap pilot into the guide. Do not cock the tap when starting the operation. Make certain the drill motor reversing switch is set for right-hand rotation. Tap the valve guide completely through.
4. Blow the metal chips from the guide with an air hose and select the bronze wall bushing that corresponds to the valve guide bore. Slide the bushing on the special inserting tool (Figure 14-12*b*). Make sure the closed tail on the bushing fits into the slot on the end of the inserting tool. Place the protruding tail into the thimble end of the inserting tool and turn to lock. Next, wind up the bushing by turning the bottom knob on the

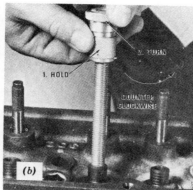

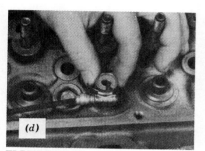
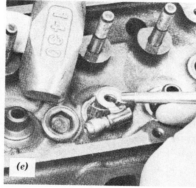
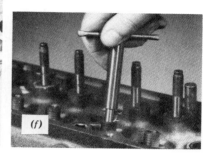

FIGURE 14-12

(*a*) The installation of bronze spiral bushings begins by tapping the guides, using a special drill motor. (*b*) Screw the bushing into the thread of the tapped guide. (*c*) Unwind the excessive portions of the bushing away from the guide, and cut. Leave a 1/32″ tail. (*d*) Install the bushing retainer to prevent the bushing from turning while reaming. (*e*) Broach the bushing. This operation presizes the bushing for finish reaming. (*f*) Cut the bushing flush with the guide top, using the special cut-off tool (Courtesy of Winona Van Norman Machine Company).

inserting tool while holding the top knob. Wind until the bottom knob stops turning. *CAUTION: Do not overwind.* Carefully start the bushing into the thread of the tapped guide. Screw into the guide until your finger can feel the inserting tool slightly recessed from the guide bottom (approximately 1/16″).

5. Hold the top knob of the insert tool and overwind the bottom knob counterclockwise until the top tail of the bushing breaks. Firmly grasp the top knob and turn the entire insert tool counterclockwise until the bottom tail breaks.

6. Unwind the protruding portion of the bushing away from and tangent to the guide (Figure 14-12*c*). With diagonal cutters, cut the bushing so that it extends 1/32″ beyond

the OD of the valve guide. *COMMENT: This is necessary to properly hold the bushing while reaming.*

7. Install the special serrated bushing retainer over the guide tower (Figure 14-12*d*). Make sure one of the serrations engages the 1/32″ tail of the bushing.

8. Lubricate the guide bore with the supplied reaming oil. Dip the bulb end of the broach in this oil. Drive the broach bulb, end first, completely through the guide (Figure 14-12*e*). Use a brass hammer to avoid "mushrooming" the broach shank.

9. Set the drill motor switch to the "ream" position. Use the cutting oil supplied for reaming. Lightly push the reamer through the guide. *COMMENT: Many shops prefer*

FIGURE 14-13
Removing a bronze spiral bushing (Courtesy of Winona Van Norman Machine Company).

to size bronze wall bushings by honing, because bronze will quickly dull a reamer.

10. Remove the special bushing retainer. Insert the cut-off tool in the guide with the "knife" under the extended portion of the bushing (Figure 14-12f). With a slight downward pressure, turn the cut-off tool clockwise. This will cut the bushing flush with the top of the guide. Thoroughly clean and lubricate the guides before you install the valves.

Removal and Replacement of Bronze Wall Bushings

If for any reason it becomes necessary to remove a bronze wall bushing, use a scratch awl to pry up the anchor end of the bushing at the top of the guide. Remove the bushing with needlenose pliers by pulling straight out (Figure 14-13). A new bushing can be easily installed again.

REAMING VALVE GUIDES

Worn guides can be reconditioned by reaming. With this method the guide ID is increased to take an oversize valve stem.

Valves with oversize stems generally are available from the manufactuer or from a parts supplier. The standard oversizes are .003″, .005″, .015″, and .030″. In some cases, there may be an available size other than one of these. Therefore, always check availability before doing any machining.

Adjustable and self-aligning reamers are available. Refer to Chart 14-1.

CHART 14-1
These reamer sizes are available from Sioux Tools Inc. They can be furnished in .001″, .002″, .003″, or .004″ oversize or undersize.

SELF-ALIGNING GUIDE REAMERS

Decimal Size & Cat. Number	Size	Length
G-250	1/4″	7″
G-312	5/16	7″
G-328	21/64	7″
G-343	11/32	7″
G-3467	11/32 + .003	7″
G-3587	11/32 + .015	7″
G-359	23/64	7″
G-3737	11/32 + .030	7″
G-373	3/8″ — .002	7″
G-375	3/8	7″
G-3780	3/8 + .003	7″
G-3800	3/8 + .005	7″
G-3850	3/8 + .010	7″
G-3900	3/8 + .015	7″
G-390	25/64	7″
G-3950	3/8 + .020	7″
G-396	.396	7″
G-4050	3/8 + .030	7″
G-406	13/32	7″
G-421	27/64	7″
G-437	7/16	9″
G-453	29/64	9″
G-468	15/32	9″
G-484	31/64	9″
G-500	1/2	9″
G-515	33/64	9″
G-531	17/32	9″
G-562	9/16	9″
G-593	19/32	9″
G-625	5/8	9″
G-687	11/16	10″
G-750	3/4	12″

Reaming Tips

1. Always clean out the guide hole to prevent the pilot end of the reamer from sticking.

2. Always advance the reamer slowly into the guide, as the lead end of the blades does all the cutting. Do not stop the reamer before going through the hole.

3. The amount of metal removed per pass should be limited to .005″. If more metal has to be removed, make progressive cuts.

4. *Do not reverse a reamer.* This will dull the cutting edges.

5. Cast iron guides should be reamed *dry* for best results in both size and finish.

6. Bronze should be reamed *wet*. Gear oil (90W) makes an excellent cutting lubricant. A very slow reamer speed (650 rpm) works best when reaming bronze.

7. The closer the tolerance needed (or the finer the finish), the slower the reaming speed must be.

8. When reaming bronze, there is the tendency for the reamer and guide to heat up and cause undersize reaming. To prevent this, use two reamers alternately, allowing the one not being used to cool in a can of cutting lubricant.

**SERVICE HINT
FOR GENERAL MOTORS 350 AND
403 CID ENGINES**

On these engines, one or more valve guide ID bores may be oversize as manufactured. The oversizes are .003", .005", .010", and .013". If an oversize condition exists, there will be a marking on the inboard side of the cylinder head (Figure 14-14). The mark is visible without removing any parts other than the air cleaner assembly. The mark will only be stamped in front of the oversize valve guide bore(s).

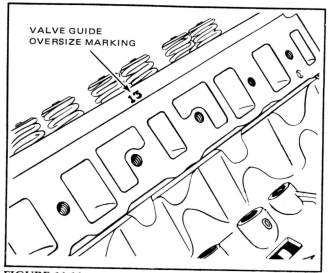

FIGURE 14-14

Oversize valve guide marking location for GM 350 and 403 CID engines.

HONING VALVE GUIDES

Valve guide honing is an ideal method for sizing cast iron or bronze replacement guides, integral guides, knurled guides, or bronze inserts. Very close tolerances (down to .0002") can be obtained with this process. When reaming, anything closer than .0005" is very difficult to obtain.

A popular guide hone is shown in Figure 14-15. A stone is expanded against the guide ID and turned with a slow-speed drill motor (approximately 350 rpm). As the head turns, metal is removed from the guide bore until the desired size is obtained.

REPLACING ROCKER ARM STUDS

A number of engines use canoe-type, individually stud mounted rocker arms in place of rocker arms pivoted on a shaft (Figure 14-16).

In standard production, the stud is normally pressed into the cylinder head. It may be a standard or "positive stop" stud (Figure 14-17). If the stud has thread damage, has started to pull out, is broken, or shows wear, it must be replaced (Figure 14-18). There are two methods to use.

1. Pressing in a replacement stud after reaming the hole oversize.

2. Threading the hole using a special threaded replacement stud.

Measure and record stud height before removal. The new studs must be installed at this height.

Several methods can be used to remove pressed-in studs. A stud removal tool set is often used (Figure 14-19a and b). Nuts and a flat washer also can be used (Figure 14-20). Tighten down on the washer so that the tension will withdraw the stud. As the stud moves up, some extra washers may have to be added. After removing the studs, it will be necessary to ream out the stud hole to accept a replacement stud (usually .003" or .006" oversize). The hole needs to be sized to provide a *.002" to .0035" negative fit.* **CAUTION:** *Do not press an oversize stud into a standard sized hole.* This can crack and ruin the head. Always measure the new stud diameter, and the ID size of the stud hole before installation.

When you install a pressed-in stud, coat the surface to be driven in with stud mount compound (Figure 14-21). Install the replacement to the correct depth.

NOMINAL FRACTIONAL SIZE	RANGE	① MANDREL	② SLEEVE	③ STONE	④ ADAPTER	⑤ HEAD & DRIVER
5/16	.308″- .316″	LG-308M	S-308	LG-13	PK10-A	
11/32	.339″- .347″	LG-339M	S-339	LG-13	PK10-A	
3/8	.370″- .385″	LH-370M	S-370	LH-13	PK12-A	
13/32	.400″- .416″	LH-400M	S-400	LH-13	PK12-A	
7/16	.432″- .447″	LH-432M	S-432	LH-13	PK12-A	
1/2	.495″- .526″	LJ-495M	S-495	LJ-13	PK16-A	
9/16	.557″- .588″	LJ-557M	S-557	LJ-13	PK16-A	P-180
5/8	.619″- .650″	LJ-619M	S-619	LJ-13	PAK16-A	
21/32	.650″- .681″	LJ-650M	S-650	LJ-13	PAK16-A	
11/16	.681″- .713″	LJ-681M	S-681	LJ-13	PAK16-A	
3/4	.744″- .775″	LM-744M	S-744	LM-13	PAK20-A	

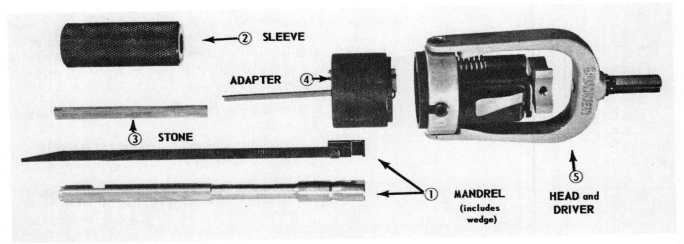

FIGURE 14-15

A valve guide hone is ideal for honing cast iron replacement guides, integral guides, or bronze-type insert guides when a precise fit is required (Courtesy of Sunnen Products Company).

Screw-in Studs

A few factory engines are equipped with threaded studs. The stud may be either a positive or non-positive stop type (Figure 14-22). Screw-in studs are made with either a jam nut or have a jamming shoulder (Figure 14-23). Screw-in studs help eliminate stud pullout. They are generally heat treated for extra strength to help reduce horsepower loss caused by flexing.

A cylinder head with pressed-in studs can be modified for the installation of threaded studs. Jam nut studs will require machining material of the stud boss. A milling tool is used to remove metal equal to the thickness of the jam nut (Figure 14-24). Then each stud hole is drilled and tapped. The two common stud sizes are 3/8″-16 and 7/16″-14. Studs with a jamming shoulder do not require boss milling. However, drilling and tapping is still required. *CAUTION: Use extreme*

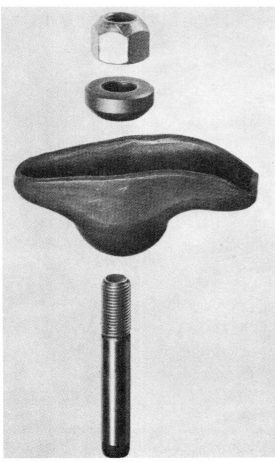

FIGURE 14-16
Individual rocker arm assembly mounted on a stud.

care to maintain correct alignment when drilling and tapping stud holes. Any stud angle other than the original will affect rocker arm geometry. Figure 14-25 shows a commercial tap machine designed to prevent misalignment.

Apply a hardening-type sealer (such as No. 1 Permatex) to the threads on the stud end. This will prevent engine coolant from working its way up the threads and mixing with the oil. Tighten the stud until the jam nut or shoulder is fully seated against the top of the stud boss.

Pinning Rocker Arm Studs

Pressed-in type studs are sometimes pinned as a precaution against pullout. This is accomplished by drilling through the boss and stud, then driving in a hardened roll pin. Commercial pinning kits are available (Figure 14-26).

Guide Plates

Some manufacturers use a push rod guide plate that is held in position by the rocker arm stud (Figure 14-27). The plate prevents push rod "whip" at high speeds. As a result, push rod guide hole wear in the cylinder head is non-existent. *COMMENT: Guide hole elongation is a frequent cause of bent push rods.*

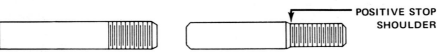

POSITIVE STOP
SHOULDER

FIGURE 14-17
A pressed stud that uses an adjustable rocker nut is shown on the left. A pressed stud that uses a "positive stop" nonadjustable rocker nut is shown on the right. This stud can only be used with hydraulic cams, since there is no provision to adjust for the lash required for solid lifter cams (Courtesy of Ford Parts and Service Division).

WEAR FROM
ROCKER ARM

FIGURE 14-18
Nicks caused by the rocker arm is reason for stud replacement.

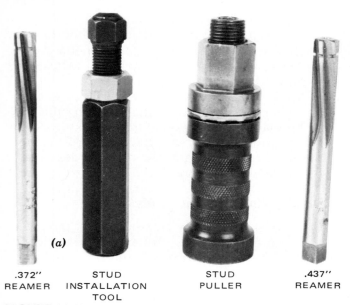

FIGURE 14-19*a*

This rocker arm stud tool set can be used for easy extracting or installing of studs.

FIGURE 14-19*b*

Installing puller on a stud. The ball bearing thrust washer helps reduce friction.

If push rod guide plates are going to be installed when converting to screw-in studs, the stud boss must be milled this additional thickness. *NOTE: When using guide plates, hardened push rods should be used because they rub on the plate.* Because of production misalignment it may be necessary to slightly increase the diameter of one or more push rod guide holes in the head when installing guide plates to prevent push rod bind.

Checking Rocker Arm-to-Stud Clearance

Sometimes, stud breakage becomes a problem when a high-lift camshaft is installed. This is due to the rocker arm stud slot binding on the stud. A minimum of .020″ clearance between the rocker slot and the stud is required (Figure 14-28). Check

FIGURE 14-20

A stud can be removed by placing a nut and flat washer over a stack of several nuts. As the top nut is tightened down, tension will withdraw the stud. An additional nut on the stack may be required as the stud comes up.

FIGURE 14-21
Loctite is recommended when installing pressed-in studs. Coat the surface to be driven in just prior to installation.

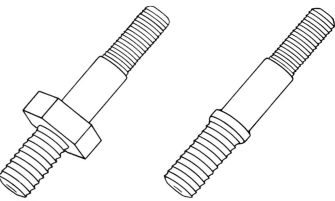

FIGURE 14-23
Jam nut and jamming shoulder type screw-in studs.

FIGURE 14-24
Milling tool for machining the top of the stud boss (Courtesy of Kwik-Way Manufacturing Company).

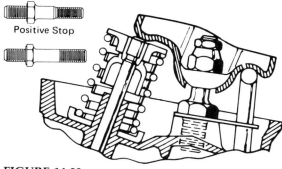

Positive Stop

FIGURE 14-22
Positive and nonpositive stop screw-in studs (Courtesy of Ford Parts and Service Division).

FIGURE 14-25
For an accurate change from factory pressed-in studs to screw-in studs, a tapping machine should be used (Courtesy of Kwik-Way Manufacturing Company).

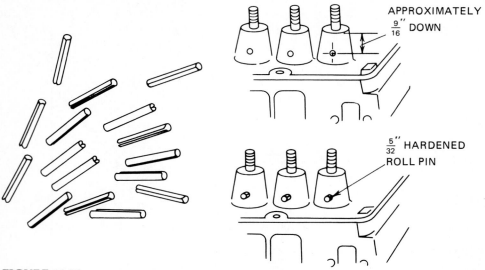

APPROXIMATELY $\frac{9}{16}''$ DOWN

$\frac{5}{32}''$ HARDENED ROLL PIN

FIGURE 14-26

Kits are made for pinning pressed-in rocker arm studs. First, mark the stud bosses with a center punch, and drill through. Then, install roll pins until flush with back side of the boss as shown.

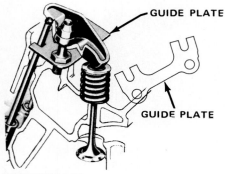

GUIDE PLATE

GUIDE PLATE

FIGURE 14-27

A stud mounted rocker arm with an attached guide plate (Courtesy of Ford Parts and Service Division).

at valve full open and closed positions. A bent paper clip can be used for a measuring tool.

To correct insufficient clearance, use a die grinder to remove metal from the rocker arm stud slot ends (Figure 14-29). *Do not grind metal from the sides of the slot.* The slot width helps maintain correct rocker alignment.

VALVE GUIDE OIL CONTROL

A running engine generally has an abundance of oil under the valve covers. To keep this oil from getting to the valve stem guide area where it can be drawn into the engine and burned, seals of various types are used.

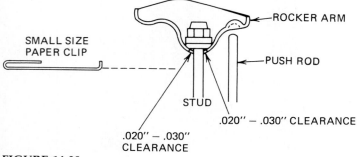

ROCKER ARM

SMALL SIZE PAPER CLIP

PUSH ROD

STUD

.020'' – .030'' CLEARANCE

.020'' – .030'' CLEARANCE

FIGURE 14-28

When canoe type stud-supported rocker arms are used with a high lift camshaft, clearance between each rocker arm and stud must be checked.

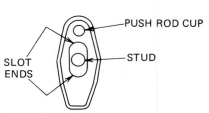

PUSH ROD CUP

STUD

SLOT ENDS

FIGURE 14-29

To increase the clearance between a rocker slot and its stud, grind material from the slot ends.

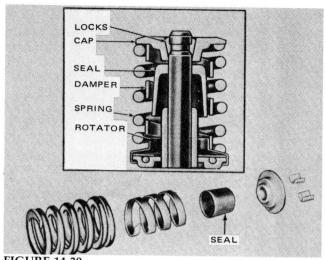

FIGURE 14-30

Deflector or "umbrella" type valve stem seal (Courtesy of Chevrolet Motor Division of General Motors Corporation).

FIGURE 14-32

Rubber "o" ring type valve stem seal (Courtesy of Chevrolet Motor Division of General Motors Corporation).

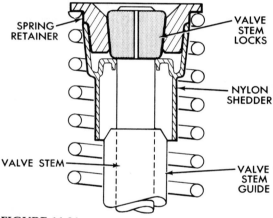

FIGURE 14-31

Nylon shedder type shield (Courtesy of Cadillac Motor Division of General Motors Corporation).

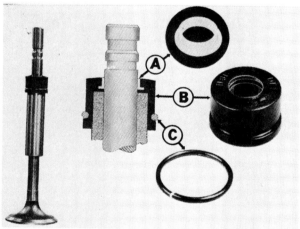

FIGURE 14-33

Perfect circle "positive" type valve stem seal. Insert *A* is attached and held to the valve guide by synthetic rubber jacket *B*. The jacket fits tightly over the Teflon seal insert and valve guide, and is further secured by steel retainer ring *C* (Courtesy of Dana Corporation—Service Parts Group).

1. Rubber "umbrella" on the valve stem (Figure 14-30). This rides up and down with the valve stem and acts as a deflector.

2. Metal shield or shedder that fits under the spring retainer (Figure 14-31).

3. Square cut rubber "o" ring seal in the keeper groove (Figure 14-32). This keeps the oil from running off the retainer and down the valve stem.

4. "Positive" insert type (Figure 14-33). This is generally considered the most effective guide seal. It is held securely to the valve guide with a snap ring and contracts around the valve stem. Sealed Power, Perfect Circle, and

Raymond are among a number of companies that sell aftermarket "positive" guide seals.

SHOP TIP

"Umbrella" seals can be installed on small-block Chevrolets without any machining to gain clearance between the guide OD and the valve spring assembly ID. Umbrellas from several other engines will fit. Check an engine parts catalog for interchangeability.

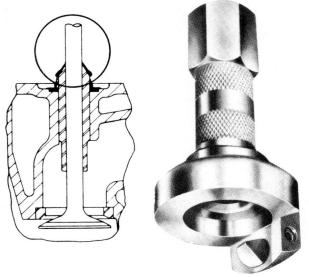

FIGURE 14-34
This valve guide chamfering tool is used in conjunction with regular valve seat grinding pilots (Courtesy of Sealed Power Corporation).

FIGURE 14-35
A Teflon and rubber "positive" type stem seal is shown on the left; an all Teflon type is shown on the right (Courtesy of Dana Corporation—Service Parts Group).

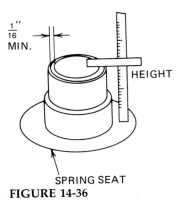

FIGURE 14-36

Valve Guide Chamfering

Some people like to chamfer (bevel) the top of the guide OD to prevent an "oil puddle" from collecting. They feel this helps to better control the amount of oil passing through the guide. Chamfering is accomplished by using a special hand-driven tool that centers on the valve guide bore (Figure 14-34).

Machining for PC (Perfect Circle) Valve Stem Seals

PC "positive" stem seals are available for most models of passenger cars, trucks, and racing engines (Figure 14-35). This widely used seal offers several excellent benefits.

1. Excess lubricating oil is prevented from reaching the combustion chambers via the valve guides.

2. By prohibiting the passage of air through the guides, PC seals help prevent changes in the ratio of air and fuel entering the combustion chamber.

3. Valve seals for an application can be used on standard through .015" oversize valve stems.

These seals are easy to install. In some cases the OD of the valve guide has a suitable surface, and no machining is required. On some engines, it is necessary to machine the top and sides of the guide by using a special cutter. This procedure is as follows.

1. Refer to the engine application information in the valve seal set. See Chart 14-2. Determine the valve guide machining requirements and the correct tool to use. Valve guide heights after machining are given in the intake (IN) and exhaust (EX) columns. This dimension is the distance from the spring seat to the top of the guide. An X in the column means the valve guide must be machined until the cutter contacts the top of the guide and produces a 1/16" wide flat top surface. A dash (—) indicates no machining is required. Refer to Figure 14-36.

2. Insert the machining tool in the chuck of a 1/2" drill. Tool speed should not exceed 450 rpm.

3. Place cylinder head on the floor. Oil the tool pilot and insert it into a valve guide (Figure

CHART 14-2

An example of machining specification information for installing PC stem seals (Courtesy of Dana Corporation—Service Parts Group).

Cylinders	Chevrolet Cars	Machining Specifications Guides IN.	EX.	Tool	Valve Seal Set
6	216, 235 Engs.				VS-1
8	396 Eng.				VS-4
8	265, 283 Engs.	$^{11}/_{16}$	$^{11}/_{16}$	VST-1811	VS-2
	283, 327 Engs.				
8	348 Eng. (single valve spring)	X	X	VST-2012	VS-4
8	283 Eng.	X	$^{11}/_{16}$	VST-1811	VS-2
4	153 Eng.	$^{11}/_{16}$	$^{11}/_{16}$	VST-1811	VS-2
6	194 Eng.	$^{11}/_{16}$	$^{11}/_{16}$	VST-1811	VS-1

14-37a). Machining tools cannot be used on guides that have been reamed oversize.

4. Apply moderate pressure to the cutting tool (Figure 14-37b). It is not necessary to lubricate the cutter. *Do not bounce the tool on the guide, as the carbide may be chipped or broken.* Machine the valve guide to specification (Figure 14-37c). **NOTE:** *If a guide is machined with the tool indicated, but does not properly clean up, the next smaller OD size tool and seal set can be used.* Sometimes, during the installation of a thin-wall guide insert the original integral guide is not entirely removed or an eccentric condition exists between the valve stem hole and the OD of the guide (Figure 14-38). Ragged edges, eccentric diameters, or portions not cleaned up, must be corrected. Otherwise, the seals will be distorted and pulled to one side when installed. This distortion opens up a path for oil to enter.

5. Machine remaining guides. Wipe and lubricate tool pilot after machining each guide.

6. Thoroughly clean cylinder head and guides.

7. Remove any burrs from the valve stems with a fine stone or crocus cloth.

8. Insert valves into guides. Lubricate lightly.

9. Place plastic installation cap over end of valve stem (Figure 14-39a). Trim cap if it extends more than $^{1}/_{16}$" below the lower keeper groove. The cap will prevent the sharp edges

(a)

FIGURE 14-37a

Insert pilot of special cutter into guide. Be sure to lubricate pilot with oil.

FIGURE 14-37b
Machine valve guides for seals.

FIGURE 14-37c
Check machined height.

FIGURE 14-38
Note the eccentricity of the valve stem hole with the guide OD (Courtesy of Sealed Power Corporation).

of the keeper grooves from cutting the seal insert.

10. Start the valve seal carefully over the cap (Figure 14-39b). On seals with rubber jackets, hold thumbs against the white insert to avoid dislodging it. Push the seal down until the jacket touches the top of the guide.

11. Use the special installation tool to push the seal over the valve guide (Figure 14-39c). If a tool is not available, use two small screwdrivers or your thumbnails on the metal retaining ring. Pull down until the top of the seal is flush with the top of the valve guide.

12. Install seals on the remaining valves.

13. Replace the valve spring assemblies. *Compress springs only enough to install keepers. Ex-*

FIGURE 14-39

(*a*) The plastic cap prevents damage to the Teflon insert during seal installation. (*b*) Push the seal down until the rubber jacket touches the top of the guide. (*c*) Using a valve seal installation tool to push the valve seal over the valve guide (Courtesy of Dana Corporation—Service Parts Group).

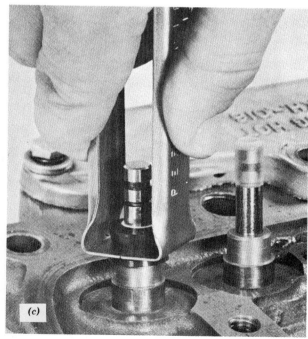

cess compression can cause the spring retainer to damage the valve seal. Do not reinstall any other seals such as original equipment metal shields or rubber "o" rings.

TRUE CASE HISTORY

Vehicle	Oldsmobile
Engine	8 cylinder, 455 CID
Mileage	39,000
Problem	1. Engine misfired badly at 2800 rpm and above (with or without a load). 2. CO and HC were low at idle. 3. With the throttle level held at 2800 rpm (transmission in neutral), the CO was .5% and HC was fluctuating around 800 ppm. 4. Partially closing the choke valve decreased the misfire a little.
Cause	Three of the cylinders had excessive carbon deposits due to defective valve stem seals. As a result, the compression ratio in three cylinders was considerably higher than normal. These cylinders became so lean at 2800 rpm that misfiring resulted.
Cure	The heads were taken off, the excess carbon removed, and the valve stem oil seals replaced.

TRUE CASE HISTORY

Vehicle	Chevrolet
Engine	8 cylinder, 305 CID
Mileage	31,000 miles since vehicle was purchased new.
Problem	Badly egg-shaped worn exhaust guides on cylinders 3 and 4. The rest of the guides were satisfactory.
Cause	The EGR valve was sticking. On this engine, exhaust ports 3 and 4 are channeled to the EGR valve. An improperly working EGR valve will cause these guides to overheat. In turn, this results in poor lubrication and rapid wear.

CHAPTER 14 SELF-TEST

1. What is the general rule-of-thumb intake valve-to-guide maximum clearance? Exhaust valve-to-guide maximum clearance?
2. Whenever possible, why should replaceable guides be driven out from the valve spring side of the head?
3. What engines use valve guides with a tapered OD?
4. On what engines are the guides removed by first breaking off the top with a brass punch and hammer?
5. Why is guide installed height important?
6. New replaceable guides have just been installed, and the valves fit the guides. Why is it important to still make a pass with the correct size reamer?
7. Certain import engines have unusual oversize OD guides. Rather than machine the bore hole in the head, what is often done?
8. What removal method is recommended when replacing VW bronze valve guides?
9. How can a head be preheated before installing new replaceable bronze guides?
10. Describe the valve guide knurling operation.
11. What is the recommended valve-to-guide clearance when knurling worn guides?
12. What tool makes an excellent drive for guide reamers?
13. It is not recommended to use positive guide seals when guides are knurled and the clearance is _____" or less.
14. Valve guide repair sleeves are available in what different styles?
15. What guide material is recommended for dry fuel application?
16. Silicon aluminum bronze valve guides cannot be run against what type of valve stems?
17. List four advantages of using the bronze walling reconditioning procedure.
18. When using the Winona bronze walling kit, why shouldn't any lubricant be used while tapping?
19. What is the reason for cutting the Winona bronze wall insert so that it extends 1/32" beyond the OD of valve guide?
20. Why do many shops prefer to size bronze wall bushings by honing?
21. How can a Winona bronze wall bushing be removed?
22. Valves with oversize stems are generally available in what oversizes?
23. When reaming a valve guide, the amount of metal removed per pass should be limited to _____".
24. Why should a reamer never be reversed?
25. Should cast iron guides be reamed wet or dry?
26. What is a good lubricant to use when reaming bronze?
27. Some General Motors V-8 engines are manufactured with oversize valve guide bores. How can this be determined without taking the engine apart?

28. What major advantage does valve guide honing have over reaming?
29. What two methods can be used to replace a broken rocker arm stud?
30. How can a stack of nuts and washers be used to withdraw a rocker stud?
31. When installing a pressed-in replacement stud, the hole in the head must be sized to provide a _____negative fit.
32. When installing a pressed-in stud, the surface to be driven in should be coated with _____ _____ compound.
33. What is the advantage of screw-in studs?
34. A cylinder head with pressed-in studs requires modification for the installation of threaded jam nut studs. Outline the procedure.
35. What is a pinning kit?
36. What is the purpose of a push rod guide plate?
37. When using guide plates, what type of push rods should be used?
38. How is rocker arm-to-stud clearance checked?
39. What is the minimum required amount of rocker arm-to-stud clearance?
40. List four types of valve guide oil control seals.
41. Name three companies that sell aftermarket "positive" guide seals.
42. Sometimes, the top of the guide is chamfered. What is the reason for doing this?
43. List three benefits of using PC seals.
44. Cutter speed should not exceed _____ rpm when machining for PC seals.
45. Why is it important to place a plastic cap over the valve stem end when installing PC seals?
46. What should be used to push PC seals over the guides?
47. What precaution must be observed when installing the valve springs on an engine equipped with PC seals?

CHAPTER 15

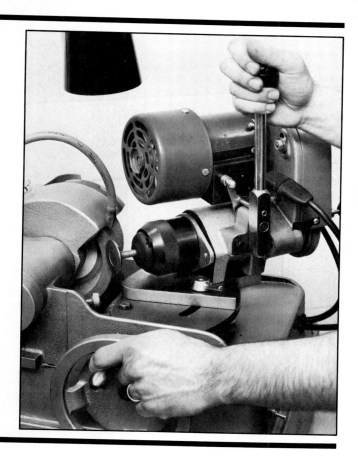

VALVE
AND
VALVE
SEAT
SERVICE

The intake and the exhaust valves in an engine absorb a fantastic amount of punishment. For instance, during a 15-minute drive to the grocery store a valve may open and close 10,000 times. The valve stem may travel a mile or more in the guide. The valve head is scorched by high temperature and stressed by spring tension. The valve face is exposed to corrosive combustion products and abrasiveness (especially when using lead-free gasoline). As a result, the valves—particularly the exhaust valves—become burned, pitted, warped, grooved, and out-of-round (Figure 15-1).

Valve seats distort, wear, and recess because of constant pounding and grinding action from the valve. Particles of hard carbon with sharp points projecting become "white hot" and cause pitting. Seat inserts can become loose or crack due to thermal variations in the engine (Figure 15-2). *COMMENT: Exhaust valves and seats are momentarily exposed to combustion temperatures in excess of 3000° F. During normal operation, the temperature all the way around an exhaust valve edge stays at about 1200° F (cherry red in color).*

The valve guide always wears in a pattern corresponding to valve stem wear (Figure 15-3). Carbon residues may form between the stem and guide and cause valve sticking. Proper heat dissipation is prevented, and warping, burning, and scoring is encouraged. To ensure top performance when doing a valve job, the valve face, valve seat, and valve guide must be closely evaluated and reconditioned accordingly. After the valve and valve seat have been reconditioned, they must be concentric with each other and with the valve guide (Figure 15-4). Otherwise, the valve will want to seat off-center, with these results.

1. Rapid valve stem and guide wear.

2. Bending stress. This increases the probability of fatigue cracking and resulting valve breakage.

3. Compression leakage.

4. Preignition.

5. Valve burning.

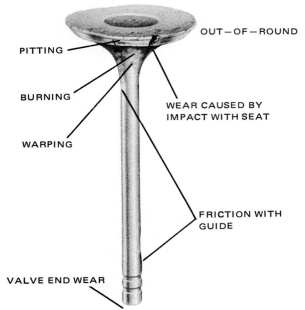

PITTING

OUT—OF—ROUND

BURNING

WEAR CAUSED BY
IMPACT WITH SEAT

WARPING

FRICTION WITH
GUIDE

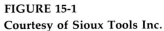
VALVE END WEAR

FIGURE 15-1
Courtesy of Sioux Tools Inc.

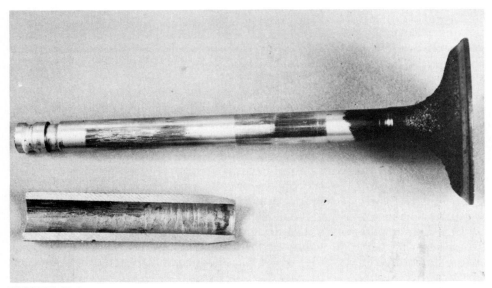

FIGURE 15-2
Seat insert cracking (Courtesy of TRW Replacement Division).

FIGURE 15-3
When replacing scored valve stems, the valve guide, which is the other 50% of the rubbing surface, has to be considered. Both the stem and guide shown in this picture are badly scored.

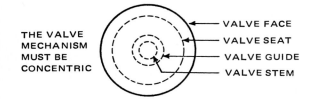

THE VALVE
MECHANISM
MUST BE
CONCENTRIC

VALVE FACE
VALVE SEAT
VALVE GUIDE
VALVE STEM

FIGURE 15-4
The valve mechanism must be concentric (Courtesy of Sioux Tools Inc.).

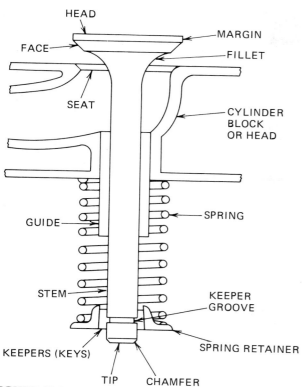

FIGURE 15-5
Valve assembly nomenclature.

FIGURE 15-6
The boss 351 Ford engine features "staggered" valves
(Courtesy of Ford Parts and Service Division).

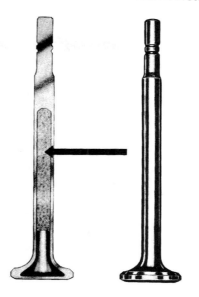

FIGURE 15-7
Sodium-filled valve (Courtesy of TRW Replacement Division).

VALVE NOMENCLATURE

Refer to Figure 15-5 and note the various sections of a typical valve assembly.

Canted Valves (or "Staggered" Valves)

Valves are generally positioned parallel to each other in a single row. On some engines, though, the manufacturer will cant the valves. This means the valves are tilted at compound angles to the intake and exhaust ports (Figure 15-6). Canting makes room for bigger valves, reduces shrouding, and reduces the sharp turns associated with engines that have valves operating in a single plane. This greatly helps engine breathing.

Sodium-Filled Valves

For extreme heat application, sodium-filled valves are often used. They have been adopted for use primarily in heavy-duty engines (Chrysler marine, Dodge trucks, Autocar, International, and White). These valves have a hollow stem, or the stem and head is hollow. The cavity is filled 60% full of metallic sodium that melts to the consistency of milk at 207° F (Figure 15-7). Liquid sodium has the ability to quickly absorb and dissipate heat. Very efficient valve cooling results as the liquid sodium is sloshed from one end of the cavity to the other. An exhaust valve with a sodium-filled stem will reduce head temperature by 10%, whereas, an exhaust valve with a sodium-filled head and stem will lower valve head

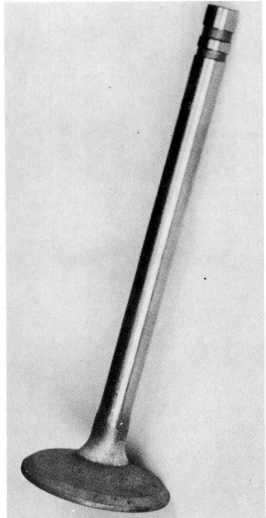

FIGURE 15-8
Valve with thin coating of aluminum on entire head and underhead area (Courtesy of TRW Replacement Division).

FIGURE 15-9
This Z/28 Chevrolet valve has a swirl-finished head.

temperature by 25%. Sometimes the internal cavity is copper plated for additional heat transfer.

However, sodium-filled valves have several disadvantages.

1. They have a greater stem diameter which makes interchange with conventional solid stem types difficult.
2. They are quite expensive to manufacture.
3. If sodium contacts water (or just moisture in the air), sparks and a small explosion can occur. Safety during removal and disposal, therefore, is a major concern.

Aluminized Valves

With the trend to smaller displacement higher rpm engines there has been an increase in valve temperature. To provide greater resistance to valve face burn and exhaust gas corrosion, aluminized valves are used, especially on turbocharged engines.

Aluminizing is a process where pure aluminum is fused to the base metal of the valve. An aluminized valve can be identified by the dull, rough looking appearance on the face area (Figure 15-8). *CAUTION: This coating must not be removed by valve grinding, as this would defeat the purpose and result in premature valve failure.*

When a postvalve job leak test (either vacuum or solvent method) is performed on a cylinder head fitted with aluminized valves, the valves will leak. This occurs because of the relatively rough aluminum finish on the valve face. After the engine has run for a few minutes, the valves will seat in and seal.

Swirl-Finished Valves

A swirl-finished valve has the fillet area ground with a special scratch pattern (Figure 15-9). The swirl scratches run down the valve stem and across the bottom of the head. The major advantage to swirl-finishing is that the valve becomes stronger (because the scratch marks line up with the inside grain of the metal).

Hard Chromed Valves

Many manufacturers are now hard chroming valve stems in order to improve wear and corrosion resistance. *NOTE: Chromium deposited for wear resistance of sliding metal surfaces is called hard chrome.* Hard chrome is a fairly thick deposit applied directly to the base metal. Do not confuse hard chroming with decorative chroming.

VALVE ROTATION

Valve rotation will reduce heat and increase valve life. On engines running leaded fuel, tests have shown that the valves last two to five times longer. *WARNING:* *Valve rotation used in conjunction with lead-free gasoline or liquified petroleum gas can cause accelerated valve seat recession.* This situation is shown in Figure 15-10. The valve seat has recessed over .100" after 4500 miles.

Besides reducing heat, valve rotation offers a number of other advantages.

1. The rotating action of the valve helps prevent sticking.
2. The valve face and seat are constantly being wiped clean of deposit buildup.
3. Localized hot spots on the valve face are eliminated.
4. Valve stem lubrication is more uniform.
5. The valve tip will not form a groove from the wear of rocker arm scrubbing.

A number of manufacturers use valve rotating devices as standard production equipment. Also, aftermarket rotators are available for many engines. There are two types of rotating devices.

1. Natural or "free" release.
2. Positive or "forced" release.

A natural rotator releases the valve from spring tension as it lifts, thus permitting engine vibration to induce an uncertain free rotation. Several "free" release devices are used.

Rotovalve

When installing this rotator (Figure 15-11), the valve tip-to-cap clearance must be gauged. If the clearance is not enough, the valve will not completely release. If the clearance is too much, shock loads will be high and valve lift not correct. Repeated shocks can cause severe damage (Figure 15-12). It is recommended to change the rotovalve cap, keepers, and spring retainer during every valve job.

FIGURE 15-10

A section of a valve seat with severe recession (Copyright, Champion Spark Plug Company, 1971. Used with Permission).

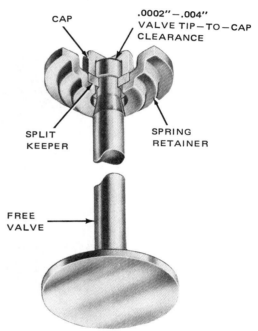

FIGURE 15-11

The Rotovalve cap provides for rotation, but not controlled (Courtesy of Ford Parts and Service Division).

Multigroove Valve Stems

Some engines use multigrooved exhaust stems to achieve valve rotation (see Figure 15-13). The grooves on the stem (four in number) are designed to have a slight interference fit with the beads on the keeper. As the valve is pushed open, each keeper bead has tension relieved on one side. This action tends to promote stem rotation. However, these type valve stems are very prone to keeper groove wear. This author has observed

FIGURE 15-12
An improperly adjusted rotater caused this valve damage.

grooves on Ford 351C and on Chrysler 383 CID engines worn to knife-edge sharpness (Figure 15-14).

Offset Rocker Arms

A few engines have the rocker arm face positioned slightly off the center of the valve stem (Figure 15-15). This results in a "wipe" action on the end of the stem that causes the valve to spin.

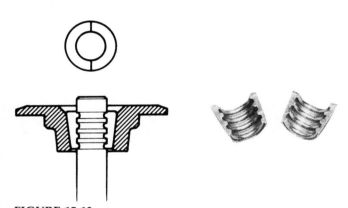

FIGURE 15-13
Four-bead keepers help rotate the valves.

FIGURE 15-14
Badly worn multigroove valve stems.

A positive type rotator will actually turn the valve a certain amount each time it opens. The turn rate is approximately five revolutions per minute. Two types of positive rotators are used—the Rotocoil and the Rotocap.

Rotocoil

This type of rotator is used on several General Motors engines (Figure 15-16). As a valve opens, a coil spring inside the rotator "lays over" and imparts a twisting action. The retainer body, keepers, and valve stem are then forced to rotate as an assembly.

Rotocap

As a valve opens, pressure is applied to the small balls inside the rotator. Each ball rolls down a ramp, applying a rotary force that results in valve rotation (Figure 15-17).

Positive valve rotators should be changed approximately every 100,000 miles. Rotators can easily be checked for proper operation. Make a

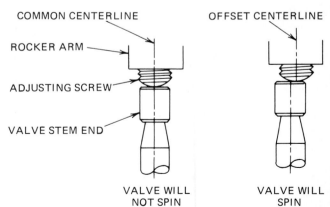

FIGURE 15-15
Volkswagen rocker arms are offset slightly from the valve stem end. This imparts a "wiping" action that causes valve rotation.

FIGURE 15-16
Cutaway of a Rotocoil (Courtesy of TRW Replacement Division).

colored felt pen mark on the valve stem tip edge. Hook up a timing light. Run the engine at 1500 rpm and strobe the mark for motion to the right or left. See Figure 15-18.

VALVE INSPECTION

Every valve must have margin (Figure 15-19). Margin helps the valve withstand pressure and control heat. If a valve has been ground to a knife edge, leaving no margin, it is extremely susceptible to warping and burning. Also, a knife-edge margin will retain heat during the compression stroke and cause preignition. *As a general rule, do not grind the valve face to a point where the margin is less than .045".*

Visually inspect the valve stems for wear, nicks, score marks, corrosion, tip wear, and keeper groove wear (Figures 15-20, 15-21, and 15-22). *Maximum allowable stem wear is .001".* Stem wear is a very important, often overlooked, measurement. Never reuse valves that are badly worn, cupped, necked, or bent. Refer to the illustrations in the valve damage analysis section of this chapter. If the valves appear to be in good condition and are to be reused, proceed with refacing.

FIGURE 15-17
Cutaway of a Rotocap (Courtesy of TRW Replacement Division).

PROPER TIP PATTERN   NO ROTATION PATTERN   PARTIAL ROTATION TIP PATTERN

FIGURE 15-18
Stem wear patterns on rotated valves (Courtesy of Oldsmobile Division of General Motors Corporation).

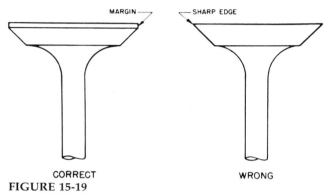

FIGURE 15-19
Correct versus wrong valve face edge.

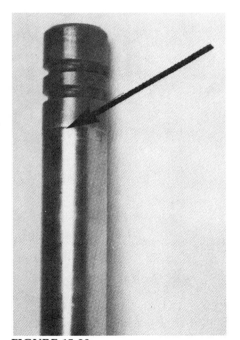

FIGURE 15-20
Badly worn valve stem (Courtesy of TRW Replacement Division).

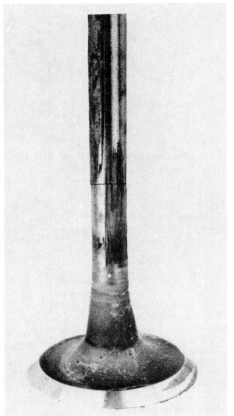

FIGURE 15-21
Stem scuffing cause by a cocked valve spring (Courtesy of TRW Replacement Division).

VALVE GRINDING

Most valve grinding equipment operates in a very similar manner. The following instructions are for a Sioux Machine (Figure 15-23) and can be considered typical.

1. Dress the left wheel (Figure 15-24). Move the chuck carriage to the extreme left. Place the dressing attachment against the stop bar on the grinding head and tighten down. The amount of diamond nib overhang should be kept at a minimum in order to maintain rigidity. Turn on the motor and coolant. Pass the diamond over the wheel while taking *cuts of .0005" or less per pass.* The diamond should be occasionally rotated to present a new cutting edge. A rapid traverse of the diamond will result in a poor finish. For hard-faced valves use a different grade wheel.

2. True valve stem ends (Figure 15-25). Dress the right grinding wheel. Square the stem ends and renew the chamfer.

3. Locate chuck head (Figure 15-26) at the exact angle you wish to refinish the valve, then

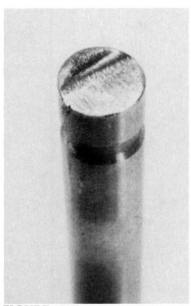

FIGURE 15-22
Badly worn valve stem tip.

FIGURE 15-23
Sioux valve grinding machine (Courtesy of Sioux Tools Inc.).

lock into position. Chuck heads are generally calibrated with markings from 15° to 90° (including 29° and 44° interference angles).

4. Chuck valve (Figure 15-27). Open the chuck sleeve and insert the valve so it is gripped just above the worn part of the stem. *NOTE: Use a rubber chuck shield to protect the inside of the chuck from wheel grit while grinding (Figure 15-28).*

5. Reface the valves (Figure 15-29). Advance the valve to a position in front of the grinding wheel. Turn the feed screw until the wheel just touches the valve. Switch on the motor and coolant. Begin grinding at the left side of the wheel, moving the valve slowly and steadily (with a "rocking" motion) back and forth across the wheel. Take light cuts by feeding the wheel up to the valve .001" to .002" at a time. Allow the valve to "spark-out" between feeds. *Do not allow the valve at any time to pass beyond either edge of the grinding wheel while grinding.* Remove just enough material to make a clean, smooth face. When inspecting the valve face during grinding, back off the stone from the valve. Do not pull the valve off the stone. Otherwise, the outer edge of the valve face can be deformed.

COMMENT: Small valves may not be of sufficient size to be traverse ground across the full width of the grinding wheel. This condition will create an interference shoulder where the traverse stops. Figure 15-30 shows the solution. A small undercut has been made by the dressing tool. The valve is then ground on the high area of the wheel.

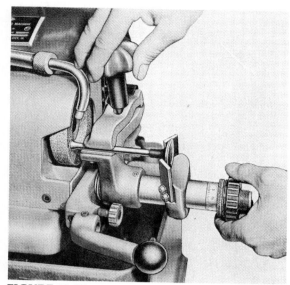

FIGURE 15-25
True stem ends (Courtesy of Sioux Tools Inc.).

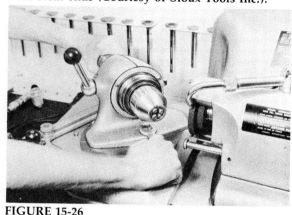

FIGURE 15-26
Locate and lock chuck head (Courtesy of Sioux Tools Inc.).

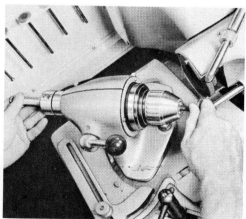

FIGURE 15-27
Insert valve into chuck (Courtesy of Sioux Tools Inc.).

FIGURE 15-24
Dress wheel (Courtesy of Sioux Tools Inc.).

FIGURE 15-28
Rubber boot protects the chuck from abrasive wear while grinding.

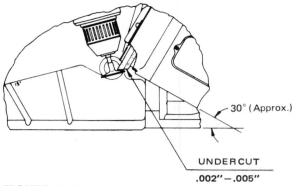

30° (Approx.)

UNDERCUT
.002" – .005"

FIGURE 15-30
It may be necessary to undercut the grinding wheel when refacing small valves (Courtesy of Sioux Tools Inc.).

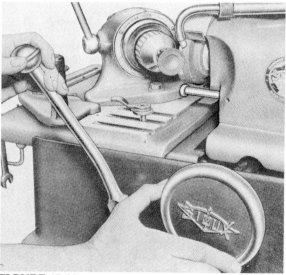

FIGURE 15-29
Grind valve faces (Courtesy of Sioux Tools Inc.).

Valve Grinding Hints

1. When dressing, have coolant flowing freely over the wheel to keep the diamond cool and wash away wheel grit.

2. When grinding the valve face, adjust the coolant to flow on the valve face rather than on the wheel (Figure 15-31).

COOLANT
RUBBER BOOT

FIGURE 15-31
Adjust coolant flow.

3. A soluble oil/water solution is a superior coolant and lubricant to use when wet grinding valves. However, more frequent machine cleaning is necessary since this type of oil becomes gummy as the water evaporates. Also, the solution can develop a rancid odor because of bacteria growing in the water.

4. One wheel dressing is usually enough for eight valves. A poor finish on valves is generally caused by a dull wheel.

5. After the valve face is cleaned up, pass the valve over the stone until the grinding sparks disappear. This procedure will give a more accurate face angle with a finer finish.

6. *The chamfer on the stem end of the valve should*

FIGURE 15-32
Chamfering valve stem tip (Courtesy of Sioux Tools Inc.).

FIGURE 15-33
A fast method for grinding valve stem ends (Courtesy of Automotive Engine Rebuilders Association).

be ⅟₃₂″ *wide*. Figure 15-32 shows renewing the chamfer using a chamfering vee.

7. After refacing, check the valve face runout. *The maximum allowable is .002″.* The face should not show any unground areas, "chatter" marks, or hairline cracks. Excessive wheel pressure or inadequate coolant flow will sometimes cause fine cracks to develop during grinding.

8. After refacing, visually check the margin around the entire valve head. An uneven width indicates bend in the head or fillet area.

9. Check the accuracy of your refacing machine at regular intervals. Here is a quick way to check it out. Grip the top end of a valve seat grinding pilot in the chuck head. Position a dial indicator on the pilot (about 2″ from the chuck jaw). Rotate the chuck by hand. *The indicator should read less than .002″ tir.* Sometimes, a thorough cleaning and oiling of the machine is all that is required to restore tolerance.

10. Here is a rapid, consecutive procedure for grinding the ends of valve stems. Clamp or hold a valve guide in the vee block attachment on the end of the grinding machine. Insert the valve and rotate by hand (Figure 15-33).

Valve Grinding with Lapping Compound

For many years this was a popular method of seating a valve. In this operation, a small amount of abrasive paste was placed on the valve face. After placing the valve in its guide, a special tool was used to turn the valve back and forth on its seat while applying slight downward pressure. This operation was repeated (changing the valve-to-seat position each time) until the valve and seat fit each other. The lapping operation could be assumed complete when the valve face and seat showed a continuous frostlike ring all around. Unfortunately, there are still some people who use the lapping method of valve seating. **This method is obsolete**. With grinding equipment, a much more accurate and effective seal can be obtained. There are other people who believe in lapping after machine grinding the valve and the seat. **This procedure is a waste of time**. As the valve heats up in the engine, its seating surface will rise (Figure 15-34). This results in the lapped

FIGURE 15-34

The illustration shows a valve and seat "ground in" with compound. When the engine is cold the valve and seat form complete contact; but when the valve is heated and has raised, the portion ground in with compound is actually not in contact with the seat at all (Courtesy of Kwik-Way Manufacturing Company).

HIGH UNIT PRESURE

WITH INTERFERENCE ANGLE

FIGURE 15-35

Interference angle between valve and seat.

FIGURE 15-36

Groove wear on the valve face.

portion of the valve no longer being in contact with the seat.

Authorities in the engine rebuilding industry agree that the use of valve lapping compound does not give effective results.

Interference Angles

In valve and valve seat reconditioning, the valve face is sometimes ground ½° to 1° less than the seat angle (Figure 15-35). Advocates of interference angle grinding give the following reasons as benefits.

1. The valve face and the valve seat will expand at uneven rates during engine warmup. An interference angle will help compensate for expansion and provide a better seal.
2. It makes a better seal when the engine is first started after a valve job.
3. The effect of distortion will be minimized when the head is torqued down.
4. Dirt or carbon particles will be cut and wedged from the seat.

Many people dislike interference angles because of the much higher loading at the seat contact point. They feel the narrow seat contact line acts likes a sharp hot chisel cutting into the valve face. This often results in a groove that shortens the valve life (Figure 15-36). Another disadvantage of interference angles is that an otherwise sloppy valve job can be temporarily covered up.

Valve Stem Grinding (or "Tipping")

"Tipping" the stem end of the valve is an extremely important part of a valve job for the following reasons.

1. Some valve grinding machines use the stem end to center the valve in the chuck. If the valve stem end has nicks, is out-of-square, or mushroomed, the valve will be cocked when ground. The valve will then operate in the engine as if bent.
2. A stem end that is not square will accelerate valve guide and rocker arm wear. This is because an out-of-square end will allow the rocker arm to apply additional side thrust while opening the valve.
3. When you do a valve job, material is removed from both the valve face and the valve

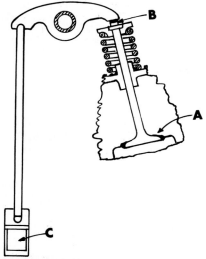

FIGURE 15-37

Valve stem height (*B*) will change when valves and seats are ground. This changes the position of the hydraulic valve lifter plunger (Courtesy of Houser Engineering and Manufacturing, Inc.).

seat at point *A* (Figure 15-37). Thus, the valve stem extends further out of the head and moves the hydraulic lifter plunger down from the midpoint tolerance position. Refer to point *C* in Figure 15-37. Certain engines use non-adjustable hydraulic lifters that have very limited travel. *For example, valve stem height on certain Buick and Chrysler engines can only vary .038" from limits.* On these engines, correcting the stem height after doing a valve job is critical. If not done, the lifters may bottom out and hold the valves open. Fortunately, most hydraulic valve lifters have a plunger travel range of about .090" to .125". However, stem height should still be adjusted in order to maintain the correct rocker arm geometry (the position of the rocker arm on the stem end of the valve at 40% opening). If the rocker arm is not in the center of its swing at 40% valve opening, extra side pressure is placed on the guide at full valve opening. Incorrect rocker arm geometry is a frequent cause of premature valve guide wear.

Valve Stem Height Specifications

Specifications for this dimension are not generally available from the manufacturer because this is a production measurement established when

TABLE 15-1
VALVE STEM HEIGHT SPECIFICATIONS

Valve stem heights shown in the table below are noted as dimension *A* in the drawing.

These dimensions have been established from numerous measurements made in the field and should be considered approximate. These may be used when manufacturer's specifications are not available.

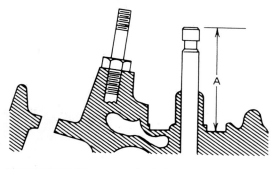

Engine	Displacement	Stem Height
AMC	360	1.965"
Buick	350	1.950
Ford	302	2.145
	351C	2.080
	351W	2.200
	360	2.125
	390	2.125
	460	2.160 intake
		2.200 exhaust
Mopar	318	1.920
	360	1.920
	383	2.130
	400	2.130
	440	2.130
Oldsmobile	455	2.050
Vega	140	2.150 intake
		2.120 exhaust

each head casting is machined. Also, the measurement is taken from different reference points (see Table 15-1).

There are several different types of gauges available for checking valve stem height (Figures 15-38 and 15-39). A specification chart is sometimes supplied with the gauge. However, the specifications are of no value without the required gauge. Hall-Toledo, UTP, Houser, and Central are companies that sell valve stem height gauges.

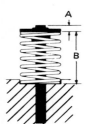

A + B = VALVE STEM HEIGHT

FIGURE 15-38

No specification sheet is needed with this height gauge. Measure the distance from bottom of valve spring retainer to end of valve stem on a new valve. Add this dimension to the valve spring installed height (given in the service manual) for correct valve stem height (Courtesy of Central, A Division of K-D Manufacturing Company).

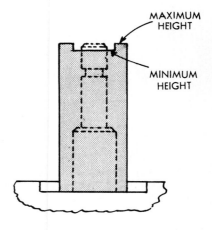

MAXIMUM HEIGHT

MINIMUM HEIGHT

FIGURE 15-39

This gauge kit indicates the minimum and maximum length of the valve stem (Courtesy of Houser Engineering and Manufacturing, Inc.).

"Tipping" Hints

1. A shop may not have valve stem height specifications information or a special measuring gauge. In this case, you must measure each valve stem length during disassembly of the cylinder head. Record these dimensions. When installing the valves, the stem lengths must match those recorded. See Figure 15-40.

2. What if a head is apart and the stem length is unknown? First, grind the stem lengths on the end valves equal. Then, use a precision straightedge to span all the valve tips. Grind as necessary.

3. When adjusting stem height, *allowance for cylinder head surfacing must be made at a ratio of about 1.5:1. In other words, if .008" has been milled from the head, the valve should be shortened .012".*

4. It is recommended to not cut more than .030" from the end of the valve stem for two reasons. First, the heat treating hardness may be removed, and, second, the valve stems may end up lower than the top of the valve spring retainer (Figure 15-41). Then the rocker arm can hit the retainer and pop out the valve keepers.

5. If a valve stem is much too long, a new valve seat may have to be installed to move the tip closer to the spring seat.

FIGURE 15-40

This gauge can be used to measure stem length before grinding either seats or valves. After valve work is completed, remeasure stem to determine the amount to grind off (Courtesy of Lisle Corporation).

CAUTION: *After adjusting a valve stem for length, keep it in the same location in the head. Do not mix the valves.*

FIGURE 15-41

The valve stems in this head are cut so short that they are lower than the top of the spring retainers. This can allow the rocker arm to hit the retainer and pop out the valve keepers (Courtesy of Automotive Engine Rebuilders Association).

FIGURE 15-42

Medium grit valve seat grinding stones (Courtesy of Black and Decker Manufacturing Company).

VALVE MATERIALS

The selection of material for intake and exhaust valves depends primarily on the temperature at which the valve will operate. As operating temperature rises, the strength of a valve decreases and the possibility of corrosion and change of shape increases. Most intake valves are a ferritic steel base material that is alloyed with chromium, nickel, and silicon. Ferritic steels utilize substantial amounts of silicon to increase resistance to oxidation. However, silicon reacts adversely with lead oxide. *Therefore, when lead corrosion is a factor, use a valve alloy that contains less silicon.*

Selection of material for exhaust valves is more critical. Intake valves operate at much lower temperatures (usually under 600° F). Austenitic steels are the most widely used for exhaust valves. This alloy contains large amounts of chromium and nickel and has excellent hot strength, and high stretch resistance.

Some valves have stellite welded onto the face. This hard-facing material is very resistant to pitting, corrosion, and burning. Stellite valves and hard seats work very well in engines that use unleaded gasoline. It is possible to tell the difference between a hard-faced valve and a standard valve by using an acid-etch test. Place several drops of battery acid on the valve faces. The standard valve will start to corrode (rust) shortly after being etched clean. The acid will have no effect on the stellite face.

VALVE SEAT GRINDING STONE SELECTION

Stones are cataloged by:

1. **Diameter size in inches** Stones are available in assorted sizes ($1\frac{1}{16}''$, $1\frac{1}{8}''$, $1\frac{3}{16}''$, $1\frac{1}{4}''$, $1\frac{5}{16}''$, $1\frac{3}{8}''$, etc.). Stones over $2''$ are generally sized in $\frac{1}{8}''$ increments. When grinding a seat, the proper size stone is approximately $\frac{1}{8}''$ larger than the valve head diameter.

2. **Angle (generally 30° or 45°)** However, special angle stones are sometimes available.

3. **Grit type** There are rough, finish, medium, stellite, and wet wheels. Roughing wheels are used to clean up steel and other hard seats. These wheels leave a rough texture that will have to be finish ground. For cast iron seats (to grind and finish) use a finishing wheel. These wheels are very fine and leave the desired texture. Medium (often called universal or general purpose) wheels are a grit compromise between the rough and finish (Fig. 15-42). Stellite wheels have a special coarse grit for cutting stellite valve seats. *NOTE: When finish grinding stellite, use a general purpose stone.* Special stones are available for wet valve seat grinding.

4. **Thread hole diameter** Seat wheels have a threaded hole through the center. The size will vary with the different makes of grinding equipment. Some of the common thread insert sizes are $\frac{1}{2}''$-20, $\frac{9}{16}''$-16, $\frac{11}{16}''$-16, and $\frac{7}{8}''$-14.

VALVE SEAT PILOTS

There are several styles of valve seat pilots. Over the years, there has been a considerable amount

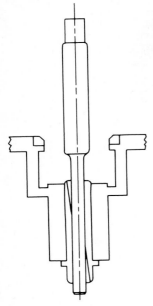

FIGURE 15-43

A slow tapered valve seat pilot (Courtesy of Kwik-Way Manufacturing Company).

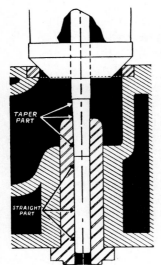

FIGURE 15-44

This valve seat pilot is only tapered in the upper portion (Courtesy of Sioux Tools Inc.).

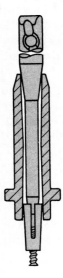

FIGURE 15-45

Self-centering expandable valve seat pilot (Courtesy of Black and Decker Manufacturing Company).

of argument and theory regarding the centering accuracy of each style. It is even a controversial subject among automotive engineers. We will not try to provide an answer, but will present instead some different points of view.

Kwik-Way says, "None but a slow tapered arbor will properly align a valve seat reconditioning operation." Kwik-Way arbors (commonly called pilots) are microscopically tapered from end to end. When inserted into a guide, alignment is taken from the least worn section. The arbor is designed to make no contact with the worn portions at the top and bottom of the guide (Figure 15-43).

Sioux makes the comment, "The greatest possible centering accuracy can only be obtained when using a pilot that is tapered in the upper portion." See Figure 15-44.

Black and Decker valve seating equipment uses self-centering (expanding) pilots that are designed on the principle of cone centering (Figure 15-45). The pilot locates on the mean or average center of the guide. Black and Decker states, "There is no straight or unworn portion to a used guide."

VALVE SEAT GRINDING

Always check and/or recondition the guides before grinding the valve seats. Check the pilot for excessive wear and bend before grinding. Also, make sure the stone holder fits on the pilot with very little side-to-side play. Valve seats are often ground off-center to the valve guide because this procedure is not followed. Under this condition, the valves will not close and seal properly. Valve seat grinding has four basic goals.

1. The seat must be the correct width. Too wide a seat will tend to trap hot carbon particles and cause pitting. Too narrow a seat will not dissipate heat fast enough and burning results.

2. The width of a seat must be uniform all around. Any variation will cause heat transfer differences which can result in distortion and burning.

3. The seat must not have excessive runout. *Try to hold seat tir to no more than .001", even though .002" is the permissible rule-of-thumb.*

4. The seat must be located in the correct position on the valve face. The valve seat should make contact with the valve just above the

face center. A valve seat-to-face location that is too high will lead to premature burning. Too low of a location can result in a leaking valve when the engine reaches operating temperature.

The actual steps and recommendations for grinding valve seats are as follows.

1. Select the proper size and grit stone.
2. Set the dressing tool head at the desired angle (Figure 15-46). Put a drop of light oil on the dressing fixture pilot to prevent sticking. Do not get oil on the grinding wheel. Adjust the diamond dressing nib until it just touches the face of the grinding stone. Bring the stone up to speed, holding the driver motor as straight as possible. Move the diamond steadily across the wheel, taking very light cuts (Figure 15-47).
3. Select the largest size pilot that will anchor snugly on the entire length of the valve guide. Pilots are available in undersize, standard, and oversize. For example, Chart 15-1 lists pilot sizes in numerical sequence that are available from Sioux. Insert the pilot into the guide and twist gently to lock (Figure 15-48). Twist in the opposite direction to remove the pilot.
4. Place the stone and holder on the pilot (Figure 15-49). Put a drop of oil on the pilot top to help reduce friction while grinding. Use a "bounce" or "lifting" spring under the grinding stone (Figure 15-50). This helps reduce stone breakage and gives the operator

CHART 15-1

A listing of valve seat pilot sizes (Courtesy of Sioux Tools Inc.).

Size, In.	Size, In.
1/4	.380
1/4 + .001	.385
1/4 + .002	
1/4 + .003	25/64
1/4 + .004	.395
9/32	.396
9/32 + .001	.397
9/32 + .002	.398
9/32 + .003	.399
9/32 + .004	13/32—.003
5/16—.002	13/32—.002
5/16—.001	13/32—.001
5/16	13/32
5/16 + .001	13/32 + .001
5/16 + .002	13/32 + .002
5/16 + .003	13/32 + .003
5/16 + .004	13/32 + .004
	7/16—.003
21/64	7/16—.002
11/32—.002	7/16—.001
11/32—.001	
	7/16
11/32	7/16 + .001
11/32 + .001	7/16 + .002
11/32 + .002	7/16 + .003
11/32 + .003	7/16 + .004
11/32 + .004	
.358	15/32
23/64	1/2
3/8—.003	1/2 + .001
3/8—.002	1/2 + .002
3/8—.001	1/2 + .003
	1/2 + .004
3/8	
3/8 + .001	17/32
3/8 + .002	17/32 + .001
3/8 + .003	17/32 + .002
3/8 + .004	17/32 + .003
	17/32 + .004

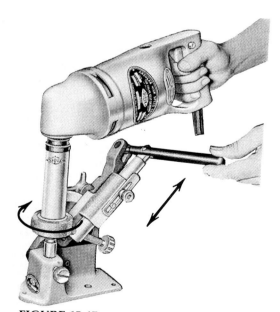

FIGURE 15-47
Dressing a valve seat grinding stone (Courtesy of Sioux Tools Inc.).

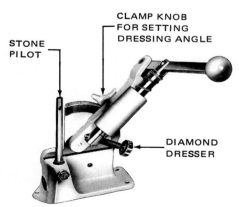

STONE PILOT

CLAMP KNOB FOR SETTING DRESSING ANGLE

DIAMOND DRESSER

FIGURE 15-46
Dressing tool (Courtesy of Sioux Tools Inc.).

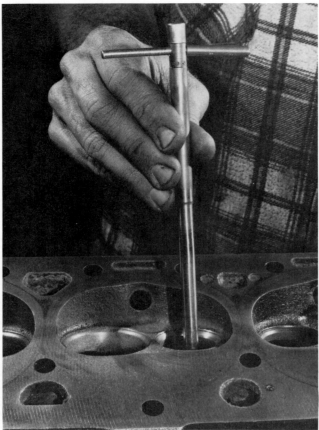

FIGURE 15-48
Installing pilot into guide. Note the pilot handle.

FIGURE 15-49
Placing the stone and holder on the pilot.

better control of the stone against the seat. Also, the spring allows the stone (while running at full speed) to be "lifted" off the seat every few seconds. This action will clear the grit and enable the stone to cut more accurately.

5. Insert the driver lug into the stone holder (Figure 15-51). Start the driver motor. When grinding, do not use heavy pressure. Use your lower hand to partially support the weight of the motor. Keep the motor held square with the other hand. Momentarily push the stone against the valve seat. *CAU-TION: It may only require several seconds of grinding to recondition a seat.* After a few revolutions of the stone, stop the motor and inspect the seat surface. Continue to grind until the seat is just free of blemishes. Then, dress and make a light finish cut to smooth up the seat surface.

6. After the seat has been satisfactorily cleaned, determine and adjust its width and valve face contact location. A good seat will show a full 360° contact pattern on the valve, and will have a *minimum of 1/32" overhang* (sometimes called rim overlap). See Figure 15-52. As a general rule-of-thumb, *automotive intake valves should have a contact pattern of 1/16" (.063"), and nonrotated exhaust valves 3/32" (.094").* Rotated exhaust valves should have the seat contact as wide as possible, while maintaining 1/32" overhang, at the top and bottom. If the valve face contact pattern is too wide, correction angles must be ground (Figure 15-53). For example, if the pattern is too wide with incorrect overhang, the seat must be "topped" *(Figure 15-54).* To accomplish this, use a 30° angle stone on 45° seats, and use a 15° angle stone on 30° seats.

FIGURE 15-50
A "lifting" spring is being used on the pilot.

FIGURE 15-51
Grinding the seat.

SHOP TIP

You can get double service from valve grinding stones by dressing both sides (Figure 15-55).

If the pattern is too wide and the overhang at the top is good, the seat must be narrowed by "throating" (Figure 15-56). This is generally done by using a 60° angle stone or cutter.

7. Check valve seat trueness with a dial indicating runout gauge. Place the indicator on the valve seat pilot, and rotate the lower section with the adjustable bar in contact with the seat (Figure 15-57). As the bar rotates, read the seat runout and note the high and low sides. Correction can sometimes be made by slightly leaning the stone and holder in the direction of the seat high side.

   Dial indicating the seats also serves as a good check on the machinist and the equipment.

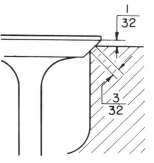

FIGURE 15-52
Correct seat location and width for a typical nonrotated exhaust valve.

Valve Seat Grinding Tips

1. Always carefully check the seat face for chatter marks and burrs if a seat cutter has been used (Figure 15-58). It may be necessary to "kiss" the seat face with a stone.

2. When machining valve seats, first rough in the seat angle. This is because the depth required to clean up the seat will determine the need for a seat ring. Then "top" with a 15° or 30° stone to establish the major seat

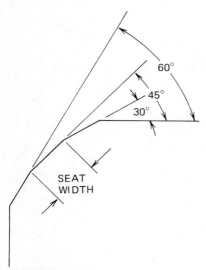

FIGURE 15-53
Valve seat correction angles.

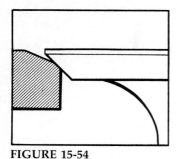

FIGURE 15-54
Narrowing the seat by "topping." Material has been removed at top of seat as shown in the illustration.

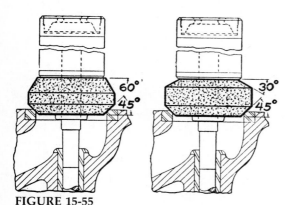

FIGURE 15-55
Both sides of a valve seat grinding stone can be dressed for double service (Courtesy of Sioux Tools Inc.).

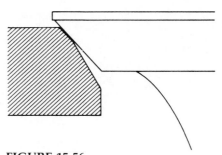

FIGURE 15-56
Narrowing the seat by "throating." Material has been removed at bottom of seat as shown in the illustration.

FIGURE 15-57
Measuring valve seat runout (Courtesy of Ford Parts and Service Division).

diameter. Next, "throat" the inside diameter to get the proper seat width.

3. After grinding the seat angle, color the surface with machinists' dye (Dycum) or with a colored felt tip marker. This will help distinguish the correction grinding angle surfaces (top and throat).

4. Valve face-to-seat contact location can be easily determined by using Prussian blue paste (Figure 15-59). Coat the valve face with paste and insert the valve in its guide. Rotate the valve several times while pushing against the seat (this will evenly distribute the Prussian blue). Remove the valve and wipe off the face. Again, insert the valve and rotate against the seat. Remove the valve and note the seating pattern on the face (Figure 15-60). *NOTE: Prussian blue will not accurately show*

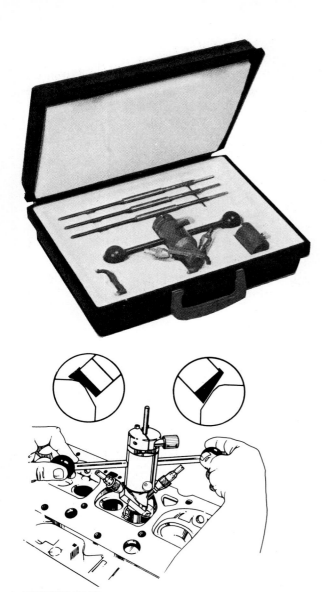

FIGURE 15-58
A carbide valve seat cutter that is designed to cut several angles at once (Courtesy of Lisle Corporation).

FIGURE 15-59
Prussian blue paste.

FIGURE 15-60
Seating pattern on valve face.

seat width if interference angles are used. Measurement will have to be made with a scale.

5. Dividers can be used to locate where the top edge of the seat is hitting the valve face. First, measure the valve head diameter with dividers. Second, reduce the divider tip spread by 1/16" (1/32" overhang on each side). Third, transfer this dimension to the valve seat (Figure 15-61). Fourth, top the seat until this seat OD is reached.

VALVE SEAT REPAIR

There are two types of valve seats.

1. **Integral** This type of seat is ground directly on the head casting material. Integral seats

FIGURE 15-61
Using dividers to determine location of valve seat on the valve face.

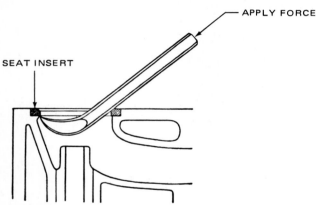

FIGURE 15-62
Removing a valve seat insert with a special tool (Courtesy of Sioux Tools Inc.).

are less costly to manufacture, and have the advantage of running cooler than the insert type.

2. **Insert** This type of seat uses a metal ring that is press-fitted into the head.

There are times when it becomes necessary to repair an integral seat. The seat may be worn beyond service limits, may be burned, or is cracked. The repair is made by machining a counterbore into the old seat and installing a seat insert. This automotive machine shop operation is commonly called *"false"* seating. The new insert is installed with a negative fit (generally about .003").

Repairing an insert-type seat begins with removing the seat ring. Sometimes the ring can be removed by prying with a hooked bar (Figure 15-62). However, it is not always possible to find a ridge under the ring to hook onto. This makes it difficult to remove the ring without damaging the bore. There is another good method to remove insert-type seats. Arc weld a bead completely around the valve seating surface of the insert. As the bead cools, it will shrink and cause the insert to loosen and fall out.

Typical "False" Seat Installation Procedure

1. Insert counterboring pilot (1) into valve guide. Mount base fixture (2) on the cylinder head. See Figure 15-63a.

2. Use a micrometer to set the cutter blades to the required counterbore size (Figure 15-63b). *COMMENT: Some machines use a fixed size cutter, so sizing is not necessary.* You just select a certain cutter for a specified seat. The press-fit clearance is built into the valve seat.

3. Place the counterboring tool on the pilot and base fixture (Figure 15-63c). Adjust the depth of cut by using the insert as a measuring gauge (3) between the sizer knob and stop ring.

4. Cut counterbore in the valve seat by turning the feed screws (4) until the sizer knob bottoms on the stop ring (5) (Figure 15-63d).

5. Remove the counterboring tool. Install the valve seat insert by using the ring driver (6) and ring driver pilot (7) (Figure 15-63e).

6. Replace the cutter blades with the proper size "spinner" tool (8). Remount the counterboring tool. Turn the feed screw to allow the "spinner" tool to "roll" (sometimes called peening) head metal over the new valve seat insert (Figure 15-63f).

Valve Seat Installation Hints

1. Table 15-2 can be used to determine the interference fit for hard cast or wrought inserts installed in a cast iron head.

TABLE 15-2
RECOMMENDED VALVE SEAT INTERFERENCE FIT

Insert OD	Insert Depth	Interference Fit
1–2″	¼–⅜″	.002–.004″
2–3″	⅜–⁹⁄₁₆″	.003–.005″
3–4″	⁹⁄₁₆–1″	.004–.006″

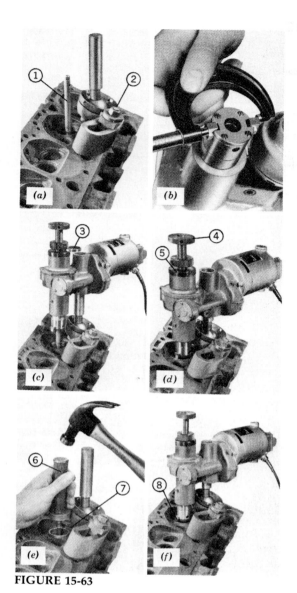

FIGURE 15-63

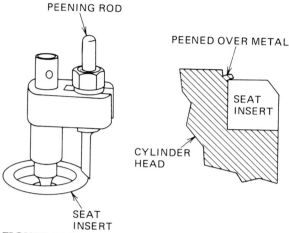

FIGURE 15-64

This equipment is used to lock a valve seat in position by "rolling" metal over the edge. Tap the top of the rod with a hammer while turning the tool on the pilot until the entire seat is peened (Courtesy of Sioux Tools Inc.).

2. Shrink inserts by freezing them in dry ice (−112° F), or by chilling with Freon. This procedure makes installation easier, and also helps prevent broaching the walls of the bore. *COMMENT: A seat ring will have a small chamfer (or sometimes a radius) on one outside edge. This chamfer must face down on installation (this minimizes shaving the counterbore wall).*

3. Aluminum heads require a .001" to .002" greater interference fit. When you install a seat in aluminum, heat the port to 300° F with a propane torch. This will expand the counterbore, allowing the insert to be tapped in.

4. If the interference fit with a standard size replacement seat is below specifications, it will be necessary to use an oversize seat. Inserts are generally available in .005", .010", and .015" oversizes. Sometimes, other oversizes are also available. Always check the availability of oversize seats for the particular engine you are working on. To randomly select a seat ring to match a certain cutter in the shop, often leads to cutting into the water jacket.

5. Very often, this question is asked, "At what point should a valve seat be replaced?" Generally, it is recommended to install a seat insert if valve spring installed height cannot be corrected with a .060" shim.

6. It is advisable to "roll" or stake insert seats in place (especially when installed in aluminum). This will prevent the seats from falling out should they loosen. See Figure 15-64.

7. After a cylinder head is disassembled, always check insert-type seats for looseness. Pry under the seat with a chisel and look for movement, or lightly hit the insert with a hammer and feel for movement with your fingers.

8. On aluminum heads, visually check the combustion chamber for coolant related erosion. If erosion is excessive, or has exposed the side of any valve seat, the cylinder head must

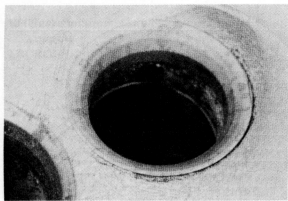

FIGURE 15-65
Unacceptable combustion chamber.

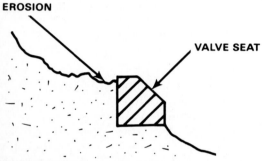

FIGURE 15-66
Unacceptable combustion chamber.

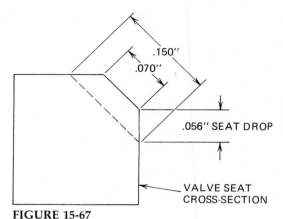

FIGURE 15-67
The valve seat will drop as its width increases (Courtesy of Automotive Engine Rebuilders Association).

be repaired or replaced (Figures 15-65 and 15-66).

9. If intake and exhaust "false" seats are being installed in the same combustion chamber, install the intake seat ring first. Intake seat rings are a softer material, whereas the exhaust ring is often a hard cast iron or alloy steel. If the two seats overlap on installation, the counterboring tool may not be able to cut through the exhaust insert.

A Quick Check Method of Determining Valve Seat Drop

A cylinder head comes into the shop for a valve job. You grind a seat with a 45° stone, and it cleans up at .150" in width. How much further down in the head is the valve seat from its original location?

This problem can be figured by simple geometry. First, we are going to assume that the original seat was around .070" wide. This means the cleaned up seat is now .080" wider (.150" − .070" equals .080"). Divide .080" by 2. This gives .040" as the width increase on each side of the seat. Multiply .040" by the diagonal constant of 1.414. This figures to be .056". The valve stem height is automatically down .056" from original location (Figure 15-67). Use this procedure as a guideline on heads for which you do not have any valve stem height specification. This will give some idea of valve "tipping" requirements.

Valve Seat Insert Materials

Valve seat inserts are cataloged according to material and dimension (Figure 15-68). In most cases, there are three groups of inserts available.

1. **High-grade iron (8000 series)** Recommended for most intake applications, and for most light- to medium-duty exhaust applications.

2. **Hi-chrome steel alloy (7000 series)** Recommended for use in engines where operating conditions are more severe than average. This includes engines operated on unleaded gasoline and liquified petroleum gas.

3. **Stellite (3000 series)** Recommended for engines subjected to extreme high temperatures, high seating forces, and adverse corrosion conditions. This includes many diesel and industrial application engines. These heavy-duty valve seats are made from a nickel base chromium-tungsten alloy which is nonmagnetic.

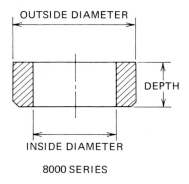

8000 SERIES

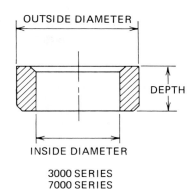

3000 SERIES
7000 SERIES

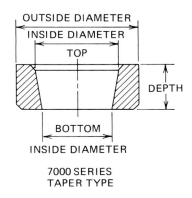

7000 SERIES
TAPER TYPE

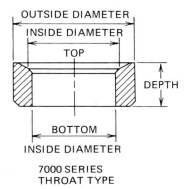

7000 SERIES
THROAT TYPE

FIGURE 15-68
Valve seat insert shapes (Courtesy of Dana Corporation—Service Parts Group).

Higher valve and seat temperatures have resulted from the many emission control devices being added to engines. Because of this, most passenger car engines now use induction hardened integral seats.

VALVE DAMAGE ANALYSIS

Basically the two types of valve failure are burning and breakage.

Burning

Burning is the most common mode of exhaust valve failure. *Burning can only occur when the temperature is raised above the melting point of the valve metal.* See Figure 15-69 for the temperature profile of an exhaust valve. Some conditions leading to burn failure are given below.

1.   **Preignition**   This will cause rapid face melting. A good example of the massive damage

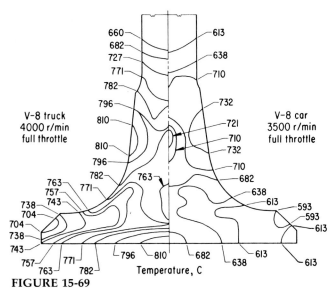

FIGURE 15-69
Temperature profile of an exhaust valve operating in a typical car and truck engine (Courtesy of TRW Valve Division).

FIGURE 15-70
Preignition damage (Courtesy of TRW Replacement Division).

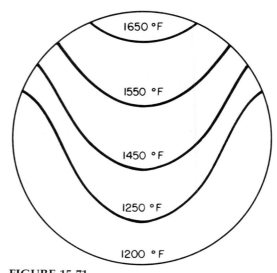

FIGURE 15-71
Effect of gas channeling on valve head temperature pattern (Courtesy of TRW Replacement Division).

FIGURE 15-72
Burned valve face.

caused by preignition is shown in Figure 15-70.

2. **Leakage due to poor seating** Poor seating will result in the leakage of hot gases at various points around the valve face. When this local leakage occurs, extreme valve face and head temperature patterns are developed (Figure 15-71). That portion of the valve face in the hot gas channel will progressively burn away (Figure 15-72).

3. **Deposit accumulation** Leakage channels also can be caused by hard combustion deposits and adhesive wear "warts" (Figures 15-73 and 15-74).

4. **Seat distortion** This is one of the most prevalent causes of burning. During engine operation, thermal and mechanical forces will mildly distort the valve seat. If gas pressure and spring load are not sufficient to conform the valve to the seat, the valve will leak. Also, bending stress in the stem-to-head blend area will result (Figure 15-75). A breakage failure due to fatigue can occur then.

5. **Valve face corrosion** Chemical erosion can create a pattern of depressions that will permit a gas escape route (Figure 15-76).

6. **Overheating stress** Sometimes excessive heat will not burn a valve. Instead, the valve head becomes deformed like a cup (Figure 15-77).

FIGURE 15-73
Closeup of exhaust valve face shows "wart" formation caused by adhesive wear. Particles of cast iron were welded to the face and were pulled away from the cylinder head. The particles then oxidized (Courtesy of TRW Valve Division).

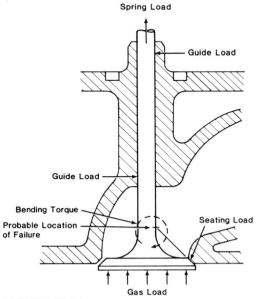

FIGURE 15-74
Face guttering due to flaked-off deposits (Courtesy of TRW Replacement Division).

FIGURE 15-76
Valve face corrosion (Courtesy of TRW Replacement Division).

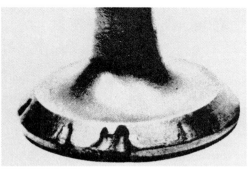

Spring Load

Guide Load

Guide Load

Bending Torque

Probable Location of Failure

Seating Load

Gas Load

FIGURE 15-75
Valve bending loads created by valve seat distortion (Reprinted with Permission, Copyright 1980 *Automotive Engineering*, Society of Automotive Engineers, Inc.).

FIGURE 15-77
Cupping (Courtesy of TRW Replacement Division).

FIGURE 15-78
"Hoop" stressing.

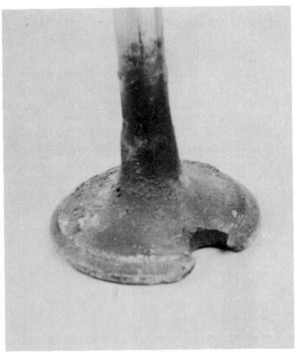

FIGURE 15-79
When an air-pump equipped engine is overheated this often results.

"Hoop" stressing is a fairly common overheating failure (Figure 15-78). It is caused by valve temperatures being much higher than the valve was designed to withstand. A radial crack in the valve face rim then results as the center of the valve head cools more slowly than the outer edge.

Figure 15-79 shows what can happen when an air-pump equipped engine is overheated. If an engine operating at high load is abruptly shut down, thermal shock can develop and cause valve head and manifold warpage and gasket failure. This is especially true of turbocharged engines. It is recommended that an engine cool-down period (at idle speed) of a few minutes be practiced prior to shutdown of these engines.

Breakage

Valve breakage failures can be of a pure mechanical nature or directly related to an inherited thermal or chemical problem. The common causes are listed here.

1. **Flexing**   Off-square seating and high seating forces are two common reasons for valve failure. The alternating stress cycles then cause a fatigue fracture (Figure 15-80).
2. **High seating forces**   Excessive valve stem-

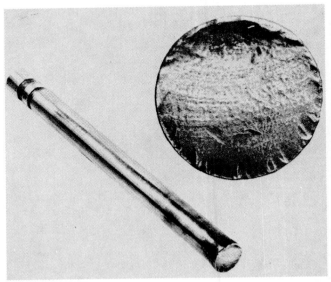

FIGURE 15-80
Valve stem failure due to fatigue from repeated flexing (with magnified cross-section of break). (Courtesy of TRW Replacement Division).

FIGURE 15-81

An improperly adjusted valve caused this stem to break at the keeper groove (Courtesy of TRW Replacement Division).

an engine that has been destroyed from a "dropped" valve, be sure to inspect the exhaust system of the vehicle—it may be full of metal. Whatever is in the exhaust system can be drawn back into the engine during high-speed closed throttle conditions.

4. **Necking** Underhead erosion is a frequent cause of valve failure. Figure 15-84 shows stem pitting and neck-down that has resulted from erosion. If the process had continued, the valve would have broken. Necking is generally an indication of elevated operating temperatures or poor seating.

5. **Valve stem "metal pick-up" and scoring (Figure 15-85).** Lack of lubrication can lead to this condition. Then the valve will either stick and be hit by the piston, or it will act like a reamer and produce excessive guide clearance. Oil flow routing to the valve stem is critical. If not enough oil gets into the valve guide to lubricate the valve stem, the valve will seize. A case in point is the Toyota 2T engine. Early production engines were plagued with valve seizures. Toyota solved the problem in an interesting manner. Valve stem diameter was reduced by .0002" for more oil passage, and the raised rib on the rocker

to-guide clearance, out-of-square valve springs, excessive valve lash, and worn rocker arm faces cause extra cocking and seating forces on valve stems. A broken stem end is often the result (Figure 15-81).

3. **Impact failure** This is the result of a high physical load suddenly imposed on the valve. Faulty keeper installation, a timing chain or belt that breaks, operating an engine beyond its normal rpm range, or hitting a foreign object are common causes of impact damage (Figures 15-82 and 15-83). *CAUTION: On*

FIGURE 15-82
Impact failure of the valve stem (with magnified cross-section of break). (Courtesy of TRW Replacement Division).

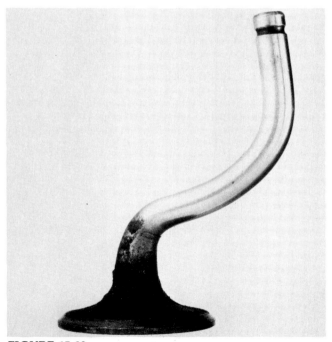

FIGURE 15-83
This bend damage resulted when a timing belt broke.

FIGURE 15-84

An example of severe necking. High temperature gases have attacked this valve neck and eroded it.

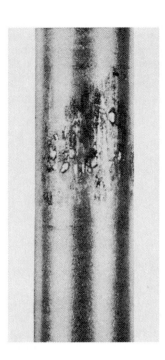

FIGURE 15-85
Scored valve stem (Courtesy of TRW Replacement Division).

arm was moved from the center to the outer edge (Figure 15-86). This provided a more direct oil drip flow on the valve stem end.

Use the preceding valve damage analysis information as a guide. Always try to isolate the cause for failure. To just replace parts does not remove the cause, and can quickly result in a repeat failure.

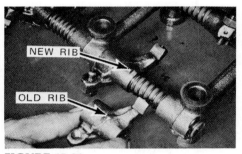

FIGURE 15-86
Rocker arm ribs are sometimes used to direct oil flow on the valve stem end.

TRUE CASE HISTORY

Vehicle	Oldsmobile
Engine	8 cylinder, 455 CID
Mileage	17,000
Problem	After being overheated, the engine started to idle very rough, shake on take-off, and smoke from the tailpipe.
Cause	An exhaust seat insert became loose, dropped out of position and wedged under the valve head. See Figure 15-87.

FIGURE 15-87
Engine overheating caused this seat ring to come out.

CHAPTER 15 SELF-TEST

1. Exhaust valves and seats are exposed to what combusion temperatures?
2. List five problems caused by off-center valve seating.
3. What are canted valves?
4. What is the advantage of using sodium-filled exhaust valves in an engine? What is a disadvantage?
5. What are aluminized valves?
6. Why shouldn't aluminized valves be ground?
7. Why will aluminized valves leak when performing a postvalve job leak test?
8. What is the major advantage to swirl-finishing valves?
9. Why are many manufacturers hard chroming valve stems?
10. List six advantages of valve rotation.
11. Why shouldn't valve rotation be used in conjunction with liquified petroleum gas?
12. What is a natural rotator? Give three examples.
13. What is a positive type rotator? Give two examples.
14. Multigroove valve stems are prone to what wear condition?
15. How often should positive type rotators be changed?
16. How can rotators be checked for proper operation with a timing light?
17. What is valve margin?
18. As a general rule, margin shouldn't be less than _____".
19. Maximum allowable valve stem wear is _____".
20. How much of a cut should be made per pass when dressing a valve refacing wheel?
21. What is the result of dressing a valve refacing wheel with a rapid traverse of the diamond?
22. How can the inside of the chuck head be protected from grit while refacing valves?
23. What is meant by "spark-out"?
24. When inspecting the valve face during grinding, why shouldn't the valve be pulled off the stone?
25. When dressing, why is it desirable to have coolant directed on the wheel?
26. Where should coolant flow be directed when grinding valve faces?
27. What is the best coolant to use when refacing valves?
28. What usually causes a poor finish on valves?
29. The chamfer on the valve stem end should be _____".
30. Maximum allowable valve face runout is _____".
31. After refacing, the margin around the valve head is uneven. What does this indicate?

32. How can the accuracy of a valve refacing machine be quickly checked out?
33. Why is the use of valve lapping compound not recommended?
34. What is an interference angle?
35. Many people do not like interference angles. Why?
36. What are three reasons for "tipping" valve stem ends?
37. What is rocker arm geometry?
38. Name three companies that sell valve stem height gauges.
39. A set of V-8 heads from an engine with nonadjustable hydraulic lifters is brought into the shop for a valve job. You do not have any valve stem height specification information for this engine. What procedure should be followed?
40. What should be done if a head is apart and the stem length is unknown?
41. When adjusting stem height, what allowance must be made if the head has been surfaced?
42. Give two reasons why it is not recommended to cut more than .030" from the end of a valve stem.
43. Most intake valves use what base material? Most exhaust valves?
44. What is stellite?
45. What simple test will determine if a valve is hard-faced?
46. How are valve seat stones cataloged?
47. What three points of view regarding the centering accuracy of valve seat pilots are presented in the chapter?
48. Why must valve guides always be reconditioned before grinding the valve seats?
49. What are the four basic goals of valve seat grinding?
50. Maximum allowable valve seat tir is _____".
51. At what location should the valve seat make contact with the valve face?
52. Too low of a valve seat-to-valve face location can result in what problem?
53. What is the reason for putting a drop of oil on the top of the valve seat pilot when grinding a seat?
54. What is a "bounce" spring? Why is it used?
55. What is rim overlap? Why is it important?
56. Intake valves should have a seat contact width of _____".
57. Nonrotated exhaust valves should have a seat contact width of _____".
58. What is "topping"?
59. What is "throating"?
60. Describe the procedure for using Prussian blue to determine valve face-to-seat contact location.
61. How can dividers be used to determine where the top edge of the seat is hitting the valve face?

62. What are the two types of valve seats?
63. What is a "false" seat?
64. What two methods are used to remove a seat ring?
65. What is the reason for peening a seat ring?
66. When installing a seat ring, what direction must the chamfer face?
67. Seat inserts are generally available in what oversizes?
68. What methods can be used to shrink seat inserts?
69. At what point is it generally recommended to install a seat insert?
70. How can insert-type seats be checked for looseness?
71. If intake and exhaust "false" seats are being installed in the same combustion chamber, what seat ring should be installed first?
72. What are the three groups of valve seat insert materials generally available?
73. What are the two basic types of valve failure?
74. List six conditions that can lead to a burn failure.
75. List five possible causes of a valve breakage failure.
76. Why is it important to check the exhaust system of the car when an engine "drops" a valve?

CHAPTER 16

VALVE COMPONENTS

Valve components discussed in this chapter include rocker arms, push rods, valve spring retainers, keepers, and valve springs.

ROCKER ARMS

Basically, a rocker arm has two purposes.

1. It acts as a reversing lever to transmit camshaft lobe lift motion to open the valve (Figure 16-1). OHC engines eliminate the need for push rods and, in some cases, the valve lifters and rocker arms (Figures 16-2, 16-3, and 16-4).

2. It provides a movement advantage during valve lift. Rocker arms generally pivot in an off-center position. This means that one end of the arm will move through a longer arc of travel than the other end. If one end rotates $1\frac{3}{4}$ the distance of the other end, the rocker arm lift ratio is 1.75:1. If a cam has a lift of .270", actual valve lift is then computed by multiplying .270 by 1.75. The valve lift would be .472". See Figure 16-5. *NOTE: If the en-*

gine does not use rocker arms, valve lift will be the same as cam lift.

Rocker arm ratios vary with different engines. Table 16-1 (Page 336) lists the rocker arm ratios for some popular domestic model engines.

Stamped-Steel Rocker Arms

Stamped-steel (die-formed) rocker arms are used on many engines. This type of rocker arm reduces valve train reciprocating weight, and is cheaper to manufacture. After the rocker arm is stamped, the wear points are surface hardened. Refer to Figures 16-6 and 16-7 for some examples of stamped-steel rocker arms.

Cast Rocker Arms

Cast type rocker arms are made from either cast iron, aluminum, or magnesium. Most cast type rockers are shaft mounted (Figure 16-8). However, some stud mounted rockers are cast (Figure 16-9). The rocker arm is sometimes bushed with a brass or bronze insert (Figure 16-10).

333

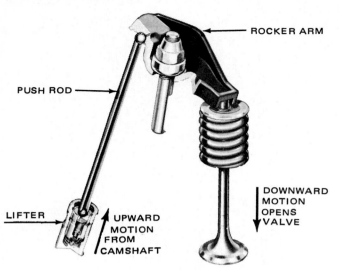

FIGURE 16-1

The lifter and push rod are lifted by the camshaft. This upward motion raises one end of the rocker arm. In turn, the other end of the rocker arm moves down to press the valve open (Courtesy of Ford Parts and Service Division).

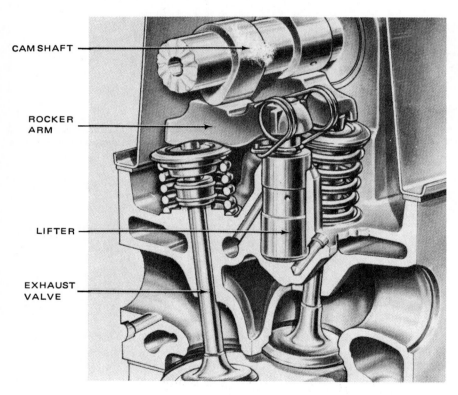

CAMSHAFT

ROCKER ARM

LIFTER

EXHAUST VALVE

FIGURE 16-2

Pinto OHC valve train system (Courtesy of Ford Parts and Service Division).

Silent Lash (or Zero Lash) Rocker Arms

Some older Ford six-cylinder engines use a novel design rocker arm (Figure 16-11). A hardened steel eccentric presses against a plunger to dampen "click" noise as valve train clearance is removed prior to valve opening.

Rocker Arm-to-Cylinder Head Mounting Methods

Rocker arms are mounted on the cylinder head by one of several methods.

1. Stud mounted with a ball pivot (Figure 16-

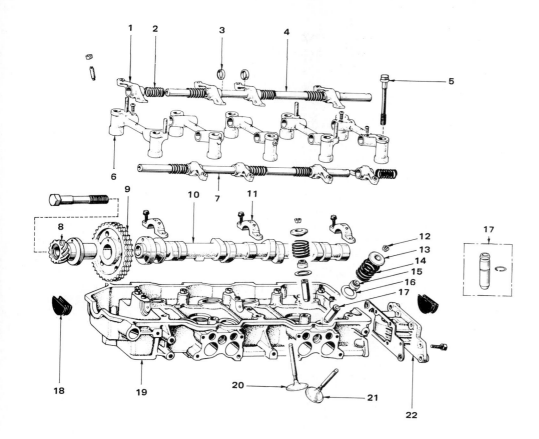

1. ROCKER ARM	12. VALVE KEEPERS
2. SPRING	13. SPRING RETAINER
3. SPACER	14. VALVE SPRING
4. ROCKER SHAFT (INTAKE)	15. VALVE SEAL
5. HEAD BOLT	16. SPRING SEAT INSERT
6. ROCKER STAND	17. VALVE GUIDE
7. ROCKER SHAFT (EXHAUST)	18. HALF—CIRCLE SEAL
8. DISTRIBUTOR DRIVE GEAR	19. CYLINDER HEAD
9. CAM SPROCKET	20. INTAKE VALVE
10. CAMSHAFT	21. EXHAUST VALVE
11. CAMSHAFT BEARING CAP	22. REAR COVER (EGR COOLER)

12). The stud is pressed or threaded into the cylinder head. On some engines, the stud supplies oil to the ball pivot through a feed hole (Figure 16-13). On other engines, the ball pivot receives oil through a hollow push rod. The rocker arm pivots on the ball, transmitting cam lift into valve lift.

2. Pedestal mounted with a fulcrum pivot (Figure 16-14). The rocker arm is bolted to a pedestal that is cast into the head. This attaching bolt holds the fulcrum inside the rocker arm. The rocker arm pivots on the bottom surface of the fulcrum.

3. T-shape support mounted. Cadillac uses a T-shape support that carries two rocker arms (Figure 16-15). Clips on a retainer plate hold the rocker arms on their supports. Oldsmobile uses a somewhat similar rocker system (Figure 16-16).

4. Shaft mounted (Figure 16-17). Shaft mounted assemblies have individual rocker arms positioned on a shaft. The assembly is suspended by stands or saddles cast in the cylinder head. Each rocker arm is correctly spaced on the shaft by thickness spacers and springs.

Rocker arms and/or stands are sometimes

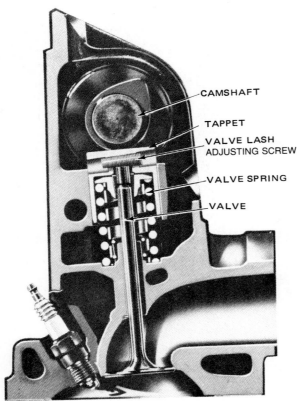

CAMSHAFT

TAPPET

VALVE LASH
ADJUSTING SCREW

VALVE SPRING

VALVE

FIGURE 16-4
OHC valve train assembly for the Chevrolet Vega engine
(Courtesy of Chevrolet Motor Division of General Motors
Corporation).

TABLE 16-1
ROCKER ARM RATIO EXAMPLES

Make/Displacement	Rocker Arm Ratio
AMC 258, 304, 343, 360, 390, 401	1.60
Buick 350	1.55
401, 455	1.60
400, 430	1.59
Cadillac 472, 500	1.72
429	1.65
Chevrolet 153, 230, 250	1.75
396, 427, 454	1.70
307, 327, 350, 400	1.50
Chrysler 225, 318, 340, 360, 383, 440	1.50
Ford 98	1.54
122	1.60
170, 200, 250	1.50
240, 302, 351W	1.61
351C	1.71 or 1.73
400, 429, 460	1.71
Oldsmobile 330, 350, 400, 425, 455	1.60
Pontiac 250	1.75
307, 350, 389, 421, 428, 455	1.50
400	1.50 or 1.65

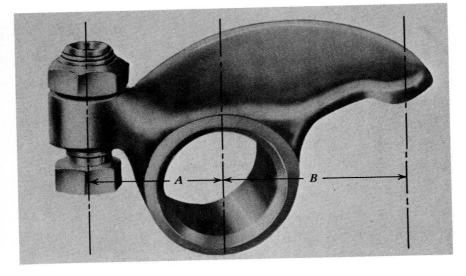

FIGURE 16-5
The distance from the center of the rocker
arm pivot to the ends is not the same.
Thus, the rocker arm becomes a lever
and has lift ratio.

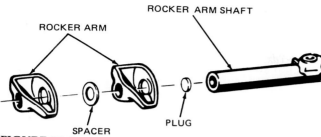

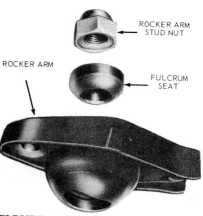

FIGURE 16-6

Shaft mounted stamped-steel rocker arms (Courtesy of Chrysler Corporation).

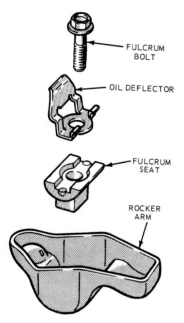

FIGURE 16-7

A stamped-steel rocker arm with related parts (Courtesy of Ford Parts and Service Division).

FIGURE 16-9

Cast rocker arm for 302 CID and 351W Ford engines (Courtesy of Ford Parts and Service Division).

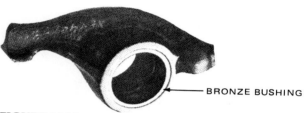

FIGURE 16-10

This rocker arm uses a bronze bushing. If there is wear, the bushing can be replaced.

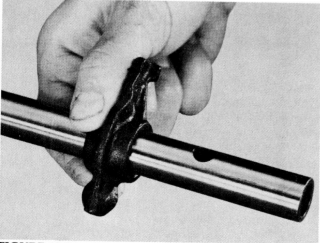

FIGURE 16-8

Removing a cast-type rocker arm from its shaft (Courtesy of Ford Parts and Service Division).

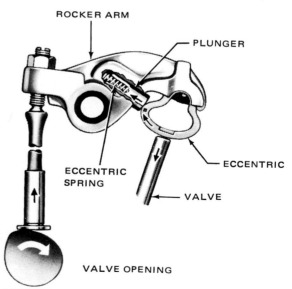

FIGURE 16-11

Operation of the silent lash rocker arm used on some older Ford six-cylinder engines (Courtesy of Ford Parts and Service Division).

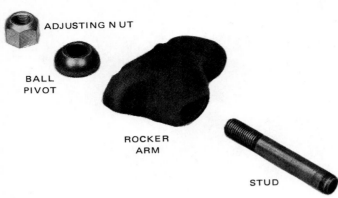

FIGURE 16-12
This rocker arm mounts on a ball pivot and stud arrangement.

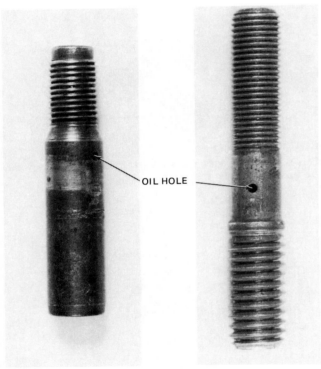

FIGURE 16-13
An oil hole is provided in these studs for ball pivot lubrication.

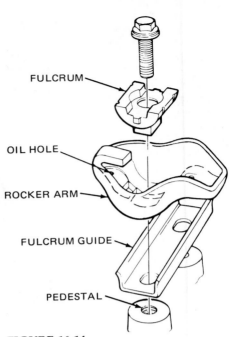

FIGURE 16-14
The 351C Ford engine uses a pedestal mounted rocker with a fulcrum pivot (Courtesy of Ford Parts and Service Division).

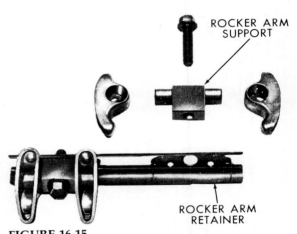

FIGURE 16-15
472 CID Cadillac rocker arm assembly (Courtesy of Cadillac Motor Division of General Motors Corporation).

designated for right- or left-hand positions (Figures 16-18 and 16-19). They must be installed in the proper location or push rod alignment will not be correct.

Rocker arm shafts usually have one side drilled with lubrication holes (Figure 16-20). *To ensure proper lubrication, these holes must face the right direction (generally down) when the shaft is installed.* The engine manufacturer may mark the end of the shaft for a visual aid in determining correct position (Figure 16-21). Note the rocker arm lubrication circuit shown in Figure 16-22.

FIGURE 16-16
Oldsmobile uses die-cast retainers for rocker alignment.

RETAINER

FIGURE 16-17
A shaft mounted rocker arm assembly. These particular rocker arms and stands are made of aluminum.

FIGURE 16-18
Left- and right-hand rocker arms are used on some engines (Courtesy of Chrysler Corporation).

ROCKER ARMS—RIGHT

ROCKER ARMS—LEFT

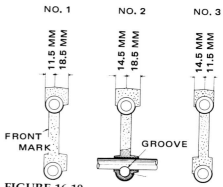

NO. 1   NO. 2   NO. 3

11.5 MM   18.5 MM   14.5 MM   18.5 MM   14.5 MM   11.5 MM

FRONT MARK

GROOVE

FIGURE 16-19
Note the rocker arm stand direction for the Toyota 2T engine (Courtesy of Toyota Motor Sales U.S.A., Inc.).

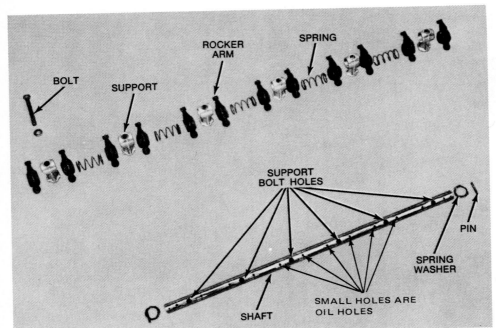

FIGURE 16-21
Installed position of rocker arm and shaft assembly on a Mercury V-8 engine (Courtesy of Ford Parts and Service Division).

Rocker Arm Lubrication

Oil flow to the rocker arm assembly usually occurs in one of four ways.

1. Through a hollow push rod (Figure 16-23).

2. Through a drilled hole in the cylinder head (Figure 16-24). *NOTE: When installing this type of rocker shaft assembly, make sure the rocker stand with the oil passage is positioned over this hole.*

3. Through a special cap screw (Figure 16-25). Oil is supplied to the rocker shaft by a hollow or undercut rocker stand screw that threads into an oil passage in the head.

4. Through an external oil supply line (Figure 16-26).

Chrysler import cars and trucks (2000 and 2600 cc engines) have recently had the rocker arms changed from cast iron to aluminum. Along with this change, longitudinal oil grooves have been added to the rocker arm shafts (Figure 16-27). *CAUTION: Aluminum rocker arms cannot be used on shafts that do not have these added oil grooves.*

Refacing Rocker Arms

The rocker arm-to-valve stem contact surface should be inspected for grooves, pits, scoring, and wear (Figure 16-28). In many cases, the contact surface can be refaced on a valve grinding machine. A swivel attachment is used to move the rocker arm back and forth against the grinding wheel until the desired surface is obtained (Figure 16-29). *NOTE: It is not recommended to reface stamped-steel or laminated rocker arms because*

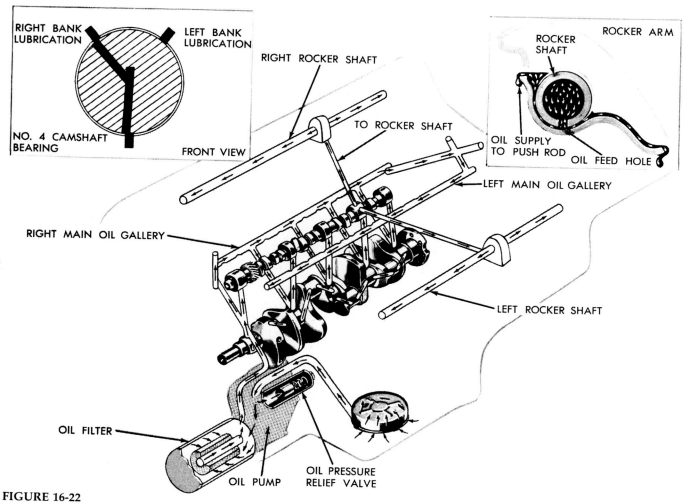

FIGURE 16-22

Engine oiling system for Plymouth 400-440 CID engines (Courtesy of Chrysler Corporation).

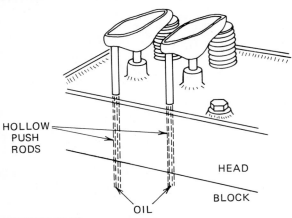

FIGURE 16-23
Oil supplied through hollow push rods.

FIGURE 16-24
Oil hole in cylinder head.

FIGURE 16-26
In the Toyota 18R engine, oil is supplied to the rocker assembly through a special bolt and crossover line. A lubrication tube with drilled nozzles is used to spray oil on the camshaft.

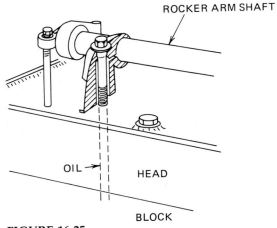

FIGURE 16-25
Oil supplied through an undercut rocker stand cap screw.

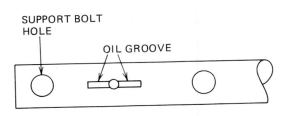

NOTE: THE LOWER SURFACE OF THE ROCKER SHAFT IS SHOWN

FIGURE 16-27
Longitudinal oil grooves on the lower rocker arm shaft surface (Courtesy of Chrysler Corporation).

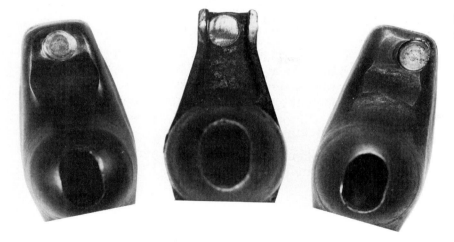

FIGURE 16-28
Worn rocker arm faces (pads).

FIGURE 16-29
Grinding rocker arm face (Courtesy of Sioux Tools Inc.).

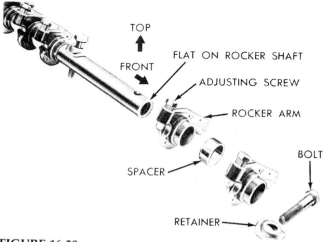

FIGURE 16-30
The rocker arms used on Chrysler Corporation 225 CID slant-6 engines are laminated. Resurfacing is not recommended (Courtesy of Chrysler Corporation).

they are only surface hardened. They should be replaced if excessive wear is present. See Figure 16-30. Some shops like to reface rocker arms on a sanding disc, or use the side of a grinding wheel. This is not good practice, because the proper face radius cannot be restored by "free-hand" grinding.

Inspecting Overhead Camshaft Rocker Arms

Rocker arms for OHC engines are subjected to heavy scuff loads. Carefully check for scoring or wear on the rubbing surfaces. *CAUTION: OHC rocker arms will wear-mate on the camshaft lobes. If the rocker arms are being reused, it is very important to return them to original position.* Failure to do this may cause quick lobe damage. Replace those rocker arms that are scored deeply enough to catch a fingernail, or that have a definite "wear line" across the camshaft contact surface (Figure 16-31). Such wear will eventually flatten camshaft lobes. Replace the camshaft if scoring on any lobe is deep enough to catch a fingernail (Figure 16-32).

Inspecting Rocker Arm Pivot Balls

Carefully examine rocker arm pivot balls. When evidence of excessive wear, galling, or burn is found, replacement is necessary.

As a pivot ball wears, it may form a near perfect seal against the inside of the rocker arm. This can cause excessive oil consumption through the valve guides due to oil being retained and spilling over the rocker lip. Also, the rocker arm and/or pivot can be badly scored because of improper lubrication. Some shops install only grooved pivot

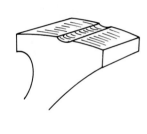

FIGURE 16-31
Unacceptable rocker.

balls when replacement is necessary (figure 16-33). The grooves carry oil and allow for better lubrication.

Inspecting Rocker Arm Stud Adjusting Nuts

When removing a stud nut, notice the turning resistance required. If the nut turns free with little or no effort, replace it. On positive stop nuts, carefully inspect the beveled stop surface (Figure 16-34). *Maximum wear for this surface is 1/16" wider than original.*

Inspecting Rocker Arm Adjusting Screws

If the rocker arm adjusting screw (Figure 16-35) breakaway torque is less than specified, install a new screw and recheck. If the torque is still too low, the rocker arm must be replaced.

Rocker Arm Breakage

Broken rocker arms are generally caused by one of the following reasons.

1. **Excessive valve spring tension** When tension exceeds the design limit of the arm, breakage will occur (Figure 16-36).

2. **Improper rocker shaft assembly removal procedure** When removing the rocker shaft support attaching bolts, loosen in sequence in two or three steps. Otherwise, binding may result and cause rocker arm and/or shaft breakage.

3. **Valve spring coil bind** When using an inner-outer valve spring arrangement, coil bind sometimes occurs. This is a condition where the coils compress together and stack solid. As the rocker arm attempts to move the valve against the stacked spring, broken parts result. *NOTE: At full open valve spring position*

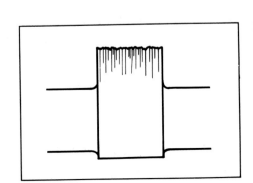

FIGURE 16-32
Unacceptable cam lobe.

FIGURE 16-33
Grooved rocker arm pivot balls aid top end lubrication.

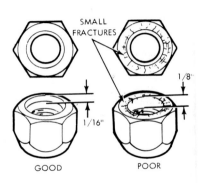

FIGURE 16-34
Inspection of rocker arm stud nut (Courtesy of Ford Parts and Service Division).

ADJUSTING
SCREW

you should be able to insert a .010" feeler gauge between at least five coils. See Figure 16-37.

4. **Valve spring-to-retainer interference** Any contact between the bottom of the valve spring retainer and the top of the guide will be disastrous to valve train parts (Figure 16-38). *Allow a clearance margin of .100".*

5. **Improper valve clearance** With excessive valve clearance, the impact between the valve tip and rocker arm pad will produce higher than normal wear, which can lead to eventual breakage.

6. **Poor lubrication** A shaft mounted rocker arm will show the greatest wear on the bottom surface of the pivot hole. The rocker shaft will also show the greatest wear on the bottom (Figure 16-39). *NOTE: As a general rule, rocker arm-to-shaft clearance should not exceed .005".* As a rocker arm and shaft wear, an oil hole ridge starts to form. If this ridge becomes high enough, oil flow will be shut off. Rocker arm seizure and breakage will then follow.

7. **Rocker arm-to-stud interference** This type of problem is generally associated with stud mounted rockers and a high-performance camshaft. With higher than normal cam lift, the clearance between the ball stud and the rocker slot end is reduced at full valve lift. If contact occurs in this area, stud and/or rocker arm damage will result. When using a high-performance camshaft, always check for interference in this location. (See Chapter 14.)

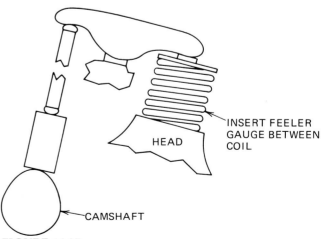

FIGURE 16-36
Rocker arm breakage caused by excessive valve spring tension (Courtesy of Dana Corporation—Service Parts Group).

INSERT FEELER
GAUGE BETWEEN
COIL

HEAD

CAMSHAFT

FIGURE 16-37
Check the clearance between each active coil with a feeler gauge when the valve is full open.

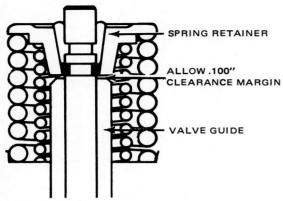

FIGURE 16-38
Valve guide-to-retainer clearance must be checked when using a high-lift camshaft.

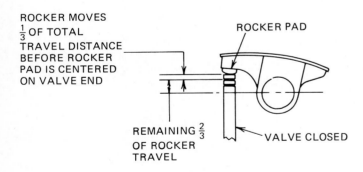

FIGURE 16-39
Rocker arm shaft wear (Courtesy of Dana Corporation—Service Parts Group).

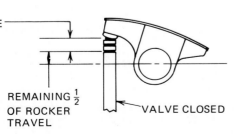

Rocker Arm Geometry

Basically, rocker arm geometry can be defined as the contact position of the rocker arm on the valve stem. Geometry is considered correct when the center of the rocker arm tip radius coincides with the centerline of the valve stem at 40% to 50% of full valve lift (Figure 16-40). If the rocker arm tip does not center on the valve stem just right, excessive side thrust will cause rapid valve guide wear. Incorrect rocker arm geometry has several causes.

1. Excessive block and/or head surfacing.
2. Excessive valve seat grinding.
3. Installing a high-lift camshaft.
4. Worn rocker arms.
5. Worn push rods.
6. Screwing in a rocker arm adjusting screw to the limit of its travel (applies to engines where the rocker arm pivot point is fixed by a rocker shaft). The cure is a longer push rod.

Correcting rocker arm geometry can be accomplished in several ways.

1. Using a longer or shorter push rod.
2. Grinding material from the valve stem end.

FIGURE 16-40
The angular relationship of the top surface of the valve stem to the rocker arm, when the valve is closed, is important when assembling an engine. Most OEM engines use a $\frac{1}{3}:\frac{2}{3}$ relationship (top illustration). For camshafts with more than .550" lift, a split setup is recommended (lower illustration). This angular relationship is what determines rocker arm geometry.

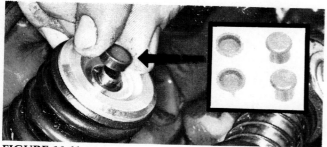

FIGURE 16-41

Lash caps can be used to correct rocker arm geometry and to help prevent valve stem "mushrooming". These caps are .050" thick, ground and heat-treated.

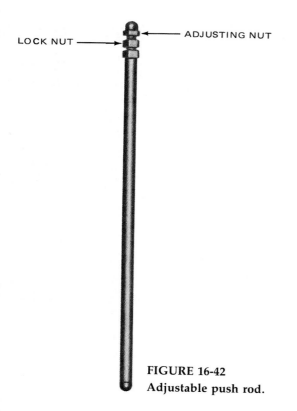

LOCK NUT —
ADJUSTING NUT

FIGURE 16-42
Adjustable push rod.

3. Adding material to the valve stem end by using a lash cap (Figure 16-41).
4. Milling material off the rocker arm stand bases.
5. Adding shims under the rocker arm stand bases.

PUSH RODS

Push rods are made in a variety of types to match the many different engines. They differ according to several dimensions and material used.

1. **Length**   Most push rods have a fixed overall length. However, adjustable types are sometimes used (Figure 16-42). ***NOTE:*** *Some Ford Motor Company big block engines equipped with hydraulic valve lifters may use push rods of varying lengths. In the same engine, push rods may be standard length, oversize by 1/16", or undersize by 1/16".* When reassembling these engines, it is very important to return the push rods to their original positions. Failure to do this can result in noise, or valves being held open.

2. **Diameter**

3. **Tubular or solid construction**   Many manufacturers use tubular (hollow) push rods in order to reduce valve train reciprocating weight. Often, both ends are drilled in order to supply oil to the rocker arms. Solid construction push rods are gradually being phased out because of the extra weight.

4. **End shape**   Tubular push rods have hardened tip inserts, either concave or convex (Figure 16-43). The upper and lower ends may have a different spherical radius.

5. **Type of material**   Generally, seamless steel tubing is used for hollow construction push rods. Performance engines may use chrome-

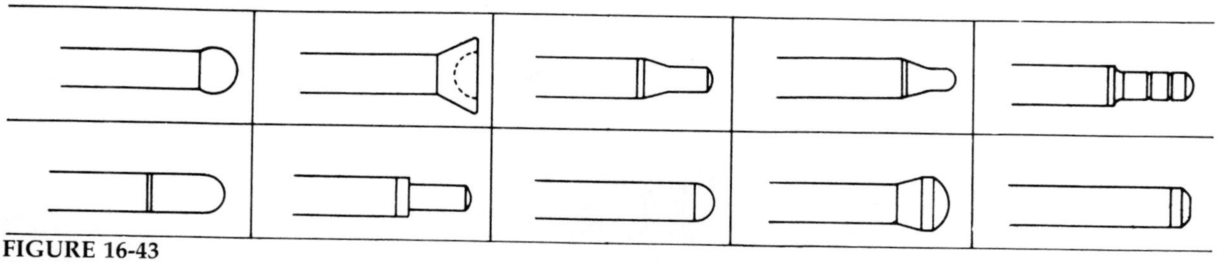

FIGURE 16-43

Typical push rod end configurations (Courtesy of Dana Corporation—Service Parts Group).

moly material for added protection against fatigue, bending, and breakage.

Inspect the push rod ends for ridges, chips, and "onioning" (Figures 16-44 and 16-45). Be sure to inspect the push rod socket in the rocker arm for wear.

Roll the push rods on a flat surface and visually check for bend. *Maximum allowable deflection is .010".* Replace any push rods that exceed this tolerance. Do not attempt to straighten them.

Tightening down rocker arm assemblies too fast (especially when using an impact wrench) on engines equipped with hydraulic lifters can cause trouble. Tightening down too rapidly does not give time for the lifters on the open valves to leak down. Consequently, valve lift will be excessive for a moment. Valves may then contact piston heads with enough force to bend or break push rods (Figure 16-46).

Because of the preceding, always tighten rocker assemblies slowly and uniformly to prevent damage.

In some engines, the upper end of the push rod runs in a guide plate (Figure 16-47). This

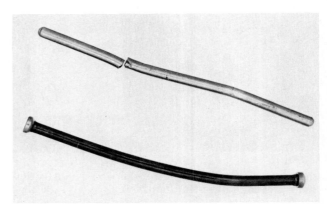

FIGURE 16-46
Damaged push rods.

prevents sideways movement of the push rod and rocker arm. Other engines use guide slots in the head. When inspecting push rods, check for wear from guide rub (Figure 16-48). *NOTE: This is fairly common on 265 to 400 CID Chevrolets, and on 260 to 351 CID Fords.* If any push rod shows guide rub, inspect the guide slot for wear. Recondition the guide slot, or install a guide plate if wear is evident. Remember to use hardened push rods when installing guide plates.

VALVE SPRING RETAINERS

A valve spring retainer (sometimes called washer) serves several purposes.

1. It provides a square setting surface for the valve spring end.
2. It serves as part of the lock mechanism that holds the valve spring in position.
3. It provides a convenient way of disconnecting and connecting the valve from the spring.

Retainers are generally manufactured from steel. The high-performance industry has them available in aluminum and titanium (42% lighter than

FIGURE 16-44
This push rod wore through the rocker arm. Note the severe wear on the push rod end.

FIGURE 16-45
Worn push rod ends.

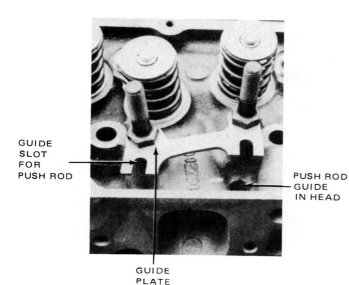

GUIDE
SLOT
FOR
PUSH ROD

PUSH ROD
GUIDE
IN HEAD

GUIDE
PLATE

FIGURE 16-47
Push rod guide plate installed on head.

FIGURE 16-48
Excessive sideways movement of the push rod or running guide plates with nonhardened push rods will cause this wear.

FIGURE 16-49
Multiple spring setups require stepped retainers.

steel). When multiple springs are used, the retainer will be stepped (Figure 16-49).

Retainers should be visually checked for cracks, wear on the spring seat surface(s), and scoring inside the hole (Figure 16-50). Replace all damaged retainers.

KEEPERS

Split-type keepers (often called keys) are generally used in the automotive engine. Each keeper has a tapered outside surface, while the inner surface is generally provided with bead locks. The keepers are pinched tightly against the valve stem by upward pressure from the spring retainer (Figure 16-51).

A thorough visual inspection of the keeper OD, and the bead locks should be made. Inspect for wear, metal shearing, and nicks. Discard all damaged keepers. *CAUTION: New keepers must be installed in pairs.* If an unworn keeper is mated with a used keeper, the spring retainer may cock and break off the valve tip (Figure 16-52).

VALVE SPRINGS

The valve spring functions in the opening and closing motion of the valve.

1. Its primary job is to close the valve.
2. It maintains proper valve train clearance. Without the correct spring tension, correct "lash" for an engine cannot be maintained. If springs are weak, the valves can bounce

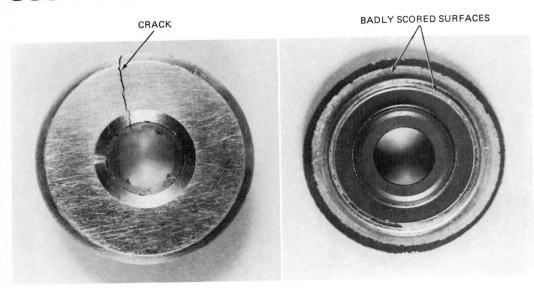

CRACK

BADLY SCORED SURFACES

FIGURE 16-50
Damaged retainers.

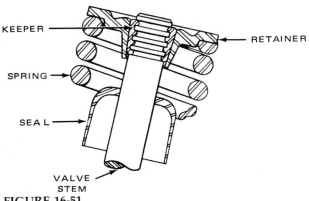

KEEPER

RETAINER

SPRING

SEAL

VALVE STEM

FIGURE 16-51
Keeper position when installed. Keeper ends should not touch one another. This allows the retainer, keepers, and valve stem to lock together completely. (Courtesy of Chrysler Corporation).

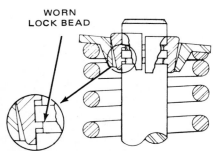

WORN LOCK BEAD

FIGURE 16-52
To prevent valve tip breakage, always install new keepers in pairs (Courtesy of TRW Replacement Division).

on the seat during closure, resulting in noise, poor performance, and valve train damage.

3. It must be able to control valve float. This is a high-speed condition where valve train inertia can sometimes be strong enough to force the valve beyond its designed lift opening. Float can destroy an engine.

4. It must control surge. Surge is an unwanted harmonic vibration that can occur in the spring coils during the opening and closing action of the valve. Surge can cause spring breakage, seat recession, and rocker arm failure.

Valve Spring Surge Control

Except for the uniform pitch design, most valve springs are built to eliminate or dampen surge. Several approaches are used.

1. Variable pitch coils (springs more closely coiled at one or both ends). This prevents vibration from occurring over the entire length of the spring (see Figure 16-53).

2. Counterwound inner springs (coils wound in opposite direction of outer spring). Harmonic vibration will occur at a different rpm for each spring. Therefore, the valve will never be unsprung from surge (see Figure 16-54).

3. Mechanical dampeners (a flat coil or a finger insert that fits tightly inside the outer spring). The friction fit reduces the tendency of the coils to vibrate (see Figure 16-55).

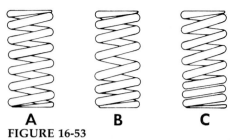

FIGURE 16-53
Spring A is uniform pitch. Springs B and C are variable pitch (Courtesy of Dana Corporation—Service Parts Group).

FIGURE 16-54
Counterwound spring setup for Boss 302 Ford (Courtesy of Ford Parts and Service Division).

FIGURE 16-55
Mechanical vibration dampeners (Courtesy of Dana Corporation—Service Parts Group).

FIGURE 16-56
Varying OD (tapered) valve spring (Courtesy of Oldsmobile Motor Division of General Motors Corporation).

4. Varying OD (one end of the spring has a smaller outer diameter than the other end). The coil tension is greater at the small end. This variable strength rate means there is spring control at all times (see Figure 16-56).

Valve Spring Inspection

After disassembling the head, lay the springs out on a bench. Visually inspect the springs for collapse, nicks, pitting, corrosion, etching, breakage, and end wear (Figures 16-57, 16-58, and 16-59). *All springs should have a free length height within 1/16".* Spring ends that show wear or polish indicate surge or rotation. Tension problems caused

FIGURE 16-57
Both springs are from the same engine. The one on the left has collapsed.

FIGURE 16-58
Spring end wear (Courtesy of TRW Replacement Division).

FIGURE 16-59
Spring breakage.

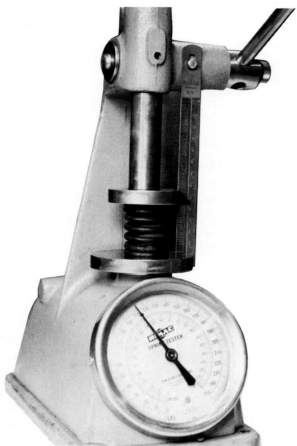

FIGURE 16-60
Checking valve spring tension.

by heat and fatigue cannot generally be detected visually. In order to determine if any tension has been lost, use a valve spring tester (Figure 16-60). *Spring tension should be at least 90% of the OEM specifications.* Valve bounce, valve float, and pre-

mature valve burning can be caused by weak springs. Springs with too much tension can cause accelerated valve train wear, valve stretch, and broken stems. Spring tension should be measured at the two length extremes at which the spring operates.

1. "Open" pressure, with spring pressure at full open valve position. Compress the spring to the specified valve open test height and read the pressure.
2. "Seat" pressure, with spring pressure when the valve is on the seat. To check seat pressure, install the spring in a tester and compress to the installed spring height dimension. See Table 16-2.

NOTE: Sometimes, the listed pressure specification is given at a compressed length different than either valve open height or installed height.

Out-of-square valve springs will cause extra side load on the valve. Premature valve guide wear or valve breakage can result. As a rule-of-thumb, *out-of-squareness should not exceed 1/32" per inch of free spring length.*

SHOP TIP

If a shop encounters a valve guide wear failure problem with no apparent reason, dial indicate the spring seats. The seats may not be perpendicular to the valve guides. This will cause the springs to be installed out-of-square. Then the springs will thrust sideways on the valves, resulting in quick stem and guide wear. The problem can be eliminated by spot facing the spring seats.

Valve Spring Identification

The use of incorrect valve springs can contribute to engine damage or failure. When rebuilding a cylinder head, determine the correct valve spring by part number and size. Do not rely on color striping. Varnish buildup on the springs can make color identification difficult. As a point of illustration, let's take a Chrysler 318 CID engine with exhaust valve rotators. On this engine, the intake valve springs have a wire diameter of .192" with 6.4 coils, and are striped blue. However, the exhaust valve springs have a wire diameter of .185" with 5.8 coils, and are striped brown.

TABLE 16-2

Typical valve spring pressure information (Courtesy of Silver Seal Products Company).

Year	Number of Cylinders	Engine or Model		Closed Pressure	Closed Height	Open Pressure	Open Height
CHEVROLET							
1972–75	4	110 CID(Luv)	(Outer)	47.4	1.580		
		(1800 cc)	(Inner)	17.8	1.500		
1976–80	4	Chevette 1398 cc, 1599 cc		64–67	1.250	173	.886
1977–80	8	305 (2 bbl.), 350	Int.	76–84	1.700	200	1.250
			Exh.	76–84	1.610	200	1.160
1977–78	6	250		78–86	1.656	175	1.260
1975–80	V-6	231, 196	Int.	59–69	1.727	164	1.340
			Exh.	59–69	1.727	182	1.340
1978–80	4	151 (2.5 L)		78–86	1.660	176	1.254
1978–80	V-6	200		76–84	1.700	200	1.250
1980	V-6	173 (2.8 L)			1.570	195	1.810
1976–80	4	110 CID (1800 cc)	(Outer)	74–82	1.390	114–126	1.190
		(Luv)	(Inner)	31–35	1.382	78–86	.961
1980	8	350 Diesel		78–86	1.670	151	1.300

Valve Spring Base Cups

Some engines use a spring base cup that mounts into position on the OD of each valve guide (Figure 16-61). The bottom of the valve spring fits into the cup. This eliminates spring "walk" and the resulting side thrust on the valve guide.

Replacing Valve Springs Without Removing the Cylinder Head

Broken valve springs, defective valve stem seals, or damaged retainers may be replaced without removing the cylinder head. The basic procedure is given below.

1. Install an air hose with an adapter in the spark plug hole (Figure 16-62).
2. Remove (or slide to the side) the rocker arm from the affected cylinder.
3. Position the piston at TDC.
4. Turn on the air supply to hold the valves closed. *NOTE: Instead of using air to hold the*

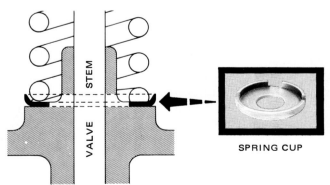

FIGURE 16-61

Spring cup position when installed (Courtesy of Silver Seal Products Co., Inc.).

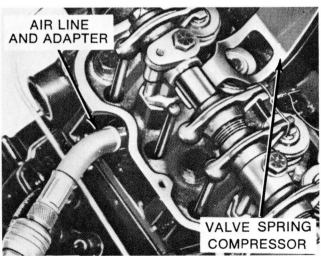

FIGURE 16-62

In-chassis valve spring assembly removal. Remove push rod and insert loop end of tool under rocker shaft. Apply pressure to compress spring (Courtesy of Ford Parts and Service Division).

valves, you can use rope stuffed into the cylinder through the spark plug hole.

5. Compress the valve spring (many special tools are available) and remove the keepers. See Figures 16-63, 16-64, and 16-65).

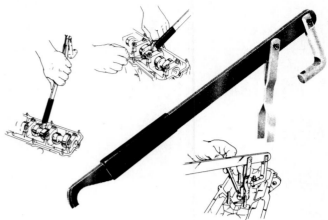

FIGURE 16-64
Valve spring compressor and lash adjustor for water cooled VW Rabbit, Scirocco, Dodge Omni, Plymouth Horizon, and Audi Fox (Courtesy of K-D Tools).

FIGURE 16-63
Valve spring removal (Courtesy of American Motors Corporation).

FIGURE 16-65
Jaw-type spring compressor designed to remove keepers and spring without removing the head (Courtesy of K-D Tools).

TRUE CASE HISTORY

Vehicle	Corvette
Engine	8 cylinder, 454 CID
Mileage	1100 miles on vehicle after purchasing new.
Problem	A noticeable ticking noise when the engine was running.
Cause	Several push rod guide plate slots had accelerated wear. This permitted the push rods to operate in a misaligned position, causing noise in the valve train.
Comment	A number of these engines were OEM built with improperly hardened guide plates. As a result, they were prematurely worn by the heat-treated push rods.

CHAPTER 16 SELF-TEST

1. What are the two purposes of a rocker arm?
2. An engine has a rocker arm ratio of 1.5:1. The cam lift is .280″. What is the actual valve lift?
3. What are the advantages of stamped-steel rocker arms?
4. Cast-type rocker arms are made in what materials?
5. What engine uses silent lash rocker arms?
6. List four rocker arm-to-cylinder head mounting methods.
7. When installing a rocker arm shaft, what direction do the drilled lubrication holes generally face?
8. List four ways in which oil flow to the rocker arm assembly occurs.
9. Why is it not recommended to reface stamped-steel or laminated rocker arms?
10. The rocker arms from an OHC engine are being reused after a valve job. Why is it very important to return them to original position?
11. What are the advantages of using grooved pivot balls?
12. What is the maximum allowable wear limit on the beveled stop surface of a positive stop nut?
13. List five conditions that can cause rocker arm breakage.
14. What is the correct procedure for loosening rocker shaft support attaching bolts?
15. How is coil bind checked?
16. What should be the clearance margin between the bottom of the valve spring retainer and the top of the guide?
17. Generally, rocker arm-to-shaft clearance should not exceed _____ ″.
18. On what surface will a shaft mounted rocker arm show the greatest wear?
19. Where does a rocker shaft show the greatest wear?
20. List four conditions that will cause incorrect rocker arm geometry.
21. How can rocker arm geometry be corrected?
22. What manufacturer uses oversize length push rods?
23. How can push rods be checked for bend?
24. Why should rocker assemblies be tightened down slowly?
25. What is a guide slot? What is its purpose?
26. What are two purposes of a valve spring washer?
27. Why must new keepers always be installed in pairs?
28. List four functions of a valve spring.
29. What is valve float?
30. What is valve spring surge?
31. Describe four ways to control valve spring surge.
32. All valve springs should have a free length height within _____ ″.

33. What do polished valve spring ends indicate?
34. Valve spring tension should be at least _____ of the OEM specifications.
35. What can be the result of out-of-square valve springs operating in an engine?
36. Spring out-of-squareness should not exceed what amount?
37. Why are spring base cups used in some engines?
38. How can the valve stem seals in an engine be replaced without removing the cylinder head?

CHAPTER 17

ASSEMBLING THE CYLINDER HEAD

After the cylinder head and its related parts have been inspected and replaced, or reconditioned, assembly can begin.

INSTALLING CORE PLUGS

Sometimes a new core plug (often called freeze plug, expansion plug, welsh plug, or soft plug) leaks after being installed by a mechanic (Figure 17-1). To help eliminate this possibility, the following installation procedures are suggested.

1. Use emery paper to thoroughly clean the inside edge of the core plug hole. Remove all burrs, rust, oil, grease, dirt, and old sealer.

2. Make sure the replacement plug is clean and free of distortion (nicks, grooves, and bent edges).

3. Lightly coat the plug OD with a hardening-type sealer (Permatex No. 1 is often used). See Figure 17-2.

4. When replacing a cup type core plug, *drive the plug 1/32" past the bottom edge of the hole chamfer (Figure 17-3). Only drive the plug into*

DISC TYPE   SHALLOW CUP TYPE   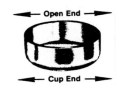DEEP CUP TYPE

FIGURE 17-1

Designs of core plugs (Courtesy of Dana Corporation—Service Parts Group).

358

FIGURE 17-2
Hard setting sealer.

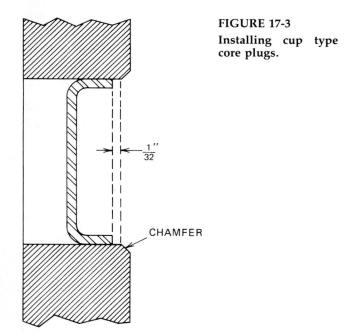

FIGURE 17-3

Installing cup type core plugs.

$\frac{1}{32}''$

CHAMFER

the machined bore using a properly designed tool (Figure 17-4). *CAUTION: Do not drive a cup plug into the bore using a tool that contacts the flange or the center portion of the plug.* This method will distort the sealing edge and may cause leakage and/or plug "blowout."

5. Most cup type plugs are installed with the open side out. However, some have a reverse taper sealing edge, and must be installed with the open side facing in (Figure 17-5). Always use the correct installation tool.

6. Disc (flat) type plugs must be installed with the dished side facing in. On this type of plug, it is very important to use a driver that contacts the center crown of the plug. *Drive on the plug until the crown just becomes flat.* This procedure will properly expand the plug, ensuring a tight fit.

7. Core plugs are generally fractional sized. Figure 17-6 shows a typical listing in an engine parts catalog. Notice that some plugs are sized apart by 1/64" or less. For this reason, it is important to check the size marking of the old core plug against the new one. Otherwise, the new plug may "blowout" under pressure. *NOTE: Some engines (Chrysler and Ford) have been produced with one or more oversize core hole plugs. Ford oversize (OS) plugs are identified by the OS stamped on the cup side of the plug. OEM Chrysler oversize plugs can be identified by a cadmium coating.*

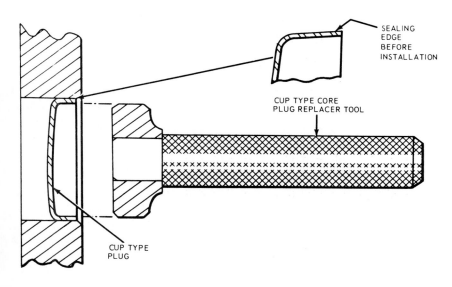

SEALING EDGE BEFORE INSTALLATION

CUP TYPE CORE PLUG REPLACER TOOL

CUP TYPE PLUG

FIGURE 17-4

Cup type plug and installation tool (Courtesy of Ford Parts and Service Division).

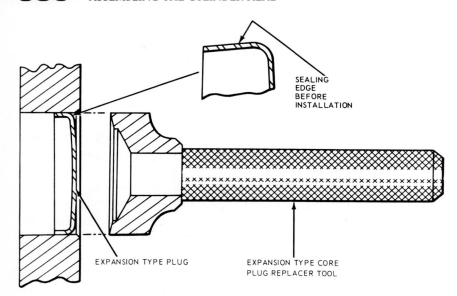

FIGURE 17-5
Expansion type plug and installation tool (Courtesy of Ford Parts and Service Division).

NOTE: All cup type plugs are installed with open end out with the exception of those numbers with an asterisk (*) which are reverse tapered plugs and installed with cup end out.

Size	Part No.	Size	Part No.	Size	Part No.	Size	Part No.	Size	Part No.	Size	Part No.
.270	219-3105	7/8	•219-6006	1-1/4	•219-6011	1-5/8	219-2126	2	•219-6026
5/16	219-3128	7/8	219-3029	1-1/4	219-3018	1-5/8	•219-7126	2	219-2066
.339	219-3101	7/8	•219-8029	1-1/4	•219-8018	1-45/64	219-3070	2	•219-7066
7/16	219-1037	29/32	219-3072	1-1/4	*219-3099	1-45/64	219-2076	2-1/64	219-3069
7/16	219-3093	29/32	•219-8072	1-1/4	219-2110	1-45/64	•219-7076	2-1/64	•219-8069
.448	219-3088	59/64	219-3077	1-1/4	•219-7110	1-47/64	219-1103	2-1/64	219-2083
1/2	219-1046	15/16	219-1007	1-5/16	219-1023	1-3/4	219-1015	2-1/16	219-2059
1/2	219-3169	15/16	219-3039	1-21/64	219-2127	1-3/4	•219-6015	2-1/16	219-1068
1/2	219-3095	15/16	•219-8039	1-3/8	219-1012	1-3/4	219-2057	2-1/16	219-3112
1/2	•219-8095	31/32	219-1104	1-3/8	•219-6012	1-3/4	219-3075	2-3/32	219-3131
33/64	219-3087	1	219-1008	1-3/8	219-3080	1-3/4	•219-8075	2.105	219-1129
.540	219-3097	1	•219-6008	1-3/8	•219-8080	1-3/4	219-3082	2-7/64	219-2078
.540	219-2098	1	219-3016	1-31/64	219-1102	1-3/4	•219-8082	2-1/8	219-1044
9/16	219-1047	1	•219-8016	1-31/64	•219-6102	1.788	219-3076	2-9/64	219-2170
9/16	219-3048	1-1/16	219-1009	1-1/2	219-1013	1-51/64	219-2096	2-9/64	219-3084
37/64	219-3094	1-1/16	219-3086	1-1/2	•219-6013	1-7/8	219-1020	2-9/64	•219-8084
19/32	219-3054	1-1/8	219-1010	1-1/2	219-3133	1-7/8	•219-6020	2-5/32	219-2071
5/8	219-1002	1-1/8	•219-6010	1-1/2	•219-8133	1-7/8	219-3050	2-5/32	•219-7071
5/8	•219-6002	1-1/8	219-3025	1-1/2	219-3030	1-7/8	•219-8050	2-3/16	219-2074
5/8	219-3052	1-1/8	•219-8025	1-1/2	•219-8030	1-15/16	219-3040	2-7/32	219-2154
11/16	219-1045	1-3/16	219-1021	1-1/2	219-3085	1-15/16	•219-8040	2-1/4	219-1022
3/4	219-1003	1-3/16	•219-6021	1-1/2	•219-8085	1-15/16	219-1043	2-1/4	219-3081
3/4	•219-6003	1-13/64	219-3134	1-5/8	219-1014	1-61/64	219-3042	2-1/4	•219-8081
3/4	219-3038	1-7/32	219-3017	1-5/8	•219-6014	1-61/64	•219-8042	2.262	*219-3090
3/4	•219-8038	1-7/32	•219-8017	1-5/8	219-3041	1-31/32	219-2058	2.396	*219-3092
7/8	219-1006	1-1/4	219-1011	1-5/8	•219-8041	2	219-1026	2.573	*219-3089

• Brass Plugs

FIGURE 17-6

(Courtesy of Dana Corporation—Service Parts Group).

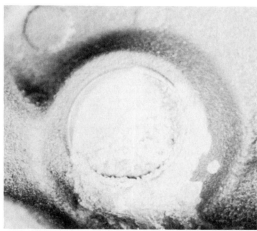

FIGURE 17-7

This rusted through cylinder head core plug is located next to the firewall. In-chassis replacement is difficult. Such problems can be avoided by using brass plugs.

It is recommended to use brass core plugs whenever possible since they will not rust through (Figure 17-7).

Special Application Expansion Plugs

Several companies make special expansion plugs that can be easily installed in those "hard-to-get-at" places (Figures 17-8 and 17-9). They are de-

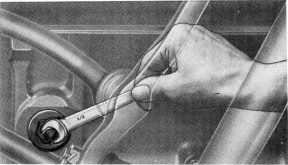

FIGURE 17-9

Solid copper expansion plug (Courtesy of Dorman Products, Inc.).

signed to be installed in minutes, even when the steering linkage, exhaust manifolds, motor mounts, or frame interferes.

CHECKING INSTALLED SPRING HEIGHT (OR ASSEMBLED HEIGHT)

The loss of material from the valve face, the valve seat, or the spring seat has the effect of lengthening the valve stem (Figure 17-10). The length-

FIGURE 17-8

Neoprene rubber expansion plugs can save time and effort. They must be installed dry with no sealing compound (Courtesy of Hunckler Products, Inc.)

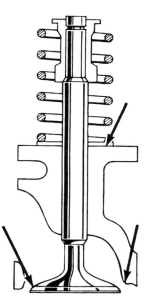

FIGURE 17-10

Loss of material at any of these points will affect installed spring height and reduce tension (Courtesy of Houser Engineering and Manufacturing Company).

ening of the valve stem reduces valve spring tension because of the increased installed spring height. ***COMMENT:*** *As little as .035" metal removal can result in the loss of 20% of original spring tension.* Valve spring shims inserted under the valve spring can compensate for loss of tension from change in installed spring height due to grinding and wear.

Valve springs sometimes encounter a tension drop during break-in. This tension drop (also called *torsion relax*) is due to heat and flexing at the time of break-in. Because of this, some shops will arbitrarily use spring shims without taking time to measure installed spring height. This practice is not suggested, as overshimming often results. ***NOTE:*** *Valve spring shims are not a cure for weak springs.* Also, shimming in excess of .060" is not recommended.

All car manufacturers recommend that installed spring height be accurately measured. Incorrect valve spring height can have several effects.

1. Incorrect valve seating.
2. Hot running valves.
3. Rough idling.

4. Reduced gas mileage.
5. Stress on springs and valve train parts.
6. Valve spring coil bind.

Valve spring shims are usually available in three sizes (Figure 17-11).

1. Type A (.060" thick) for springs in service.
2. Type B (.030" thick) for new springs.
3. Type C (.015" thick) for balancing.

Many spring shims have a serrated side (Figure 17-12). The shim should be installed with the serrated side toward the valve head. The serrated louvers retard heat transfer to the spring, thereby reducing torsion relax.

Proper installed valve spring height is usually determined by one of these methods.

1. Measure the height of the assembled valve spring from the surface of the cylinder head spring seat to the underside of the spring retainer with dividers (Figure 17-13). Check the dividers against a scale. If the assembled

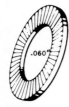  

TYPE A   **TYPE B**   **TYPE C**

FIGURE 17-11
Valve spring shims (Courtesy of Silver Seal Products Co., Inc.).

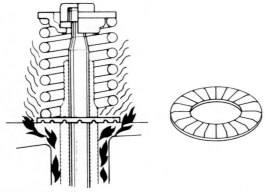

FIGURE 17-12

Serration on spring shims help dissipate heat away from the valve spring (Courtesy of Silver Seal Products Co., Inc.).

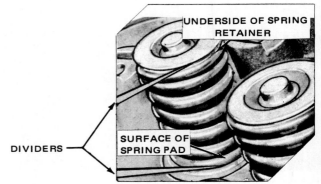

FIGURE 17-13

Measuring valve spring assembled height with dividers (Courtesy of Ford Parts and Service Division).

height is greater than specification, install the necessary shims. *NOTE: Although this method is usually shown in shop manuals, it is not always practical.* Also, it is not the most accurate method.

2. Use a narrow, thin scale and measure from the spring seat in the head to the top of the valve spring (Figure 17-14).

3. Use a caliper (Figure 17-15).

4. Install the valve in the head with the spring retainer and keepers. Holding the valve against the seat, measure the distance between the spring seat and the spring side of the retainer with a telescoping gauge (Figure 17-16).

5. Use a commercial spring shim gauge kit (Figure 17-17). Determine the specified setting by referring to the selector chart in the kit.

With the valve held snugly on the seat, and the gauge held against the valve, measure to the top edge of the top keeper groove. The distance above the proper marking on the gauge indicates the shim thickness needed to restore original tension.

Measuring Installed Spring Height on Chevrolet Engines

Many small block Chevrolet engines use a metal oil shield between the top of the retainer and valve spring. See Figure 17-18. Shops often make

FIGURE 17-16

Using telescoping gauge for measuring assembled height after valve grinding. This method is fast and accurate (Courtesy of Silver Seal Products Co., Inc.).

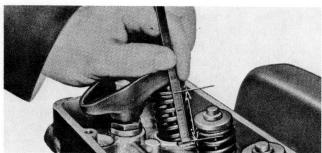

FIGURE 17-14

Checking installed height of valve springs with a scale (Courtesy of Chevrolet Motor Division of General Motors Corporation).

1.417 IN

FIGURE 17-15

Checking valve spring assembled height with a caliper (Courtesy of Ford Parts and Service Division).

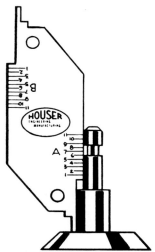

FIGURE 17-17

Checking valve spring assembled height with a special gauge plate (Courtesy of Houser Engineering and Manufacturing Company).

FIGURE 17-18

Checking valve spring installed height on a Chevrolet engine equipped with metal oil shields. Measure from the bottom of the retainer to the spring seat surface (Courtesy of Chevrolet Motor Division of General Motors Corporation).

the mistake of including the oil shield thickness when measuring installed height. *The installed height specification given by Chevrolet for small block engines does not include shield thickness.*

Factory Shims

Aluminum heads generally have a hardened steel shim placed under each valve spring to protect the soft spring seat surface. *Unless otherwise specified, these factory shims must be in place when checking installed spring height.*

Valve Spring Load Balance (Static)

The purpose of this analysis is to determine ultimately what has to be done to have equal seating pressure and loading force on each valve. Then, final adjustment can be made by selective shimming to balance all of the valve preload conditions. The procedure is given below.

1. Measure and record valve spring free length (FL).
2. Measure and record spring assembled height (SAH).

3. Calculate initial spring deflection (ISD). This is determined by subtracting SAH from FL.
4. Install the camshaft, lifters, push rods, rocker arms, and assembled head(s) on the block.

   Valve displacement is fixed by the valve lifter-push rod-rocker arm relationship for each valve. However, spring loads can vary depending on initial spring deflection and valve movement as the camshaft rotates.
5. With a dial indicator, measure the displacement of each valve at the spring retainer while rotating the cam. This will take into account all variations in cam lobes, lifters, push rod deflection, and rocker arm ratio. This measurement gives you static spring deflection (SSD). *The total height deflection (THD) on each spring equals ISD plus SSD.*
6. On a spring tester, compress each valve spring to THD and record the pounds of tension. See Figure 17-19.

   Use the highest individual tension figure as a baseline. Determine the pounds of variation for all other springs by comparing to the baseline. Balancing the tension of all the springs is the objective.
7. Determine the shim thickness required to adjust the tension variation of each spring to the baseline. This can be done on the spring tester. Place shims under the spring, compress to the THD and measure the tension increase. ***NOTE:*** *A four lb increase per .015" of shim thickness is a general rule-of-thumb guideline.*

FIGURE 17-19

A locked telescoping gauge can be used to show spring pressure at preassembled height (Courtesy of Silver Seal Products Co., Inc.).

Valve Spring Load Balance (Dynamic)

To correlate static load balance with dynamic spring behavior, the following procedure can be used.

1. Remove valve cover(s).
2. Hookup a strobe light (timing light).
3. Hold the engine at a constant 2000 rpm.
4. Strobe each spring retainer and compare the apparent steadiness. If one or more are out of phase (not in dynamic balance), the retainer will appear to wander up and down. Correction is made by adding or subtracting shim thickness.

TIPS WHEN ASSEMBLING CYLINDER HEADS

Listed below are some precautions and component installation tips.

1. When installing valve keepers, compress the springs just enough to complete the job. Umbrella seals can be crushed if the valve springs are excessively compressed. Be extra careful if using an air-operated compressor (Figure 17-20). *COMMENT: Chrysler Corporation 318, 340, 360, 400, and 440 CID engines are especially prone to this problem.*

2. Some engines use oil shields that are designed to "snap" inside of the valve spring retainer cap (Figure 17-21). They must not be installed on the valve stem and allowed to move up and down with the valve.

3. The valve face-to-valve seat seal can be bench checked by using solvent. Position the head with the combustion chamber facing up, and install the spark plugs. Set the valves into position without the spring assembly. *The valve face and valve seat must be absolutely clean.* Fill the combustion chambers with solvent and observe for leakage.

4. Valve assembly vacuum gauges are available to check the valve-to-seat seal in seconds (see Figure 17-22). The correct vacuum cup is placed over the valve head or combustion chamber. A handle is pulled to expel air from the vacuum cup. The rate of loss of vacuum is read directly on the vacuum gauge.

5. When installing variable pitch valve springs,

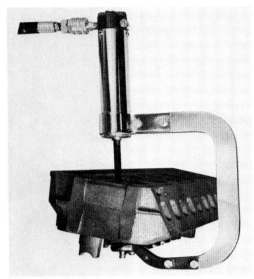

FIGURE 17-20
Air-operated valve spring compressor (Courtesy of Hall-Toledo, Inc.).

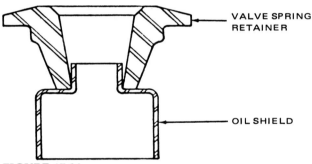

VALVE SPRING RETAINER

OIL SHIELD

FIGURE 17-21
International Harvester oil shields must be "snapped" in the retainer to assure a tight seal (Courtesy of Sealed Power Corporation).

the close coil end must be placed against the cylinder head (Figure 17-23).

6. Prelubricate the valve stems and guides. Use an engine oil supplement (EOS) or a thick assembly lube (Figure 17-24a). Thinner lubricants (such as 30-weight engine oil) tend to drain off before the engine is run.

SHOP TIP

90-weight hypoid gear oil will give excellent results when used for a valve stem and guide prelubricant.

FIGURE 17-22
A commercial tester for checking the valve face-to-seat seal (Courtesy of Hall-Toledo, Inc.).

7. Before inserting the valves into the head, polish the stems with crocus cloth that has been saturated with solvent. This will give added protection against scuffing and scoring.

New valves sometimes have the part number stamped on the stem. As a result, there is a slight raising of metal around the edges of the number. To prevent possible valve sticking, this area must be smoothed down.

For many years, engine requirements allowed the use of a one-piece steel valve stem. As valve temperatures increased, welded tips were introduced to resist wear and scuffing. When the tip wafer is applied, the welding process may leave a visible gap on the stem OD. As shown in Figure 17-24b, this gap can be a maximum of .016" and extend around the entire stem circumference. However, the valve can still be considered satisfactory and functionally acceptable.

8. After the head has been assembled, tap the valve stem tips with a soft-faced hammer (Figure 17-25) to ensure seating of the valve keepers, or to cause them to fly out if they were improperly installed (Figure 17-26). *CAUTION: If a valve keeper drops out when the engine is running, the valve will hit the piston and cause major damage (Figure 17-27).*

9. Check each assembled valve spring for looseness (Figure 17-28). The spring should not turn under finger grip pressure with the valve in the closed position. Low spring pressure allows hydraulic lifter pump up.

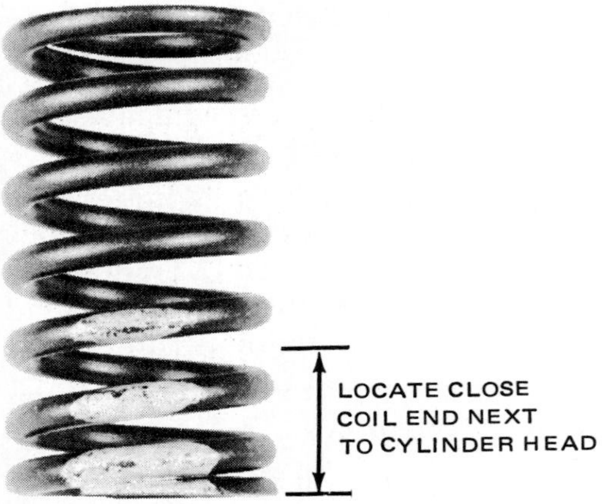

LOCATE CLOSE
COIL END NEXT
TO CYLINDER HEAD

FIGURE 17-23
Basket type variable pitch valve springs must be installed in a certain direction.

Installing Chevrolet Valve Stem "O" Ring Seals

Sometimes the wrong procedure is followed when installing Chevrolet valve stem "o" ring seals (Figure 17-29). *It is a mistake to install the "o" ring*

FIGURE 17-24*a*

HRL assembly lube is blended with a "tacky" grease carrier that sticks to metal wherever it is applied.

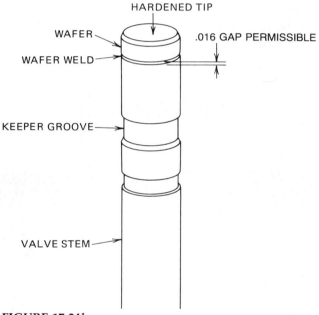

FIGURE 17-24*b*

Wafer welded exhaust valve (Courtesy of Sealed Power Corporation).

on the valve before the spring and the retainer. Then the "o" ring is pushed out of place when the retainer is moved down to install the keepers. The correct procedure is explained next.

1. Insert the valve in the cylinder head. Position the spring and retainer assembly over the valve stem and compress.

FIGURE 17-25

Tap valve stem lightly to check keeper installation (Courtesy of Toyota Motor Sales U.S.A., Inc.).

FIGURE 17-26

This engine was running when it came into the shop. One set of keepers had almost completely come out.

2. Install the "o" ring seal in the second keeper groove (Figure 17-30).
3. Place the keepers in the first groove and release the spring and retainer assembly (Figure 17-31). Slowly release the spring compressor, keeping the retainer hole centered on the valve stem as it comes up. Otherwise, the "o" ring may be pinched and it will leak oil.

Installing Toyota Valve Stem "O" Ring Seals

On certain model Toyota engines, the valve stem "o" ring seal is in an unusual location (Figure 17-32). It fits in a special groove on the valve stem, just above the keepers.

FIGURE 17-27
Here is the result of faulty valve keeper installation.

FIGURE 17-28
Checking valve spring looseness (Courtesy of Sealed Power Corporation).

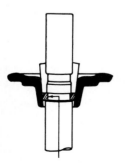

FIGURE 17-29
Chevrolet valve stem "o" ring seal location.

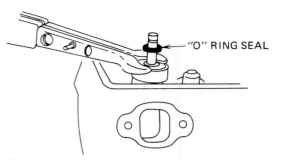

FIGURE 17-30
After compressing valve spring, install "o" ring in second keeper groove.

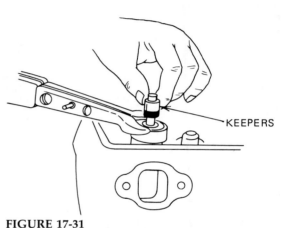

FIGURE 17-31
Hold keepers in first groove and slowly release the spring compressor.

Testing Valve Stem "O" Ring Seals for Leakage

After "o" ring seals are installed, test for leakage. This can be done several ways.

1. Place a rubber bulb from a battery hydrometer over the spring retainer and squeeze (Figure 17-33). The bulb must fit tightly on the retainer. If the bulb will not hold pressure, air is leaking past the seal. The assembly must be disassembled, and the seal and/or retainer replaced.

2. Fill the recess around the hole in the center of the spring retainer with 30-weight oil. Let the oil stand and observe for leakage down the valve stem.

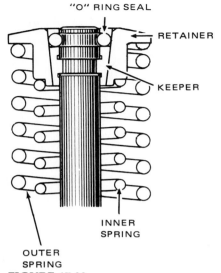

FIGURE 17-32

Toyota valve stem "o" ring seal location (Courtesy of Toyota Motor Sales U.S.A., Inc.).

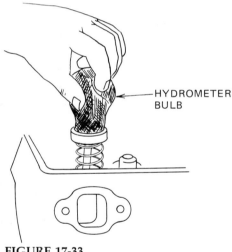

FIGURE 17-33
Test Seal.

Sealing Cylinder Head Attaching Bolts

Some engines have certain head bolt holes that are open to the water jackets. Use a nonhardening sealant on the threads to prevent coolant leakage.

The following engines have cylinder head bolts that enter the water jackets.

1. Chevrolet small blocks (265, 283, 305, 307, 327, 350, and 400 CID engines).

2. AMC six-cylinder (199, 232, and 258 CID engines).
3. AMC V-8 (290, 304, 343, 360, 390, and 401 CID engines).
4. Some 318 CID Chrysler Corporation engines.

When removing head bolts, always look for any sealer previously used. Check for any visible evidence that the bolt has been in contact with engine coolant. If uncertain as to which bolts require sealing, examine the bolt hole with a flashlight to see if it enters a water passage.

Check the removed head bolts for evidence of bottoming. This condition can be identified by the friction shine on the threaded end. If necessary, the end of the bolt can be ground down to eliminate bottoming. *Always check bolt lengths prior to cylinder head installation.*

Head Bolt Installation Tips

1. Always clean old sealer from the bolts before new sealer is applied.
2. Run the proper size tap into all head bolt holes in order to clean the threads. Discard any bolt that shows visible corrosion on the fillet or threads. Reject any bolt that is necked down from stretch (Figure 17-34).
3. Do not put sealer on bolts that enter blind holes. This can form a hydraulic lock and cause an incorrect torque reading.
4. Before installing bolts in a blind hole, lightly lubricate the threads and the bottom of the bolt head face with 30-weight oil.

Head Gasket Installation Precaution

Inspect the head gasket to be sure it is being installed properly. If the gasket is in a reversed position (with the front end to the rear), coolant passages could be restricted and the engine would overheat. Also, if the gasket is reversed on some engines, the oil feed hole to the rocker shaft will be blocked.

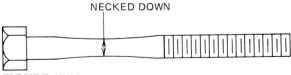

FIGURE 17-34
Head bolts are often stretched in the shank area.

TRUE CASE HISTORY

Vehicle	Toyota
Engine	4 cylinder, 3K model (1200 cc)
Mileage	Zero mileage. An exchange short block was being installed. The original cylinder head was being used.
Problem	All the head bolt holes in the block stripped during installation of the head. The mechanic had started all the head bolts by hand. After they were engaged several threads, the mechanic ran them down with an impact wrench. While the head bolts were being final tightened with a torque wrench, the stripped threads were discovered.
Cause	The old head bolts (non-ISO thread) and the bolt holes in the new block (ISO standard thread) were a different thread pitch. The threads were not matched, even though they could be initially started together.
Comment	Until only a few years ago, there were five metric fastener systems. To conform with International Standardization Organization (ISO) standards, there is now a unified metric system, called the ISO standard. *Only install isometric nuts, screws and bolts in ISO parts. Otherwise, serious thread damage will result.* ISO parts often carry distinctive identification marks (Figure 17-35).

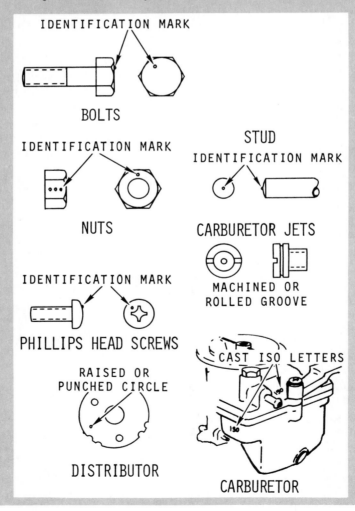

FIGURE 17-35

CHAPTER 17 SELF-TEST

1. What are three other names for core plugs?
2. How far into the bore should a cup type core plug be driven?
3. What type of driver should be used on a flat type core plug?
4. What engines may have oversize core hole plugs?
5. List four problems that can occur from incorrect valve spring height.
6. What is the recommended limit for spring shimming?
7. What is torsion relax?
8. Valve spring shims are usually available in what three sizes?
9. Why do some spring shims have a serrated side?
10. Name five ways to determine installed valve spring height?
11. What precaution must be observed when measuring installed spring height on small block Chevrolet engines that use metal oil shields?
12. Why do aluminum heads generally have a steel shim placed under each valve spring?
13. Outline the procedure for determining static spring balance.
14. How is dynamic spring balance correction made?
15. What spring compresser precaution must be observed when assembling a head that uses umbrella seals?
16. How can solvent be used to check the valve face-to-valve seat seal?
17. Variable pitch valve springs must be installed with the close coil end facing what direction?
18. What lubricant is recommended for valve stems and guides when assembling a head?
19. Why must stems be smooth before installing the valves?
20. Why should the valve stem tips be tapped with a soft-faced hammer after the head has been assembled?
21. What is indicated if an assembled valve spring can be turned with finger grip pressure?
22. Why is it wrong to install a Chevrolet "o" ring seal on the valve stem before the spring and the retainer?
23. Valve stem "o" ring seals on some Toyota engines are found in what unusual location?
24. How can valve stem "o" ring seals be tested for leakage?
25. What should be used on the head bolt threads of small block Chevrolet engines to prevent coolant leakage?
26. Why shouldn't sealer be used on head bolts that enter blind holes?
27. What problems can be caused by installing a head gasket with the front end to the rear?
28. What are ISO standard threads?

CHAPTER 18

CAMSHAFTS AND LIFTERS

In this chapter we discuss camshafts, lifters, and drive mechanisms and the trade-related service procedures for each.

PURPOSE OF CAMSHAFT

Basically, the camshaft controls the entry and exit of the air/fuel mixture at points that are timed to piston position. Also, the camshaft links to the distributor in such a way that the compressed mixture is fired at the right time.

A typical overhead valve engine's camshaft is located in the center of the block directly above the crankshaft (Figure 18-1). A number of engines use the overhead cam design (Figure 18-2).

Review of Basic Valve Timing

From a theoretical point of view the intake valve would open when the piston is at TDC beginning its intake stroke. It would close when the piston reached BDC so no intake mixture could escape

when the piston started upward. The exhaust valve would open at BDC after the power stroke, and close at TDC after the exhaust stroke. However, an engine with this valve timing would not be very efficient.

Intake mixtures and valve train parts have inertia. In simple language this means that they are hard to start in motion. Once started, they are difficult to slow down, stop, or change direction. As a result, valve timing events are started early in the cycle. The intake valve gets a head start and opens before the piston begins the intake stroke (then it will be open even further when the intake stroke actually starts). The intake valve does not close until the piston passes BDC, because air/fuel mixture can still fill the cylinder (due to inertia). These same principles are true for the exhaust valve.

During high-speed engine operation the air/fuel charge has a packing force (kinetic energy). By holding the intake valve open longer (duration), this packing force can be used to improve the engine's power output. This is why high-per-

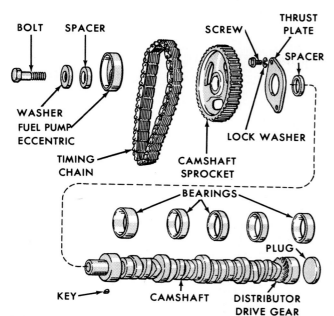

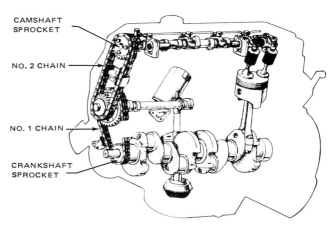

FIGURE 18-1

Typical OHV camshaft and related components (Courtesy of Ford Parts and Service Division).

FIGURE 18-2

The top picture shows a phantom view of the OHC Toyota 8R engine. The bottom picture shows the same engine mounted upside down on an engine stand.

formance camshafts are designed with more duration.

Overlap

The period of crankshaft rotation when both the intake and exhaust valves are unseated is called *valve overlap*. The primary function of valve overlap is to scavenge (remove) remaining gases from the cylinder. This is done by taking advantage of the pressure variations that exist when the piston reaches TDC on the exhaust stroke. Additionally, the outgoing exhaust gases help to "pull" in the next intake charge. Study Figure 18-3.

VALVE LIFTERS

Valve lifters (sometimes called cam followers or tappets) follow the contour of the cam lobe and produce reciprocating motion in the valve train. Lifters are either *solid* (mechanical) or *hydraulic*. A solid lifter transfers motion as a rigid mechanical piece from the cam. A hydraulic lifter operates like a solid lifter when the engine valve is opened. However, the hydraulic lifter is designed to automatically maintain zero lash clear-

ance in the valve train at all times. In so doing, the hydraulic valve lifter offers certain advantages over the solid lifter.

1. The elimination of periodic valve clearance adjustments.
2. The elimination of tappet clearance noise.
3. Longer valve and cam lobe life by the elimination of pounding.
4. Smoother engine operation due to more precise control of valve timing.
5. Automatic compensation for the expansion and contraction of the valve train caused by temperature changes.

Figure 18-4 shows the construction of a typical hydraulic valve lifter.

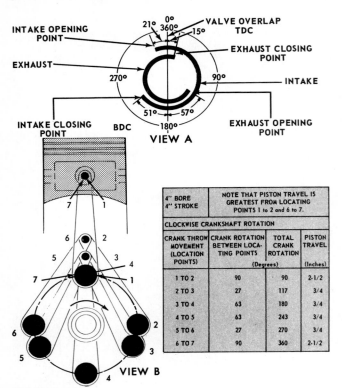

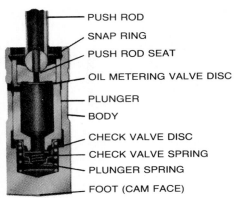

FIGURE 18-3

View A is a typical valve timing schematic. View B relates piston travel to valve timing (Courtesy of Ford Parts and Service Division).

FIGURE 18-4

Hydraulic lifter nomenclature (Courtesy of Dana Corporation—Service Parts Group).

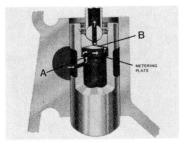

FIGURE 18-5

Metering valve mechanical tappet (Courtesy of Ford Parts and Service Division).

Metering Valve Solid Lifters

There is a type of solid lifter that is often confused with a hydraulic lifter because it uses a "metering plate" that rattles when the lifter is shaken. Here is how this lifter works.

Oil enters the tappet from the main gallery bore in the block (Figure 18-5). The oil travels, unmetered, up through the body of the lifter, passing through slot *A* on the underside of the metering plate. The oil is then precision-metered around the OD of the metering plate to slot *B* on the top face of the plate. Oil travels through the push rod, and up to the rocker arms. The amount of oil that is allowed to flow up the push rod is governed by the fit of the metering plate OD to the ID of the tappet body.

Hydraulic Valve Lifter Operation

Let's now take a look at what happens inside a hydraulic valve lifter during a complete cycle of operation. Figure 18-6 shows the lifter on the base circle of the cam. In this position the plunger spring takes up all the clearance in the valve train. Oil enters the lifter body through feed holes and flows into the inside of the plunger. The oil continues to flow down through the hole in the bottom of the plunger, around the check valve, through the holes in the check valve retainer, to completely fill the cavity below. In Figure 18-7 the lifter is raised up on the cam lobe nose. The oil below the plunger tries to escape past the check valve. This sudden rush of oil forces the check valve to seat, which seals the hole at the bottom of the plunger. The lifter then acts as a relatively solid unit. At this time, the full load of the valve train is being applied on the lifter. A predetermined and closely held clearance between the plunger and the body permits a minute amount of oil to escape from below and past the plunger. *This relative movement of the plunger with respect to the body after the check valve is seated is termed leakdown.* As the lifter returns to the cam

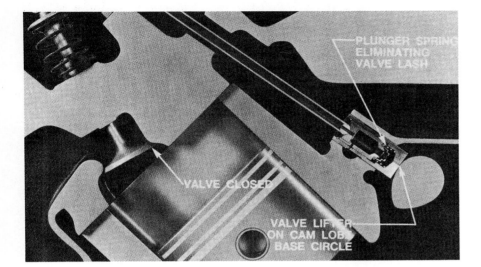

FIGURE 18-6
(Courtesy of Dana Corporation—Service Parts Group).

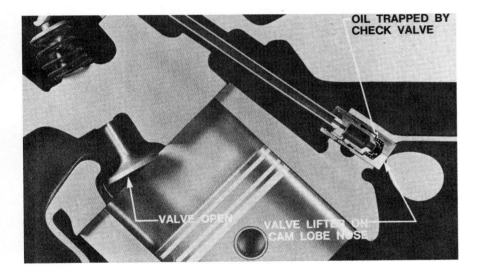

FIGURE 18-7
(Courtesy of Dana Corporation—Service Parts Group).

base circle, oil fills the high pressure cavity, and the cycle begins again (Figure 18-8). When engine temperature changes require shortening of the lifter length, the engine valve spring forces the plunger down (because of leakdown), thus automatically correcting. When lengthening of the lifter length is required, the return spring raises the plunger, causing more oil to flow into the spring cavity.

Hydraulic Valve Lifter "Pump-Up"

Generally, this condition is associated with higher engine rpm. At high speed, valve train inertia may open the valves further than they are designed to open. This results in additional valve train clearance. The lifter "senses" this clearance and the plunger begins to lengthen. The plunger may extend far enough to prevent the valve from closing. This can cause valve-to-piston interference, resulting in extensive engine damage.

Lifter "pump-up" can be caused by sticky valves, weak valve springs, or by excessive revving of the engine. Many people attribute "pump-up" to faulty hydraulic lifters. This is a misconception. The lifter automatically reacts to valve train separation and eliminates it.

Replacement Hydraulic Valve Lifters

Some mechanics think that hydraulic valve lifters used in a particular engine application must look identical. This is a common error. Actually, only four dimensions are critical.

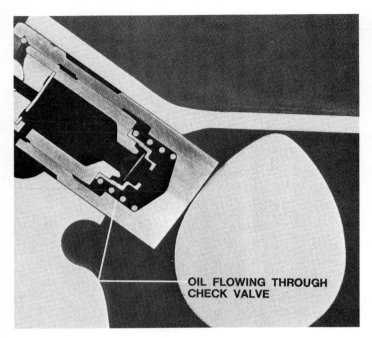

OIL FLOWING THROUGH CHECK VALVE

FIGURE 18-8
(Courtesy of Dana Corporation—Service Parts Group).

1. The outside diameter.
2. The operating height (Figure 18-9).
3. The width of the oil feed groove on the valve lifter OD.
4. The position of the oil feed groove.

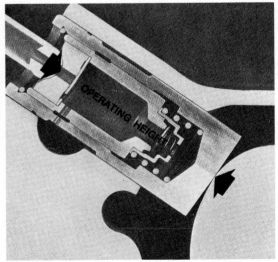

OPERATING HEIGHT

FIGURE 18-9
Always compare operating height when replacing a set of hydraulic valve lifters (Courtesy of Dana Corporation—Service Parts Group).

In Figures 18-10 and 18-11 four hydraulic valve lifters are shown. Each is a different brand, but all are used in the same engine. To compare the four lifters is interesting.

1. Each body is different in appearance.
2. The plungers vary in length and configuration.
3. The plunger springs vary in size and coil number.
4. Lifter *B* uses a conical spring under the check valve, while *A*, *C*, and *D* use a straight coil spring.
5. Lifters *C* and *D* use a ball-type check valve. *A* and *B* use a disc-type check valve. *COMMENT: Laboratory tests show that the ball check valve provides a better seal, but is more prone to leakage after extended use.*
6. Each push rod seat is recessed a different amount.

Oversize Hydraulic Valve Lifters

Over the years, a limited number of engines have been released from the factory with oversize OD lifters. Such engines are usually identified by a special mark. For example, Chrysler Corporation V-8 engines have a diamond mark on the engine number pad that indicates one or more lifters are .008" oversize. Some Oldsmobile engines use .001"

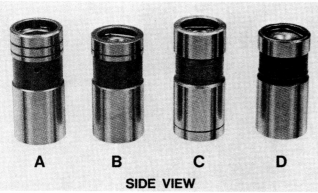

FIGURE 18-10

Outside comparison of four hydraulic valve lifters. Even though each looks different, they are for the same engine. (Courtesy of Dana Corporation—Service Parts Group).

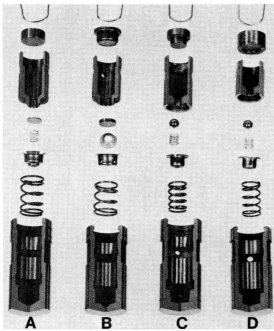

FIGURE 18-11

Internal comparison of the lifters shown in Figure 18-10 (Courtesy of Dana Corporation—Service Parts Group).

and .010″ oversize lifters. The cylinder block is marked "1" or "10" on the gasket rail under the valve cover.

Hydraulic Valve Lifter Noise Analysis

An important point to remember is that several conditions will give a noise identical to that caused by an inoperative lifter. Some of these conditions are:

1. Lack of lubrication between the rocker arm and the push rod.
2. Lack of lubrication between the rocker arm and the end of the valve stem.
3. Excessive valve guide-to-valve stem clearance.
4. Badly worn valve tips and/or rocker arm pads.
5. Worn cam lobes.
6. A loose valve seat.
7. A warped valve.
8. Excessive clearance between the lifter and the lifter bore hole in the block.
9. Loose rocker arm shaft.
10. A broken valve spring.

If you determine that the noise is not caused by one of the preceding conditions and is definitely caused by a hydraulic valve lifter, isolate the inoperative lifter. A simple way is to use a piece of three-foot long garden hose. Place one end of the hose near the spring retainer of each intake and exhaust valve and the other end of the hose to your ear. It will then be very easy to locate the offending lifter or lifters.

SHOP TIP

An inoperative lifter can also be found by stopping the engine and pushing down on each of the rocker arms (on the lifter side). If the rocker arm is free to move or there is a spongy feeling, it is a good indication the lifter is leaking down too fast or not receiving oil from the engine. See Figure 18-12.

Once you locate the lifter that is causing the noise, try to establish the reason for the malfunction. Basically, there are four types of noise that result from an inoperative lifter.

1. **Loud, rapping noise**   This can be caused by the plunger being "frozen" in the body, usually due to excessive varnish between the plunger and body, or by foreign matter (dirt or metal chips) wedged between the plunger and body. Another cause of a loud rapping noise is an excessively worn or spalled foot on the lifter (Figure 18-13).

2. **Moderate clicking noise**   As was the case with the loud, rapping noise, moderate click-

FIGURE 18-12
This hydraulic lifter was completely inoperative. The bottom face wore completely through to the pressure cavity.

FIGURE 18-13
Badly worn lifter foot.

ing can be caused by varnish or a worn lifter bottom. The noise intensity depends on the amount of varnish and the degree of wear. Two other causes of a moderate clicking noise are fast or slow leakdown. Slow leakdown generally will cause the engine to be noisy only when the engine is cold. With fast leakdown, the valve train will become noisy when the engine is warm. Fast leakdown will also occur if the ball check fails to seal.

3. **Intermittent clicking noise** This type of noise is hard to locate because there will be a few clicks, and then the noise will disappear and reappear after a short period of time. The usual cause of intermittent clicking is a very minute piece of dirt that holds the ball check off the seat for a few seconds and then passes through. In rare cases, the cause of the intermittent clicking is a pitted or flat spotted ball check. Figure 18-14 shows an unusual lifter failure that caused intermittent noise.

4. **General noise throughout the valve train** When the noise is throughout the en-

FIGURE 18-14
A cracked hydraulic lifter body.

tire valve train, the cause will usually be found in the oil or the oil supply. Too much oil in the crankcase will cause foaming and aeration. When the air gets into the lifters, they will fail to operate properly.

Insufficient oil supply to the lifters can also cause general valve train noise. This could be the result of not enough oil in the crankcase, an oil pump not operating correctly, or clogged main oil gallery lines.

SERVICE HINT FOR BIG BLOCK CHEVROLET ENGINES

Valve train noise may be caused by an oil gallery air lock. Sometimes an air pocket will develop in the oil galleries where they dead end at the front of the block. This causes oil flow to be restricted to the front hydraulic lifters. The lifters will leak down and partially collapse at higher engine speeds. The formation of the air lock can be prevented by drilling a 1/32" bleed hole in the middle of each front oil gallery plug. Do not drill the hole any larger or a significant engine oil pressure loss can result.

CAMSHAFT/LIFTER RELATIONSHIP

The mating relationship of the lifter bottom to the camshaft lobe is very critical. From visual appearance, the cam lobes look straight across. Actually, most automotive camshafts (with the exception of some older Chevrolet and Buick engines) are ground with a face taper of .0007" to .002" (Figure 18-15). The bottom of the lifter is spherically ground with a .002" crown in the center (Figure 18-16). When installed in the block, the lifters are offset to the centerline of the cam lobe toward the low side of the lobe. The mating between the lobe and lifter will then occur at a single point somewhere near center (Figure 18-17). This prevents loading on the cam lobe edge, and aids in rotating the lifter to distribute the load (normal loading between a cam lobe and valve lifter often reaches 100,000 psi).

Clearance Ramps

A solid lifter cam lobe has a clearance ramp that allows the cam to rotate a few degrees off the base circle in order to gradually take up the valve train clearance (Figure 18-18). A hydraulic lifter

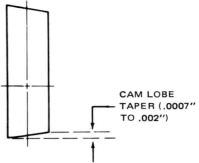

FIGURE 18-15
Camshaft lobes are generally ground to a slight taper. The taper is usually less than .001".

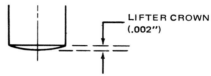

FIGURE 18-16
Lifter bottoms are radius ground so that they do not ride on the edge of the camshaft lobes.

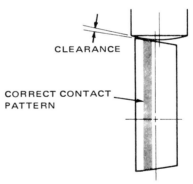

FIGURE 18-17
Correct contact between lifter and camshaft lobe.

maintains zero lash. Therefore, a hydraulic lifter camshaft has no clearance ramps ground into the lobes. Using solid lifters on a hydraulic camshaft will result in severe valve train pounding, severe wear to the lobes and lifters, and will change valve timing.

Camshaft/Lifter Wear

The correct mating path between a lifter and lobe will show a defined track around the entire lobe

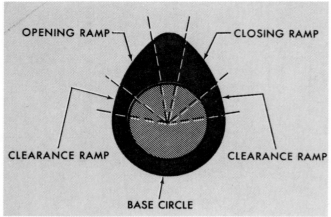

FIGURE 18-18
Names of solid lifter cam lobe lift areas (Courtesy of Ford Parts and Service Division).

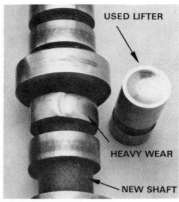

FIGURE 18-20
This is what happens when used lifters are used on a new camshaft (Courtesy of TRW Replacement Division).

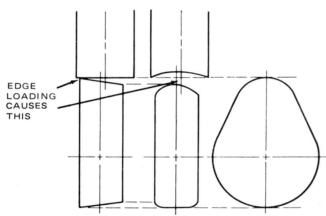

FIGURE 18-19
Point contact of the cam lobe and lifter will cause spalling of the lifter and early failure of both parts.

FIGURE 18-21
Badly sludged lifter (Courtesy of Dana Corporation—Service Parts Group).

as shown in Figure 18-17. The majority of camshaft wear difficulties occur when this mating pattern takes place on the edge of the cam lobes rather than on the track near center (Figure 18-19).

A lifter with a worn bottom will cause cam lobe edge loading. Proper rotation will not occur, and the camshaft and lifter will soon be destroyed (Figure 18-20).

When the lifters are severely varnished or filled with sludge, they should be replaced (Figure 18-21). Varnish and sludge can cause sticking and improper lifter operation. If either the valve lifters or cam lobes are worn at all, both should be replaced. *Do not install worn valve lifters with a new*

camshaft. Do not install new valve lifters on a worn camshaft. See Figure 18-22.

Some other, often overlooked, reasons for camshaft and lifter failure are listed here.

1. Insufficient idle speed or lubrication during the break-in of a new camshaft.
2. Dilution of the oil by fuel or coolant.
3. Connecting rods installed backward (oil spit hole facing away from the camshaft).
4. Introduction of foreign material on the lobe surface.
5. Excessive valve spring pressure.
6. Incorrect valve clearance adjustment.

FIGURE 18-22
Closeup of a "flat" camshaft.

FIGURE 18-23
EOS additives.

7. Excessive camshaft end play.
8. Incompatibility of lifter and cam material.
9. Reinstalling used lifters on different lobes.
10. Distributor gear failure due to excessive drag.

Camshaft/Lifter Installation and Break-In

When installing a new set of lifters and/or a camshaft, the following procedures are recommended.

1. Thoroughly clean the lifter bores.
2. Prelube the cam lobes generously with a lubricant containing an EP (extreme pressure) additive. Lubricants such as engine oil, white grease, wheel bearing grease, or STP are not recommended.

   This author recommends using an engine oil supplement (EOS) additive for camshaft and lifter prelubrication. EOS is a friction reducing additive recommended as an aid to initial "sea- soning" after engine teardown and installation of new parts (camshaft bearings, main and rod bearings, valve lifters, and so on).

   The regular use of EOS, as with each engine oil and filter change is not recommended. EOS contains a high percentage of zinc. Engineering data (published by SAE) concludes that over an extended period, too much zinc can cause pitting and premature deterioration of engine bearing surfaces.

   AMC, Chrysler, Ford, and General Motors dealership parts departments generally stock EOS (Figure 18-23). The part numbers are as follows—AMC #8991634, Chrysler #4-185, Ford #D2AZ-19579-A, General Motors #1051396 or #105004.

3. The most critical time in a camshaft's life is the first 15 minutes. The characteristics of the lifter and camshaft material allow "wear in" to each other after a few minutes of operation. To enable the camshaft and lifters to wear in, always fast idle (1200 to 1800 rpm) the engine for the first 30 minutes of operation. *Do not rev the engine higher or allow it to idle.*

4. Early camshaft failure can also result from incorrect installation. Many V-8 engines use a bolt or bolts to hold the timing sprocket, thrust plate or washer, and other parts in place. If these items are assembled in the wrong order, or if the bolt is not tightened sufficiently, the camshaft may be free to slide toward the rear of the engine. This may let the cam lobes collide with adjacent lifters (Figure 18-24). Lobe and/or lobe edge chipping result.

FIGURE 18-24
A camshaft free to slide toward rear of the engine lets lobes contact adjacent lifters and cause damage (Courtesy of TRW Replacement Division).

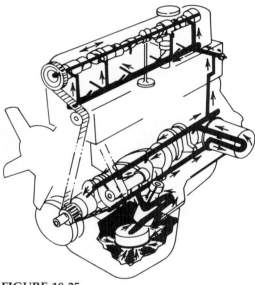

FIGURE 18-25
Ford Pinto 2300 cc engine lubrication circuit (Courtesy of Sealed Power Corporation).

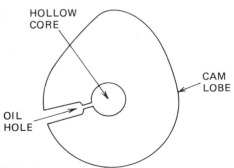

FIGURE 18-26
Oil holes in the Ford Pinto 2300 cc cam lobes (Courtesy of Sealed Power Corporation).

Ford Pinto Camshaft Failure

Camshaft lobe failures on the 2300 cc engine are not uncommon. The problem is usually related to lubrication (lack of oil, dirty oil, or the wrong oil). To help avoid a repeat failure after installing a new camshaft, the following is recommended.

1. Use API Service Classification SF oil.
2. Check all the overhead oil holes for restriction (Figure 18-25). Note that the camshaft is fed oil through the no. 2 journal, then through the hollow center of the camshaft to each cam lobe. The rest of the journals feed oil to their respective cam bearing only.
3. The oil metering hole in each cam lobe should be checked for blockage (Figure 18-26).
4. Cam journal-to-bearing clearance should be measured. Clearance greater than specifications will result in an oil pressure drop to the cam lobes.
5. Install new camshaft fingers (followers).
6. Change oil and replace the filter.

> **SHOP TIP**
> Removing a stuck or broken valve lash adjuster from a Pinto OHC engine can be a very difficult job (they are too hard to drill out). The following method of removal can be used.
> 1. Enlarge the lifter boss oil hole with a size D drill.
> 2. Tap the hole with a 1/16" pipe tap.
> 3. Install a grease fitting.
> 4. Use a grease gun on the fitting. The lash adjuster will "jump up" to a point where it can be easily removed with vise grips.

CAMSHAFT SERVICE

Because the camshaft turns at half crankshaft speed, its bearing journal surface speeds are less than those of the crankshaft. This is one reason why camshaft bearings generally wear less than connecting rod bearings. However, camshaft bearings tend to wear out-of-round. *Cam bearing clearance of .003" or more will result in more than normal internal oil leakage and a drop in engine oil pressure.* Clearance less than specifications can cause a spun cam bearing.

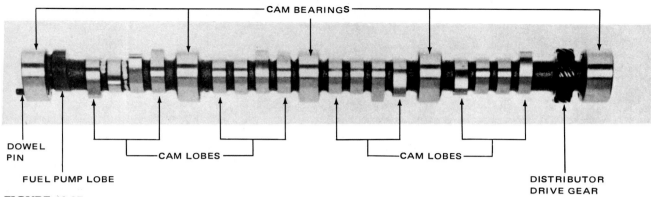

FIGURE 18-27
Camshaft terminology.

Inspect Camshaft

Replace a camshaft that has scored, chipped, pitted, worn, or badly discolored lobes. Replace a camshaft that has visibly worn, broken, chipped, or badly scored drive gear teeth. Refer to Figure 18-27. Check the camshaft for straightness.

Measuring Lobe Lift with Camshaft in Engine

Figure 18-28 shows a dial indicator setup to use for measuring lobe lift wear on an OHV engine without removing the camshaft.

On an OHC engine, this measurement is much simpler. Using a vernier or dial caliper, measure the lobe height and the base circle diameter (Figure 18-29). The difference between these two measurements is the amount of lobe lift.

Camshaft End Play

In order for the camshaft lobes to properly line up with their valve lifters, the camshaft has to be restricted in its endwise movement. One method is to bolt a thrust plate or washer to the front of the cylinder block (Figure 18-30). This thrust plate allows the camshaft to move only a slight amount from its correct position.

Some engines control camshaft end play by the side stiffness of the chain. The small block Chevrolet is one such example.

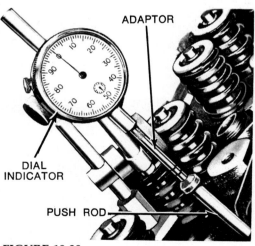

FIGURE 18-28
Checking lobe lift on an OHV engine (Courtesy of Ford Parts and Service Division).

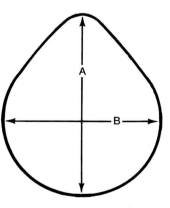

A — B = LOBE LIFT

FIGURE 18-29
$A - B$ = lobe lift (Courtesy of Ford Parts and Service Division).

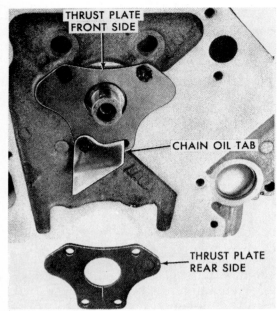

FIGURE 18-30
Camshaft thrust plate installation (Courtesy of Chrysler Corporation).

Camshaft Straightening

Once the amount of bend has been measured (and the high side determined), the camshaft can be straightened. Camshafts are generally straightened on centers by applying moderate pressure with a blunted air chisel (Figure 18-31).

A camshaft can also be straightened by using

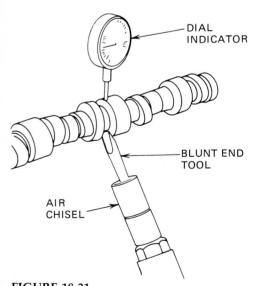

FIGURE 18-31
Straightening a camshaft. Hold the camshaft on centers and peen the low side.

two 2 × 4 wooden blocks placed on the floor by a workbench, and your body weight. Set the blocks under each end bearing journal and position the camshaft with the high side facing up. Keep your balance by holding on to the workbench and apply pressure to the camshaft by standing with both feet on the high side. Maintain this position for approximately 30 seconds. Recheck the camshaft tir and repeat the straightening procedure as required. Although this method may sound primitive, it works very well in the absence of a commercial straightening fixture.

Always check a camshaft for straightness before installation.

Camshaft Surface Treatment

After a camshaft is ground it is given a special surface treatment that aids the break-in of the lobes and lifters. This treatment is usually a molybdenum disulphide spray coating or a manganese phosphate etching process (Figure 18-32). The surface treatment turns the camshaft a dirty

FIGURE 18-32
Grinding the camshaft and treating it in a manganese phosphate solution soak tank.

black color (Figure 18-33). Because of this, a mechanic may want to clean the lobe surfaces. *However, do not wash or sand off this coating, as it is put on for the engine's protection.*

Checking Camshaft End Play

To check camshaft end play, mount a dial indicator against the timing sprocket or gear (Figure 18-34). Then proceed as follows.

1. With your hand, push the camshaft to the rear of the engine as far as it will go.
2. Zero the indicator hand.
3. Use a large screwdriver and *lightly pry* behind the sprocket or gear to move the camshaft forward.
4. Compare the indicator reading with specifications. If correction is necessary, install a new or different thickness thrust plate.

Excessive camshaft end play can change distributor timing as the cam thrusts back and forth during engine operation. In addition, cam lobe damage can result. However, insufficient end play can cause bind and subsequent camshaft breakage.

Measuring Camshaft Base Circle Runout

Cam lobe base circle runout is a very important, seldom performed, camshaft measurement (Figure 18-35). Excessive base circle runout can cause hydraulic pump-up to occur at a lower rpm limit. Also, correct valve train clearance adjustment is impossible to obtain. *Maximum total indicated runout on the cam lobe base circle is .001".*

Base circle runout can be checked with vee blocks and a dial indicator. Rotate the cam lobe until the dial indicator needle shows the start of the base circle (this can be determined by watching until the needle just reaches its lowest counterclockwise point). Then, continue to slowly rotate the camshaft until the lift ramp on the other side of the lobe is encountered.

Camshaft Bearing Installation

Camshaft bearing installation is a very critical operation, and it is essential that certain procedures be observed.

1. Check the cam bearing bores for rust, burrs

FIGURE 18-33
A surface treated camshaft.

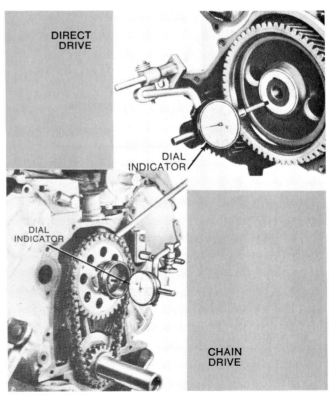

FIGURE 18-34
Measuring camshaft end play (Courtesy of Ford Parts and Service Division).

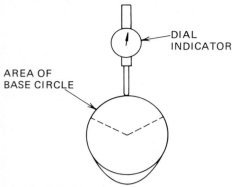

DIAL INDICATOR

AREA OF BASE CIRCLE

FIGURE 18-35
Checking concentricity (runout) of the lobe base circle.

CONSISTS OF	POSITION
2—SH 287	3-4
1—SH 288	2
1—SH 289	5
1—SH 290	1

FIGURE 18-36
Typical information usually boxed with a set of cam bearings. The left column shows the quantity of each bearing in the box and its identification number (SH 287, SH 288, and so on). The numbers to the far right show the position the bearings must be installed. Number 1 bearing is at the front of the engine, with the rest of the bearings numbered consecutively from this point.

nicks, scratches, and scores. The bores can be cleaned up by using a Flex-Hone®.[1] In order to reach all the bores with the hone it may be necessary to braze on an extension.

2. With a piece of crocus cloth, break the sharp lead-in edge of each bearing bore. This will prevent shaving off material from the bearing OD during installation.

3. Check the bearing bores for correct size. *NOTE: Some OEM blocks have been produced with oversize cam bearing bores.*

4. Make certain that the cam bearings are correct for the application.

5. Determine the correct sequence position for each bearing. This is very important because in many engines each bearing is a different size. The largest is in the front, and the smallest is in the rear.

Cam bearing location is numerical in order from the front of the engine to the rear. Number 1 is at the front, then number 2, number 3, and so on. Each bearing will generally have the number position stamped on its back. Also, the bearing box usually contains location information (Figure 18-36).

6. Select the right driver(s). When installing new cam bearings, you usually start at the rear and work up to the front.

7. Install the cam bearings dry. For easier installation, each bearing should be installed with the OD chamfered edge facing into the bore. Do not let the driver "bounce" against

the bearing during installation. This can "upset" material on the bearing edge and cause binding.

8. Make sure that each cam bearing is driven in straight to the proper location, and that the cam bearing oil holes are indexed (in line) with the oil holes in the block.

9. After the cam bearings are installed, check the alignment by installing the camshaft. Slowly rotate the camshaft, as it is inserted. Be very careful not to damage the cam bearings. There should be no binding. If binding is encountered, *do not force or drive.*

SHOP TIP

Fabricate a camshaft holding tool from a piece of round steel stock. This will provide leverage to hold and support the cam during installation (Figure 18-37).

Binding can be caused by several conditions—a bent camshaft, an improperly installed cam bearing, or cam bearing bore misalignment. In locating the reason for the bind, start by rechecking the camshaft straightness. Then, try to install the camshaft. Turn it by hand against any high spots. The point of binding will show as a shiny spot on the cam bearing. These high spots can be removed by using a bearing scraper or by honing (Figure 18-38). The fit of the camshaft is considered acceptable if it can be turned by moderate hand pressure. In the case of severe misalignment, the cam bearing tunnel will have to be trued by boring and installing oversize OD bearings or by installing semifinished bearings and machining to size.

---

[1]Flex-Hone® is the registered trademark for the hones manufactured by the Brush Research Manufacturing Co., Inc., Los Angeles, CA.

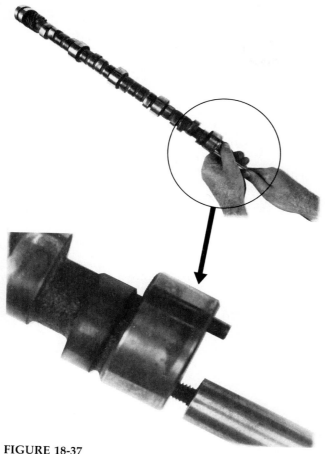

FIGURE 18-37
Camshaft installation tool.

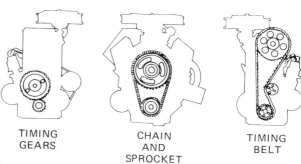

TIMING GEARS

CHAIN AND SPROCKET

TIMING BELT

FIGURE 18-39
Different cam drive arrangements (Courtesy of TRW Replacement Division).

VALVE TIMING MECHANISMS

Proper engine valve timing is critical. The valve action has to be kept in correct relationship to the piston movement. This is done by transmit-ting power from the crankshaft to the camshaft by means of timing gears, chain and sprockets, or belt and sprockets (Figure 18-39).

Excessive wear or improper installation of these parts will greatly affect the efficiency and performance of the engine. For example, a crank gear incorrectly meshed with a cam gear by only one tooth to the right or left will cause a lag or advance of valve opening or closing of from 3° to 7° of crank travel.

Timing Gears

Timing gears are made in aluminum, fiber, cast iron, and case hardened steel. The teeth are shaved at the crown to avoid concentrating the load on the gear teeth ends (Figure 18-40). Also, this helps to distribute the load equally across the gear face.

FIGURE 18-38
Bearing scraper tool.

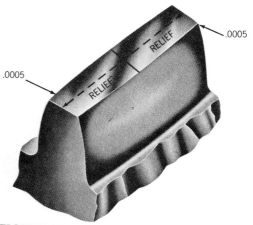

.0005

.0005

RELIEF

RELIEF

FIGURE 18-40
Crown shaved tooth profile (Courtesy of TRW Replacement Division).

Timing Gear Installation Tips

1. It is recommended to install a pressed-on camshaft gear with the camshaft out of the engine. Trying to drive it on in the engine could drive out the cam plug at the rear of the block.

2. Clean the gear and apply a daub of Loctite to the ID of the gear before pressing it on.

3. Prior to pressing on an aluminum or all metal timing gear, heat the bore to approximately 200° F. Use a hot plate, hot oil, or boiling water. Be sure the gear is installed hot to permit easy fitting. Do not heat fiber gears.

4. Remove all burrs from the gear shaft.

5. Note the location of the gear timing marks. The gears are usually installed with the marks facing out. The timing marks should be aligned after gear installation (Figure 18-41).

6. Align gear keyways with the shaft. Start the gear squarely into position.

7. Avoid exerting pressure against any part of the gear, except the steel hub (Figure 18-42). This is especially true with fiber gears. After the gear is pressed on, check thrust plate clearance (Figure 18-43).

8. Use a feeler gauge to check backlash (clearance) between the mating gear teeth at several points (Figure 18-44). As a general rule, the backlash should be between .003" and

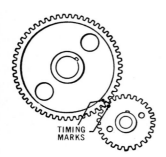

FIGURE 18-41

Typical gear timing mark alignment (Reprinted from *Motor Handbook* © 1983, The Hearst Corporation).

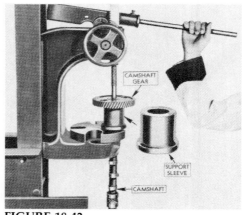

FIGURE 18-42

Pressing off camshaft gear with an arbor press. Support sleeve is positioned against the gear hub. (Courtesy of Chevrolet Motor Division of General Motors Corporation).

FIGURE 18-43

Measuring camshaft thrust plate clearance (end play) (Courtesy of Chevrolet Motor Division of General Motors Corporation).

.010". If not, check on the availability of oversize (+) and undersize (−) camshaft gears.

9. Cam and crank gears should have less than .003" face runout when checked with a dial indicator (Figure 18-45).

Timing Gear Failure Analysis

Timing gear failure is usually the result of improper backlash. Excessive backlash can cause noise, high impact loads, and often breakage. Insufficient backlash puts a bind on the gears. The lubrication film between the teeth is rup-

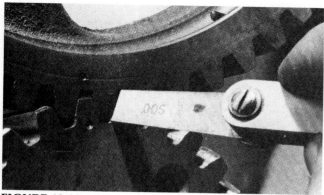

FIGURE 18-44

Measuring timing gear backlash (Courtesy of TRW Replacement Division).

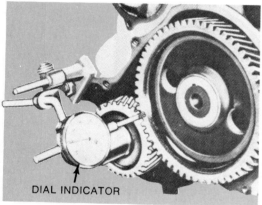

DIAL INDICATOR

FIGURE 18-45

Checking camshaft gear face runout (Courtesy of Ford Parts and Service Division).

tured, and the result is severe wear (Figure 18-46).

Timing Chain and Sprocket Assembly

Today, there are basically two types of timing chains in use—the silent tooth and the roller type (Figure 18-47). Timing chain sprockets are manufactured in a variety of materials (Figure 18-48).

Timing Chain and Sprocket Wear

There are more bearings in a timing chain than in the entire engine. Wear in the chain or sprockets means a timing lag that results in poor engine performance. Follow this rule-of-thumb to determine when to replace a timing chain.

1. Turn the cam sprocket by hand to get all the slack on one side of the chain.

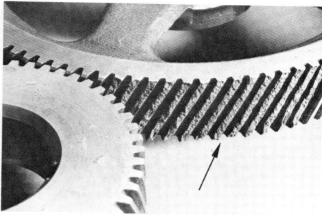

FIGURE 18-46

Badly worn cam gear (Courtesy of TRW Replacement Division).

2. Grip the slack side of the chain with your thumb and index finger midway between the sprockets.

3. If you can move the slack side in and out more than $\frac{1}{2}$", the chain should be replaced (Figure 18-49).

4. Before replacing the chain, run your fingernail along the teeth of both sprockets. If your fingernail catches at any point there is sufficient wear to warrant replacing the sprockets. Worn sprockets may cause a damaging valve time lag.

Timing Chain and Sprocket Installation Tips

1. Make a preinstallation check of the cam sprocket. In some cases, the replacement sprocket may incorporate an integral spacer on the back. This eliminates the need for the separate factory spacer. To determine if the spacer is to be reused, always read the instruction sheet in the parts package.

2. Install the chain and sprockets as an entire assembly, keeping keyways and timing marks aligned (Figure 18-50). Press both sprockets on evenly, making sure they stay parallel. Do not hit the chain with a hammer—this can bend the pins and links in the chain. Use a sleeve and tap gently (Figure 18-51).

    Some engines run dampers or tensioners against the timing chain (Figure 18-52). Visually inspect the rub contact surface(s) before installing a new chain.

ROLLER CHAIN

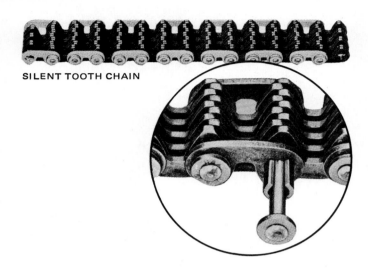

SILENT TOOTH CHAIN

FIGURE 18-47
(Courtesy of TRW Replacement Division).

CAST IRON
SPROCKET

ALUMINUM—WITH—NYLON
CAMSHAFT SPROCKET

POWDERED METAL AND
STEEL CRANK SPROCKET

FIGURE 18-48
(Courtesy of TRW Replacement Division).

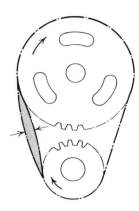

FIGURE 18-49

If over ½″ movement, replace timing chain (Courtesy of Cy-lent Timing Gears Corporation).

FIGURE 18-51

Using a sleeve to tap on a cam sprocket (Courtesy of TRW Replacement Division).

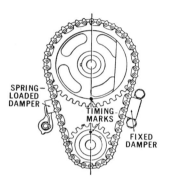

FIGURE 18-52

Buick V-6, Oldsmobile V-6, and Pontiac four-cylinder engines use this chain and sprocket setup. The timing mark alignment shown is typical of most OHV engines (Reprinted from *Motor Handbook* © 1983, The Hearst Corporation).

FIGURE 18-50

Do not pry or stretch chain over its sprockets. Install the parts as an assembly (Courtesy of Ford Parts and Service Division).

Timing Chain and Sprocket Failure Analysis

Chain and/or sprocket failure is generally the result of misalignment, overload, or hammering on the sprockets.

1. **Misalignment** If the cam and crank sprockets are not properly aligned, the chain is forced to operate in a twisted position. This causes

FIGURE 18-53

Side wear on sprocket teeth (Courtesy of TRW Replacement Division).

heavy wear on the sides of the sprocket teeth (Figure 18-53). Because of the high stress placed on the chain links, the chain may break. Breakage due to misalignment usually can be identified by a diagonal break across the links (Figure 18-54).

Failure to properly tighten sprocket mounting bolts, and excessive wear on the cam sprocket thrust surface are two common reasons for misalignments.

2. **Overload** If a foreign object lodges between the chain and sprocket, the chain can break. Figure 18-55 shows a chain broken by improper installation. A screwdriver was

FIGURE 18-54

This chain broke because of misalignment (Courtesy of TRW Replacement Division).

FIGURE 18-55

This chain broke because of overload (Courtesy of TRW Replacement Division).

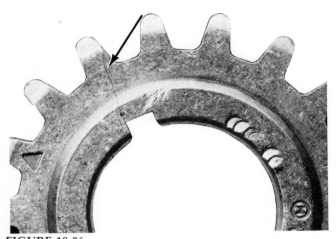

FIGURE 18-56

Improper installation caused this crank sprocket to crack (Courtesy of TRW Replacement Division).

being used to stretch the chain over the cam sprocket while another person cranked over the engine. *This method of installation is never recommended.* Breakage due to overload usually can be identified by a straight break across the links.

3. **Hammering on the sprockets** Hammering often causes cracking through the keyway (Figure 18-56). Sometimes a tooth will break off. Use a sleeve (a piece of pipe or tubing) to tap on the sprockets.

OHC Timing Belts

The OHC design has proven to be very desirable for many subcompact four-cylinder engines. Since there is less valve train weight, the engine will generally rev higher and produce a decent amount of power even with small displacement. The cogged (toothed) rubber timing belt is being used in many of these engines (Figure 18-57). This belt offers several advantages when compared to a long drive chain.

FIGURE 18-57

Timing belt (Courtesy of TRW Replacement Division).

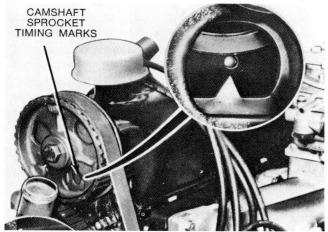

FIGURE 18-58

Pinto engine camshaft sprocket timing marks (Courtesy of Ford Parts and Service Division).

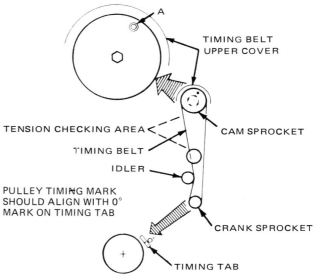

FIGURE 18-60

Chevette engine valve timing and timing belt tension (Courtesy of Chevrolet Motor Division of General Motors Corporation).

1. A belt stretches very little in service (due to its fiberglass cord reinforcement).

2. Since a belt requires no lubrication, it can be replaced without disturbing oil tightness.

3. A belt is relatively cheap and almost noiseless.

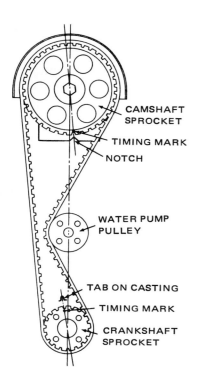

FIGURE 18-59

Chevrolet Vega engine valve timing marks (Reprinted from *Motor Handbook* © 1983, The Hearst Corporation).

Installation Tips for OHC Timing Belts

1. Handle belts carefully. **NOTE:** *The belt used on the German built two-liter Pinto engine has "Nicht Knicken" (Do not crimp) printed on it.* This warning applies to all timing belts.

2. Keep belts free of oil, grease, and dirt.

3. Timing belt sprockets are relatively soft. Do not turn them with a pipe wrench. A gouge can easily tearup a belt.

4. Never turn the crankshaft backward with the belt installed. The belt could jump several teeth and change valve timing.

5. Valve timing for most OHC engines is correct when the #1 piston is at TDC and the cam sprocket timing marks are aligned (Figures 18-58, 18-59, and 18-60).

6. Always follow the factory recommendations when adjusting a timing belt (Figures 18-61 and 18-62).

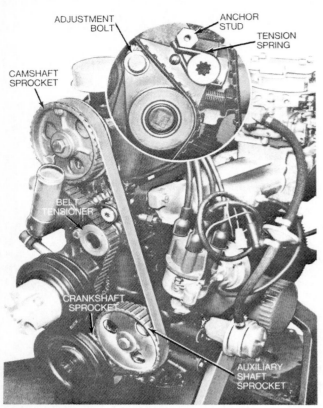

FIGURE 18-61

Pinto engine timing belt adjustment (Courtesy of Ford Parts and Service Division).

Warning for Some OHC Engines

Several model OHC engines use a timing belt cover that is open at the bottom. There have been a number of reported cases during the winter of these engines "jumping time." The problem occurs when a driver noses the vehicle against a pile of snow while parking. The snow wedges between the timing belt and crank sprocket. Several hours later the driver returns to start the vehicle. As soon as the engine is cranked the belt jumps several cogs and there is a no-start condition.

FIGURE 18-62

Vega engine timing belt adjustment. Tension is set by moving the water pump (which acts as an idler pully). The special tool shown in the inset will tighten the belt to correct tension when 15 ft/lbs. are applied to it. Hold the torque and tighten the water pump bolts (Courtesy of Chevrolet Motor Division of General Motors Corporation).

TRUE CASE HISTORY

Vehicle   Toyota Corona

Engine   4 cylinder, 18R-C model (1968 cc)

Mileage   86,000

Problem
1. Lack of power, backfiring, and hard starting.
2. Widely fluctuating timing mark readings.
3. Excessive distributor rotor back and forth movement.
4. Acted just like a worn timing chain and sprocket assembly.

Cause   The distributor drive gear teeth were badly stripped due to a lubrication failure (Figure 18-63).

FIGURE 18-63
Damaged distributor drive gear.

Cure   The distributor drive gear, and its mating gear (the oil pump driveshaft) were replaced. On the later model 18R-C engine, oil flow to these gears was improved.

CHAPTER 18 SELF-TEST

1. Basically, what is the purpose of the camshaft?
2. The intake valves on most engines open before the piston begins the intake stroke. Why?
3. Define valve overlap.
4. What is the primary function of valve overlap?
5. What advantages do hydraulic valve lifters have over solid lifters?
6. What is the purpose of the metering plate used in some solid lifters?
7. Explain the term "leakdown."
8. What is "pump-up?"
9. What are the critical dimensions to consider when replacing hydraulic valve lifters?
10. How can a Chrysler Corporation engine equipped with factory oversize OD lifters be identified?
11. Name five conditions that will make a noise identical to that caused by an inoperative lifter.
12. How can an inoperative lifter in an engine be isolated?
13. What can cause a lifter to make an intermittent clicking noise?
14. What can cause a lifter to make a loud, rapping noise?
15. What is sometimes done to eliminate valve train noise in big block Chevrolet engines?
16. The bottom of a new lifter should have what configuration?
17. Why doesn't a hydraulic lifter camshaft have clearance ramps ground into the lobes?
18. The correct mating path between a lifter and lobe will show what kind of pattern?
19. What is edge loading?
20. List twelve possible reasons for a camshaft and lifter failure.
21. When installing a new camshaft and lifters, what is the recommended prelubricant?
22. Why is regular use of EOS not recommended?
23. How should an engine with a new camshaft and lifters be operated for the first 30 minutes?
24. What can cause lobe chipping?
25. You have just replaced the camshaft and fingers in a Ford 2300 cc engine. What type of engine oil should be used to help avoid a repeat failure?
26. How can a broken valve lash adjuster from a Pinto OHC engine be removed?
27. Why do camshaft bearings generally wear less than connecting rod bearings?
28. What can happen if cam bearing clearance is less than specifications?
29. How can lobe lift wear on an OHV engine be measured without removing the camshaft?

30. What is the purpose of a camshaft thrust plate?
31. How can a camshaft be straightened by using your body weight?
32. What is the reason for surface treating a camshaft after it is ground?
33. What problems can be caused by excessive camshaft end play? Insufficient end play?
34. How is camshaft base circle runout checked?
35. Maximum base circle total indicated runout is _____ ".
36. Excessive base circle runout on an engine equipped with hydraulic valve lifters can cause what problem?
37. Prior to installing camshaft bearings, why should you "break" the sharp lead-in edge of each bearing bore?
38. How can the correct sequence position for each camshaft bearing be determined?
39. Should cam bearings be installed dry?
40. New cam bearings are installed in an engine. Binding is encountered as the camshaft is inserted. This could be caused by what conditions?
41. Timing gears are made of what materials?
42. Why are timing gear teeth shaved at the crown?
43. Why is it risky to install a pressed-on camshaft gear when the camshaft is still installed in the engine?
44. How should an aluminum timing gear be heated when pressing it on?
45. When pressing on a fiber gear, where should the pressure be exerted?
46. Generally, timing gear backlash should not exceed _____ ".
47. What is the maximum allowable cam and crank gear face runout?
48. What problems can be caused by excessive timing gear backlash? Insufficient backlash?
49. What are the two basic types of timing chains in use?
50. What is the timing chain slack limit?
51. What can happen if a timing chain is hit with a hammer?
52. How can timing chain breakage due to misalignment usually be identified?
53. How can timing chain breakage due to overload usually be identified?
54. What advantages does a timing belt offer over a timing chain?
55. Why shouldn't a crankshaft be turned backward with the timing belt installed?
56. Valve timing is correct for most OHC engines when the #1 piston is at _____ and the cam sprocket timing marks are aligned.
57. What is the warning regarding OHC engine timing belt covers that are open at the bottom?

CHAPTER 19

CRANKSHAFTS AND FLYWHEELS

When a crankshaft is removed from an engine, it is often only measured for size. If measurement indicates reusable dimensions, the crankshaft sometimes is assumed to be ready for reinstallation. To assume this is dangerous shop practice. The crankshaft should receive further inspection and service.

RECOMMENDED CRANKSHAFT INSPECTION PROCEDURE

1. Inspect keyway condition. Damaged keyways are often the result of lower than specified crankshaft damper bolt torque. Due to the unsatisfactory clamping force, the crankshaft sprocket is then driven by the key rather than by bolt friction.
   Always inspect the flat washer under the damper bolt. If the washer is distorted or dished, it should be replaced.

2. Visually inspect the snout threads. If there is damage, determine if repair can be made by chasing the threads, or by installing a thread repair insert.

3. Visually inspect the flywheel flange bolt holes. Chase the threads or install thread repair inserts as required.

4. Inspect the rear main oil seal surface for wear. Wear sleeves and special "oversize" seals are available for certain model engines.

SERVICE TIP FOR CHEVROLET 153, 194, 230, AND 292 CID ENGINES

These engines often develop a serious leak at the rear main oil seal that does not respond to conventional repair procedures. To correct this problem, Fitzgerald Gasket Company manufactures a .010" smaller ID seal (part no. CB 136). This special seal is designed to cure rear main oil seal leak problems on these engines.

---

5. Check the main bearing journals and crankpins with a micrometer (see Figure 19-1).

6. With the crankshaft mounted in vee blocks, check for bend, snout trueness, and flywheel flange runout. Correct as necessary. Some shops straighten crankshafts by striking the fillet area with a rounded chisel. *This method*

398

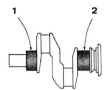

TWO MAIN BEARING CRANKSHAFT
(2 CYLINDERS)

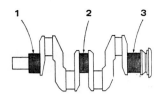

THREE MAIN BEARING CRANKSHAFT
(4 CYLINDERS)

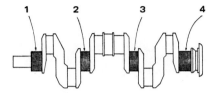

FOUR MAIN BEARING CRANKSHAFT
(6 CYLINDERS)

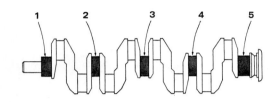

FIVE MAIN BEARING CRANKSHAFT
(8 CYLINDERS)

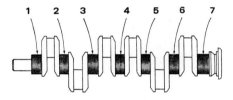

SEVEN MAIN BEARING CRANKSHAFT
(6 CYLINDERS)

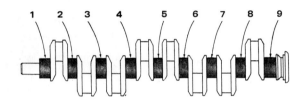

NINE MAIN BEARING CRANKSHAFT
(8 CYLINDERS)

FIGURE 19-1

Crankshaft classifications. The shaded areas are main bearing journals with position numbers. The crankpins are unshaded (Courtesy of Federal-Mogul Corporation).

is not recommended by any engine manufacturer. The best way to straighten crankshafts is with a specially designed press (Figure 19-2).

7. Visually inspect the entire crankshaft for any deviation from normal (Figure 19-3). Nicks, dents, or pits in the journal OD must not exceed 1/8", with no raised metal (Figure 19-4). Any journal OD score marks must be:

- At least 1/8" from the fillet tangent point.
- A maximum of .010" wide.
- Circumferential, not axial.

8. Magnetically inspect the shaft after visual inspection.

9. If the shaft requires machining, check the fillets, thrust, and all crankpin widths to determine if enough stock remains to warrant regrinding. See Figure 19-5. *Maximum thrust wall runout is .0015" tir.*

MACHINING RECOMMENDATIONS

1. If possible, have the crankshaft ground between centers (as opposed to holding it in two universal chucks).

2. Grind main bearing and crankpin OD surfaces to specified undersize (Figure 19-6). All journals should be ground to the same undersize.

3. Grind the rod journal side walls just wide enough to clean up the basic face surface. Do not exceed the allowable crankpin width (Figure 19-7).

4. Restore the fillets. *The radius should never be less than 1/64" below the original radius (Figure 19-8).*

5. Re-radius and repolish the edges of all oil holes and grooves (Figure 19-9).

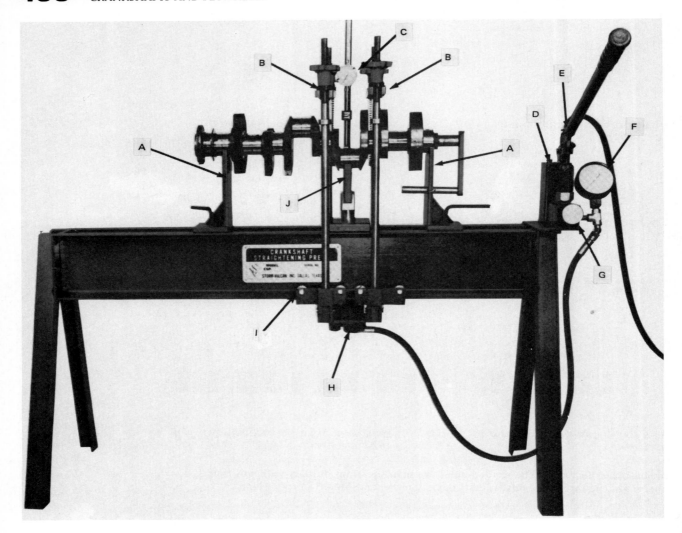

A. V BLOCKS
B. OVER—THE—TOP CLAMPS
C. DIAL INDICATOR
D. HYDRAULIC PUMP
E. AIR VALVE
F. HYDRAULIC PRESSURE GAUGE
G. PUMP VALVE
H. HYDRAULIC CYLINDER UNIT
I. STEEL ROLLERS
J. RAM NOSE

FIGURE 19-2
Crankshaft straightening press (Courtesy of Storm Vulcan, A Division of the Scranton Corporation).

ACCEPTABLE ROD JOURNAL

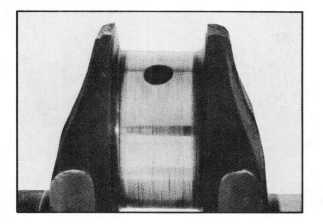

ACCEPTABLE ROD JOURNAL

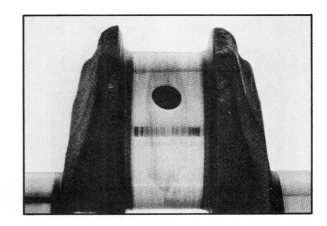

UNACCEPTABLE ROD JOURNAL

UNACCEPTABLE ROD JOURNAL

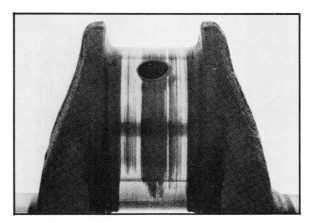

FIGURE 19-3

FIGURE 19-4
The nicks on this crankpin were caused by failing to use protective covers on the rod bolts when removing the piston.

FIGURE 19-5
Badly worn thrust face.

FIGURE 19-6
Grinding a crankshaft.

FIGURE 19-7
Checking crankpin width (Courtesy of *Popular Hot Rodding* Magazine).

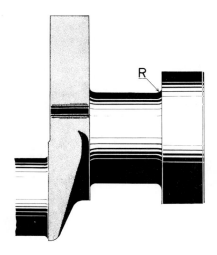

FIGURE 19-8
The fillets at the ends of the journal should have a specified radius.

FIGURE 19-10
Micro finishing (polishing) a crankshaft.

6. After grinding has been completed, inspect the crankshaft by magnetic particle inspection to determine whether the grinding operation has caused cracks.

7. Remove all burrs on the counterweights with a rotary file.

8. Polish all journals, fillet radii, and seal surfaces to a 15 microinch finish. The crankshaft should be rotated in the same direction it operates (usually clockwise looking from the front of the engine), with the polishing cloth pulling the material away from rotation (Figure 19-10). Sometimes a strip of fine grain abrasive cloth is used to hand lap the journals (Figure 19-11). *CAUTION: Nodular cast iron crankshafts should be final polished in the same direction as normal engine rotation.*

FIGURE 19-9
Ford 429 BOSS crank with "forged-in" oil groove (Courtesy of Ford Parts and Service Division).

FIGURE 19-11
Hand lapping a crank journal.

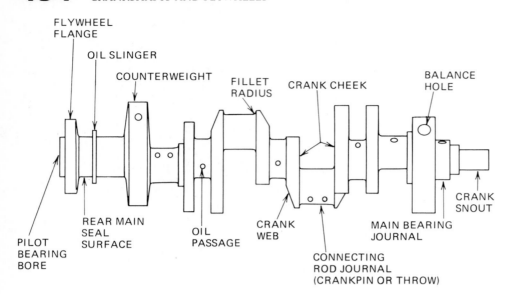

FIGURE 19-12

Crankshaft nomenclature. Note the oil passages.

9. When grinding and polishing is completed, the crankshaft surfaces and all oil passages (Figure 19-12) must be thoroughly cleaned.

10. Dynamically balance the crankshaft (Figure 19-13). After regrinding, the excess weight is generally removed by light grinding on the counterweights.

CRANKSHAFT REBUILDING

There are several methods used to add material to damaged crankshaft bearing surfaces. Listed below are a few descriptions.

FIGURE 19-13

Balancing a crankshaft. Bob weights are used to simulate the total rod/piston assembly weights.

1. **Hard chroming** This is a plating process where pure hard chromium is electrolytically applied to journal or crankpin surfaces (Figure 19-14). A nonporous, well bonded coating is obtained. Grinding and polishing then follow.

2. **"Submerged arc" welding** An alloy or a high carbon steel wire material is deposited in a continuous weld. This process involves covering the electric arc with flux powder to keep the weld clean and free of porosity (see Figure 19-15).

3. **Metal spraying** In this operation, molten steel is sprayed on the bearing surfaces while the crank is rotating.

4. **Oxyacetylene welding** This method has limited use for journal or crankpin repair because of shaft warpage from the high heat. However, worn thrust surfaces are often built up by oxyacetylene welding with good success if heated and cooled correctly.

Tufftriding[2]

Tufftriding is a critically controlled, low-temperature salt bath treatment based on carbon and nitrogen diffusion. It is applicable to all ferrous metals. It has several primary advantages.

1. Significantly increased wear resistance (wear properties of all cast irons are typically improved 300 to 400%).

2. Greatly increased fatigue resistance (fatigue

---

[2]Tufftriding is a registered trademark of the Kolene Corporation, Detroit, MI.

CHROME TANK

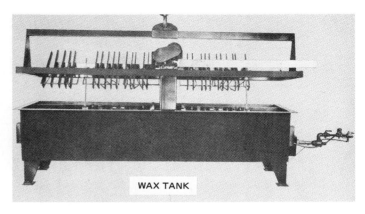

WAX TANK

ROTATING FIXTURE

FIGURE 19-14
A crankshaft hard chrome plating system. The system consists of a chrome tank, wax tank, and rotating fixture (Courtesy of Storm Vulcan, A Division of The Scranton Corporation).

endurance is increased by as much as 80%).

3. Improved hardness.
4. Freedom from galling and seizing.
5. Improved corrosion resistance.
6. No significant dimensional change.
7. Short processing time (about three hours).
8. Economical.

Sources of Tufftriding

See Appendix D for a listing of companies that will Tufftride crankshafts. Several of the companies also offer Melonite, a new nonpolluting nitriding process.

COMMENT: Several years ago certain manufacturers released bulletins that Tufftrided crankshafts should not be reground, because the hardened surface would be removed. Since that time, tests have shown that Tufftrided crankshafts can be successfully reground, provided they are rehardened by Tufftriding.

Procedures to Follow Prior to Tufftriding

1. Check the crankshaft for cracks by magnetic particle inspection.
2. Rebuild any worn areas as necessary.
3. Stress relieve at 1100° F. Hold this oven temperature for a minimum of one hour per inch of the largest journal diameter size.
4. Shot peen and straighten as necessary.
5. Grind and polish.
6. Check for cracks again.
7. Send the crankshaft out for Tufftriding. *NOTE: Tufftriding has a stress relieving effect. Distortion will occur unless the shaft is properly stress relieved before Tufftriding (refer to Step 3).*
8. Check for straightness. *WARNING: Do not bend to straighten a Tufftrided crank. You can break the shaft, or start microcracking. The crankshaft will have to be ground and re-Tufftrided.*

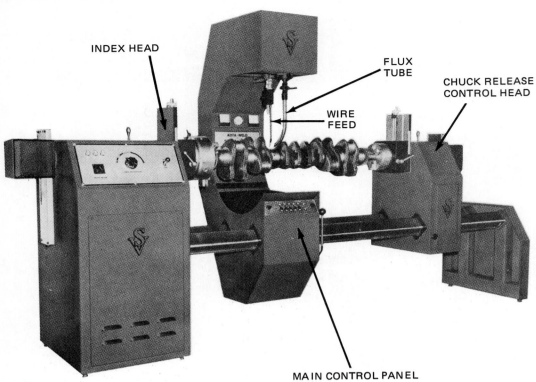

INDEX HEAD

FLUX TUBE

CHUCK RELEASE CONTROL HEAD

WIRE FEED

MAIN CONTROL PANEL

FIGURE 19-15

A machine for welding crankshafts with the submerged arc process. With submerged arc welding there is no visible arc and no distortion of metal (Courtesy of Storm Vulcan, A Division of The Scranton Corporation).

Determining if a Crankshaft Has Been Tufftrided

1. Perform a file test. If any part of the crankshaft can be easily filed, you can assume that it has not been Tufftrided. *NOTE: Make the file test on an area other than the journal ODs.* When a crankshaft has been Tufftrided, the entire surface is hardened, while other methods of hardening are generally confined to the bearing surfaces only.

2. Make a chemical test. Obtain a 10% aqueous solution of copper ammonium chloride at a pharmacy. With an eye dropper, apply a drop to the crankshaft journal surface. If the drop turns to a copper color in less than 10 seconds, the crankshaft has not been Tufftrided. If the crankshaft has been Tufftrided, the drop will take a much longer time span to turn copper.

Shot Peening

The basic purpose of this process is to improve fatigue properties of a part. Laboratory testing of shot peened crankshafts has shown crankshaft fatigue life improvement of up to 100% over the original.

The part to be shot peened is placed on a turntable within a cabinet (Figure 19-16). Fine steel shot is then blasted at the piece (Figure 19-17). This action introduces compressive stress (preloading) on the metal and causes a slight hardening effect. The actual service stress is reduced.

Cross-Drilling

For the high rpm needed in competition application a crankshaft is sometimes cross-drilled (Figure 19-18). This adds a second source of oil during each 360° of crank rotation. For example, standard V-8 cranks usually have one oil outlet

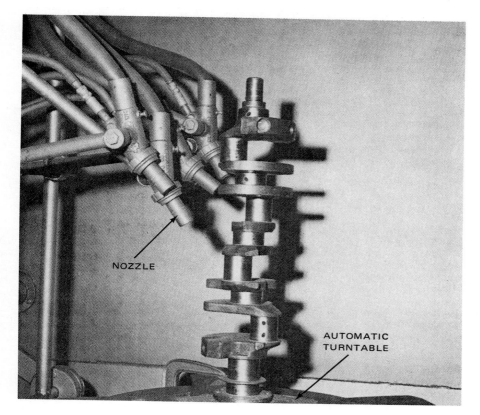

NOZZLE

AUTOMATIC
TURNTABLE

FIGURE 19-17
Steel shot peen pellets, .023" in diameter.

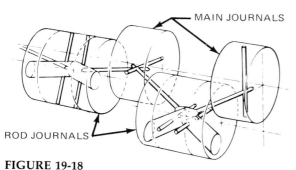

MAIN JOURNALS

ROD JOURNALS

FIGURE 19-18
Cross-drilled crank (Courtesy of Ford Parts and Service Division).

on nos. 1 and 5 main journals. Main journals nos. 2, 3, and 4 have two outlets. All rod journals have two outlets (one for each rod). Cross-drilled cranks have two outlets on all mains and four on the rod journals, 180° apart.

CRANKSHAFT FAILURE ANALYSIS

The main function of the crankshaft is to convert the up-and-down motion of the pistons into rotary motion. In performing this function, the crankshaft is subjected to torsional and bending stresses. On today's engines, these stresses are becoming higher (due to increased emission requirements, more accessory devices, reduced weight versus horsepower, and turbocharging).

Torsional Stress Failure

Torsional stress is a condition inherent to engine design. The inertia of the crankshaft is speeded up on each power stroke, and slowed down on each compression stroke. This creates a twisting and untwisting action on the crankshaft. To help

FIGURE 19-19
This crank failure was caused by a faulty vibration damper.

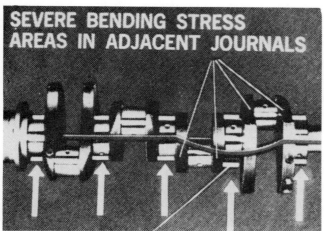

FIGURE 19-20
(Courtesy of Automotive Engine Rebuilders Association).

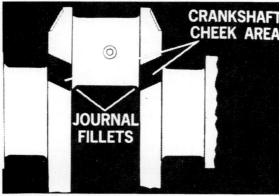

FIGURE 19-21
Most bend failures cause breakage at the crank cheeks (Courtesy of Automotive Engine Rebuilders Association).

minimize this stress, the engine manufacturer generally adds a vibration damper that is tuned to the torsional twist amplitude.

A crankshaft torsional fatigue failure can almost always be recognized by a 45° angle fracture across the bearing journal and through the oil hole (Figure 19-19). Torsional failures can be caused by several factors.

1. The addition of improperly balanced pulleys or driven accessories.
2. A loose flywheel.
3. The wrong vibration damper.
4. A faulty vibration damper.
5. The wrong torque converter.
6. Slapping belts.
7. Damaged crankshaft keyways.
8. Excessive gear train backlash.

Bending Stress Failure

The combustion force on the piston is equally divided between adjacent main bearings (Figure 19-20). This support must be maintained with each revolution of the crankshaft. Otherwise, excessive cyclic bending and fatigue failure will result.

Bending failures generally occur in the crank cheek area (Figure 19-21). The fracture lines will be at right angles to the crankshaft. *COMMENT: The fracture lines are not always straight and may look "hammered" because the engine ran after the fracture occurred.* Bending failures can be caused by:

1. Breakdown of oil clearances.
2. Bearing caps loose.
3. Excessive bearing wear.
4. Saddle bore misalignment.
5. Reduced fillet radii.
6. Uneven journal finish (this creates many notch type surface interruptions that act as stress raisers).

Figure 19-22a shows a bend failure caused by overheating. Hairline heat check cracks progressing from the fillet radius can be seen under section magnification (Figure 19-22b).

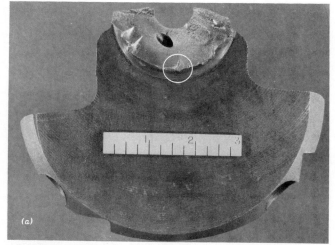

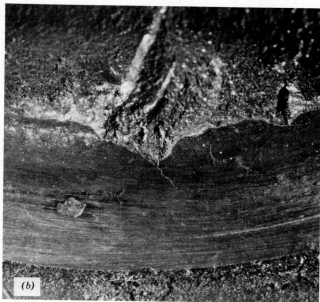

FIGURE 19-22

(*a*) Overheating in the crankpin fillet radius area caused this failure. Poor lubrication or excessive friction caused by connecting rod side thrust could have been the reason. (*b*) Magnification of the circled area in Figure 19-22*a* shows heat cracks (6.5× section). Burnishing over the cracked areas point out that the engine was still running while cracks had already developed. The failures were ductile, indicating the crank casting was okay and that it was not too hard or brittle for proper service.

CRANKSHAFT MATERIAL

Crankshafts used in automobile engines are either forged or cast. Forged crankshafts (made from AISI 1049 or similar steel) are stronger than cast crankshafts (made from molded steel alloy, nod-ular iron, or malleable iron). However, forged crankshafts are more expensive.

Forged crankshafts can be identified by the wide separation lines on the crank throws (Figure 19-23). Cast crankshafts are visually identified by the narrow mold parting lines (Figure 19-24). *NOTE: A forged steel crankshaft will produce a sharp "ringing" sound when tapped lightly with a ball-peen hammer on one of the counterweights. A cast crankshaft will produce a dull "thud" sound.* Table 19-1 shows the type of crankshaft material used in some current and recent vintage standard production domestic engines.

FACTORY UNDERSIZE CRANKSHAFT DIAMETERS

In past years the automobile manufacturer has produced new engines that have one or more undersized crankshaft bearing journals. These crankshafts can generally be identified by special markings. Some recent examples are listed next.

1. **Chrysler products 318 and 360 CID engines** Engines with .001″ or .010″ undersize main or rod journals will have a stamped mark on the crankshaft. A crank with .001″ undersize rod journals will have a R1, R2, R3, or R4 stamped on the #8 counterweight. All .001″ undersize main journals will have a M1, M2, M3, M4, or M5 stamped on the #8 counterweight. A crankshaft with .010″ undersize main and/or rod journals will have RX, MX or RX-MX stamped on the #1 and #8 counterweights along with a spot of yellow, light brown, or black paint.

2. **Chrysler products 225 CID engines** A R1, R2, R3, R4, R5, or R6 stamped mark on the center counterweight indicates a crankshaft with .001″ undersize rod journals. Main journals .001″ undersize will use the mark M1, M2, M3, M4, or M5 stamped on the center counterweight. Cranks with journals .010″ undersize will be stamped R10 or M10 on the center counterweight.

3. **Chrysler products 400 and 440 CID engines** These engines with .001″ undersize bearing journals will be marked in the same manner as the 225 engines, except on the #3 counterweight. A .010″ undersize shaft will be marked on the #2 counterweight.

TABLE 19-1

Make/Displacement	Crankshaft Material
AMC 199,232	Nodular cast iron or malleable iron
290,343	Malleable iron
258,304,360	Nodular cast iron
390,401	Forged steel
Buick 225,300	Malleable iron
250,350,400,430	Nodular cast iron
401	1145 or 1053 steel
Cadillac 420	Malleable iron
472,500	Nodular cast iron
Chevrolet 153,250,283,327,350,396	Nodular cast iron or forged steel
140,194,230,250,307,400	Nodular cast iron
302,427,454	Forged steel
Chrysler 170,198,225,340,	
383,413,426,440	Forged steel
273	Forged steel or nodular cast iron
318,360	Nodular cast iron
*400	Forged steel or nodular cast iron
Ford 98,122,170,200,240,250,289,	
302,351,352,390,400,427,428,429,430,460	Cast iron alloy
302 Boss, 361 Truck, 391 Truck, 427 Boss	Forged steel
Oldsmobile 225	Malleable iron
250	Nodular cast iron
330,425	1049 Steel
350,400,455	Nodular cast iron or 1049 steel
Pontiac 326,389,421	Malleable iron
215,250,307,350,400,455	Nodular cast iron

*The 400 CID forged crank requires a different torque converter and damper than the one used with the cast crank. Otherwise, there will be serious engine vibration.

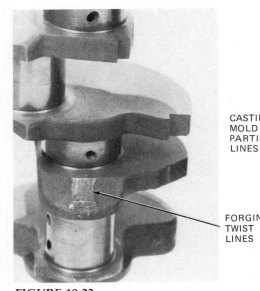

CASTING
MOLD
PARTING
LINES

FORGING
TWIST
LINES

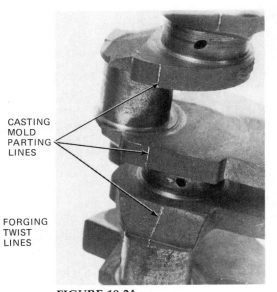

FIGURE 19-23

Forged crankshaft. Note the wide separation lines where the crank has been twisted to index the throws.

FIGURE 19-24

Cast crankshaft. Note the straight narrow mold parting lines.

4. **Chrysler products 1.7 liter engines** This engine may have .25 mm undersize bearings. If the crankshaft is undersized, the engine serial number will have the last digit circled. Check for a 6 mm to 8 mm stamped marking on the #1 counterweight. R indicates all crankpins are .25 mm undersize. M indicates all main journals are .25 mm undersize. RM indicates all crankpins and main journals are .25 mm undersize.

5. **American Motors products 232, 258, 304, and 360 CID engines** All engines with undersize crankshaft journals can be identified by a letter code stamped on the boss between the distributor and coil. The code is as follows.

F or P All connecting rod journals are .010″ undersize.

M All main bearing journals are .010″ undersize.

6. **American Motors products two-liter engines** Some two-liter, four-cylinder engines have been built with undersize rod and/or undersize main bearings. A letter code is stamped on the engine build date flange (located at the left rear corner of the block by the oil pressure sending unit). The code is given below.

P Rod bearings .25 mm undersize.

M Main bearings .25 mm undersize.

PM Rod and main bearings .25 mm undersize.

7. **Pontiac 231, 350, and 400 CID engines** These engines may be equipped with crankshafts having .010″ undersize rod journals. These crankshafts can be identified by a spot of orange paint on the counterweights. If one rod journal is .010″ undersize, all others will be the same size.

8. **Chevrolet and Pontiac 305 and 350 CID engines** These engines may be equipped with crankshafts having .009″ undersize main journals. When this was done, the oil pan rail was stamped ''009'' and the bearing inserts were marked with a ''009'' on the back. *COMMENT: Instead of the pan rail marking, some engines may have a ''9'' stamped on one side of the undersize journal along with a spot of light green paint.*

FLYWHEEL INSPECTION

The flywheel face is a friction surface that must be in good condition to give proper clutch performance. Flywheels must be precision measured. Visual inspection is not enough. Remember that clutch components are designed to work on parallel surfaces. A dished, crooked, or worn flywheel can cause clutch release problems (especially on late model clutches that have small travel).

Flywheel problems can be grouped into the following categories.

1. **Face runout** Check with a dial indicator while the flywheel is mounted on the crankshaft (Figure 19-25). The Society of Automotive Engineers specification (SAE J6186) is *.0005″ maximum tir per inch of diameter.* Runout in excess of specification can cause chatter, vibration, and premature clutch failure. Many clutch complaints are caused by flywheel related problems.

    Excessive flywheel runout is often caused by a previous misaligned resurfacing method. Resurfacing operations should always reference off the crankshaft mounting flange of the flywheel.

FIGURE 19-25

Checking flywheel face runout (typical). Courtesy of Ford Parts and Service Division .

FIGURE 19-26
Badly worn ring gear teeth.

FIGURE 19-27
Removing ring gear (Courtesy of Ford Parts and Service Division).

FIGURE 19-28
Measuring clutch pilot bushing ID. If the bushing is worn bell mouthed, if it is loose, or if the diameter is larger than specified, install a new one (Courtesy of Ford Parts and Service Division).

2. **Dishing** Today, most flywheels are made of forged steel rather than cast iron. Steel flywheels are much stronger, but they are more prone to warpage and distortion. A flywheel may be dished .010″, yet will look perfect. A simple way to detect dishing is by using a straightedge and feeler gauge.

3. **Hardspots and discoloration** Excessive clutch slippage or localized overheating will cause the nucleation and growth of carbide hardspots on the flywheel surface. Hardspots are best removed by grinding. Sometimes, cutting with a grade 999 carbide at a slow speed will work.

4. **Groove wear** This condition is evidenced by grooves on the clutch friction disc and on the flywheel face.

5. **Ring gear wear** Inspect the condition of the flywheel ring gear. Worn teeth mean that weight has been lost on a large diameter. This metal loss affects engine balance (Figure 19-26). *COMMENT:* *The practice of reversing a worn ring gear is not recommended.*

    To remove a worn gear, heat it with a torch on the engine side of the gear (Figure 19-27). When the gear is thoroughly hot, knock it off the flywheel with a hammer and punch. *Never heat a replacement gear hotter than 500° F, or the teeth will loose their temper and soften.*

Flywheel Installation Tips

1. Before attaching any flywheel, check the flange end of the crankshaft and the flywheel contact surface. Lightly use a stone to remove any fretting or raised metal.

2. Make sure the bolt threads in the flange are clean and dry before attaching the flywheel.

3. Thoroughly clean the flywheel attaching bolts.

4. Check the condition of the clutch pilot bearing or bushing (if used). See Figure 19-28.

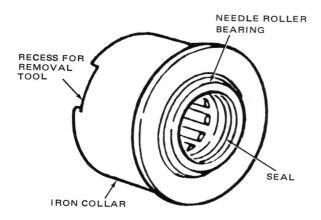

CLUTCH PILOT BEARING
ASSEMBLY

NEEDLE ROLLER
BEARING

RECESS FOR
REMOVAL
TOOL

SEAL

IRON COLLAR

SLIDE HAMMER
PULLER

FIGURE 19-29

Removing pilot bearing assembly (Courtesy of Ford Parts and Service Division).

Replace if worn or damaged (Figure 19-29).

Sometimes a pilot bushing can be removed by packing the ID with wheel bearing grease. Then, select a bolt that just fits the bushing ID. Position the bolt against the grease and strike the end with a hammer. The bushing will be forced out hydraulically.

Use an accurately calibrated torque wrench to tighten the flywheel bolts. *WARNING: Make sure the bolt head corners do not overlap the pilot bearing bore in the flywheel. If so, slightly turn the bolts.*

TRUE CASE HISTORY

Vehicle	Ford Truck
Engine	8 cylinder, 6.6 liter
Mileage	Unknown; truck was just two years old
Problem	1. Repeated flywheel (flex-plate) replacement in subject vehicle because of cracking.
	2. Unusual sound while cranking.
Cause	Mislocated rear block face dowel pins.
Comment	A special fixture is made by OTC (Owatonna Tool Company) to check the dowel pin location on the rear block face of any 5.8 liter or 6.6 liter Ford engine. The fixture is OTC #44853.

CHAPTER 19 SELF-TEST

1. What is often the cause of crankshaft keyway damage?
2. What special procedure can be used to cure rear main oil seal problems on Chevrolet 153, 194, 230, and 292 CID engines?
3. What is the preferred crankshaft straightening method?
4. Nicks, dents, or pits in the crankshaft journal OD must not exceed _____ ".
5. What is the maximum allowable thrust wall runout?
6. Crankshaft fillets should never be less than what amount?
7. Journals, fillet radii, and seal surfaces should be polished to what microinch finish after machining?
8. How should nodular iron crankshafts be final polished?
9. What four methods can be used to add material to damaged crankshaft bearing journals?
10. What is Tufftriding?
11. List several advantages of a Tufftrided crankshaft.
12. What can happen if you attempt to straighten a Tufftrided crankshaft?
13. How can you determine if a crankshaft has been Tufftrided?
14. What is the basic purpose of shot peening?
15. What is a cross-drilled crankshaft?
16. How is a crankshaft torsional fatigue failure usually recognized?
17. List five items that can cause a crankshaft torsional failure.
18. In what area of the crankshaft do bending failures generally occur?
19. List five causes of crankshaft bending failure.
20. How can forged crankshafts be visually identified?
21. How can cast crankshafts be visually identified?
22. A Plymouth 318 CID engine has RX-MX stamped on the #1 and #8 counterweights along with a spot of yellow paint. What does this indicate?
23. Dodge and Plymouth 1.7 liter engines may have been produced with _____ mm undersize crankshaft bearings.
24. An American Motors 304 CID engine has the letter M stamped on the boss between the distributor and coil. What does this indicate?
25. What engines may be equipped with crankshafts having .009" undersize main journals?
26. What problems can be caused by excessive flywheel runout?
27. How can flywheel dishing be detected?
28. How are hardspots on the flywheel surface best removed?
29. What is the procedure described in the chapter for removing a worn ring gear?
30. How can wheel bearing grease and a bolt be used to remove a pilot bushing?

CHAPTER 20

CONNECTING RODS AND PISTON PINS

A list of the subjects covered in this chapter is given below.

1. Why is the connecting rod big-end bore dimension so critical?
2. What is the rod reconditioning process?
3. Connecting rod alignment.
4. Connecting rod breakage analysis.
5. What is a proper pin fit?
6. Wrist pin failures and causes.
7. Assembling the piston to the rod.

CONNECTING RODS

The load and stress demands on present-day connecting rods are hard for most people to visualize. For instance, the connecting rod must support four to five tons of pressure during combustion. Also, the connecting rod is required to reverse its vertical motion many times a second—at 3000 rpm this motion is being reversed 100 times in one second. These load and stress forces, combined with the engine heating and cooling cycles, will eventually deform the big-end bore of the rod. Mechanics need to be aware of this when an engine comes into the shop for a complete rebuild or just an overhaul. When assembling an engine, there are only two approaches to assure the proper functioning of rod bearing inserts.

1. Install new connecting rods.
2. Measure the big-end bore of the rod; have it reconditioned if beyond limits.

Stretch

Connecting rod big-end bore out-of-roundness (often called concentricity, stretch, or elongation) is usually the greatest in a vertical direction (Figure 20-1). Most of the stretch will occur in the rod cap of the rod (especially if the cap is weak).

Measuring the Big-End Bore

Several methods can be used to check the out-of-roundness of the big-end bore.

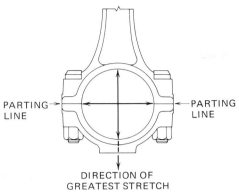

FIGURE 20-1
Rod big-end bore out-of-roundness almost never occurs across the parting lines. An exception would be a rod that was "burned out."

1. **A special bore gauge (Figure 20-2)** The back of this gauge has two rigid pins and one floating pin. The floating pin is adjustable to fit the particular rod bore being checked. When the gauge is properly adjusted to suit the rod bore, rotate the gauge around the bore. The dial will show out-of-roundness in thousandths of an inch.

2. **Use a precision dial gauge (Figure 20-3)** Measure the rod in both directions. The dif-

ference in readings gives the out-of-roundness.

3. **Measure with either an inside micrometer or a telescoping gauge** By taking a series of readings across the bore, a determination of the size and roundness can be made (Figure 20-4).

CAUTION: Always torque up the rod caps before measuring the big-end. Otherwise, the reading may be inaccurate.

Always remember that connecting rod big-end bore elongation will cause the rod bearing insert assembly to be out-of-round when installed.

Rod Reconditioning

Very often this question is asked, "How can a connecting rod be reconditioned when the big-end bore is out-of-round beyond original size?" The answer becomes apparent when you understand the steps of the rod reconditioning process. There are basically three steps.

1. Reduce the big-end bore size slightly below its standard size. The rod bore is made up of two parts (the rod and the bearing cap). By removing stock from the parting line of these parts, you reduce the vertical dimen-

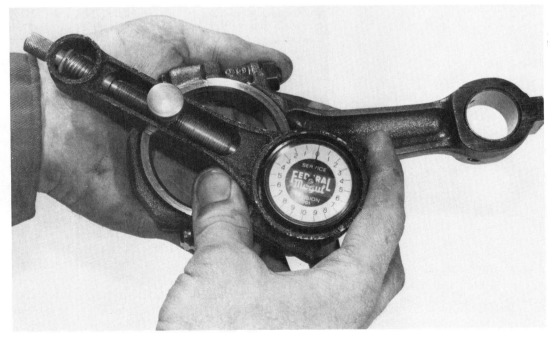

FIGURE 20-2
Federal-Mogul gauge in position for measuring rod big-end bore out-of-roundness.

FIGURE 20-3

The Sunnen AG-300 Precision Gauge measures in .0001" increments. Note the .002" out-of-round reading on the dial. Manufacturers now hold rod bores to a maximum out-of-round tolerance of .0003" (Courtesy of Sunnen Products Company).

FIGURE 20-5

Parting face of rod being ground square on a cap and rod grinder. A micrometer dial controls exact amount to be removed (Courtesy of Sunnen Products Company).

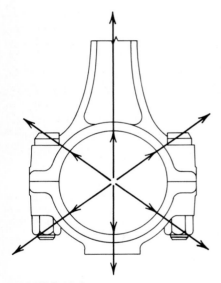

FIGURE 20-4

Recommended positions for checking big-end bore out-of-roundness.

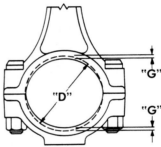

FIGURE 20-6

Reassembled rod after grinding parting faces. The rod now has a smaller vertical dimension (by amounts G) than diameter D of original true-round factory rod (Courtesy of Sunnen Products Company).

sion of the rod bore. Stock removal is held to a minimum (generally about .002") with equal amounts removed from both the rod and cap parting surfaces. This operation is very critical and requires precision equipment (Figure 20-5). Do not attempt to freehand grind or file the parting faces. This will result in misalignment of the rod assembly.

2. Attach the bearing cap and torque to specifications. The reassembled rod now has a smaller vertical dimension than originally (Figure 20-6).

3. Hone (or machine) just enough material from the rod bore to bring it back to the same diameter size as originally specified by the engine manufacturer (Figure 20-7). In addition, the finished bore must be round, straight, and have the proper surface finish (Figure 20-8).

Manufacturers' specifications on new rods call for .0003" maximum tolerance for roundness and straightness, .001" or less allowable size tolerance, and a surface finish of 30 mi-

FIGURE 20-7

Two V-8 rods being honed simultaneously on a Sunnen honing machine. Note the double bar for absorbing honing torque. Larger rods are generally honed one at a time. (Courtesy of Sunnen Products Company).

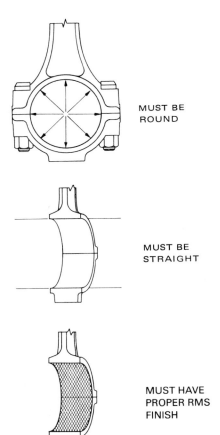

MUST BE ROUND

MUST BE STRAIGHT

MUST HAVE PROPER RMS FINISH

FIGURE 20-8

Finished rod bore requirements (Courtesy of Sunnen Products Company).

croinches. A properly reconditioned rod will often surpass these requirements.

In some cases after a rod has been reconditioned, you may notice a small area at the parting line that did not clean up. This is usually of no consequence, and will have no effect on the life or load carrying capacity of the rod bearing. As a matter of fact, some bearing manufacturers even thin their inserts at this area to avoid "pinching" (this ensures sufficient oil clearance at the parting line).

Connecting Rod Alignment

Rods are normally checked for bend and twist (Figure 20-9) which can cause abnormal wear and stress on the rod bearings, rings, pins, and pistons. See Figure 20-10a and b. Alignment is generally checked with the piston assembled to the rod (Figure 20-11). However, on rods that use floating pins this is not necessary.

Sometimes, bend and twist measurements are made using dial indicators with mandrels positioned in each end of the rod (Figure 20-12).

Straightening (aligning) a rod is generally done by using a bending bar. A slightly bent or twisted connecting rod can be straightened (Figure 20-13). *Always straighten in the area of the original bend to avoid bending an offset in the rod.* Severely misaligned rods should be replaced (Figure 20-14).

Connecting Rod Breakage Analysis

Connecting rod breakage can occur at any time during the life cycle of the engine, but it usually happens early in the life. Breakage can be caused by many factors. The following reasons are listed in the order of likelihood.

1. **Improper bolt tightening** Most connecting rod failures occur from undertightening or overtightening the rod bolts. The rod bolts must be prestressed to eliminate breakage and to hold the cap firmly in position. If overtightened, a rod bolt will often crack and snap off at the base of the threads. If undertightened, a rod bolt will stretch under load. This can cause the rod nut to back off, permitting a hinge action on the opposite side. Then, breakage from metal fatigue will soon occur (Figure 20-15).

   For maximum safety and performance, rod nuts should never be reused. Nuts are deformed each

BEND MUST BE
WITHIN .001" IN 6"

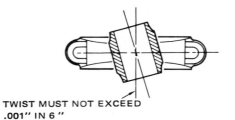

TWIST MUST NOT EXCEED
.001" IN 6"

FIGURE 20-9
Exaggerated sketches of recommended limits for connecting rod alignment (Courtesy of Federal-Mogul Corporation).

(a)

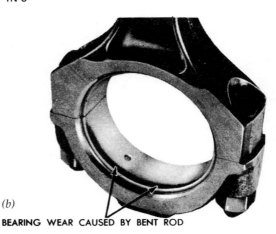

(b)

BEARING WEAR CAUSED BY BENT ROD

FIGURE 20-10
(a) Align or replace rods if piston skirts show crooked (diagonal) wear pattern (Courtesy of McQuay-Norris Manufacturing Company).
(b) The upper half of bearing would show similar wear on opposite side (Courtesy of Ford Parts and Service Division).

FIGURE 20-11
Checking connecting rod alignment with a special fixture (Courtesy of TRW Replacement Division).

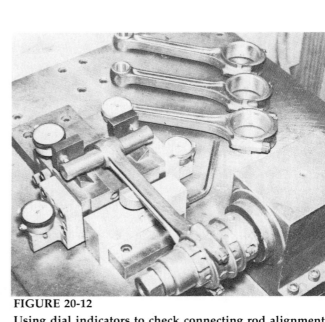

FIGURE 20-12
Using dial indicators to check connecting rod alignment.

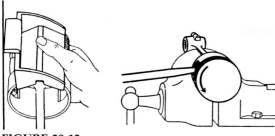

FIGURE 20-13
Straightening a twisted rod (Courtesy of Sunnen Products Company).

FIGURE 20-14
Badly misaligned rod.

FIGURE 20-15
A loose rod nut caused this failure.

SHOP TIP
Some engine builders prefer to tighten rod bolts using the stretch method. First, measure the overall length of the bolt with a micrometer. Then gradually tighten the nut. The bolt is properly torqued when it has stretched .006" ± .0005".

---

2. **A big-end bore that is out-of-round** During high engine rpm, the bore will flex in shape. This can cause bending stress and breakage of the beam section (Figure 20-16).

time they are reused. This causes increased assembly friction and decreased loading strength. Even though a torque wrench shows proper loading, the rod cap will actually be loose when the old nuts are used. Laboratory tests on 3/8" nuts, tightened to 60 ft/lbs., show a 20% tension loss the first time the nuts were reused. Increasing losses are shown in Table 20-1.

TABLE 20-1

FOR MAXIMUM SAFETY AND PERFORMANCE, A ROD NUT SHOULD BE DISCARDED ONCE IT IS REMOVED.

Times Nut Was Reused	Percentage of Tension Lost
1st	20%
2nd	33%
5th	47%
9th	60%

FIGURE 20-16
Rod beam failure is generally caused by an out-of-round big-end bore (Courtesy of TRW Replacement Division).

FIGURE 20-17
Special vise guarantees alignment of rod and cap during torque-up prior to honing (Courtesy of Sunnen Products Company).

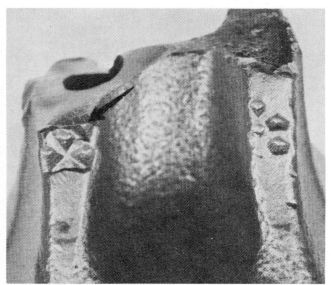

FIGURE 20-18
Vise jaw marks on the surface of this rod (see arrow) created a stress raiser and caused breakage (Courtesy of TRW Replacement Division).

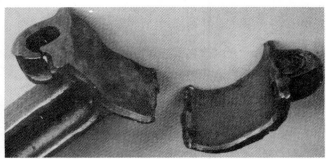

FIGURE 20-19
This connecting rod failed because the wrong clamp bolt was used (Courtesy of Sunnen Products Company).

3. **Breakage due to bearing failure** When a rod bearing starts to "knock", an excessive oil pressure loss exists at that rod journal. Continued running of the engine, when these conditions are present, will allow the rod to "pound" the big-end bore. This results in distortion and then breakage, usually through the cap.

4. **A pressed-in pin that seizes in the piston** Such seizure prevents the connecting rod from oscillating properly. This will cause the rod to bend and break, usually just below the pin bore.

5. **Nicking the rod** A nick or scratch mark on the surface of the rod forging creates a stress raiser. A point of stress concentration (weakness) exists in the rod at that point. During heavy loading or high rpm, breakage is then possible at that point.

    Handle a connecting rod with care. Special vises are available for clamping the rod (Figure 20-17). If a bench vise is used instead, protect the rod by using pieces of wood or soft metal on each side. Any vise jaw marks can lead to an early rod failure (Figure 20-18).

6. **Breakage due to a poor quality clamp bolt** This connecting rod broke at the small end because the mechanic installed a plain no-line bolt on the piston pin clamp (Figure 20-19).

PISTON PINS

Piston pins (often called *wrist pins*) attach the piston to the small end of the connecting rod. The combustion chamber thermal force is carried

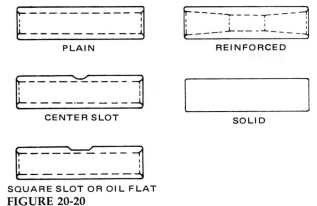

PLAIN

REINFORCED

CENTER SLOT

SOLID

SQUARE SLOT OR OIL FLAT

FIGURE 20-20

Piston pin shapes. The reinforced (tapered hole) tube provides extra strength (Courtesy of Dana Corporation—Service Parts Group).

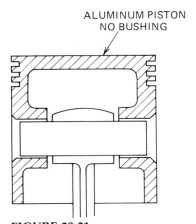

ALUMINUM PISTON NO BUSHING

FIGURE 20-21

Press fit in rod-oscillating in piston pin attachment (Courtesy of Sunnen Products Company).

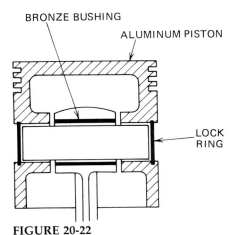

BRONZE BUSHING

ALUMINUM PISTON

LOCK RING

FIGURE 20-22

Full floating pin attachment (Courtesy of Sunnen Products Company).

through the pin into the connecting rod. This thermal pressure reaches 600 lb./sq.in. or higher in many engines. On a 4″ diameter piston that figures to 7500 lbs.—that is, $3\frac{3}{4}$ tons of load on the pin.

The pin is made from high quality steel (for strength) in the shape of a tube (for lightness). Sometimes, the inside hole of the pin is manufactured with a taper (Figure 20-20) to provide extra load strength.

Piston Pin Attachment Methods

The pin must be fastened so that it stays centered in the piston. Otherwise, the pin would move endwise and gouge the cylinder wall. There are four different attachment methods.

1. **The pin is pressed in the rod and oscillates only in the piston bosses (Figure 20-21)** This is the method used in the majority of passenger car engines. In this design, the pin is press fit through the "eye" (small-end bore) of the connecting rod with a −.0008″ to −.0012″ interference. This means the hole in the rod is smaller in diameter than the outside diameter of the piston pin. The pin usually has a .0003″ to .0005″ clearance fit in the piston bosses.

2. **Full floating (Figure 20-22)** This design allows free pin rotation in the "eye" of the connecting rod and in both bosses of the piston. The fit of the pin in the connecting rod ranges from .0003″ to .0005″. Normal clearance between the pin and the piston bosses is .0001″ to .0003″.

A lock ring is installed at the end of each piston boss to retain the pin and hold it centered in the piston. Some engines have used aluminum buttons or plastic plugs at both ends of the pin in order to to retain it.

3. **The pin is clamped in the rod and oscillates only in the piston bosses (Figure 20-23)** This style attachment method uses a bolt to clamp the rod around the pin. The pin fit in the piston should be .0003″ to .0005″.

4. **The pin is locked in the piston by a set screw (Figure 20-24)** On some older engines a cap screw is located on one piston boss. This screw, when tightened, locks the pin in po-

ALUMINUM PISTON
NO BUSHING

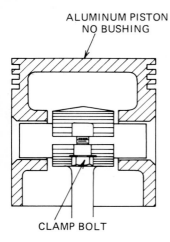

CLAMP BOLT

FIGURE 20-23

Oscillating in piston-clamped in rod pin attachment (Courtesy of Sunnen Products Company).

BRONZE BUSHING

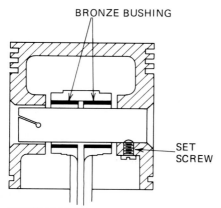

SET SCREW

FIGURE 20-24

Pin attachment using set screw in piston (Courtesy of Sunnen Products Company).

FIGURE 20-25

The dark area in this reamed pin hole shows where the pin has contacted the bushing. Pin fit has excess clearance at top and bottom of hole (Courtesy of Sunnen Products Company).

FIGURE 20-26

This pin hole is tapered. The dark area in front shows where the pin made contact. The light areas have excessive pin clearance (Courtesy of Sunnen Products Company).

sition. The screw side of the piston should have −.0002″ to −.0003″ interference fit, while the free side should have zero to .0001″ clearance. The pin should have .0007″ to .0009″ clearance in the connecting rod.

Proper Pin Fit

To withstand the tremendous impact loads, a pin must have a 100% bearing surface. Yet, it still has to oscillate without drag. In order to achieve these requirements, a precision pin fit is necessary. The days of fitting pins by reaming to a thumb or palm push fit is history. Today, pin fitting is one of the most critical operations in engine service.

When fitting pins, certain criteria must be met.

1. The pin hole in the piston and rod must be round, and free from high spots (Figure 20-25).
2. The pin hole must be straight, and free from taper and bell mouth (Figure 20-26).
3. The bore in the rod and piston must be in perfect alignment, and free from bind and deflection (Figure 20-27).
4. There has to be a precise amount of oil clearance, as recommended for each pin attachment design (Chart 20-1).
5. The surface finish must be correct in order to sustain and support an oil film. See Figure 20-28.

Checking the Quality of a Pin Fit

To accurately check a given pin fit, a precision pin hole gauge is needed. However, in the ab-

CHART 20-1

Recommended clearances when piston pins are fit by honing (Courtesy of Sunnen Products Company).

Precision Pin Fits on Engines with $\frac{3}{4}''$ to $1\frac{1}{4}''$ Diameter pins			
Description	Aluminum Piston	Cast Iron Piston	Connecting Rods
Full floating	.0001 to .0003" clearance	.0003 to .0005" clearance	.0003 to .0005" clearance (all pressure feed, .0005 to .0007" clearance)
Oscillating in bushed piston		.0003 to .0005" clearance	clamped in rod
Oscillating in piston (no bushing)	.0003 to .0005" clearance	.0006 to .0008" clearance	clamped in rod
Oscillating in piston—press fit in rod	.0003 to .0005" clearance		− .0008 to − .0012" press fit
Set screw type piston	Screw side − .0002 to − .0003" press fit Free side 0 to .0001" clearance	Screw side − .0001 to − .0002" press fit Free side 0 to .0001" clearance	When locked in piston—and all pressure feed, .0007 to .0009" clearance

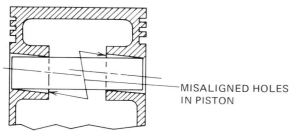

—MISALIGNED HOLES IN PISTON

FIGURE 20-27

Cross-section of piston with misaligned pin holes. This is generally caused by finishing each pin hole separately (Courtesy of Sunnen Products Company).

sence of such a gauge you can make some simple tests to show whether the fit of a new pin meets certain necessary standards.

1. Rotate the pin, first in one boss of the piston and then in the other. Examine the surfaces for total contact.

2. Push the pin from one boss to the other. The pin should enter the second boss without force or bind.

3. On full floating pins, check for taper and bell-mouthing of the rod "eye" bushing. Insert

FIGURE 20-28

Some shops diamond bore the pin bushing. This gives a perfect 360° bearing surface.

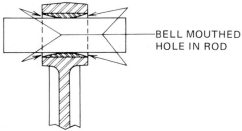

FIGURE 20-29

Bell mouthed pin hole in rod (Courtesy of Sunnen Products Company).

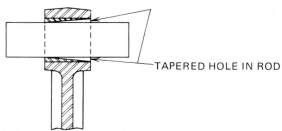

FIGURE 20-30

Tapered pin hole in rod (Courtesy of Sunnen Products Company).

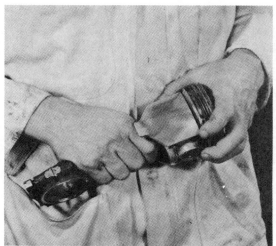

FIGURE 20-31

Checking piston pins for excessive clearance in this manner is not accurate.

the pin from either end of the bushing. If the pin starts easily from either end, but becomes tight in the center, the hole is bell-mouthed (Figure 20-29). If the pin is tight on one end, but free on the opposite end, the hole is tapered (Figure 20-30).

Some people "feel" for excessive wrist pin clearance by rocking the piston against the pin (Figure 20-31). *This method is not always valid*. It is very difficult to determine clearance when the pin is installed. A pin that "feels" normal may actually have more clearance than specifications. This situation often causes a problem after a ring job. The mechanic did not replace the pins because they "felt" tight and were not noisy prior to disassembling the engine. However, the engine now has noisy pins because the pistons slide less freely in the cylinders (new rings cause increased friction).

How Temperature Affects Pin Fit Clearance

We should realize by now, pins must be fit with sufficient and precise clearance in order to allow for temperature variation. What is not generally known, however, are the expansion and contraction rates.

As a general rule, *a 1" diameter steel wrist pin will expand .0003" for every 50°F increase in temperature, while the pin hole (in an aluminum alloy piston) will expand .0006" for every 50°F increase in temperature.*

Pin Seizure Analysis

Under certain engine operating conditions, the pin temperature can be much higher than the bosses in the piston. When this happens, the oil clearance is reduced. If the pin fit becomes too tight, a seizure will result.

A press fit pin will sometimes seize after a prolonged heavy load. What happens is during heavy load conditions, the engine is generating extra heat. The crankshaft "soaks" up this heat. Under light load (a downhill run) the piston pin-holes cool rapidly and shrink. However, the heat that has been stored in the crankshaft continues to travel up the connecting rod directly into the pin. This prevents the pin from cooling and shrinking in size. As a result, pin clearance becomes tight and a seizure occurs (Figures 20-32 and 20-33).

Full floating pins often experience seizure in the rod bushing when fit with insufficient clearance (Figure 20-34). *NOTE: Replacement bushings should always be expanded and seated (Figure*

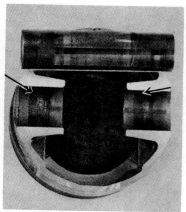

FIGURE 20-32
Pin seizure marks can be seen in the middle of both piston bores as well as on pin (Courtesy of Sunnen Products Company).

FIGURE 20-33
Piston and rod damage caused by a rod seizure. The failure was due to insufficient pin clearance (Courtesy of Sunnen Products Company).

FIGURE 20-34
Pin seizure in rod bushing (Courtesy of Sunnen Products Company).

20-35). This assures 100% contact for perfect heat transfer.

In the last several years, many shops have reported cases where a rebuilt engine has run very well for several thousand miles, then suddenly it tightens up and knocks out a rod bearing. When the engine is torn down for inspection, the pins are so tight that the connecting rods can barely be moved when the piston is held. When the pins are removed, there is no evidence of scoring. However, the bearing area of the pin is black and shiny. Investigation has revealed that this black discoloration is a layer of hardened colloidal graphite (found in certain prelubricants and oils). Deposits do not have to be very heavy to seize a pin (a properly fit pin may only have .0001" clearance).

To prevent a recurrence, keep the oil temperature around 180° F, use a straight high detergent oil, and avoid using additives that contain a high percentage of graphite.

WARNING: After rod bushings or piston bosses are honed, the grit must be completely washed out. Figure 20-36 shows a pin that scored because all the grit was not removed.

PRESSED-IN BUSHING FITS ON THE HIGH SPOTS ONLY

EXPANDED BUSHING FILLS IN VALLEYS AND IS LOCKED IN PERFECTLY TO ROD EYE

FIGURE 20-35
Using a special mandrel to expand a rod "eye" bushing (Courtesy of Sunnen Products Company).

FIGURE 20-36
Badly scored pin (Courtesy of TRW Replacement Division).

Piston Pin Lock Ring Failures

Occasionally, a piston pin lock ring will come loose or fall out. This can result in severe cylinder wall damage. Mechanics should be aware of the possible causes for such failure. Improper installation is the most common reason for lock ring failure. A lock ring should be compressed just small enough to slip into the pin boss groove. Never close it entirely, as this will cause overstressing and a loss of tension. *The open end of the lock ring should be installed toward the bottom of the piston (Figure 20-37).* The reason for this is that a piston changes its direction more rapidly in the upper half of its stroke than in the lower half. By installing in this manner, the lock ring has a

full bearing area at the top, which is the point of greatest motion stress.

Always install one eared lock rings with the tang pointing out from the pin (Figure 20-38) and with the open end facing the bottom of the piston.

Misaligned, bent, or twisted connecting rods, tapered crankshaft journals, and excessive crank end play all tend to cause excessive side movement on a floating pin. Such action pushes on the lock ring, wears the groove, and eventually results in failure (Figure 20-39).

ASSEMBLING THE PISTON TO THE ROD

First, determine the position of each piston and connecting rod. This is best accomplished by consulting the appropriate shop manual. However, here are some general guidelines.

1. Most pistons are marked with a *notch*, an *arrow*, the letter *F*, or the word *Front*. The

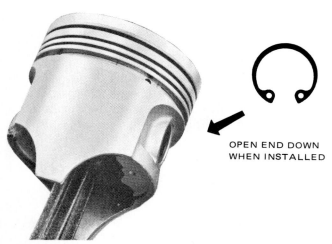

OPEN END DOWN
WHEN INSTALLED

FIGURE 20-37
Piston pin lock rings should be installed with the open end down. This Tru-Arc (brand name) ring should be placed with the rounded edge facing towards the center of the piston. This positions the sharp edge in the groove toward the direction of pin thrust and helps insure retention.

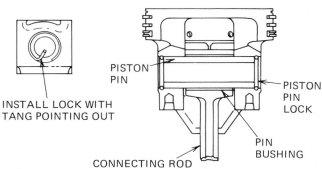

PISTON PIN

PISTON PIN LOCK

PIN BUSHING

CONNECTING ROD

INSTALL LOCK WITH TANG POINTING OUT

FIGURE 20-38
Cross-sectional view of a full floating pin in place (Courtesy of Buick Motor Division of General Motors Corporation).

FIGURE 20-39
This piston damage was caused by the end of the pin lock breaking off. Do not reuse pin lock rings (Courtesy of Dana Corporation—Service Parts Group).

piston must be installed with this mark facing the front of the engine. If the piston is designed with valve reliefs, incorrect assembly can cause piston-to-valve interference. See Figure 20-40. If the piston is installed backward (and it has an offset pin), noise and slap can result.

2. Connecting rods are generally positioned with respect to an oil spurt or bleed oil, factory stamped number markings, bearing tangs, a special marking on the rod beam, a chamfer on the big-end bore, or an offset shoulder.

Figures 20-41, 20-42, and 20-43 show examples of correct rod and piston installation for several common engines.

Spurt Holes and Bleed Holes

Many connecting rods have a spurt hole that throws out oil from the connecting rod journal (Figure 20-44). The oil is aimed to help lubricate the cylinder wall and piston pin. When assembling the piston to the rod, make sure the spurt hole is facing the right direction and is not clogged. Some connecting rods have a hole similar to a spurt hole (Figure 20-45). However, its only purpose is to help control oil flow through the rod bearing assembly.

Reassembling Pistons and Rods with a Press Fit

The assembly of press fit piston pins can be accomplished by using a hydraulic press and special pin inserter tools and adaptors (Figure 20-46). This tooling prevents piston distortion and "rod cocking" during assembly. It also automatically centers the pin in the piston bosses. *CAUTION: Some aftermarket standard size pins*

FIGURE 20-40
These piston head top views show the correct installation position for the various Chevrolet engines in the left column (Courtesy of Chevrolet Motor Division of General Motors Corporation).

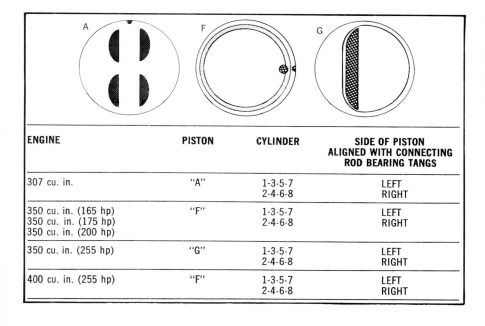

ENGINE	PISTON	CYLINDER	SIDE OF PISTON ALIGNED WITH CONNECTING ROD BEARING TANGS
307 cu. in.	"A"	1-3-5-7 2-4-6-8	LEFT RIGHT
350 cu. in. (165 hp) 350 cu. in. (175 hp) 350 cu. in. (200 hp)	"F"	1-3-5-7 2-4-6-8	LEFT RIGHT
350 cu. in. (255 hp)	"G"	1-3-5-7 2-4-6-8	LEFT RIGHT
400 cu. in. (255 hp)	"F"	1-3-5-7 2-4-6-8	LEFT RIGHT

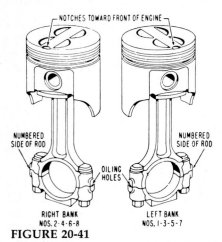

FIGURE 20-41

AMC 343 CID engine (Reprinted from *Motor Handbook* © 1983, The Hearst Corporation).

FIGURE 20-42

Chevrolet, Buick, and Oldsmobile six-cylinder 250 CID engine (Reprinted from *Motor Handbook* © 1983, The Hearst Corporation).

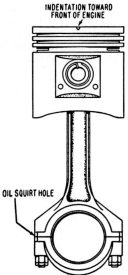

FIGURE 20-43

Ford six-cylinder 223, 240, and 300 CID engines. Notice that the pin lock is incorrectly installed (Reprinted from *Motor Handbook* © 1983, The Hearst Corporation).

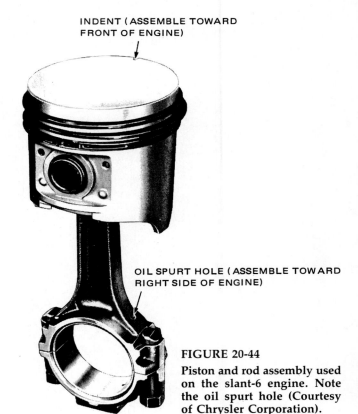

FIGURE 20-44

Piston and rod assembly used on the slant-6 engine. Note the oil spurt hole (Courtesy of Chrysler Corporation).

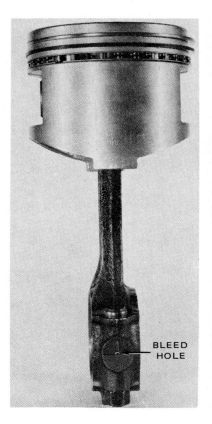

FIGURE 20-45

Connecting rod bleed hole. This rod is offset in order to provide crankshaft cheek clearance (Courtesy of American Motors Corporation).

FIGURE 20-46
Removing piston pin.

(a)

FIGURE 20-48*a*
Electric pin furnace.

FIGURE 20-49
Fixture for centering rods on interference fit piston pins.

FIGURE 20-47
Countersink for chamfering the wrist pin end of connecting rods (Courtesy of Sunnen Products Company).

are up to .0005" larger than the standard factory size. Before pressing in these pins it will be necessary to slightly open up (enlarge) the "eye" of the rod for the correct negative fit. Otherwise, the pin may be galled during installation.

When installing pins using the hydraulic press method, lubricate the pin. Also, chamfering the rod "eye" makes assembly of the pin easier and reduces the possibility of galling. Figure 20-47 shows a special countersink tool designed for that purpose.

An alternate way used in many shops, is the electric pin furnace method (Figure 20-48 *a* and *b*). This furnace will install press fit pins without the fear of scoring or cracking pistons. The "eye" of the rod is expanded by heat (approximately 425° F) so the pin can be pushed into place by hand pressure. When pushing, a special fixture is used to center the rod on the pin (Figure 20-49).

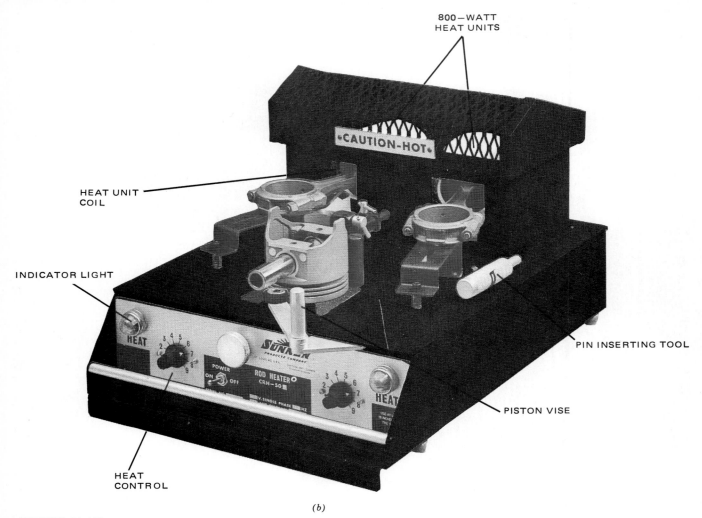

800—WATT
HEAT UNITS

CAUTION-HOT

HEAT UNIT
COIL

INDICATOR LIGHT

PIN INSERTING TOOL

PISTON VISE

HEAT
CONTROL

(b)

FIGURE 20-48b
Rod heater (Courtesy of Sunnen Products Company).

TRUE CASE HISTORY

Vehicle	Ford Truck
Engine	6 cylinder, 223 CID
Mileage	Unknown
Problem	1. Extremely rough idle after installing new rings and rod bearings.
	2. Severe engine vibration while driving. As road speed increased, the condition became worse.

Cause   The mechanic lost one of the connecting rod bearing caps when the rod bearings were being replaced. He went to the local auto salvage yard and obtained a cap from an identical engine. This cap, however, had a much larger balancing pad (Figure 20-50). The extra weight caused the described out-of-balance condition.

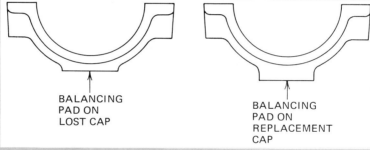

BALANCING PAD ON LOST CAP

BALANCING PAD ON REPLACEMENT CAP

FIGURE 20-50

Cure   Metal was ground off the balancing pad of the replacement cap until it visually matched the other caps.

Comment   Even though the problem was solved, the cure was risky. A rod cap should never be interchanged unless the rod big-end bore is reconditioned. Also, the rod should have been weight matched to the others by using a gram scale.

CHAPTER 20 SELF-TEST

1. List the only two assembling approaches that can be used to assure the proper functioning of rod bearing inserts.
2. Connecting rod big-end bore stretch is usually the greatest in what direction?
3. What are three methods to check the out-of-roundness of the rod big-end bore?
4. Describe the procedure for reconditioning a rod.
5. What problems can be caused by a connecting rod that has twist?
6. How can a slightly bent connecting rod be straightened?
7. What causes most connecting rod failures?
8. Why should rod nuts never be reused?
9. Explain the stretch method of rod bolt tightening.
10. What happens to engine oil pressure when a rod bearing starts to "knock"?
11. What is a stress raiser?
12. How can vise jaw marks on a connecting rod lead to an early failure?
13. Why is the inside hole on some wrist pins manufactured with a taper?
14. What are the four different piston pin attachment methods?
15. What method of pin attachment is used in the majority of passenger car engines?
16. What pin attachment design uses a lock ring installed at the end of each piston boss?
17. What should be the clearance between the pin and the piston bosses on a a full floating pin?
18. What five criteria must be met when fitting pins?
19. On a full floating pin, the pin starts easily from either end of the rod "eye" bushing, but becomes tight in the center. What does this indicate?
20. Why is "feeling" for excessive wrist pin clearance by rocking the piston against the pin not always a valid test?
21. As a general rule, what is the expansion rate of the pin hole in an aluminum alloy piston?
22. What can sometimes cause a press fit pin to seize after a prolonged heavy load?
23. Why is it important to expand replacement rod "eye" bushings?
24. What engine problem can be caused by using an oil that contains a high percentage of graphite?
25. A piston pin lock ring must be installed with the open end facing what direction?
26. List three markings used on pistons to determine correct installation position.
27. Connecting rod installation position can be generally identified with respect to what?
28. What two methods can be used to install press fit pins?

CHAPTER 21

PISTONS AND RINGS

A piston in an automobile engine is more than just a slug of aluminum that fits into a cylinder and converts the combustion pressure to a force on the crankshaft. A piston must be lightweight, yet strong in the areas of the head, the ring belt, and the pin bosses. It must allow for thermal expansion and provide precise operating clearances. It must confine and support the rings. Additionally, the piston must have durability to withstand the extra heat buildup caused by non-leaded gasoline and emission control units.

PISTON NOMENCLATURE

The important piston parts are described below.

1. **Piston head** The top piston surface that receives the force of combustion. The head configuration may be flat, domed, dished, or irregularly shaped (Figure 21-1).
   Some engines use chamfered head pistons in order to eliminate hydrocarbons from forming in the top ring land-to-cylinder bore crevice (Figure 21-2).

2. **Pin bosses** The holes through the piston skirt that carry the piston pin (Figure 21-3).

3. **Compression height** This is the distance from the center of the pin boss to the top of the piston head (Figure 21-3).

4. **Ring belt** That area between the piston head surface and the pin bore where grooves are cut for the installation of the rings (Figure 21-3).

5. **Ring lands** That section of the piston above the top ring and between the ring grooves (Figure 21-3). The lands support the rings in their grooves.

6. **Oil ring groove** A groove cut into the piston around its circumference at the bottom of the ring belt (Figure 21-3). There are generally holes or slots through the bottom of the groove for oil drainage to the interior of the piston.

7. **Compression ring grooves** Grooves (generally two) cut into the piston circumference at the upper part of the ring belt (Figure 21-3).

8. **Skirt** That section located between the

435

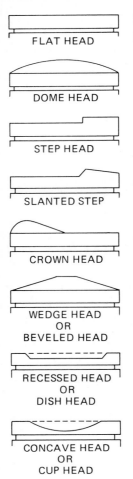

FIGURE 21-1
Some piston head shapes.

FLAT HEAD

DOME HEAD

STEP HEAD

SLANTED STEP

CROWN HEAD

WEDGE HEAD
OR
BEVELED HEAD

RECESSED HEAD
OR
DISH HEAD

CONCAVE HEAD
OR
CUP HEAD

FIGURE 21-2
Chevrolet 400 CID chamfered head piston (Courtesy of Automotive Engine Rebuilders Association).

bottom of the piston and the first ring groove above the pin bore (Figure 21-3).

Most automotive passenger cars use pistons with an open-type slipper skirt (Figure 21-4). Heavy-duty pistons generally use a solid-type trunk skirt. The undulated or wavy skirt is used to provide extra strength in certain high-performance engines (Figure 21-5).

9. **Skirt taper** This is the difference between the diameter of the piston at the top of the skirt and at the bottom of the skirt (Figure 21-6).

10. **Piston cam** The circumferential contour in which a skirt is manufactured (Figure 21-7). This provides for proper cylinder wall con-

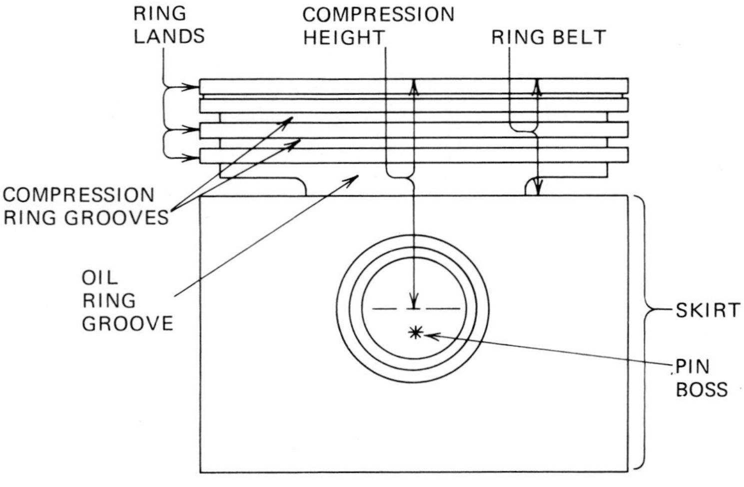

FIGURE 21-3
Typical piston nomenclature. This piston uses a solid-type skirt.

RING
LANDS

COMPRESSION
HEIGHT

RING BELT

COMPRESSION
RING GROOVES

OIL
RING
GROOVE

SKIRT

PIN
BOSS

FIGURE 21-4
Forged piston with a slipper skirt (Courtesy of TRW Replacement Division).

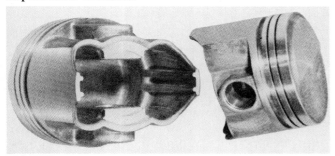

FIGURE 21-5
Piston with an undulated skirt.

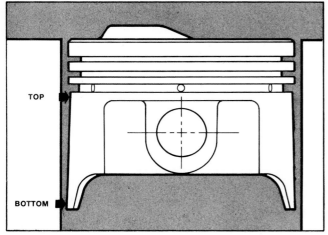

FIGURE 21-6
Skirt taper is provided to give additional clearance at top of the piston skirt (Courtesy of TRW Replacement Division).

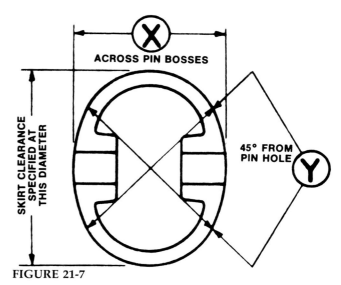

FIGURE 21-7
Cam-ground piston skirt contour (cross-section through pin hole).

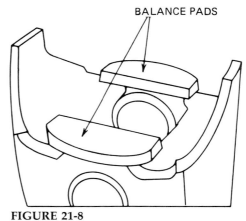

FIGURE 21-8
Pistons are balanced by removing material from balance pads (see arrows).

tact and running clearance under normal temperature and load changes.

11. **Balance pads** Extra aluminum material that is just below the wrist pin bosses (Figure 21-8). This provides material that can be removed when the piston is balanced to a specified weight.

12. **Wrist pin oil holes** Holes drilled from the outside of the piston into the pin bosses allow the entrance of oil for lubrication between the pin and the pin bosses (Figure 21-9). On some aftermarket pistons, another method of wrist pin oiling is used.

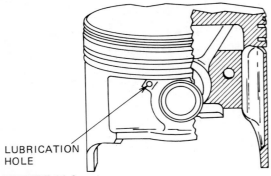

LUBRICATION
HOLE

FIGURE 21-9
Hole for wrist pin lubrication.

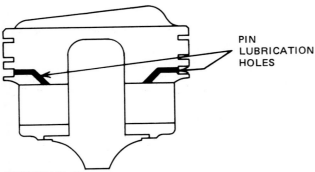

PIN
LUBRICATION
HOLES

FIGURE 21-10
Oil ring drain hole provides pin lubrication on this piston.

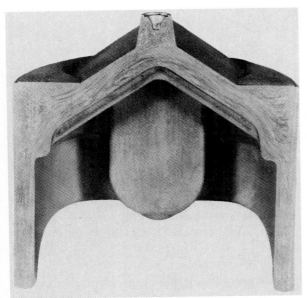

FIGURE 21-11
Grain flow in a forged piston gives it extra strength to withstand forces that could normally crack the conventional cast aluminum types (Courtesy of TRW Replacement Division).

Lubrication is provided through an oil ring drain hole that intersects a drilled hole that runs to the pin bore (Figure 21-10).

PISTON MANUFACTURE

Traditionally, most pistons are made of cast aluminum. Molten metal is poured into a mold, cooled, and then machined to size.

Some pistons are forged. Here a slug of aluminum alloy bar stock is preheated and squeezed under pressure into a piston blank, then later machined. A forged piston has a controlled grain flow in critical areas (Figure 21-11) which makes it stronger and more ductile than the conventional cast aluminum types. Also, a forged piston normally operates up to 100° F cooler than a cast piston (Figure 21-12).

Control Buttons

Teflon control buttons are sometimes used in racing engines to prevent piston rock (Figure 21-13).

Piston Pin Location

The wrist pin hole does not always run through the exact center of the piston. Some manufacturers offset the pin hole by as much as .090" from the piston centerline toward the major thrust face (Figure 21-14). This helps to quiet down "piston slap" noise. Sometimes in racing engines the pin hole is offset toward the minor thrust face because it is felt that a slight torque gain results.

Compensated/Noncompensated Oversize Pistons

Briefly stated, noncompensated oversize pistons have deeper ring grooves than compensated oversize pistons.

The original equipment manufacturers supply oversize pistons of a compensated design. The groove depths of these oversize pistons have the same dimensions as those of the original standard size. To maintain the same groove depths throughout the range of oversize pistons, the manufacturer compensates for the increases in piston diameter by increasing the groove root diameter of each oversize piston. Therefore the groove depths of the oversize pistons remain the same as those of the standard size pistons (Figure

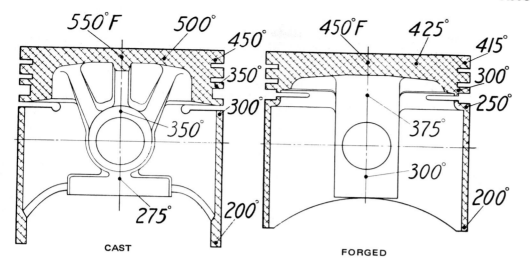

CAST

FORGED

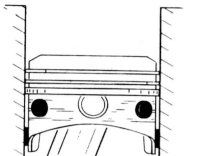

FIGURE 21-13

Piston buttons can be installed in both new and used pistons. Interference fit is achieved by cooling buttons before insertion into hot piston.

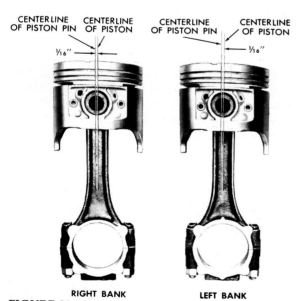

RIGHT BANK   LEFT BANK

FIGURE 21-14

Piston pin is sometimes offset toward the major thrust surface to help control "piston slap" (Courtesy of Pontiac Motor Division of General Motors Corporation).

21-15a). **NOTE:** *When insufficient piston ring back clearance is found, the piston is almost always a compensated type.*

A noncompensated oversize piston has deeper groove depths than a standard size piston. These pistons have deeper grooves because the manufacturer machines the groove of the raw or semifinished piston the same dimensions as would be found in a standard size piston (Figure 21-15b). Since the groove root diameter is not increased for an oversize diameter, this method of piston design and manufacture is termed noncompensated. Independent piston manufacturers normally supply a noncompensated type of piston.

PISTON CAM ACTION

Most engines use pistons that are cam-ground, that is, the diameter across the pin side is less than the diameter across the thrust face side.

The purposes of cam-grinding are twofold.

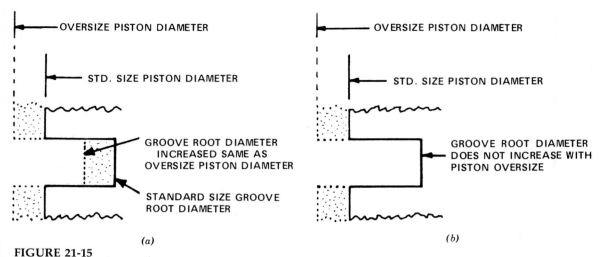

FIGURE 21-15

(*a*) Compensated piston ring groove design. (*b*) Noncompensated piston ring groove design.

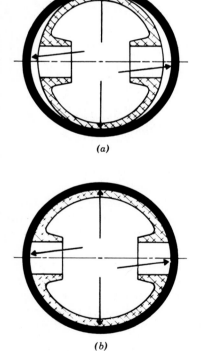

FIGURE 21-16

(*a*) Diagram of cam-ground piston in cylinder at beginning of "full-cam" position. Note that piston has cylinder wall contact in direction of thrust, and clearance along the axis of the pin bosses. (*b*) Diagram of cam-ground piston at engine operating temperature. The piston is in "expanded-cam" position. It has expanded till it is now practically round in the cylinder (Courtesy of Sunnen Products Company).

1. To maintain proper clearance between the piston thrust faces and the cylinder wall when the engine is cold. This helps avoid piston slap.

2. To allow room for expansion as the engine warms up to operating temperature to prevent piston seizure.

As a cam-ground piston warms up, it will expand until it is practically round. Because expansion of the skirt contact surfaces is restricted by the cylinder wall, the piston will expand along the axis of the pin bosses (Figure 21-16*a* and *b*). The pin bosses move outward on the pin as the amount of cam decreases. When the engine cools off, the pin bosses move inward as the piston resumes its cam shape.

Proper cam action is necessary in order to maintain correct piston-to-cylinder wall clearance at all temperatures and to stablize the rings in the cylinder.

Piston Clearance

Piston clearance (sometimes called skirt clearance) is the difference between the piston skirt diameter (measured in a plane perpendicular to the pin), and the cylinder diameter. The proper clearance varies with the type of piston, piston material, bore diameter, and the type of engine service. *Generally, passenger car piston clearance ranges from .001" to .004".*

FIGURE 21-17

Measuring piston diameter (Courtesy of Ford Parts and Service Division).

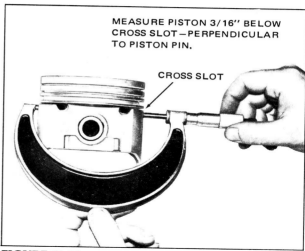

MEASURE PISTON 3/16" BELOW CROSS SLOT—PERPENDICULAR TO PISTON PIN.

CROSS SLOT

FIGURE 21-18

The manufacturer specifies that this piston diameter be measured above the pin (Courtesy of Cadillac Motor Car Division of General Motors Corporation).

The piston skirt-to-bore clearance specification is usually indicated as the minimum clearance. This clearance is for normal operation and conditions. A faulty cooling system, improper timing, and abusive operating conditions can cause higher than normal piston temperatures. Under these conditions, minimum piston clearance may not be sufficient. Pistons used for racing, marine, or special applications may require additional clearance.

Pistons and cylinder bores should be measured at room temperature (70°F). Piston sizes may vary, because of production tolerance, from the minimum to an additional .001" clearance. Measure pistons across thrust faces to check for clearances. Slipper type pistons are generally measured at the centerline of the pin, but at a right angle to the pin (Figure 21-17). In some cases, the manufacturer specifies that the measurement be taken at a point midway between the balancing pad and the tail (bottom) of the skirt. Occasionally, clearance specifications are based upon piston measurement above the pin centerline (Figure 21-18).

The standard size bores of some engines vary in size and initially the pistons were selectively fit to obtain the correct clearance. Standard size pistons for these engines are supplied to fit the average bore size. It may be necessary to hone the smaller bores for minimum piston clearance. Use the basic bore diameter listed to calculate oversize bore diameters. Add the amount of oversize to the basic bore diameter to determine the overbore size. Factory finished oversized pistons will fit with the proper clearance in cylinders bored to this size.

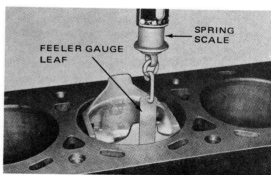

SPRING SCALE

FEELER GAUGE LEAF

FIGURE 21-19

Measuring piston-to-cylinder fit with a feeler gauge leaf. The rings are removed from the piston when making this measurement.

Pistons are available in oversizes larger than those supplied by the engine manufacturer. However, the use of such pistons beyond the manufacturer's bore limit is not recommended.

An old-fashioned (but still viable) method of determining skirt clearance is to use a feeler gauge and a spring scale (Figure 21-19). Push the piston all the way into its bore upside down. Select a feeler gauge leaf that requires 5 to 10 lb. of pull to remove it from between the skirt and the wall. The thickness of this leaf is the same as the skirt clearance.

PISTON FAILURES

A piston can fail from a number of direct or indirect causes. Some common failures and how they develop are analyzed below.

1. **Piston failure due to abnormal combustion** At low cruising speeds the flame front takes about 1/264 of a second to travel across the combustion chamber of a 4" bore engine. During high-speed wide open throttle operation, the flame travel may be three times as fast. What is important is that the burning rate be controlled in order to exert a steady downward force on the head of the piston. With abnormal combustion, the air-fuel mixture is either burned too rapidly (detonation), or the charge is consumed at the wrong time (preignition). This results in sudden pressure rise, overheating, loss of power, rough engine operation, and the possible destruction of head gaskets, valves, pistons, and rings.

Detonation or preignition often burns a hole through the head of a piston (Figure 21-20). Sometimes the burning will be on the top edge of the piston, and extend down through and past the rings (Figure 21-21). Most authorities feel that preignition generally burns pistons, while detonation causes breakage (Figure 21-22). *COMMENT: During detonation, the flame front often reaches the supersonic speed of about 3000 ft per second. This results in a violent combustion chamber explosion. When the combustion chamber walls vibrate, an audible "ping" is heard.*

Deposit induced runaway ignition (DIRSI) is an abnormal combustion condition generally found in engines that use oil with a high concentration of metallic additives. When oil consumption is high, yellowish brown

FIGURE 21-20
The result of preignition on a piston head (Courtesy of TRW Replacement Division).

FIGURE 21-21
The abnormal condition of preignition (higher than normal temperature and pressure) caused melting and burn-through on this piston head (Courtesy of TRW Replacement Division).

FIGURE 21-22
This piston was broken by the continued use of fuel having a lower octane rating than required (Courtesy of TRW Replacement Division).

deposits that look like small corn flakes form on the piston head and on the combustion chamber walls. These deposits start to glow and cause runaway preignition (surface ignition which occurs earlier and earlier in the cycle). Using a low ash or ashless oil helps protect against DIRSI.

2. **Piston scuffing caused by lack of lubrication** Some piston scuffing in the skirt thrust area is acceptable (Figure 21-23a). However, if the piston is scuffed toward the pin side area, this indicates the piston was overheated due to lack of proper lubrication and should be replaced (Figure 21-23b).

   Piston scuffing is often the result of using engine oil of the wrong viscosity for the prevailing temperature. Another item that contributes to piston scuffing is a plugged connecting rod squirt hole.

   COMMENT: In the mid 1970s Ford Motor Company failed to drill squirt holes in the connecting rods and bearings of some 2.3 liter four-cylinder, 200 CID six-cylinder, and 250 CID six-cylinder engines. The result was scuffed pistons and cylinder walls along with high oil consumption and noise. When overhauling one of these engines, you must install rods and bearings with oil squirt holes.

3. **Piston scoring due to loss of cam action** If a pin fit is too tight, the piston is prevented from moving outward on the pin axis as it wants to expand. As a result, the piston is forced to expand across the skirt surfaces. Extra pressure, friction, and heat develops against the cylinder wall which causes the piston to score on the contact surfaces (Figure 21-24) or to seize in the cylinder. Sometimes the piston skirt will distort a slight amount and take a permanent "set." This results in a collapsed piston that slaps badly and passes oil.

4. **Piston breakage caused by insufficient pin clearance** A pin that binds or seizes will often break out the piston boss. Figure 21-25 shows the results of a pin seizure failure caused by incorrectly fitting an oversize pin.

5. **Piston failure due to a drifting pin** In full floating pin type pistons, the lock ring is sometimes pushed out of the piston by the pin. This breaks off the edge of the pin boss and ruins the piston (Figure 21-26). Accurate alignment of the piston and rod assembly usually prevents this trouble from happening. In pistons where the pin is retained by a press fit in the rod "eye," the correct interference fit is absolutely essential. Otherwise, the pin can come loose and drift against the cylinder wall.

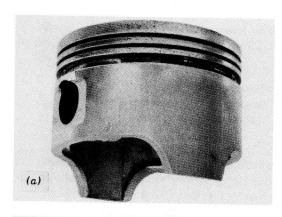

(a)

(b)

FIGURE 21-23

(a) Acceptable piston scoring. (b) Unacceptable piston scoring.

FIGURE 21-24

Too tight of a pin fit caused this piston skirt to score (Courtesy of Dana Corporation—Service Parts Group).

FIGURE 21-25
Results of a pin seizure (Courtesy of Sunnen Products Company).

FIGURE 21-26
The pin lock in this piston was forced out. Note destruction of retaining groove.

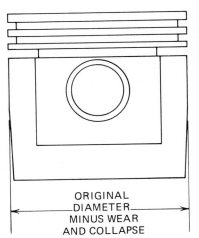

ORIGINAL
DIAMETER
MINUS WEAR
AND COLLAPSE

FIGURE 21-27
Piston collapse occurs along the thrust surface of the skirt. Collapse can be measured with a micrometer.

Overheated Pistons

Today's engines are more subject to overheating (because of lead-free gasoline, lean air-fuel mixtures, and emission control units). As a result, greater numbers of pistons are found to be scored or collapsed (Figure 21-27).

Laboratory tests show that such pistons often lose their original hardness. *NOTE: Knurling a collapsed piston is not recommended in order to establish original piston-to-cylinder wall clearance. Instead, the piston should be discarded and replaced with a new one, because it may not hold size when returned to service.*

COMMENT: Knurling pistons (that are not collapsed) in order to reduce clearance and retain more oil on the skirt is still done in the trade, with usually good results.

Skirt Cracking

Pistons sometimes crack on the skirt or near the pin boss (Figure 21-28). Such cracks can result from overloading, high mileage, or improper design.

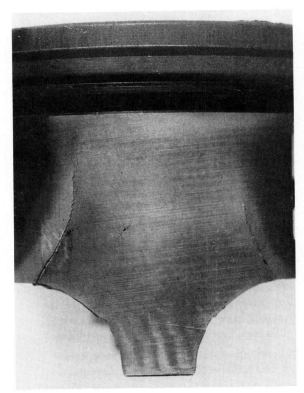

FIGURE 21-28
A cracked piston skirt.

PISTON RINGS

Basically the rings function as a moving seal between the piston and the cylinder wall. This seal must prevent excessive escape of combustion gases (blow-by) during the power stroke. Also, the rings must prevent engine oil from entering the combustion chamber via the crankcase.

Piston Ring Arrangement

Most current pistons are fitted with three rings: top compression, second compression (sometimes called a scraper ring), and a segmented oil ring. Figure 21-29 shows a three ring piston arrangement.

Each ring serves a different function. The top compression ring is used primarily for gas sealing. During the power stroke, gas pressure and ring tension force the face against the cylinder wall (Figure 21-30). The second compression ring acts as a final control for both oil and blow-by. The oil ring must meter just the right amount of oil to the two compression rings and scrape off the rest. If even a small amount of oil gets by the compression rings, the engine can be considered an oil burner. To illustrate this point, consider that one quart of oil contains approximately 36,500 drops. If an engine was driven at 55 mph and consumed 1/864 of a drop during each piston stroke, this would mean a one quart oil loss every 1000 miles. Ring makers use different materials and coatings, and offer dozens of basic designs.

Piston Ring Material

Several materials are used in the manufacture of piston rings. Most top compression rings use gray cast iron as the base material. Plain cast iron is used in the majority of automotive applications for the second compression ring. The oil ring spacer-expander is generally made of die-formed stainless steel. The rails are usually steel with a prelapped chrome face.

Nodular cast iron is the compression ring material used in some heavy-duty gasoline, diesel, and high-performance engines. Nodular cast iron has a rupture strength of over 200,000 psi, and is virtually unbreakable when used as a piston ring. It has two to three times the tensile strength of gray cast iron, and can be bent without breaking (Figure 21-31). It will withstand the impact loading generated in excess of 10,000 rpm. Nodular cast iron is so termed due to its nodular

FIGURE 21-29
Typical ring position on piston.

TOP GROOVE
TORSIONAL COMPRESSION ID BEVEL UP

SECOND GROOVE
WIPER RING OD CHANNEL DOWN

THIRD GROOVE
CIRCUMFERENTIAL CHROME OIL RING

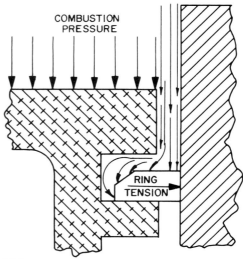

COMBUSTION PRESSURE

RING TENSION

FIGURE 21-30
Compression ring forces (Courtesy of TRW Replacement Division).

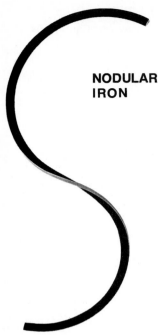

FIGURE 21-31

The ductility of nodular iron rings allows them to be bent into an *S* shape. A regular iron ring would break (Courtesy of TRW Replacement Division).

NODULAR IRON

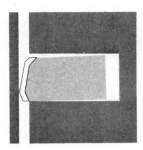

FIGURE 21-33

A chrome faced top compression ring.

dispersion of graphite. Figure 21-32 shows the dispersion of graphite in nodular and in plain cast iron. The increase in rupture strength due to the interval of cast iron between particles of graphite is readily apparent.

Pistons Ring Coatings

Early ring face coatings were tin or electroplated cadmium. Most modern engines use an electroplated chrome or plasma-sprayed molybdenum coating on the top compression ring. Phosphate or ferrax coated plain cast iron is still widely used as the number two compression ring. Some oil

ring spacer-expanders are coated with Teflon to prevent contaminants from sticking to the surface.

During manufacture, chrome plated rings are beveled at the outer edges to aid separation after plating. About .005″ of chromium is then applied to the ring face. A finished chromium face is shown in Figure 21-33. Moly rings are channeled (grooved) prior to spraying. The moly coating is actually an inlay (Figure 21-34).

Chrome faced rings were developed as a result of World War II. The dust and sand conditions of the African campaign created an acute piston ring and cylinder wall wear problem. This urgent demand for an abrasive resistant material prompted the production of chrome rings.

Chromium is extremely hard, dense, and has high resistance to abrasive wear. Also, its surface has less tendency to become impregnated with abrasives than does a softer, less dense material (such as cast iron). Plain cast iron has a melting point of about 2800° F, while chrome melts at 3212° F. Thus, chrome plated rings are very resistant to scuffing and scoring.

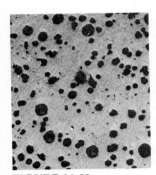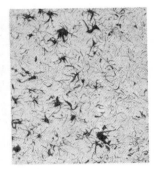

FIGURE 21-32

Microphotographs of nodular graphite (left) and flake graphite (right) ring material (Courtesy of TRW Replacement Division).

MOLYBDENUM

FIGURE 21-34

Cast iron compression ring with a moly inlay.

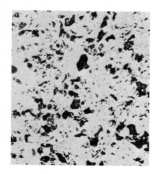

FIGURE 21-35

This microphotograph shows the porous characteristics of a moly ring face (Courtesy of TRW Replacement Division).

Molybdenum is a face material designed to make a ring live under severe engine performance conditions. The physical properties of molybdenum are more impressive than chrome's—a lower coefficient of friction (less drag), greater hardness (longer life), and high resistance to abrasion.

Moly's unusual resistance to scuffing is due primarily to its high melting point of 4750° F. Piston ring scuffing occurs only when the boundary layer of the ring face reaches the melting point. Therefore, the use of a metal such as molybdenum provides a ring face that is beyond the reach of operational engine temperatures and effectively eliminates scuffing.

Moly is thermally bonded to the cast iron in a process that leaves a 15% open porosity facing (Figure 21-35). The microporosity allows the moly to act as an oil reservoir, maintaining upper cylinder wall lubrication at all times. For heavy-duty and high-performance applications, a ceramic ring coating has been recently developed that is composed of aluminum oxide and titanium oxide.

Piston Ring Design

The most visible trend in compression ring design over the past years has been that of ring width and configuration. At first, 1/8″ and 3/32″ widths were the standard; now 5/64″ is the rule, with 1/16″ rings starting to be used.

Ring width reduction, however, has some trade-offs as described below.

1. A thinner ring reduces mass and cuts the ring's inertia. A reduced mass means less groove pounding (an important feature with aluminum pistons).
2. Ring flutter is easier to control with a thin ring.
3. Since a thinner ring presents less material to the bore face, scuff is minimized.
4. Machining the ring grooves during piston manufacture becomes more difficult as the width narrows. The practical limit seems to be 1/16″.
5. A thinner ring is less rigid and, therefore, more difficult to install.

There are a number of cross-section configurations used in compression ring design. Here are the ones in use today.

1. Plain compression.
2. Taper face.
3. Torsional twist.
4. Taper face torsional twist.
5. Taper face reverse torsional twist.
6. Scraper torsional twist.
7. Barrel face.
8. Head land.

The plain compression ring is, in most cases, being replaced by other designs. For instance, by putting a slight taper on the ring face, ring-to-cylinder wall contact is reduced to a narrow line (Figure 21-36). This narrow line contact helps to accelerate ring seating and ensures excellent oil control. On piston downstrokes, the bottom edge of the taper scrapes oil from the cylinder wall. During piston upstrokes, the taper rides

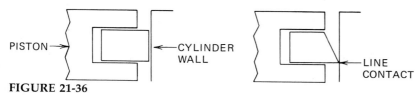

PISTON → ←CYLINDER WALL

←LINE CONTACT

FIGURE 21-36

A plain compression ring design is shown in the left drawing. A taper face compression ring design is shown in the right drawing. Note the narrow cylinder wall contact line of the taper face ring.

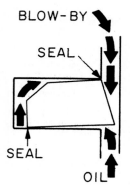

FIGURE 21-37

A positive torsional twist ring shown in its groove. Note seal points.

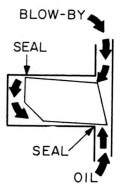

FIGURE 21-38

A reverse torsional twist ring is often used as the second compression ring.

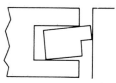

FIGURE 21-39
Scraper torsional twist ring.

FIGURE 21-40
Barrel face chrome compression ring (Courtesy of TRW Replacement Division).

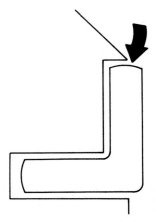

FIGURE 21-41

Head land ring relies on gas pressure for seal. Tension and inertia play only minor roles (Reprinted with Permission, Copyright 1977 *Automotive Engineering*, Society of Automotive Engineers, Inc.).

over the oil. Occasionally, a taper face ring is installed upside down. When this happens, oil is scraped into the combustion chamber and high oil consumption is the result.

The torsional twist ring is designed to twist in its groove because the cross-section is intentionally unbalanced. Material is generally machined away from the inside upper corner of the ring (Figure 21-37). This changes the internal stress of the ring, so that when it is compressed in the cylinder, it tilts on its axis. This twist causes a line contact seal at the top and bottom of the groove and at the ring-to-cylinder wall surface. This minimizes leakage through the groove and makes for good oil control.

The reverse torsional twist ring first patented by Muskegon has come into industrywide use as the second compression ring (Figure 21-38). It is designed to stop oil pumping under high intake manifold conditions. The chamfer that makes the ring tilt is on the bottom inside corner. This makes a seal at the upper inside and bottom outside edge of the ring groove. Oil is, therefore, blocked from entering the groove and trying to make the ring float up.

The scraper torsional twist ring has the lower outside corner removed (Figure 21-39). This causes the same action as in the torsional twist ring.

Barrel or radius faced rings are gaining popularity as the top compression ring (Figure 21-40). The ring-to-cylinder wall contact is along the center of the face. This compensates for piston rocking and any misalignment in the ring groove. Barrel faced rings offer fairly high unit pressure, good oil control, and quick seating.

The head land ring (often called a Dykes ring because it was invented by Dr. Paul de k. Dykes of Cambridge University) are popular in high-performance engines. The piston groove, like the ring, is L shaped (Figure 21-41). Because of this design, combustion gas pressure has a clear path into the piston groove where it can effectively force the ring against the cylinder wall.

A conventional oil ring is made up of two rail segments and an expander-spacer (Figure 21-42). Its fundamental job is to wipe the bore, leave just the right amount of oil, and direct the rest into the inside cavity of the piston. The expander-spacer is designed to produce the correct amount of radial pressure around the bore. *NOTE: When installing an expander-spacer, it is very important to avoid overlapping the ends. Different designs have evolved to ensure proper installation (Figure 21-43).*

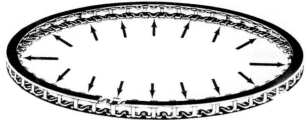

FIGURE 21-42
Oil ring showing uniform radial pressure (Courtesy of Dana Corporation—Service Parts Group).

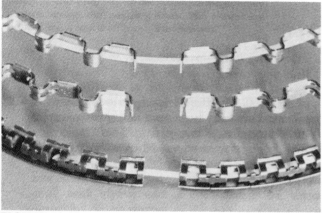

FIGURE 21-43
Guide strip nylon blocks are often used to prevent over-lapping ends of oil ring expander-spacer.

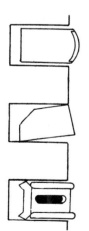

FIGURE 21-44
This factory ring set uses a barrel face top compression ring, positive twist second ring, and a three piece oil ring (Reprinted with Permission, Copyright 1977 *Automotive Engineering*, Society of Automotive Engineers, Inc.).

A Modern Piston Ring Set

Figure 21-44 shows a ring set used on a new Ford engine. The top compression ring is barrel faced, with either a chrome or moly coating. The second

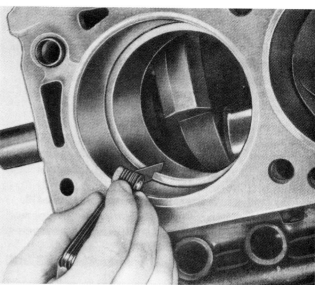

FIGURE 21-45
Measuring ring end gap clearance (Courtesy of Ford Parts and Service Division).

compression ring is taper faced with torsional twist. ***COMMENT:*** *Ford tends to favor regular twist rings, while General Motors seems to prefer reverse twist.* The oil ring is a three piece rail-expander unit.

PISTON AND RING INSTALLATION TIPS

1. Thoroughly clean all carbon from ring grooves and drainage holes. Check top ring grooves for excessive wear and replace or groove insert (GI) the pistons, if necessary.

2. Check end gap clearance on all the compression rings. Measure each ring in the proper cylinder below the bottom of ring travel (Figure 21-45). *Use an inverted piston to push the ring down square with the cylinder.* Minimum end clearances are shown in Chart 21-1.

   Maximum end clearance is not shown because it is not critical. Engine tests have proven

CHART 21-1

Cylinder Bore Diameter	Minimum End Clearance
Less than 3"	.007"
3–3$\frac{31}{32}$"	.010"
4–4$\frac{31}{32}$"	.013"
5–6$\frac{31}{32}$"	.017"

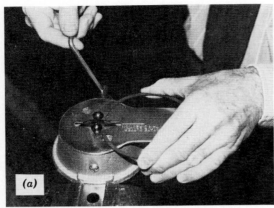 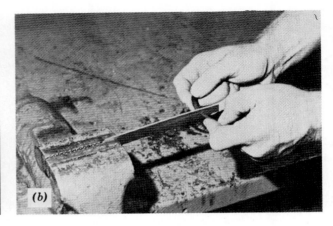

FIGURE 21-46

(*a*) Filing end gap. (*b*) If ring gaps are filed as shown above, always file from outside ring face toward inside diameter. Filing this way will not chip the face coating or leave burrs on OD edges. Break all edges at the gap with a fine stone after filing (Courtesy of Sealed Power Corporation).

that the end clearance may be as much as .040" greater than the recommended minimum, without affecting engine performance. *When the end clearance is more than .040" greater than the specified minimum, a larger oversize ring set should be used.*

Figures 21-46*a* and *b* show methods that can be used to increase ring end clearance when it is too small.

3. Check for worn top grooves. Install a new ring in the piston top groove, and hold the face of the ring flush with the land. Insert a feeler gauge between the groove and the upper side of the new ring to determine side clearance (Figure 21-47). *If the clearance is greater than .006", the groove has excessive wear or the rings are the wrong width.*

FIGURE 21-47
Checking side clearance.

4. Check ring back clearance (sometimes called ring depth clearance). To check compression ring back clearance, place a small straightedge against the piston lands and measure the clearance between each ring and the straightedge (Figure 21-48*a*). *The minimum clearance is .005".*

To check oil ring back clearance, place the straightedge against the piston skirt and measure the clearance between the face of the ring rails and the straightedge (Figure 21-48*b*). *The minimum clearance is .006".*

An alternate method to determine the correct ring-to-piston back clearance is to use a depth micrometer using the skirt of the piston as a reference point (Figure 21-49).

If the clearances are less than minimum the piston grooves will have to be machined deeper in order to ensure proper functioning of the rings.

5. When installing a segmented oil ring, begin by placing the expander-spacer in the piston groove. *The ends must not overlap.* Carefully wind one of the two steel rails over the piston, into the top of the groove. Then, wind the other rail into position at the bottom of the groove. See Figures 21-50*a*, *b*, and *c*.

6. Use a good ring installing tool when placing compression rings in the piston grooves. Pay careful attention to the instructions in the ring box, so all rings are correctly installed in grooves. Compression rings generally have the word *Top* or other mark facing toward the combustion chamber (Figure 21-51). If

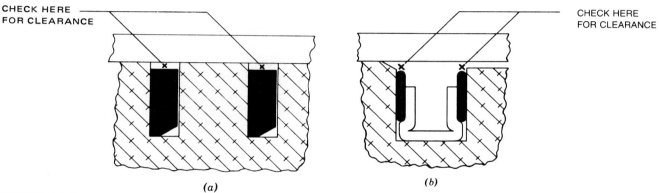

CHECK HERE FOR CLEARANCE

CHECK HERE FOR CLEARANCE

(a) *(b)*

FIGURE 21-48

(*a*) Checking compression ring back clearance. (*b*) Checking oil ring back clearance.

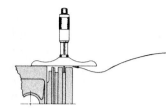

A SLIGHT CLEARANCE WILL BE FOUND HERE WHEN THE GAUGE ANVIL IS HELD FLAT AND IN PLANE WITH THE PISTON THRUST FACE. THIS IS DUE TO THE GROOVE LAND DIAMETER BEING SMALLER THAN THE SKIRT DIAMETER.

FIGURE 21-49

Making groove depth measurement.

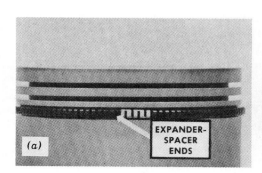

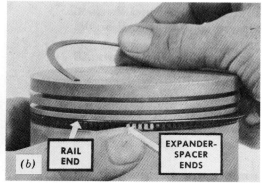

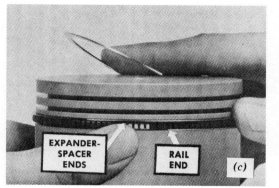

FIGURE 21-50

(*a*) Place expander-spacer in oil ring groove with ends of expander-spacer above solid portion of groove bottom. (*b*) Holding ends of expander-spacer butted, install steel rail on top side of expander-spacer. Locate rail gap approximately one inch to left of ends of expander-spacer. (*c*) Install remaining steel rail on lower side of expander-spacer with gap approximately one inch to right of ends of expander-spacer (Courtesy of Dana Corporation—Service Parts Group).

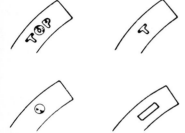

FIGURE 21-51

Typical compression ring installation marks (Courtesy of American Motors Corporation).

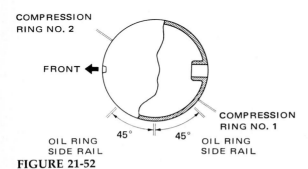

COMPRESSION RING NO. 2

FRONT

OIL RING SIDE RAIL

45°   45°

COMPRESSION RING NO. 1

OIL RING SIDE RAIL

FIGURE 21-52

Position ring end gaps on piston.

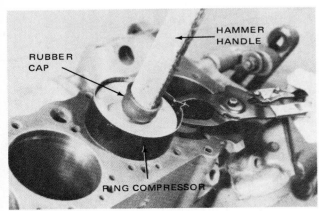

HAMMER HANDLE

RUBBER CAP

RING COMPRESSOR

FIGURE 21-53

Installing piston in bore.

installed upside down, rings will scrape oil up, causing smoking and plug fouling.

7. Dip the entire ring belt and pin area of the piston in clean, light motor oil just before installing. This is very important. There have been cases where new pins have seized and

FIGURE 21-54

Assembling connecting rod and piston assembly to crankshaft using boots to prevent nicking of crankshaft (Courtesy of D.A.B. Industries, Inc.).

locked an engine during initial cranking because of no prelubrication.

8. Stagger the ring end gaps as shown (Figure 21-52). *Do not position any ring gap at the piston thrust surfaces. Do not position any ring gap in line with the pin hole.*

9. Install each piston assembly into the cylinder with the rod journal on the crankshaft at bottom center. Recheck to be sure the piston is facing the right direction. With the rings properly squeezed, "bump" the piston head with a wooden hammer handle until the top ring is just inside the cylinder (Figure 21-53). *CAUTION: No rod bolt should ever be allowed to touch the crank journal.* A short piece of fuel line makes an excellent protector and will help guide the rod onto the crank journal. Better still, use a pair of commercially available protective boots. (Figure 21-54).

ANALYSIS OF PISTON RING FAILURE

Piston rings are now very durable—they usually last thousands of miles. However, a small percentage fail prematurely because of abrasive wear, abnormal rotation, scuffing, detonation, or incorrect installation. Let's take a look at these causes.

1. **Abrasive wear** This is the major cause of premature piston ring failure. The scrubbing action of foreign material will rapidly remove material from the rings and cylinders. Abrasive wear is recognized easily by the dull gray vertical scratches on the ring faces (Figure 21-55).

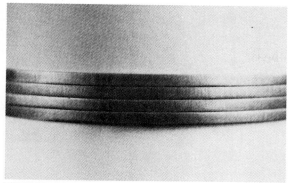

FIGURE 21-55

Abrasive wear (Courtesy of TRW Replacement Division).

Abrasives may enter an engine through many sources, some of which are listed below.

- The air intake system.
- Through dirt left in the engine during a repair or rebuild.
- A vacuum accessory.
- The crankcase breather system.
- Through gasoline.
- A worn carburetor throttle shaft.
- Defective intake manifold gaskets.
- Through the cooling system.
- Through the lubrication system.
- Through the exhaust system.

SHOP TIP

Worn chrome rings can be determined by a simple chemical test. Obtain a 15% copper sulphate solution (available at most pharmacies). Smear a few drops of the solution on the ring face. After several seconds, the copper sulphate will react with any exposed iron and form a layer of copper. If the chrome on the ring face is not worn away, there will be no reaction.

---

2. **Abnormal rotation** Piston rings tend to "creep" during engine operation. "Spinning" of the rings, however, is considered abnormal. This condition contributes to side wear on the rings and piston grooves (Figure 21-56). Ring spinning may be caused by a bent connecting rod or by a one-directional honing pattern on the cylinder wall.

3. **Scuffing** This is a fairly common cause of early ring failure. Scuffing will occur when-

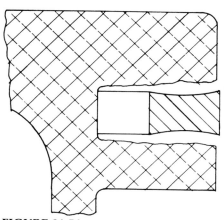

FIGURE 21-56

Abnormal rotation (Courtesy of TRW Replacement Division).

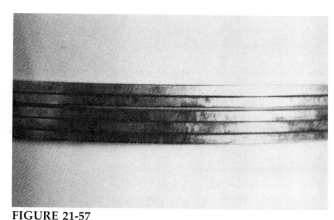

FIGURE 21-57

Badly scored ring faces (Courtesy of TRW Replacement Division).

ever the ring-to-cylinder wall oil films breaks down. This causes excessive heat to be generated and a momentary welding of the ring to the cylinder wall. As the piston moves, these welds break loose leaving scores and voids on the ring face (Figure 21-57). See Chart 21-2 for causes of scuffing and scoring.

4. **Detonation damage** Ring damage caused by detonation is dramatic. When detonation occurs, high pressure shock waves set up vibrations in the pistons and rings. This can break the ring lands, and in turn, cause the rings to break. If the compression rings are moly, the inlay material usually fractures in segments (Figure 21-58).

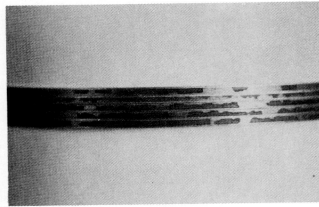

FIGURE 21-58

Detonation damage (Courtesy of TRW Replacement Division).

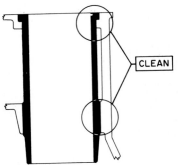

FIGURE 21-59

When installing a wet sleeve, thoroughly clean at points shown. This will prevent cylinder wall distortion and coolant leakage (Courtesy of Dana Corporation—Service Parts Group).

CHART 21-2

Scuffing and Scoring Can Be Caused by:
1. Lubrication System
   a. Worn oil pump
   b. Clogged oil screen
   c. Oil pump pressure relief valve stuck open
   d. Incorrect bearing clearance
   e. Contaminated oil
   f. Low oil pressure
   g. Low oil level
2. Cooling System
   a. Internal leaks
   b. External leaks
   c. Clogged radiator
   d. Deposits in the block
   e. Defective thermostat
   f. Defective radiator pressure cap
   g. Overheating
   h. Corroded water pump impeller
3. Carburetor
   a. Too rich a mixture
   b. Too lean a mixture
   c. Stuck automatic choke
4. Incorrect Ignition Timing
   a. Wrong setting
   b. Worn distributor
5. Detonation and Preignition
6. Improperly Fitted Parts
   a. Not enough piston-to-cylinder wall clearance
   b. Too tight of a piston pin fit
   c. Rings with insufficient end gap
7. Incorrect Engine Break-in
   a. Slow idle
   b. Coolant not permitted to reach operating temperature.
8. Lugging or overloading

5. **Incorrect compression ring installation** When either a tapered face or torsional twist design compression ring is installed upside down, the ring will be tilted on its axis in the wrong direction. This will quickly cause the sharp sealing edge on the ring face to be rounded off and oil will be pumped upward into the combustion chamber.

PISTON/RING PROBLEMS IN WET SLEEVE ENGINES

These important checks should always be made when installing wet sleeves (or liners) in order to prevent scuffed or scored pistons and rings and excessive oil consumption.

1. Thoroughly clean the sleeve counterbore so the flange of the sleeve will rest flat on the block. Also, clean the lower cylinder block bore at the end of the water jacket. Refer to Figure 21-59.
2. Thoroughly clean the cooling system.
3. Fit sleeves in the block without excess clearance. This will prevent vibration of the sleeve during engine operation and help reduce cavitation and corrosion (Figure 21-60).
4. Lubricate and install sealing rings with care.
5. Use a water filter to soften the water and dampen electrochemical action.

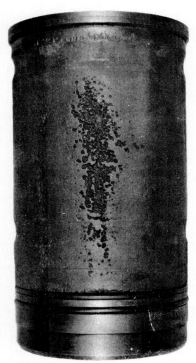

FIGURE 21-60

Cavitation (rhythmic vibration) causes small water bubbles to burst on the sleeve OD. The bubbles burst with enough force to break the granular iron structure until a series of rough appearing cavities is created. Damage is usually found on the sleeve thrust sides as shown (Courtesy of Sealed Power Corporation).

TRUE CASE HISTORY

Vehicle	Buick
Engine	V-6, 231 CID
Mileage	New vehicle; just purchased
Problem	Faint knocking noise heard only during warmup.
Cause	The no. 6 piston was striking the crankshaft counterweight.
Cure	A small amount of material (about .010") was carefully filed off the point of contact on the crankshaft. This did not have any noticeable effect on engine balance.
Comment	Some Buick V-6 engines (found also in other GM vehicles) were built with inadequate clearence in this area.

CHAPTER 21 SELF-TEST

1. Why are some engines using chamfered head pistons?
2. Sketch a section of the piston and identify the ring lands.
3. What is compression height?
4. What is the purpose of the holes or slots through the bottom of the oil ring groove?
5. What is the function of the skirt cam?
6. What methods are used to oil the wrist pin?
7. What are the advantages of a forged piston over a cast piston?
8. Why do some manufacturers offset the pin hole toward the major thrust face?
9. Describe the difference between a compensated and a noncompensated oversize piston?
10. What is a cam-ground piston?
11. Why are pistons cam-ground?
12. Generally, what is the range of passenger car piston clearance?
13. Pistons are usually measured across what point?
14. How can a feeler gauge be used to determine skirt clearance?
15. What problems can be caused by engine "ping"?
16. What can be done to help protect against DIRSI?
17. List three causes of piston scuffing.
18. How can a pin fit that is too tight cause piston scoring?
19. What can happen if a pressed-in pin does not have the correct interference fit?
20. Why shouldn't a collapsed piston be knurled?
21. Most current automobile pistons are fitted with how many rings?
22. What is the advantage of using nodular cast iron as a ring material?
23. What two coatings are used on the top compression ring?
24. What are the advantages of chrome plated rings?
25. What are the advantages of moly rings?
26. What seems to be the practical ring width limit?
27. What are the cross-section configurations used in current compression ring design?
28. What advantages are offered by barrel faced top compression rings?
29. Why is it important to avoid overlapping the ends of an expander-spacer?
30. Explain the difference between a regular twist ring and a reverse twist ring.
31. Where in the cylinder should ring end gap clearance be measured?
32. Ring end gap clearance is .050" greater than the specified minimum. In this case, what should be done?
33. How can ring end clearance be increased when it is too small?
34. What is the maximum allowable top ring side clearance?

35. How is compression ring back clearance checked? What is the minimum allowable?
36. What is the procedure for installing a segmented oil ring?
37. What will happen if a compression ring is installed upside down?
38. How should the piston assembly be prelubricated before installing?
39. How should ring end gaps be staggered?
40. What can be used to protect the crank journal when installing a piston assembly?
41. What is the major cause of premature piston ring failure?
42. What is indicated by dull gray vertical scratches on the ring faces?
43. List 10 ways abrasives can enter an engine.
44. What chemical test can be used to determine worn chrome rings?
45. What conditions can cause ring spinning?
46. Why is it important to fit wet sleeves in the block without excess clearance?

CHAPTER 22

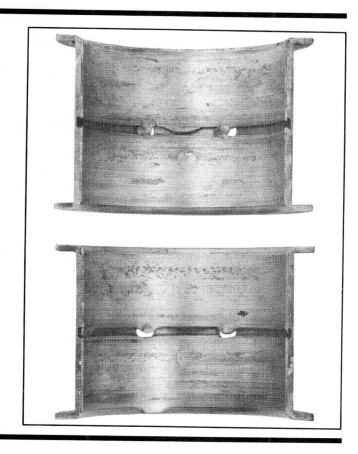

BEARINGS

Engine bearings look deceptively simple to most people. Actually, they are complex in design, structure, and function.

Bearings must support thousands of pounds of force. Yet, at the same time, they must allow parts to rotate freely and noiselessly. Bearings must be capable of performing in the normal presence of heat, dirt, and carbon. They must aid in protecting the crankshaft, cylinder block, connecting rods, and camshaft. Finally, bearings are designed to reduce friction by maintaining a film of oil between mating surfaces.

OIL WEDGE THEORY

When an engine is not running, the crankshaft rests on the bearings. When the engine starts, the crankshaft moves up on the bearing and pulls some oil under it. As crankshaft rotation continues, the oil is wedged between the shaft and bearing, keeping them separated (Figure 22-1).

As engine speeds and loads change, the position of the crankshaft with respect to the bearing changes also. Under certain conditions, the mating surfaces may break through the oil wedge. There is still enough of an oil film to prevent scoring, but not enough to keep the shaft and bearing completely separated. *COMMENT: Do not confuse oil clearance with the space between the shaft and bearing made by the oil wedge.* The oil wedge film is about .0001" (less than a grain of talcum powder). Oil clearance for main and rod bearings is in the range of .001" to .003".

PRECISION INSERTS

Modern automotive engines use thin-shell, precision insert bearings. The bearings are manufactured to very close tolerances so there is proper running clearance right from the start. When wear takes place, the old inserts are simply replaced with new ones. Figure 22-2a and b shows some insert bearings.

The names of the basic parts of a plain and flange type bearing are shown in Figure 22-3. Learning the names will help you understand later information in this chapter.

459

FIGURE 22-1

An oil wedge forms when the crankshaft rotates (Courtesy of Imperial Clevite, Inc., Engine Parts Division).

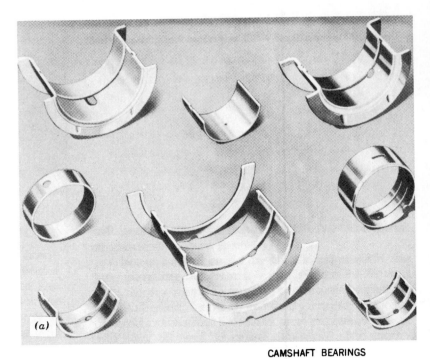

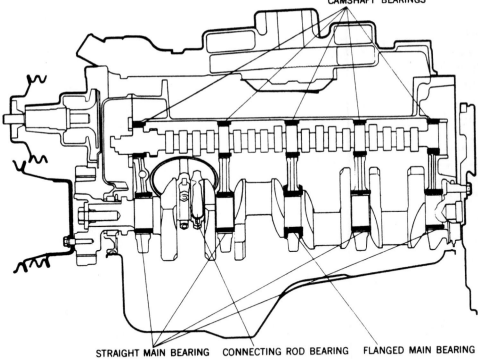

FIGURE 22-2

(*a*) Types of engine bearings. (*b*) Typical bearing location on a 5 main V-8 engine (Courtesy of Dana Corporation—Engine Parts Group).

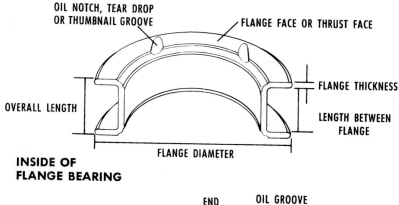

OIL NOTCH, TEAR DROP OR THUMBNAIL GROOVE

FLANGE FACE OR THRUST FACE

FLANGE THICKNESS

LENGTH BETWEEN FLANGE

OVERALL LENGTH

FLANGE DIAMETER

INSIDE OF FLANGE BEARING

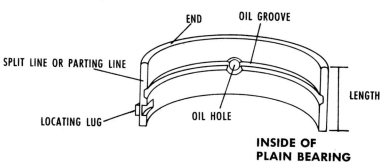

END

OIL GROOVE

SPLIT LINE OR PARTING LINE

LENGTH

LOCATING LUG

OIL HOLE

INSIDE OF PLAIN BEARING

FIGURE 22-3

Plain and flange bearing nomenclature (Courtesy of McQuay-Norris Manufacturing Company).

Oil Grooving

Oil grooves sometimes are added to a bearing to help distribute the oil and maintain a lubrication film. The groove location is critical. One groove properly located is much more effective than any number improperly placed.

As a point of interest, tests show that bearings that can function without oil grooves and still provide proper oil distribution will outlast similar bearings that are grooved. This is due to the fact that oil grooves reduce oil film hydrostatic pressures (Figure 22-4).

Sometimes "thumbnail" grooves are located on flange bearing faces to help distribute oil evenly over the thrust surfaces. A drain groove may be added to a bearing to cause extra oil to drain back into the pan rather than running out the end (Figure 22-5). In some engines, lateral oil grooves are placed on bearings to feed oil to a selected area (Figure 22-6).

Observe the small chamfer on the bearing shown in Figure 22-6. This is a standard design

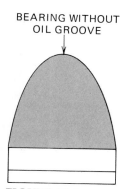

BEARING WITHOUT OIL GROOVE

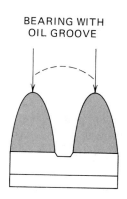

BEARING WITH OIL GROOVE

FIGURE 22-4

How an oil groove effects oil film hydrostatic pressure. Pressures are shown by shaded areas (Courtesy of Federal-Mogul Corporation).

OIL DRAIN GROOVE

FIGURE 22-5

A drain groove may be part of a bearing (Courtesy of Federal-Mogul Corporation).

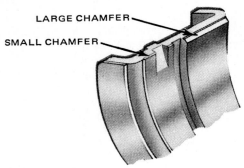

FIGURE 22-6

The parting edge on this bearing has a large chamfer for feeding oil to flange thrust face (Courtesy of Federal-Mogul Corporation).

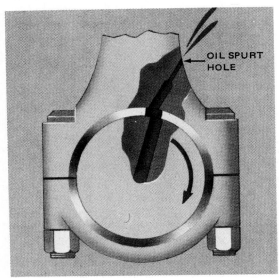

FIGURE 22-7

This connecting rod and bearing assembly allows for extra oil to be directed onto the cylinder wall (Courtesy of Federal-Mogul Corporation).

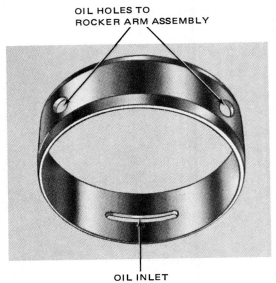

FIGURE 22-8

Camshaft bearing (Courtesy of Federal-Mogul Corporation).

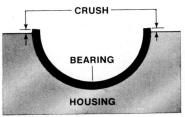

FIGURE 22-9

Crush is manufactured into each half of a split bearing (Courtesy of TRW Replacement Division).

feature on OEM and on replacement bearings. It is done for several reasons. First, it allows for oil bleed-off which helps flush out wear particles. Second, the small chamfer is added assurance against any slight mismatch at the parting line. Third, the chamfer helps prevent raising burrs on the inside edge of the bearing at the parting line during installation.

Oil Holes

Oil holes are an extremely important requirement on many sleeve type bearings. Holes in the bearings must register (be in line) with corresponding holes in the crankcase or other parts to assure proper oiling. Oil holes are sometimes added to direct extra oil to another area (Figure 22-7). In OHV engines, oil holes are often incorporated in the camshaft bearings to direct a metered oil supply to the rocker arms (Figure 22-8).

Crush

Each half of a split bearing is made so it is slightly higher than an exact half (Figure 22-9). *This extension (as little as .00025") is called crush or pinch.* When the two bearing halves come together in assembly, the crush sets up a radial pressure on the bearings so they are forced tightly into the housing bore. This assures good contact between the bearing and housing, which is vital to the proper conduction of heat away from the bear-

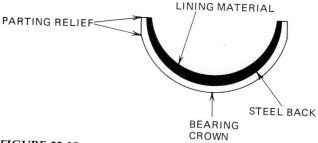

FIGURE 22-10

Exaggerated wall thickness of a new bearing. Note that the thickness near the parting line is slightly thinner.

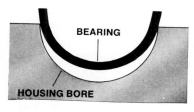

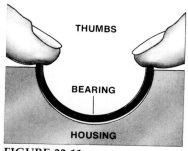

FIGURE 22-11

Bearing spread (Courtesy of TRW Replacement Division).

ing. Crush also keeps the bearing from turning when the engine runs.

Bearing crush is engineered to exert a force that ranges from 12 psi to 40,000 psi. Because of crush, the bearing wall thickness is not manufactured uniform. It is about .001" less at the end than at the crown (Figure 22-10). Note that each end of the bearing has a parting relief so that the inner corners do not curl up when crush is applied.

Spread

Most connecting rod and main bearings are made with spread. This means that the distance across the parting faces of the bearing insert is slightly greater than the actual diameter of the housing bore (Figure 22-11). *Spread may vary in the range*

of .005" to .030". Spread enables a positive fitting of the bearing against the total housing bore. It also keeps the bearing in place during assembly operations.

Sometimes after bearings have been in service, a loss of spread occurs. If it is necessary to reuse the bearings, the halves can be respread. *Gently tap the back of the bearing half with a soft face mallet while the parting faces are resting on a hard flat surface.*

BEARING PROPERTIES

For engine bearings to function properly in a given application, these important material properties must be considered.

1. **Fatigue strength (load capacity)** This is the ability of the bearing to carry the load to which it is subjected without deteriorating. This characteristic is sometimes defined as the *maximum unit load* (measured in psi) for a specified length of time.

2. **Conformability** This is the ability of the bearing surface to creep or flow slightly to compensate for tiny high spots or unavoidable misalignment.

3. **Embedability** This defines the behavior of the bearing material in the presence of hard, foreign particles. The bearing material must be soft enough to allow these particles to embed themselves in it. Otherwise, the shaft which the bearing is supporting will be scratched.

4. **Corrosion resistance** Today, corrosion resistance is more important than ever. Extended oil drain periods, higher engine operating temperatures, and sulphur fuel can cause high concentrations of acid to form in the crankcase oil. This acid can then attack the bearings (especially the copper-lead type). ***NOTE:*** *If you feel that corrosion is going to be a problem, use aluminum bearings. They are highly resistant to acid attack.*

5. **Surface action (resistance to scoring)** The bearing material must be able to resist scoring if the bearing and shaft make contact. This can occur when a heavy load breaks through the oil film that is wedged between the shaft and bearing.

Unfortunately, there is no one bearing material best suited for all applications. Because of this, bearing manufacturers offer a variety of material groups. Each has a different combination of the properties just covered.

For example, one material might rate high on fatigue strength but exhibit poor embedability properties. Another material might have good conformability and embedability, yet have poor fatigue strength.

BEARING MATERIALS

The majority of bearings in use today are made up of two or more layers (Figure 22-12*a* and *b*). The lining material is supported by a low carbon steel back.

Engine bearing materials can be grouped into the following classifications.

1. **Babbitt** This is the most universally used bearing surface material. Babbitt is a soft alloy composed of 83% lead, 15% antimony, 1% tin, and 1% arsenic. This material can be divided into two categories, conventional and thin babbitt.

Conventional babbitt bearings have excellent conformability, embedability, corrosion resistance, and surface action. They have poor fatigue strength.

Thin babbitt bearings provide good fatigue strength, corrosion resistance, and surface action. However, they have poorer conformability and embedability than conventional babbitt. Figure 22-13 shows the effect of babbitt thickness on bearing life.

Babbitt is a light duty bearing material, used in limited applications by some OEM's.

2. **Sintered copper-lead** This bearing material is made by sintering (heating) metal powders on a steel strip to produce an approximate 50% copper and 50% lead surface. See Figure 22-14 for a photomicrograph of a typical bimetal sintered copper-lead bearing.

Sintered copper-lead bearings have much better fatigue strength than conventional or

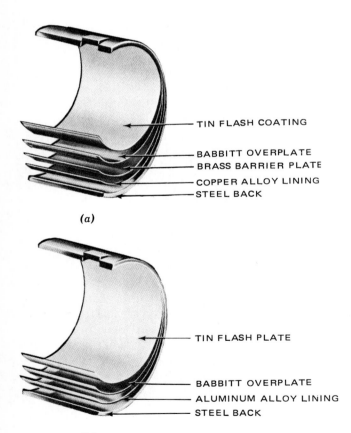

TIN FLASH COATING
BABBITT OVERPLATE
BRASS BARRIER PLATE
COPPER ALLOY LINING
STEEL BACK

(a)

TIN FLASH PLATE

BABBITT OVERPLATE
ALUMINUM ALLOY LINING
STEEL BACK

(b)

FIGURE 22-12

(a) A five-layer overplated copper alloy bearing. *(b)* A four-layer overplated aluminum alloy bearing (Courtesy of Sealed Power Corporation).

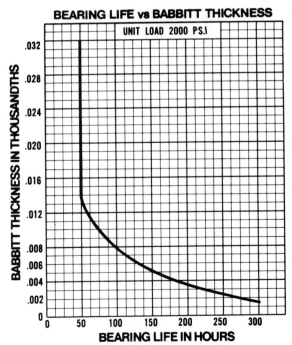

BEARING LIFE vs BABBITT THICKNESS

UNIT LOAD 2000 P.S.I

BABBITT THICKNESS IN THOUSANDTHS

BEARING LIFE IN HOURS

FIGURE 22-13

Bearing life versus babbitt thickness (Courtesy of Imperial Clevite, Inc., Engine Parts Division).

TIN FLASH PLATING ▶

COPPER-LEAD-TIN MATRIX ▶

STEEL BACK ▶

FIGURE 22-14

Cutaway and photomicrograph of typical bimetal sintered copper-lead bearing (Courtesy of TRW Replacement Division).

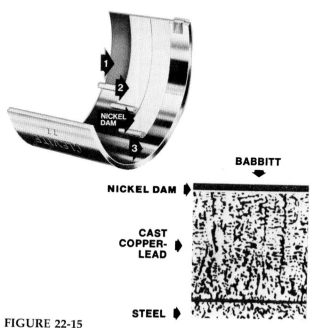

BABBITT ▶

NICKEL DAM ▶

CAST COPPER-LEAD ▶

STEEL ▶

FIGURE 22-15

Cutaway and photomicrograph of trimental cast copper-lead bearing (Courtesy of TRW Replacement Division).

thin babbitt bearings (especially at high temperatures). This gain, though, causes some sacrifice in corrosion resistance and surface action.

3. **Cast copper-lead** These bearings have an electroplated babbitt overlay (about .001" thick) and contain a higher percentage of copper than sintered types. A photomicrograph of a trimetal cast copper-lead bearing is shown in Figure 22-15.

    Cast copper-lead bearings are particularly suited for heavy-duty application (original equipment for Cummins, Ford, Mack, Volkswagen, and Toyota). They offer excellent fatigue strength and corrosion resistance. Conformability, embedability, and surface action are good.

4. **Tin-aluminum** These bearings have found wide acceptance within the American automotive industry. Tin-aluminum bearings are not affected by unleaded fuels. Nor do they suffer the same lead leaching at high oil temperatures as do copper-lead bearings.

    At present, some new passenger car engines (Chrysler, Fiat, Renault, and Peugeot) are using a 20% tin, 80% aluminum composition for crankshaft and connecting rod bearings. A .5% tin, 6% lead, 4% silicon, 89.5% composition is being used as original equipment on General Motors products.

BEARING FAILURES AND CAUSES

Whenever bearings fail, the damage generally can be classified under one of the following five headings.

1. **Fatigue (lining break-out)** This failure appears as fine cracks or cavities in the surface material (Figure 22-16). The major cause of fatigue in the bearing lining is load concentration on one section of the surface rather than overall. Figure 22-17 shows fatigue along the bearing edge that was caused by too large a fillet radii. Fatigue distress can also be caused by an out-of-round housing, bearing cap shift, "hot rodding," lugging, or improper bearing material selection.

2. **Scoring** The main cause of bearing scoring is the presence of hard foreign material in the lubricating oil. Figure 22-18 shows a bear-

FIGURE 22-16
Fatigue failure (Courtesy of Mc-Quay-Norris Manufacturing Company).

FIGURE 22-18
Badly scored bearing caused by improper cleaning of engine parts before assembly (Courtesy of McQuay-Norris Manufacturing Company).

FIGURE 22-17
Fatigue failure caused by fillet radii that were too large (Courtesy of Federal-Mogul Corporation).

ing cut by metal debris left from a machining operation.

Scoring may originate from a piston failure. Aluminum particles are compressed by shaft pressures and scar or indent the bearing surface.

Scored bearings are often caused by sand entering the engine through the air cleaner, or by failure to thoroughly clean parts before reassembly.

In recent years, there has been a widespread use of nodular iron crankshafts (especially in passenger cars and light trucks). The polishing of these shafts after grinding is critical due to their microstructure. Ex-

amine the surface with a magnifying glass. There must be no burrs or raised ferrite peaks on the surface exposed graphite. Otherwise, the engine bearings will score and wear.

3. **Wiping or seizure** The babbitt lined connecting rod bearing in Figure 22-19 failed because of insufficient clearance. The "wiped" look is the result of picking up some of the lining material and redepositing it on another area of the bearing.

Wiping is the result of metal-to-metal scrubbing action by the crankshaft. It can be caused by oil dilution, lack of oil, stoppage of the oil supply, excessive heat, or insufficient oil clearance space.

A prevalent cause of wiping or seizure during early engine running is a "dry start." This is a condition where the engine lubrication system has not been primed with oil before starting the engine for the first time after an overhaul.

Sometimes, severe heat in copper alloy lined bearings will "melt out" the lead. The copper rich areas are then torn away by seizure. Pronounced cases can create such severe force that the bearing locking lips are

FIGURE 22-19
Badly wiped bearing caused by lack of oil and clearance (Courtesy of Federal-Mogul Corporation).

FIGURE 22-20
Corrosion failure looks like fatigue failure (Courtesy of Federal-Mogul Corporation).

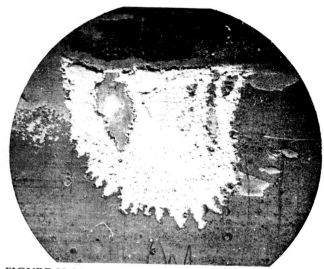

FIGURE 22-21
Bearing lining damage from erosion (Courtesy of Federal-Mogul Corporation).

sheared, thus allowing the bearing to "spin" in its bore.

4. **Corrosion** This condition takes place when corrosive media in the oil attacks certain elements in the bearing lining material. Sulphuric acid, "sour" gas, high sulfur fuels, and antifreeze are the most responsible media for corrosive attack. Bearing manufacturers report that corrosion failure can be caused by using upper cylinder lubricants containing fatty oils or acids.

What generally takes place is that the corrosive media attacks the lead and changes it to a lead soap. This is then washed out of the lining, leaving the other elements wide open to attack. Copper-alloy bearings are the most vulnerable.

A corroded bearing surface, in the final stages, looks like a fatigue breakdown. The bearing in Figure 22-20 shows a loss of overplate, exposing the lining. The dark areas is where the lead has been lost and corrosion is severe.

One method of diagnosing corrosion attack is to probe the bearing surface with a pointed tool under magnification. If oil oozes around the probe, corrosion is the likely suspect.

5. **Cavitation erosion** In today's engines, more and more of this condition is being seen. This is where the oil film (under vibration) pulls out lining material, grain by grain. It is like washing away a dirt bank with a stream of water. The bearing illustrated in Figure 22-21 has been affected by erosion. Cavitation erosion can be sudden and profuse in certain engines if aluminum bearings are used. This is why copper-lead bearings are used in GM passenger car diesels. For some peculiar reason, cavitation erosion is generally confined to the unloaded rod bearing half.

BEARING REPLACEMENT DATA

In most bearing replacements, the original manufacturer's oil clearance is the desired result. *Oil clearance is the difference between shaft OD and the*

CROWN WALL THICKNESS

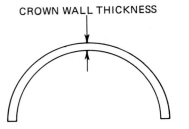

FIGURE 22-22

Bearing crown wall thickness.

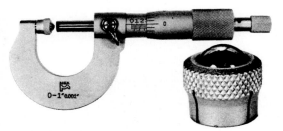

FIGURE 22-23

Micrometer with steel ball attachment (Courtesy of Yamazen U.S.A., Inc.).

bearing ID (after installation). The bearing, the shaft, and the housing bore must always be considered when determining oil clearance.

Bearing Sizing

Replacement bearings for most engines are available in standard, .001" undersize, .002" undersize, .010" undersize, .020" undersize, .030" undersize, and precision resizable .060". The precision resizable .060" bearing can be machined to size when there is no listed undersize bearing available to fit the crankshaft involved. These bearings can be bored to standard size, if necessary, to correct saddle bore misalignment.

Crown Wall Thickness

Incorrect bearing fit is sometimes the result of faulty packaging or mislabeled boxes. For this reason, it is a good idea to measure the bearing wall thickness at the crown prior to assembly (Figure 22-22). A ball anvil micrometer should be used to make this measurement due to the arc of the bearing (Figure 22-23).

To check a bearing, measure the thickness of the wall at the crown and compare to specifications (Figure 22-24). The specification listing is for a standard size bearing. *A .010" undersize bearing would measure .005" larger than the listing.* This is because only half of bearing is being measured, and 2 × .005" = .010" undersize.

SHOP TIP

If a ball anvil micrometer is not available, this measurement can be made by placing a roller bearing on the insert (Figure 22-25). Measure the roller and the insert, then subtract the diameter of the roller to get the thickness of the bearing insert at the crown.

Crankshaft Sizing

The original factory standard size of crankpins and main journals must be known when regrinding the shaft for use with undersize replacement bearings.

FORD MOTOR CO.   ENGINE BEARINGS

CYLINDER BORE&STROKE CUBIC DISP. LINE NO.	JOURNAL & POSITION	NO REQ'D	STANDARD PART NO. & MATERIAL	UNDERSIZES	DIAMETER OF STANDARD SHAFT	HOUSING BORE	OIL CLEARANCE	MAX.WALL THICKNESS AT CROWN	OVERALL LENGTH
6 3.680X3.910 250	ROD BRG	6	2380CP	1-2-10-20-30-40,CA60	2.1232/2.1240	2.2390/2.2398	.0005/.0026	.0575	.810
	MAIN SET	1	4916M	1-10-20-30					
	1-2-3-4-6-7	6	3151CPA	1-10-20-30	2.3982/2.3990	2.5902/2.5910	.0005/.0028	.0955	.970
	5	1	▲ 3153CPA	1-10-20-30	2.3982/2.3990	2.5902/2.5910	.0005/.0028	.0955	1.195
	CAM SET	1	1400M	20-30					
	RODS	6	R25AW	(Frg. No. C90E) NUT 9975N CRANK FRG. NO. (3.910)C9DE-A					

Where both **CATALOG INFORMATION** and **SHOP SPECIFICATIONS** head the table.

FIGURE 22-24

Crown wall thickness information can be found in most bearing catalogs (Courtesy of Federal-Mogul Corporation).

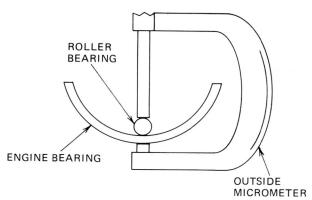

FIGURE 22-25
Using an outside micrometer and a roller bearing to determine crown wall thickness.

For example, the diameter of a 400 CID Pontiac crankpin is given as $2\frac{1}{4}''$ on most specification charts. In reality, this shaft as originally produced is 2.2487/2.2497". The ID of a standard bearing for this crankpin is 2.2505/2.2510".

In order to achieve the correct oil clearance, the regrinder would grind the shaft to 2.2387/2.2397" for .010" undersize bearings, 2.2287/2.2297" for .020" undersize bearings, and 2.2187/2.2197" for .030" undersize bearings. *To grind from the nominal diameter size of $2\frac{1}{4}''$ would remove the necessary oil clearance.*

Main and Rod Bearing Clearance

The oil clearances shown in Chart 22-1 can be generally recommended in replacement situations.

Crankshaft Thrust Bearing Clearance

Crankshafts must have a certain amount of endwise thrust. The thrust is kept within limits by several means. A common way is to use a flange face on the bearing. Another method uses washers (Figure 22-26).

Recommended crankshaft end clearance is generally as shown in Chart 22-2 (Page 470).

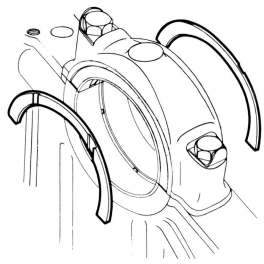

FIGURE 22-26
Some engines use separate thrust washers to control crankshaft end play.

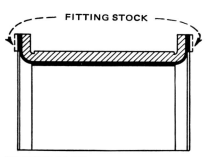

FIGURE 22-27
Undersize crankshaft thrust bearing with fitting stock on flange faces (Courtesy of Federal-Mogul Corporation).

Some flange thrust bearings carry extra stock on the flange faces so that they can be fitted to the crankshaft (Figure 22-27).

NOTE: A complaint sometimes heard is that the thrust bearings on an import engine cannot be installed. The reason is that in most import engine applications, the

CHART 22-1

Recommended replacement bearing oil clearances (Courtesy of Federal-Mogul Corporation)

Shaft Size	Babbitt (SB)	Copper Alloy (CA)	20% Tin Aluminum (RA)	Overplated Lining (AP/CP)
2–$2\frac{3}{4}''$	.001–.002"	.002–.003"	.001–.002"	.001–.002"
$2\frac{13}{16}$–$3\frac{1}{2}''$	.0015–.0025"	.0025–.0035"	.0015–.0025"	.0015–.0025"
$3\frac{9}{16}$–$4\frac{1}{2}''$	.002–.003"	.003–.004"	.002–.003"	.002–.003"

CHART 22-2

Recommended crankshaft end clearance (Courtesy of Federal-Mogul Corporation)

Shaft Size	Recommended End Clearance
2–2$\frac{3}{4}$"	.004–.006"
2$\frac{13}{16}$–3$\frac{1}{2}$"	.006–.008"
Over 3$\frac{1}{2}$"	.008–.010"

flanged bearing length is increased with the undersize of the bearing. This means those grinding the crankshaft must grind the thrust face of the crank to accommodate the increased bearing length and secure proper end thrust clearance.

Camshaft Bearing Clearance

Most camshaft bearings have multilayer metal construction. When made, they start out flat and then are rolled and secured with clinch locks. Solid-type camshaft bearings are fabricated from a tube, therefore making it impractical to apply layered metal. With both types of bearings, the correct ID is automatically obtained when pressed into place provided that the bores are true.

The OHC engine pictured in Figure 22-28 uses split half precision insert cam bearings. The arrow is pointing to the thrust face on the lower bearing. Undersize camshaft bearings are available for engines when the camshaft journals have been ground to an established undersize.

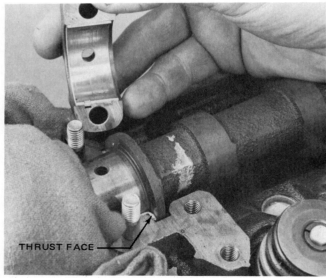

THRUST FACE

FIGURE 22-28

Split half precision insert cam bearings are used on this OHC Toyota engine. Note the thrust face on the lower bearing.

TABLE 22-1

BEARING MATERIAL CODING

Material Code	Meaning
AP	Steel back bearing, aluminum alloy, babbitt overplate
AT	Solid aluminum alloy bearing
CA	Steel back bearing, copper alloy lined
CP	Steel back bearing, copper alloy lined, lead-tin overplate
RA	Steel back bearing, aluminum alloy
SA or SB	Steel back bearing, babbitt lined

Determining Bearing Material

Some of the bearing manufacturers use a material code that is adjacent to the part number on the box (Figure 22-29). By knowing this code, you can readily determine the type of bearing material. See Table 22-1 for code information.

BEARING INSTALLATION TIPS

1. Bearing and bore must be *clean and dry*. Wipe the bearing back and bore surfaces with a lint-free cloth (Figure 22-30). Never put anything on the bearing back. Keep fingers off of bearing surfaces.

2. Check oil hole alignment. Oil holes in the block must line up with the hole or slot in the bearing (Figure 22-31).

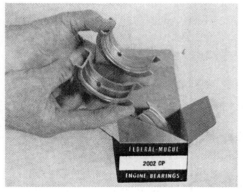

FEDERAL-MOGUL
2002 CP
ENGINE BEARINGS

FIGURE 22-29

The *CP* suffix on this box designates the bearing material as overplated copper alloy (Courtesy of Federal-Mogul Corporation).

3. Wipe cam bearing bores before installing cam bearings. Make certain the right driver is used, and each bearing is driven in straight and into the correct bore. Check oil hole alignment (Figure 22-32).

4. Rod and main bearing oil clearance must be checked. The Plastigage method is an accurate, recommended technique.

FIGURE 22-30

Make sure bearing backs and bores are clean and dry before installing bearing halves (Courtesy of D.A.B. Industries, Inc.).

FIGURE 22-31

Check oil hole alignment to insure proper oil flow (Courtesy of D.A.B. Industries, Inc.).

FIGURE 22-32

Checking camshaft bearing oil hole alignment (Courtesy of Ford Parts and Service Division).

5. Always check thrust bearing clearance (Figure 22-33). If the clearance is not enough, dress down the bearing flanges or install a new thrust bearing. If there is too much clearance, install a new thrust bearing, or repair or replace the crankshaft.

6. Lubricate all bearing lining surfaces and seal lip surfaces with a good supply of oil or assembly lubricant (Figure 22-34).

7. When installing bearing caps, tighten the nuts or cap screws finger tight first. Use a lead hammer and tap each cap so it moves to its natural position. At this time, force the crankshaft forward with a large screwdriver until the rear faces of the thrust flanges are properly aligned (Figure 22-35).

8. Avoid interference when using a socket to tighten rod bearing caps. In Figure 22-36 the socket is crowding against the cap. This can cause it to shift, producing excessive pressure areas on opposite sides of the upper and lower bearing halves.

FIGURE 22-33

Checking end clearance between crankshaft thrust and bearing flange by using a feeler gauge of proper thickness (Courtesy of D.A.B. Industries, Inc.).

FIGURE 22-34

Using an assembly lubricant on bearing surfaces (Courtesy of D.A.B. Industries, Inc.).

FIGURE 22-35
Aligning thrust bearing flange faces (Courtesy of Imperial Clevite, Inc., Engine Parts Division).

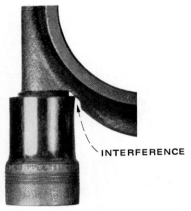

FIGURE 22-36
When torquing rod nuts, slightly lift the socket to clear the cap and gain clearance. Socket interference can give a false torque reading and/or push the cap sidewise (Courtesy of Federal-Mogul Corporation).

Caution on All Small Block Chevrolet V-8 Engines

This very popular engine has two different styles of rear main bearing caps (Figure 22-37). The old style main cap requires OEM oil pump bolt #3744998, $\frac{7}{16}$ − 14 × $2\frac{3}{16}$". The new style cap re-

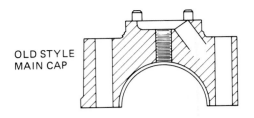

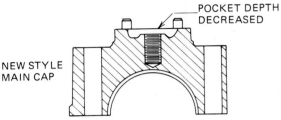

FIGURE 22-37
Old and new style main cap used on small block Chevrolet V-8 engines.

quires OEM oil pump bolt #3892678, $\frac{7}{16}$ − 14 × 2".

If the 2" long bolt is used with the old style cap, there is not sufficient thread engagement and the cap can be stripped. If the $2\frac{3}{16}$" long bolt is used with the new style cap, the bolt will bottom in the hole and push against the main bearing and cause destruction.

Selective Fitting of Bearings

Certain vehicles equipped with air-conditioning may develop a slight knock at the front main bearing, due to the additional load placed on the crankshaft by the drive belts. The noise intensity can be decreased by loosening the drive belts.

This noise condition does not necessarily mean a bearing failure. Often, the noise can be eliminated by select fitting the no. 1 main bearing set. By using a standard bearing half along with a .001" or .002" undersize bearing half, oil clearance can still be kept within specifications and the knock removed. ***NOTE:*** *Install the larger undersize shell in the upper position.*

CHART 23-1

These Ford oil pump springs are available to increase oil pressure over stock specifications.

Relief Valve Spring Part Number	Rating	Engine
C8FE-6670-A	80 psi	302 Boss (fits all 289-302)
C5AZ-6670-A	80 psi	351W
D2ZX-6670-A	100 psi	351C (fits 400-429-460)
C9ZZ-6670-A	80 psi	428 (used in H.D. pump C9ZZ-6600-A)
C1AE-6670-A	100 psi	427 (used in H.D. pump C3AZ-6600-B)

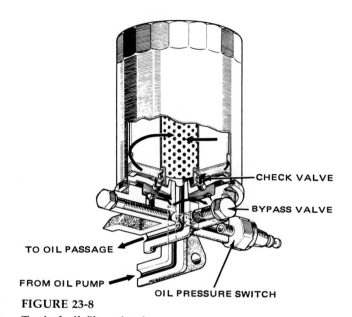

FIGURE 23-8
Typical oil filter circuit.

All filters or filter adapters contain a bypass valve in case the filter plugs (Figure 23-8). The bypass valve is set to open at about a 10 psi differential. Note that the oil filter in the picture is supplied with a check valve. It prevents oil in the filter from draining back into the pan when the engine is stopped. This keeps the oil pump primed for startup.

Sometimes a spin-on replacement filter has been on the shelf long enough for the check valve to stick to the filter end. When this happens, oil will not pass through the filter. To check, feel the filter with your hand. It should be hot when the engine is at normal operating temperature.

Oil Coolers

Some engines use an oil cooler. Most look like a small radiator mounted near the front of the engine (Figure 23-9). Heat is removed from the oil as air flows through the cooler core.

Normal maximum engine oil temperature is considered to be 250° F. Hot oil combining with oxygen will break down (oxidize) and form carbon and varnish. The higher the temperature, the faster these deposits build. An oil cooler helps keep the oil at its normal operating temperature.

WARNING: If there is any internal metal failure on an engine equipped with an oil cooler, it is imperative to replace the cooler core when the engine is repaired. Due to the construction of cooler cores,

FIGURE 23-9
Oil cooler and attaching parts (Courtesy of Ford Parts and Service Division).

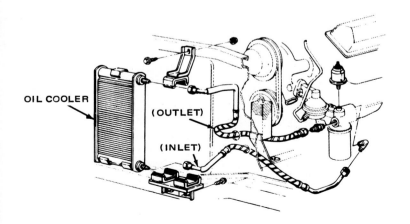

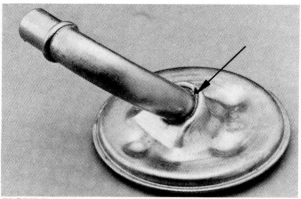

FIGURE 23-10
Cracked oil pump pickup tube (Courtesy of TRW Replacement Division).

FIGURE 23-11
Casting breakage caused by failure to hold pump squarely against the block when tightening mounting bolts (Courtesy of TRW Replacement Division).

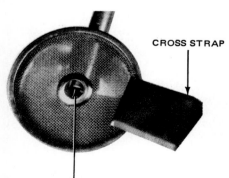

CROSS STRAP

BYPASS VALVE
FIGURE 23-12
The cross strap on this pickup screen has been bent out of position in order to show the bypass valve (Courtesy of TRW Replacement Division).

it is almost impossible to clean out trapped metal particles. Failure to replace the cooler core may allow this metal to be released later into the oil, thus ruining the engine.

OIL PUMP PROBLEMS

Oil pumps usually last for the life of the engine. There are times, however, when problems develop (often caused by another parts failure). Some more common problems include aeration, broken casting, broken intermediate shaft, stuck pressure relief valve, worn gears or rotors, and excessive internal clearance. Let's examine each of these problems.

1. **Aeration** This means the oil pump is moving air bubbles and oil through the engine. Aerated oil is a poor lubricant that can cause problems. *One giveaway to aerated oil is noise in all the hydraulic valve lifters.*

   An air leak on the suction side of the pump can cause a "flickering" oil pressure indicator light. In some cases, the air can cause the pressure relief valve to hammer back and forth and fail.

   Air leakage can come from a cracked seam in the pickup tube, loose pickup tube, cracked pump housing, or a worn pump (Figure 23-10). Too high or too low an oil level in the crankcase can lead to bubbles in the oil. A dented oil pan is an often overlooked reason for high oil level.

2. **Broken casting** Pump casting breakage is usually caused by mounting the pump in a cocked position.

   Figure 23-11 shows an oil pump with a fractured mounting neck and split drive slot. This was caused by the installer who unknowingly allowed the mounting neck to drop slightly while torquing the outward mounting bolt.

   Proper knowledge of oil pump installation is a must. *Always read the caution sheet included with most aftermarket pumps.*

3. **Broken intermediate shaft** Foreign particles (portions of valve stem seals, bearing fragments, valve lifter lock rings, and plastic cam sprocket segments) can pass through the pickup screen into the pump chamber. Figure 23-12 shows a typical domestic oil pump screen and how this can happen. Un-

derneath the metal cross strap there is a bypass valve. The valve is seated on the cross strap most of the time. However, when there is a demand for a large amount of oil (such as with sudden heavy acceleration), or the oil is cold and thick, the valve will unseat. This allows oil to bypass the screen and go directly into the pump.

As oil rushes through the bypass valve, debris lying in the bottom of the pan can be picked up and passed into the pump. If a large particle gets trapped between the gears or rotors, the pump will physically lock up. The intermediate shaft is then twisted and broken off (Figure 23-13).

4. **Stuck pressure relief valve** The oil pump pressure relief valve is fit to a very close tolerance. Foreign material entering a pump has been known to jam the pressure relief valve. If the relief valve sticks in the open position, there will be little or no oil pressure. If the relief valve sticks in the closed position, a blown oil filter is the usual result.

COMMENT: *The relief valve on Jeep V-6 engines frequently becomes stuck in the closed position when driving off-road. This is caused by the front axle housing rising and denting the aluminum timing case cover which contains the oil pump assembly. The result is a burst oil filter.*

5. **Worn gears or rotors** Abrasive particles (dirt, casting slag, machining residue, wear grit) can cause scratched rotors or gears. Figure 23-14 shows severe abrasive wear in an oil pump.

6. **Excessive internal clearances** Increased pump clearances will cause a drop in oil pressure.

FIGURE 23-14
Badly worn oil pump gears (Courtesy of TRW Replacement Division).

Pump covers must be flat within .0015" (Figure 23-15). Rotors or gears must fit together with the proper clearance (Figures 23-16, 23-17, and 23-18).

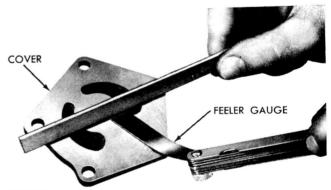

COVER

FEELER GAUGE

FIGURE 23-15
Measuring oil pump cover flatness (Courtesy of Chrysler Corporation).

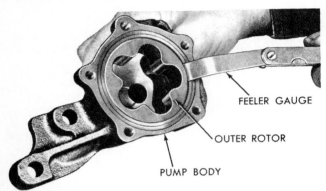

FIGURE 23-16

Measuring outer rotor clearance (Courtesy of Chrysler Corporation).

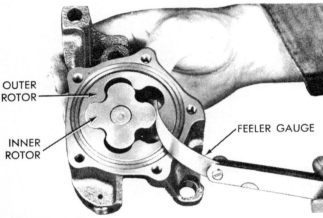

FIGURE 23-17

Measuring clearance between rotors (Courtesy of Chrysler Corporation).

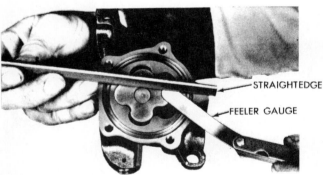

FIGURE 23-18

Measuring clearance over rotors (Courtesy of Chrysler Corporation).

OIL PUMP INSTALLATION TIPS

1. Special attention is needed on engines where the oil pump is an integral part of the timing chain cover. The pump for these engines is not self-priming. *The pump gear cavity must be filled with petroleum jelly if the pump cover has been removed for any reason.*

2. If a new press-in type screen assembly is to be installed, *avoid hammering on the oil pump.* This can distort the pump body and cause immediate internal seizure.

   One installation procedure is to install the pump in the engine block; grip the inlet tube with channel-lock pliers or an open-end wrench placed just ahead of the tube bend; drive the screen assembly into place by tapping on the pliers or wrench with a plastic face hammer.

   Special service tools are available for installing the pickup screen assembly (Figure 23-19).

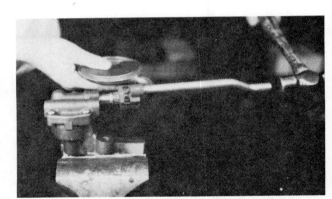

FIGURE 23-19

This illustration shows two different Chevrolet oil pumps and special Kent-Moore tools used for the pickup screen installation (Courtesy of TRW Replacement Division).

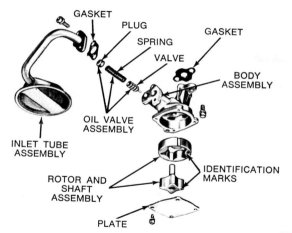

FIGURE 23-20
Oil pump and screen mounting details (Courtesy of Ford Parts and Service Division).

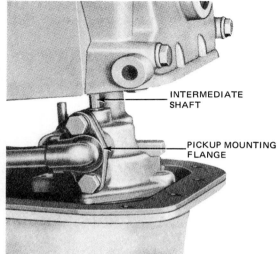

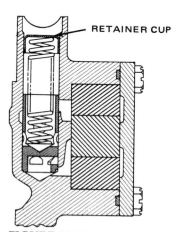

3. *Use of old pickup screens is not recommended.* Frayed wire may be hidden under the cross strap or baffle. There are cases where a piece of this wire has released and locked up a new oil pump. Also, the screen could be distorted, keeping the bypass valve under the strap open.

4. Carefully check the mounting surfaces of bolt-on type inlet tubes. Check for old gasket material. A mismounted flange can cause aeration (Figure 23-20). Make sure the center hole of the new gasket is punched out.

5. Two different pressure relief valve springs are supplied with some aftermarket pumps. The spring may have to be changed (depending on engine model). If a retainer cup is used, removal can be difficult (Figure 23-21).

FIGURE 23-21
Retainer cup installation (Courtesy of Chrysler Corporation).

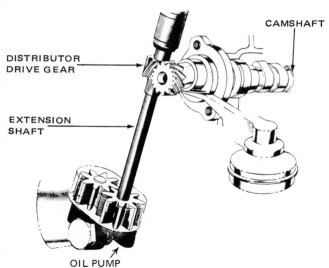

FIGURE 23-22

A gear on the end of this camshaft drives the distributor. An extension from the distributor drives the oil pump (Courtesy of Buick Motor Division of General Motors Corporation).

Some mechanics drill a 1/8″ hole in the cup, insert a self-threading sheet metal screw, and then pull on the screw.

6. The oil pump is driven by an extension (intermediate shaft) from the distributor on many engines (Figures 23-22 and 23-23). Examine the intermediate shaft for any visible damage or rounding of hexagonal ends.

7. Be sure to check intermediate shaft endplay. Failure to do so can cause the oil pump gears or rotors to jam against the inside of the pump cover, ruining the oil pump.

8. Determine if there is any interference as the intermediate shaft is turned. *NOTE: Lack of flatness of the oil pump mounting surface on*

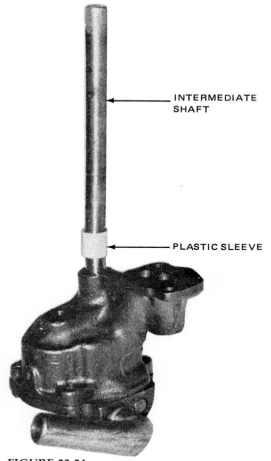

FIGURE 23-24

A plastic sleeve is used to connect this Chevrolet oil pump to the intermediate shaft.

the rear main bearing cap may cause this problem. If the mounting surface is not flat, it must be trued.

9. Some engines use a retaining clip or sleeve on the intermediate shaft (Figure 23-24). In-

FIGURE 23-23

Oil pump driveshaft used on certain model Chrysler Corporation engines. Note broken end.

stall with the oil pump from the bottom side of the engine.

SHOP TIP

On certain Ford engines, the intermediate shaft may fall into the oil pan when the distributor is removed. Usually this occurs if the shaft retaining clip was left off or installed incorrectly. A new intermediate shaft can be installed without removing the oil pan. Use a length of copper tubing as an extension and guide the new shaft into position. Leave the old shaft in the pan.

---

10. Final pump tightening should be done with a torque wrench.

TRUE CASE HISTORY

Vehicle	Dodge Dart
Engine	6 cylinder, 225 CID
Mileage	Zero mileage; new short block just installed
Problem	No oil pressure.
Cause	The oil pickup tube and screen were incorrectly positioned and no oil was being pumped. A measurement is required when the pickup assembly is installed on this engine (Figure 23-25).

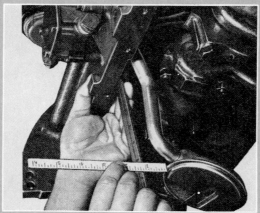

FIGURE 23-25
Positioning oil pickup tube and screen (Courtesy of Chrysler Corporation).

TRUE CASE HISTORY

Vehicle	Buick Wildcat
Engine	8 cylinder, 401 CID
Mileage	Unknown
Problem	1. Severe hydraulic lifter noise at driving speeds above 30 mph. 2. The noise would disappear after about a minute of rough idling. 3. The engine would run fine as long as the speed was kept below 30 mph. 4. The noise developed immediately after the vehicle owner replaced the rear main seal (the oil pump was disassembled and cleaned in the process).
Cause	The oil pump had developed a varnish film between the driveshaft and body. Cleaning the pump removed this varnish and increased the clearance enough to allow the entry of air.
Cure	Installing a new oil pump solved the problem.

TRUE CASE HISTORY

Vehicle	Chevrolet
Engine	8 cylinder, 283 CID
Mileage	Zero mileage; engine was just overhauled
Problem	1. As soon as the engine started, oil pressure read maximum (80 psi) and bent the gauge hand.
	2. After a few seconds of engine operation, oil pressure dropped to zero. At the same time, a "grinding" sound was heard in the area of the timing chain cover.
Cause	The mechanic had installed a 1/2" high stack of steel washers on the end of the pressure relief valve spring in order to "beef" up engine oil pressure. The oil pressure that developed was excessively high and pushed out both front oil gallery plugs (Figure 23-26). The "grinding" noise was caused by the pushed out plugs catching in the timing chain assembly.

FIGURE 23-26
These front oil gallery plugs came out and caught in the timing chain assembly.

Comment	Shimming the relief valve spring for more pressure is not recommended normally. The only time it may be necessary is when a remote filter or cooler is used. In this case, *a 1/8" shim is maximum.*

TRUE CASE HISTORY

Vehicle	Dodge Omni
Engine	4 cylinder, 105 CID
Mileage	24,500
Problem	Seized oil pump
Cause	A piece of the fuel pump rocker arm damper spring broke off and worked through the oil pump strainer and up into the gear cavity.

CHAPTER 23 SELF-TEST

1. What is the major advantage of the pressure feed lubrication system?
2. Name and describe the operation of the two types of oil pumps discussed.
3. Explain the operation of the oil pressure relief valve.
4. What factors influence oil pressure in an engine?
5. What is the purpose of the oil filter bypass valve?
6. What is the purpose of the oil filter check valve?
7. Normal maximum engine oil temperature is considered to be ___ .
8. Why must an oil cooler be replaced when there has been a bearing failure in the engine?
9. What is one giveaway to aerated oil?
10. List six ways oil can become aerated.
11. What is the usual cause of oil pump casting breakage?
12. Many oil pump screens have a bypass valve. Under what conditions will oil bypass the screen and go directly into the pump?
13. Name some foreign particles that often get picked up and passed into the pump.
14. What will happen if a large foreign particle gets trapped between the gears or rotors in an oil pump?
15. What will be the result if the oil pump pressure relief valve sticks in the open position? In the closed position?
16. Oil pump covers must be flat within _____ .
17. What must be done to prime an oil pump that is not self-priming?
18. What methods can be used to install a press-in type oil pump screen assembly?
19. Why is it not recommended to reuse a pickup screen?
20. The pressure relief valve spring in an aftermarket pump has to be changed. What method can be used to remove the spring retainer cup?
21. What can happen if there is insufficient intermediate shaft end play?
22. What can cause the intermediate shaft to fall into the oil pan when the distributor is removed on certain Ford engines?

CHAPTER 24

SEALING THE ENGINE

Gaskets or seals are used between engine parts to contain a variety of fluids and gases under conditions of heat and strain. A gasket may appear to be simple, but in reality there are many design considerations and material options.

GASKET PROPERTIES

All of the following properties must be considered when a gasket is designed.

1. **Conformability** Gaskets must be able to conform to slight surface imperfections. This includes warpage, corrosion, erosion, and roughness left from machining.

2. **Creep relaxation** As a gasket is continually stressed, its interporosity undergoes a collapse change over a period of time. This leads to a lowering of the bolting force and flange pressure. If flange pressure falls below its critical value, the gasket will fail.

3. **Impermeability** A gasket must not permit passage of the medium it is designed to seal.

4. **Elasticity** This property allows the gasket to maintain sealing pressure regardless of normal vibration or temperature change.

5. **Wear resistance** For long-term performance, a gasket must be able to resist the conditions of age, fluid attack, pressure, and heat.

GASKET MATERIALS

A multitude of gasket materials, alone or in combination, are commercially available. The more common ones are discussed here.

1. **Cork** Pure cork is the inner bark of the Mediterranean live oak tree. For gaskets, granules of this cork (Figure 24-1) are held together with binders (phenolic resin, latex, and animal glue). It is this binder material that gives cork gaskets certain undesirable characteristics.

   Cork gaskets have several disadvantages. Shelf life of cork gaskets is relatively short (they lose moisture, shrink, and become brit-

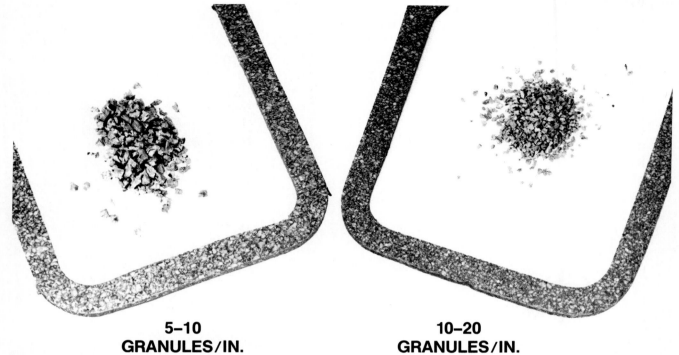

5–10 GRANULES/IN. **10–20 GRANULES/IN.**

FIGURE 24-1
Cork composition gaskets are granules of cork sealed together by a binder material. Granule size will vary with the gasket manufacturer (Courtesy of Dana Corporation—Service Parts Group).

tle). They also have a tendency to "wick." This is caused by capillary action pulling oil through the voids between the cork granules. A kerosene lantern operates on this same principle.

COMMENT: When many mechanics see dirt and oil film on the outside surface of an oil pan or valve cover gasket, they think the gasket is leaking and tighten the hold-down bolts. This will not stop the "wicking" process. It only causes distorted oil pan and valve cover flanges.

Automotive cork gaskets do offer, however, some advantages. They are easily installed on wet or dry surfaces (due to their high coefficient of friction), have excellent compressibility, and have good torque retention.

Cork is used in applications where oil is to be contained, where bolt loadings are low, and where there are no pressure and temperature extremes.

Some manufacturers coat cork gaskets with aluminum paint to improve heat resistance.

2. **Rubber** This gasket material is more expensive than cork, but has an unlimited shelf life and no "wicking" problem. Heat resistance is excellent.

Rubber gaskets have the shortcoming of displacing under pressure rather than compressing. Studies have shown this displacement can result in a 50% loss of flange pressure after 500 to 1000 hours of engine operation.

The installation of rubber gaskets is often a problem. For example, you have probably experienced a rubber valve cover gasket squirt out of position as torque is applied. This occurs because oil wet rubber has almost zero coefficient of friction. To make sure that mating surfaces to be sealed are absolutely *clean and dry* is the key to a successful installation.

3. **Cork/rubber** Basically, this type of gasket material is cork with rubber used as the binder. It provides all the good features of both cork and rubber gaskets with none of the bad qualities of either.

4. **Paper** These gaskets are designed to be used in a low-pressure, low-temperature environment. Paper gaskets usually are treated to resist oil, coolant, and gasoline. The mounting gasket for a mechanical fuel pump is an example.

5. **Fiber** Cellulose, asbestos, or a combination of both materials often is used in high-pressure, high-temperature situations. The fiber may be bonded to a steel core.

6. **RTV (room temperature vulcanizing) silicone rubber** This material is commonly used for formed-in-place gaskets. RTV usually comes in a tube and is applied as a paste (Figure 24-2). The curing agent for this material is humidity in the air. After curing, the material has tough, rubber-like characteristics. It has temperature resistance of 450° F.

   Formed-in-place silicone gasket material works especially well around water passages or where uneven clamping pressure occurs.

7. **Anaerobic materials** These formed-in-place compounds cure opposite to that of RTV. An anaerobic adhesive/sealant is shown in Figure 24-3.

   Anaerobic materials cure when confined (anaerobic means life in the absence of air) between metal parts.

   Initially, Loctite Corporation developed anaerobic materials as a way to secure threaded fasteners. Later, the materials were made suitable for formed-in-place gaskets. After curing, the gasket is a tough polymerized plastic that has 400° F temperature resistance.

CYLINDER HEAD GASKETS

Head gaskets have a difficult sealing job. Coolant, lubricating oil, and combustion pressures must be simultaneously sealed. In addition, a head gasket must be able to perform in wide temperature ranges. The seal area may be well below zero when the engine is started, to over 400° F when the engine is running.

In-line engines generally have the most difficult head sealing problems. They have an inherently unbalanced bolt load distribution. In many cases, no head bolts are used to clamp the gasket on the outer periphery of the push rod holes (Figure 24-4). To prevent oil seepage down the side of the block, the manufacturer often coats this area of the gasket with synthetic rubber.

FIGURE 24-2

This formed-in-place gasket compound is specially formulated to stay flexible at temperatures up to 600°F. In addition it is noncorrosive and meets OEM specifications for use on oxygen-sensor equipped engines (Courtesy of Automotive Trades Division/3M).

FIGURE 24-3

Anaerobic adhesive/sealants are liquid resins that turn into a tough solid resin when confined between close fitting metal parts. They cure through the process of air elimination and the catalytic action of metallic surfaces (Courtesy of Automotive Trades Division/3M).

There are several different styles of head gaskets.

1. **Metal sandwich** This style uses a face of copper or steel over an asbestos core (Figure 24-5). These head gaskets are widely used on cars and trucks.

   Always handle sandwich gaskets with care. Avoid distorting or bending. Store them flat, never on edge. Do not attempt to straighten a bent sandwich gasket, as premature failure will result (Figure 24-6).

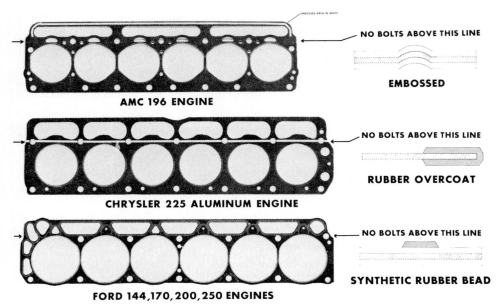

EMBOSSED AREA IN WHITE

NO BOLTS ABOVE THIS LINE

EMBOSSED

AMC 196 ENGINE

NO BOLTS ABOVE THIS LINE

RUBBER OVERCOAT

CHRYSLER 225 ALUMINUM ENGINE

NO BOLTS ABOVE THIS LINE

SYNTHETIC RUBBER BEAD

FORD 144,170,200,250 ENGINES

FIGURE 24-4

Gasket designs for some poorly clamped engines. As shown, various means are used to obtain clamp load at the push rod area (Courtesy of Fel-Pro, Inc.).

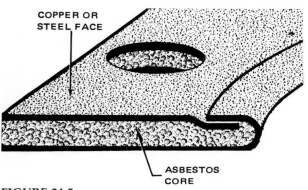

COPPER OR STEEL FACE

ASBESTOS CORE

FIGURE 24-5

Metal sandwich type head gasket (Courtesy of Fel-Pro, Inc.).

2. **Embossed steel** The embossed steel or shim style head gasket is commonly used as original equipment (Figure 24-7). The raised or embossed areas is what gives it the required resiliency.

The steel is usually around .030″ thick and coated with aluminum or tin. Under heat, the coating flows into the parting surface machine marks and provides additional sealing.

3. **Soft surface** Many aftermarket head gaskets are this style. These gaskets have treated asbestos facing attached to both sides of a perforated steel core (Figure 24-8). A stainless steel fire ring is wrapped around the

FIGURE 24-6

This is what happens when you attempt to straighten a bent sandwich gasket (Courtesy of Fel-Pro, Inc.).

THIS IS WHAT HAPPENS WHEN BENT SANDWHICH GASKETS ARE STRAIGHTENED

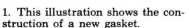

1. This illustration shows the construction of a new gasket.

2. Here, the gasket is bent, breaking the asbestos filler.

3. This gasket has been straightened and appears to be in good condition, but note the break in the asbestos filler.

4. In use, the gasket soon burns through at this weak spot. Compression and coolant loss result.

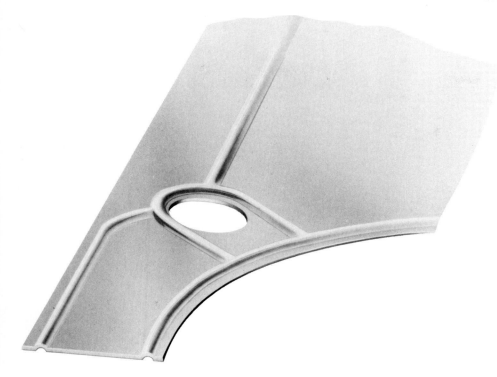

FIGURE 24-7
Embossed steel type head gasket (Courtesy of Fel-Pro, Inc.).

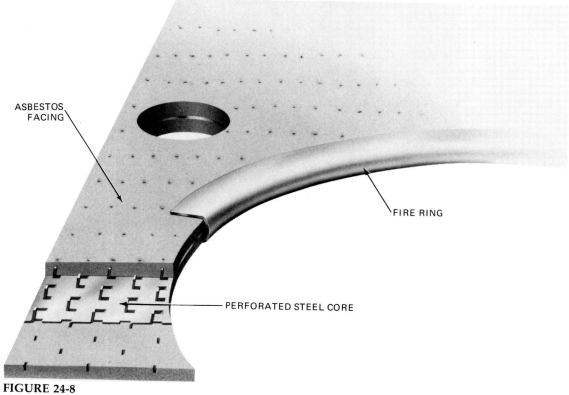

ASBESTOS FACING

FIRE RING

PERFORATED STEEL CORE

FIGURE 24-8
Soft surface type head gasket (Courtesy of Fel-Pro, Inc.).

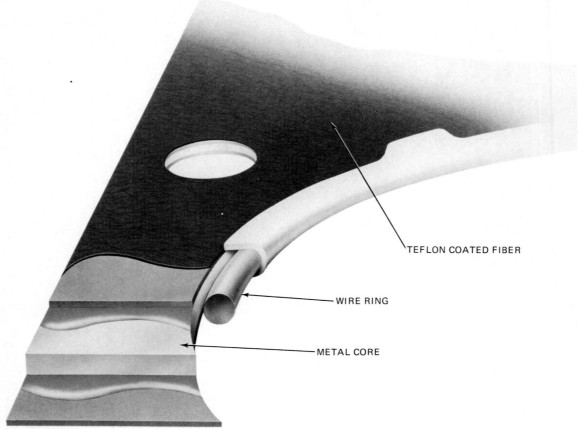

TEFLON COATED FIBER

WIRE RING

METAL CORE

FIGURE 24-9
No-retorque type head gasket (Courtesy of Fel-Pro, Inc.).

cylinder opening to seal the combustion chambers.

These gaskets are slightly thicker than the embossed steel style in order to help compensate for casting resurfacing. Also, these gaskets are less susceptible to liquid leaks than are metal face styles.

During the past few years there has been a trend toward the use of *no-retorque* head gaskets. This saves the cost, inconvenience, and time of retorquing the cylinder head after run in.

No-retorque head gaskets combine the compressibility of the metal sandwich style with the torque retaining ability of the embossed steel shim.

The very popular Permatorque head gasket manufactured by Fel-Pro, Inc. is shown in Figure 24-9. It has a metal core and fiber face that is coated with Teflon.

Cylinder head torque retention is directly related to the amount of head bolt stretch and the compressiblity of the head gasket materials (Figure 24-10). It then becomes the job of the gasket engineer to control creep relaxation, and retain the desired clamp up torque when designing a no-retorque head gasket.

Determining Head Gasket Stress Distribution

Potential head gasket problem areas can be located by making an impressions test. Unbalanced bolt load distribution is very often the reason for a leaking joint. A visual indication of the bolt load can be obtained by placing no carbon required (NCR) transfer paper in the joint.

The procedure for making this test is as follows.

1. Completely clean the block and head mating surfaces.

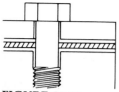

BOLT STRETCH @ .015'' @ CLAMP UP
GASKET SET .005''

BOLT STRETCH = .010'' @ END

THEREFORE $\frac{.010''}{.015''}$ OR 2/3 OF
CLAMP UP TORQUE RETAINED

FIGURE 24-10
(Courtesy of Fel-Pro, Inc.).

2. Trace all bolt and dowel pin holes on the NCR paper (Figure 24-11a).

3. Punch out all traced holes with a gasket punch (Figure 24-11b).

4. Place the paper on the block. Keep the two sheets oriented or no impression will result (Figure 24-11c).

5. Torque the head to specifications to make an impression on the paper.

6. Remove the head.

7. Carefully lift off the gasket and the top sheet of paper.

8. Observe the bolt load distribution impression (Figure 24-11d). On a comparative basis, dark areas show more compressive load than light areas.

Determining Head Gasket Compressed Thickness

By knowing the compressed thickness of an installed head gasket, piston-to-head clearance can be confirmed. Also, true compression ratio can be calculated.

Lead pellet shot (available from gun supply stores) can be used to accurately indicate the head gasket thickness after compression. The steps for this test are listed below.

1. Purchase a new head gasket. Trace the shape on blank paper.

(a)

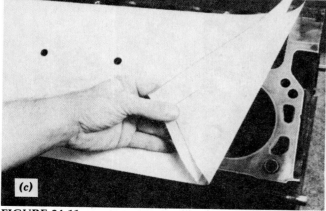

(c)

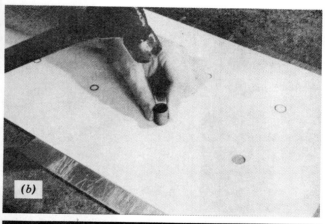

(b)

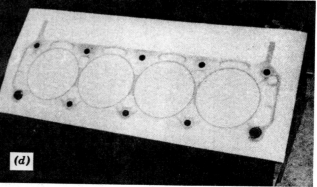

(d)

FIGURE 24-11
(Courtesy of Fel-Pro, Inc.).

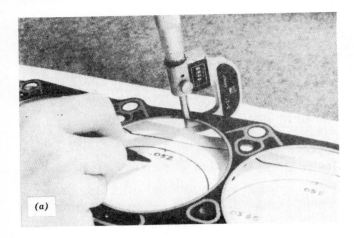

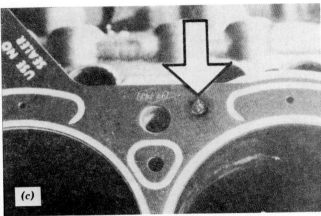

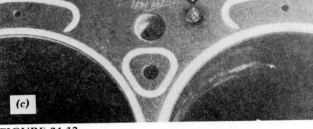

FIGURE 24-12
(Courtesy of Fel-Pro, Inc.).

2. Measure and record the gasket thickness of each fire ring (Figure 24-12*a*).

3. Obtain lead pellets. Select a size approximately twice the thickness of the head gasket.

4. Drill a hole through the gasket body near each fire ring (Figure 24-12*b*). The hole should be $1\frac{1}{2}$ to two times the pellet diameter.

5. Mount the gasket(s) on the engine and put a small amount of grease in each drilled hole (Figure 24-12*c*).

6. Place a pellet in each greased hole.

7. Mount the head and torque to specifications. The pellets are now compressed.

8. Remove the head, being careful not to lose any pellets.

9. Measure the compressed thickness of each pellet (Figure 24-12*d*).

Never Reuse Head Gaskets

A new head gasket will conform to microscopic surface irregularities when compressed (Figure 24-13*a*). If the gasket is removed and replaced, it is impossible to exactly refit it to these irregularities (Figure 24-13*b*). Early gasket failure may then result.

Head and Block Gasket Surface Finish

Sometimes, a cylinder head or block is resurfaced too smoothly or too roughly. A finish that is too smooth offers low resistance to combustion blowout. The fluid and gas seal is poor when the finish is too rough.

Casting finish limits are 60 to 125 microinches. The preferred range is 90 to 110 microinches.

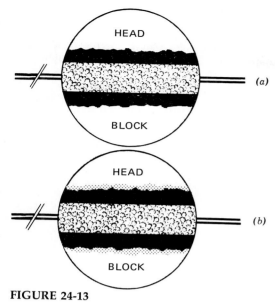

FIGURE 24-13

(*a*) New head gasket installation. (*b*) Used head gasket installation (Courtesy of Fel-Pro, Inc.).

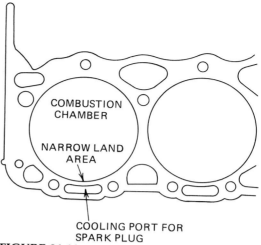

FIGURE 24-14

Before installing a reconditioned lightweight General Motors cylinder head, it is very important to check the head gasket-to-casting fit around the cooling ports (Courtesy of Fel-Pro, Inc.).

Head Gasket Installation Tips

1. Most head gaskets must be installed in a certain direction. This is critical for proper coolant flow. Look for a *Top, Front,* or *This Side Up* marking on the gasket face. If the face is not marked, install the gasket with the stamped identification numbers toward the head.

2. Check for proper gasket fit and hole alignment by placing the gasket on *both* the cylinder head and block. In some cases, a water jacket hole in the gasket may be considerably smaller than the corresponding block and head hole. This is done to meter coolant flow to a specific area. In other cases, a water jacket hole may not be punched in the gasket. This is done intentionally to aid coolant distribution.

3. General Motors lightweight cylinder heads have created head gasket sealing problems in the spark plug cooling port area (Figure 24-14). Head surfacing or an existing core shift will cause the narrow land area to move inward toward the combustion chamber and reduce gasket contact. Coolant may then leak into the cylinder. Before installing a lightweight head (Figure 24-15), carefully check this area for proper head gasket fit.

4. Use an air hose to blow out all head bolt holes in the block. If dirt or oil is in holes, a false torque reading will result when the head bolts are tightened.

5. Clean head bolt threads with a wire brush. *A dirty bolt can easily cause a torque wrench reading to be off 20 ft./lb.*

6. Lightly chamfer head bolt holes if the threads run up to the block deck. This will prevent "thread pull" when the head bolts are tightened.

7. Most gasket manufacturers say that sealers or adhesives should not be used on head gaskets. The one exception is the embossed steel type. On these, use a spray application sealer on both sides of the gasket (Figure 24-16*a* and *b*).

8. Head bolts on some engines vary in length. It is important to return all bolts to their correct location.

MANIFOLD GASKETS

Basically, there are three kinds of manifold assemblies in use—the intake, the exhaust, and the combined. Different materials are used for each.

Intake manifold gaskets must have the ability

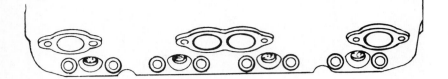

DESIGN CONTOUR (HEAVY HEAD)

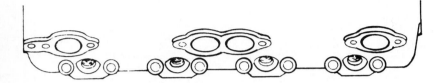

DESIGN CONTOUR (LIGHTWEIGHT HEAD)

FIGURE 24-15

Head identification for General Motors 307 and 350 CID (Type L) engines. The lightweight heads weigh approximately six pounds less than the heavy heads due to changes in internal coring of the water jackets. Lightweight heads are used on Chevrolet (passenger and truck), Buick, Oldsmobile, Pontiac, and GMC truck engines (Courtesy of Fel-Pro, Inc.).

(a)

(b)

FIGURE 24-16

(a) A special formulated spray adhesive for head gaskets (Courtesy of Automotive Trades Division/3M). (b) This spray adhesive contains fine copper particles to help dissipate heat and prevent gasket burn-out (Courtesy of K&W Products).

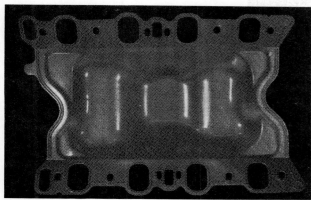

FIGURE 24-17

One piece metal embossed type intake manifold gasket (Courtesy of Fel-Pro, Inc.).

to seal vacuum, fuel vapor, and coolant pressure at the crossover. To provide the necessary rigidity and prevent wall collapse, several structure designs are used.

One popular type uses a solid steel core with chemically bonded soft fiber faces. Some are covered with a blue Teflon coating to give stick resistance and eliminate the need for gasket sealer.

Certain V-8 engines use a one-piece metal pan with both sides cut and embossed to mate with the intake manifold (Figure 24-17). This "turtle-back" gasket also serves as a deflector plate to prevent oil splash from caking on the bottom of the exhaust crossover.

Many OEM built engines are designed without exhaust manifold gaskets. They depend on casting-to-casting fit for the seal. However, repeated heating and cooling cycles often distort the exhaust manifold, and a gasket becomes necessary. The most common type has a perforated steel core with asbestos attached to each side.

The combined type manifold gasket (where intake and exhaust ports are intermixed) is used on many in-line engines. For best sealing, the gasket manufacturer often uses different materials at alternate ports.

Manifold Gasket Installation Tips

1. Some engines are very prone to exhaust manifold leaks. This procedure, if followed, often eliminates the problem.

   Before installing new gaskets, check for warping of the manifold. If dowel pins are used, the dowel pin holes should be in-

creased in diameter to enable the manifold to expand and contract without buckling. In some larger engines the exhaust manifold length increases by 1/8" as operating temperature is reached. The manifold has to be allowed to slide, or it will buckle.

   Gasket life can be increased if the exhaust manifold face contacting the gasket is given a liberal application of antiseize compound. This creates a heat resistant film with a very low coefficient of friction, allowing the exhaust manifold to expand and contract.

   When installing the exhaust manifold, tighten the center bolts to the maximum torque specification, and the bolts toward either end to the minimum torque specification.

2. When installing small block Ford intake manifold gaskets, sparingly apply sealer. Too much sealer can plug the drain slot and cause the normal coolant seepage to be drawn into the intake port or drained into the lifter galley (Figure 24-18). *These drain slots must be kept open.*

3. On small block Chevrolet engines, add a dab of RTV at the intake manifold gasket and endstrip seal intersections (Figure 24-19) to assure a positive seal.

4. Some exhaust manifold gasket designs use a two layer structure. One face is perforated steel, and the other face is asbestos. *When assembling, place the asbestos face toward the cylinder head.* This will allow the exhaust manifold to slip without gouging the asbestos face of the gasket.

5. For a positive water seal on "turtle-back" type intake manifold gaskets, use RTV. Apply a 1/8" bead all around the water passage holes on both sides of the gasket (Figure 24-20).

6. Use antiseize compound on exhaust gas recirculation (EGR) mounting bolt threads, and on exhaust pipe flange mounting bolt threads. This will prevent rusting, galling, and seizing.

7. On some Ford V-8 engines, the intake manifold serves as the lifter chamber cover and the valve cover fits over part of the intake manifold. Oil can be drawn into the intake ports from *both* the lifter chamber and rocker arm area in the event of an intake manifold gasket leak. Observe extra installation attention on these engines.

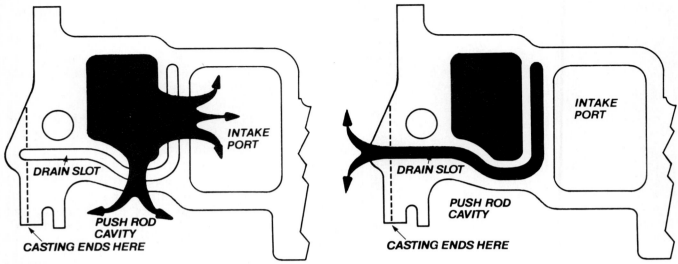

FIGURE 24-18

Left illustration shows what can happen when drain slot on small block Ford is clogged with sealer. Right illustration shows normal drain slot operation (Courtesy of Fel-Pro, Inc.).

REPLACING A VALVE COVER GASKET

Certain steps must be followed to guarantee a successful valve cover gasket installation.

1. Clean gasket surfaces of foreign materials. Old gasket particles and adhesive residue can form channels through which oil can travel.
2. Check gasket flange flatness by visual sighting or by using a straightedge. Straighten as required (Figure 24-21).
3. Clean oil return holes in the cylinder head.
4. Wipe excess oil from gasket mating surfaces.
5. Adhere gasket to valve cover. Apply a thin, even film of a good general purpose sealer

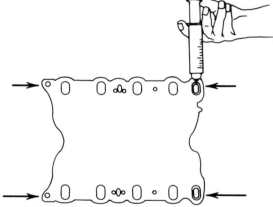

FIGURE 24-20

Prevent intake manifold water seepage by using RTV silicone sealant/adhesive (Courtesy of Fel-Pro, Inc.).

FIGURE 24-19

RTV silicone sealant/adhesive is an excellent supplementary sealer for critical areas such as V-8 intake manifold gasket ends (Courtesy of Fel-Pro, Inc.).

FIGURE 24-21

Straightening valve cover flanges (Courtesy of Fel-Pro, Inc.).

to entire cover flange. Align gasket to cover and let sealer dry until the gasket is stuck in position.

SHOP TIP

Rubber valve cover gaskets generally give the longest life, but can present installation problems on a narrow head land (Figure 24-22). Slippage can be avoided by attaching the gasket to the valve cover with 3M Super Weatherstrip Adhesive. Be sure the gasket is set before installing cover. Do not apply any sealer to the cylinder head side of the gasket.

---

6. Install valve cover. Avoid overtorquing. If load spreader washers are used (Figure 24-23), make sure they are in correct position under the head of each mounting bolt. This is very important, because valve covers have less fasteners per inch than other engine tinware.

7. Run the engine for 10 minutes and check the sealing areas for leaks.

COMMENT: *Do not ever attempt to reuse cork valve cover gaskets. Cork can be compressed up to 50% of its original thickness (Figure 24-24), but it has very poor resiliency. Reuse usually results in leakage because there is no more compressibility.*

REPLACING AN OIL PAN GASKET

The formed metal oil pan is generally used for passenger cars and light trucks. As in the case of valve covers, the thin gasket flange is easily distorted. *It is an absolute must to check the flange for dents and warpage along the length of the pan.* Check by placing the pan flange down on a flat surface. Set a flashlight inside the pan, and sight along the edge for light (Figure 24-25). Straighten as required with a hammer and block of wood.

Check the gasket fit before attempting installation. *Shrunken cork gaskets can be stretched by placing in lukewarm water.* Gaskets will be damaged if they are forced to fit.

Use an adhesive to hold the gasket in place on the pan. Allow the adhesive to set and install the pan. Tighten the bolts in uniform stages so that gasket compression is even (Figure 24-26). Complete the last stage with a torque wrench. Start the engine and check all sealing areas for leaks.

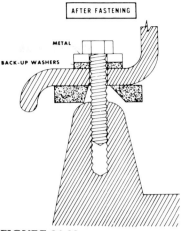

FIGURE 24-22
This valve cover gasket has slipped during installation (Courtesy of Fel-Pro, Inc.).

FIGURE 24-23
To ensure a positive valve cover seal, load spreader washers are often used under the head of each valve cover mounting bolt.

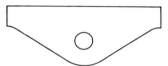

NEW CORK   USED CORK
FIGURE 24-24
Reusing a cork valve cover gasket often results in leakage (Courtesy of Fel-Pro, Inc.).

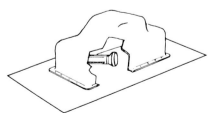

FIGURE 24-25
Checking oil pan flange straightness (Courtesy of Fel-Pro, Inc.).

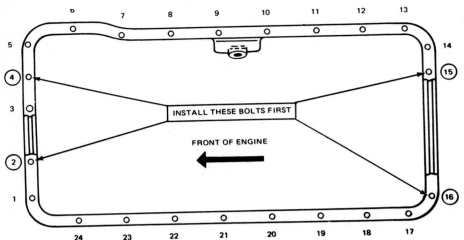

FIGURE 24-26
Typical oil pan installation and bolt torque sequence (Courtesy of Ford Parts and Service Division).

NOTE: *On some engines, the oil pan gasket must be attached first to the block. This is because the pan end strips go over the ends of the pan side strips.*

VS   Valve cover set
WP   Water pump set
T   Timing cover gaskets
F   Exhaust system gaskets

REPLACING OTHER GASKETS

In most cases, water pump, timing cover, fuel pump, and oil pump assemblies use paper gasket material treated to withstand liquids. Because these flange joints are subjected to a working load, the gasket must have high density and flexibility. This keeps bolt torque loss at a minimum, thereby reducing the possibility of fluid leakage.

Paper gasket material is affected by humidity. Therefore, always check fit before starting installation. After the fit is determined to be okay, apply a nonhardening sealer to both sides of the gasket. Clean joint surfaces, bolts, and bolt holes thoroughly. Coat bolt threads with blue Loctite. Use a torque wrench to tighten all bolts evenly. Even bolt pressure is a must for good gasketing.

GASKET PURCHASE

Engine gaskets can be purchased in one of several ways, depending on the type of work being done. A typical parts catalog would show these listings.

FS   Full set (general overhaul)
HS   Head set (valve grinding)
OS   Oil pan set
MS   Manifold set

SHOP TIP

On certain engines you can remove and replace the timing cover without removing the oil pan. To get a positive seal between the underside of the timing cover and the rails on the oil pan, you do not have have to remove the oil pan. Nor, do you have to cut and make pieces from another pan gasket. Gasket segments are made to fit this pan section (Figure 24-27).

Gasket manufacturer's generally label the contents of each package. When purchasing gaskets, read this label to make sure you have all needed gaskets. Figure 24-28a and 24-28b shows the typical location of the more important gaskets and seals in an engine.

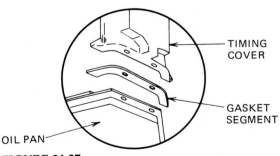

FIGURE 24-27
Oil pan gasket segments.

1. AIR CLEANER TO CARBURETOR GASKET
2. CARBURETOR BASE GASKET
3. EGR GASKET
4. WATER OUTLET GASKET
5. INTAKE MANIFOLD END SEALS
6. INTAKE MANIFOLD GASKET ("VALLEY PAN")
7. EXHAUST MANIFOLD GASKET
8. EXHAUST PIPE SEAL
9. VALVE COVER GASKET
10. VALVE STEM SEALS (INTAKE & EXHAUST)
11. CYLINDER HEAD GASKET
12. HARMONIC BALANCER SLEEVE
13. TIMING COVER SEAL
14. WATER PUMP GASKET
15. OIL PUMP COVER GASKET
16. FUEL PUMP MOUNTING GASKET
17. DISTRIBUTOR MOUNTING SEAL
18. TIMING COVER GASKET
19. OIL PUMP PICKUP TUBE GASKET
20. REAR MAIN BEARING SEAL
21. REAR MAIN BEARING CAP SIDE SEALS
22. OIL PAN SIDE RAIL GASKETS
23. OIL PAN END SEALS
24. OIL PAN DRAIN PLUG GASKET

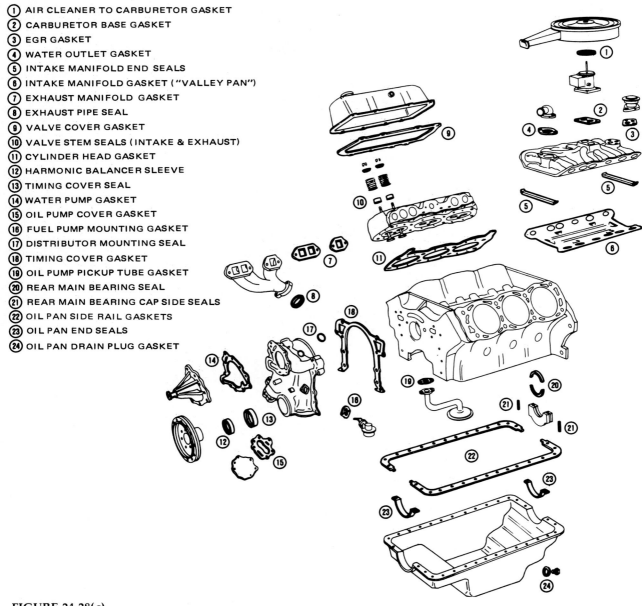

FIGURE 24-28(*a*)
Buick V-6 engine (Courtesy of Fel-Pro, Inc.).

OIL SEALS

Oil seals are classified as either *static* or *dynamic*. A static seal is used between two stationary parts. Front and rear pan end seals are examples of static seals. Dynamic seals are used between a stationary and a moving part. The timing cover seal, and the rear main bearing seal are examples of dynamic oil seals.

Oil Seal Installation Tips

1. Coat the seal wiping edge with the lubricant to be sealed.

2. Extra care should be taken to protect the sealing lip during installation.

3. Uniform pressure should be applied to assemble the seal in its bore (Figure 24-29).

1. AIR CLEANER TO CARBURETOR GASKET
2. CARBURETOR BASE GASKET
3. WATER OUTLET GASKET
4. THERMOSTAT SEAL
5. INTAKE MANIFOLD GASKET
6. EXHAUST MANIFOLD GASKET
7. EXHAUST PIPE SEAL
8. CAM COVER SEALING WASHER
9. CAM COVER GASKET
10. OVERHEAD CAMSHAFT SEAL
11. VALVE STEM SEALS (INTAKE & EXHAUST)
12. CYLINDER HEAD GASKET
13. WATER PUMP GASKET
14. INSULATOR TO BLOCK GASKET
15. PUMP TO INSULATOR GASKET
16. DISTRIBUTOR MOUNTING SEAL
17. AUXILLIARY SHAFT SEAL HOLDER GASKET (FRONT CRANKSHAFT SEAL HOLDER GASKET)
18. AUXILLIARY SHAFT SEAL
19. FRONT CRANKSHAFT SEAL
20. TIMING BELT COVER DUST SEALS
21. OIL PUMP PICKUP TUBE GASKET
22. REAR MAIN BEARING SEAL
23. OIL PAN SIDE RAIL GASKETS
24. OIL PAN GASKET SEGMENTS
25. OIL PAN END SEALS
26. OIL PAN DRAIN PLUG GASKET

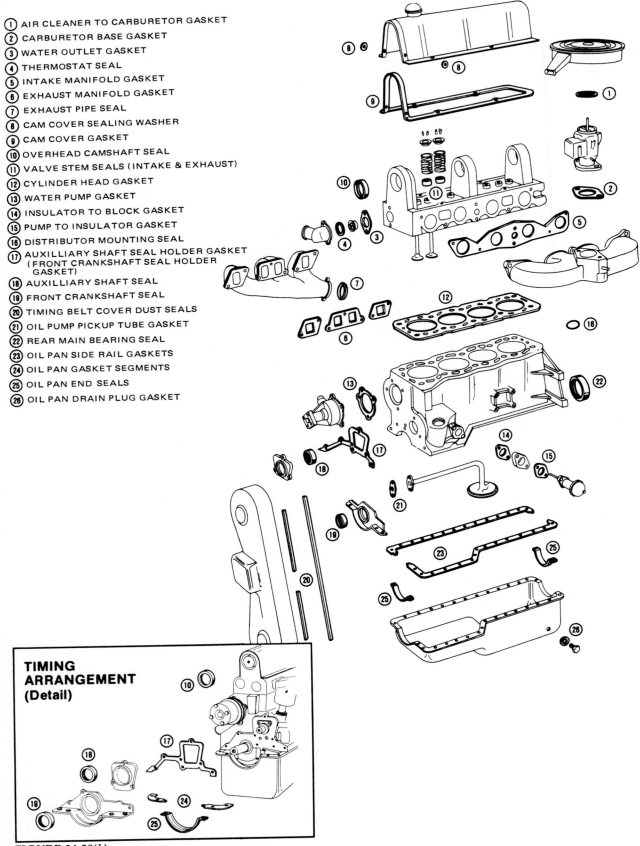

TIMING ARRANGEMENT (Detail)

FIGURE 24-28(b)
Pinto OHC engine (Courtesy of Fel-Pro, Inc.).

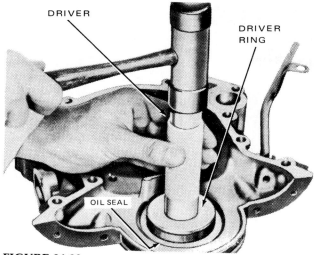

DRIVER

DRIVER RING

OIL SEAL

FIGURE 24-29
Installing crankshaft front oil seal (Courtesy of Ford Parts and Service Division).

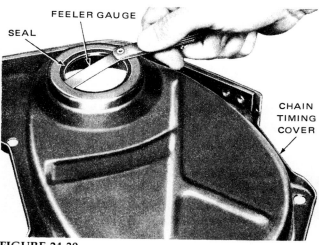

FEELER GAUGE

SEAL

CHAIN TIMING COVER

FIGURE 24-30
Inspecting seal for proper seating. Try to insert a .0015" feeler gauge between the seal body and cover. If the seal is installed properly, the feeler gauge cannot be inserted (Courtesy of Chrysler Corporation).

(a)

(b)

(c)

FIGURE 24-31
(a) Shaping a wick type rear main oil seal. (b) Trimming seal ends. (c) Installing rear main bearing cap side seals (Courtesy of Fel-Pro, Inc.).

Never apply direct hammer blows to the seal surface. Check the seal for proper seating by using a feeler gauge (Figure 24-30).

4. Soak wick (rope) type oil seals in engine oil for 10 minutes just before installing. Then shape the seal into position with a smooth round tool such as a piece of pipe (Figure 24-

31a). Use a new single-edge razor blade and trim the seal ends flush with the block machined surface (Figure 24-31b). If laminated asbestos side seals are used, coat them with mineral oil when installing (Figure 24-31c). This will cause swelling to help maintain the seals after installation.

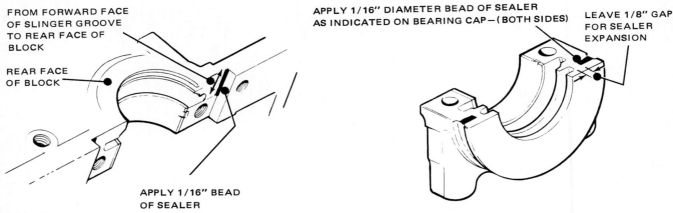

FROM FORWARD FACE
OF SLINGER GROOVE
TO REAR FACE OF
BLOCK

REAR FACE
OF BLOCK

APPLY 1/16" BEAD
OF SEALER

APPLY 1/16" DIAMETER BEAD OF SEALER
AS INDICATED ON BEARING CAP — (BOTH SIDES)

LEAVE 1/8" GAP
FOR SEALER
EXPANSION

FIGURE 24-32
Applying RTV sealer to main bearing cap and block (Courtesy of Ford Parts and Service Division).

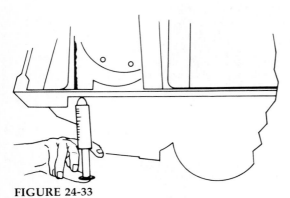

FIGURE 24-33
Forming a rear main bearing seal by injection (Courtesy of Fel-Pro, Inc.).

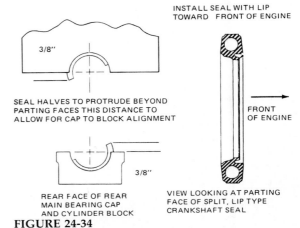

INSTALL SEAL WITH LIP
TOWARD FRONT OF ENGINE

3/8"

SEAL HALVES TO PROTRUDE BEYOND
PARTING FACES THIS DISTANCE TO
ALLOW FOR CAP TO BLOCK ALIGNMENT

FRONT
OF ENGINE

3/8"

REAR FACE OF REAR
MAIN BEARING CAP
AND CYLINDER BLOCK

VIEW LOOKING AT PARTING
FACE OF SPLIT, LIP TYPE
CRANKSHAFT SEAL

FIGURE 24-34
Crankshaft rear oil seal installation position (Courtesy of Ford Parts and Service Division).

5. When installing a rear main bearing cap, apply a thin film of anaerobic sealer to the cap parting face (Figure 24-32).

6. Certain engines (Vega, Pinto, and Capri) require "flowed-in-place" rear main bearing seals. Press the applicator plunger slowly and steadily (Figure 24-33). Pressing too fast can trap air bubbles and result in an oil leak.

7. When installing rear main split-lip seals, let the seal halves protrude beyond the parting faces (Figure 24-34). This will allow for any slight cap-to-block misalignment. Also, be sure the open (undercut) side of the seal faces toward the *front* of the engine. In other words, the seal lip must face the lubricant you wish to seal.

TRUE CASE HISTORY

Vehicle	Ford
Engine	8 cylinder, 351 Cleveland
Mileage	67,000
Problem	Engine overheating condition after doing a valve job.
Cause	The head gaskets were installed backward. On this engine, the gaskets must be installed with the square holes to the front of the block (Figure 24-35). Otherwise, serious coolant blockage will result.

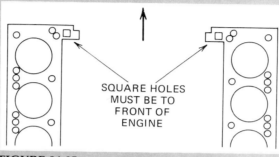

FIGURE 24-35

TRUE CASE HISTORY

Vehicle	Toyota Corolla
Engine	4 cylinder, 2T-C model (1600 cc)
Mileage	23,000
Problem	Oil dripping from the underside of the bell housing (Figure 24-36).
Cause	The leakage was from the rear area of the head gasket, and not from the rear main bearing seal. On this engine, the rocker arm supply oil passes up through the left rear head bolt hole. The gasket was not sealing, and oil seeped down the back of the block.
Cure	A new head gasket was installed. Sealer was applied to both sides of the gasket at the left rear head bolt hole area (Figure 24-37).

FIGURE 24-36

FIGURE 24-37

TRUE CASE HISTORY

Vehicle	Dodge Van
Engine	6 cylinder, 225 CID
Mileage	Unknown
Problem	Repeated exhaust manifold breakage and gasket failure.
Cause	The manifold was not being allowed to expand and contract with the engine heating and cooling cycles. Clearance was not being maintained between the stud nut washers and the manifold. This caused bind and the resulting breakage.
Cure	The stud nut washers were filed in order to obtain an edge clearance of 1/16" (Figure 24-38).

FIGURE 24-38

CHAPTER 24 SELF-TEST

1. What five properties must be considered when a gasket is designed?
2. List several disadvantages of using cork as a gasket material.
3. What advantages do cork gaskets offer?
4. Rubber gaskets have what shortcoming?
5. What are some advantages of using rubber as a gasket material?
6. Paper gaskets are designed to be used in what type of environment?
7. For what is RTV commonly used? How does it cure?
8. What is an anaerobic material?
9. List the three styles of head gaskets discussed in the chapter.
10. Why shouldn't you attempt to straighten a bent sandwich gasket?
11. Permatorque head gaskets offer what advantages?
12. Outline the procedure for making a head gasket impressions test.
13. What is the lead pellet shot test?
14. Why should a head gasket never be reused once it is installed?
15. What is the preferred microinch finish range for a head and block gasket surface?
16. Why must most head gaskets be installed in a certain direction?
17. What check should be made before installing General Motors light-weight cylinder heads?
18. What can a dirty bolt do to a torque wrench reading?
19. Why is it sometimes necessary to lightly chamfer head bolt holes before installing a head gasket?
20. Should an adhesive be used on head gaskets?
21. What are the three kinds of manifold assemblies in use?
22. What is a "turtle-back" gasket?
23. What procedure given in the chapter often eliminates exhaust manifold leak problems?
24. What can happen if the drain slots on small block Ford intake manifold gaskets become plugged with sealer?
25. On small block Chevrolet engines, what should be added at the intake manifold gasket and end strip seal intersections?
26. An exhaust manifold gasket has one face of perforated steel and the other face is asbestos. When assembling, how should the gasket be positioned?
27. What should be used on exhaust gas recirculation (EGR) mounting bolt threads?
28. How can valve cover gasket flange flatness be checked?
29. How can slippage be avoided when installing rubber valve cover gaskets?
30. Why should a cork valve cover gasket never be reused?
31. How can a flashlight be used to check oil pan gasket flange flatness?
32. What can be done to stretch a shrunken cork gasket?
33. What type of sealer is suggested for paper gaskets?

34. What is a static oil seal? Give an example.
35. What is a dynamic oil seal? Give an example.
36. Outline the procedure for installing a rope type rear main oil seal.
37. What engines use a "flowed-in-place" rear main bearing seal?
38. When installing a lip type rear main seal, what direction must the seal lip face?

CHAPTER 25

ASSEMBLY AND BREAK-IN

The following procedure is considered to be the minimum when assembling a short block. If these steps are carefully followed, any incorrect parts or machining will be revealed to the assembler. *CAUTION: Unless you thoroughly understand these steps and have the necessary tools and instruments, you should not attempt assembly.*

Assembly of a short block begins with a block that is true and has been thoroughly cleaned. All parts necessary to complete assembly are at hand.

CAUTION: It is extremely important that all work be done in a clean area. Dirt is the main enemy of an engine.

1. Check condition of every stud, bolt, and bolt hole in the block. Correct any damage.

2. Blow air through all oil holes in the block to make sure they are open (Figure 25-1). Remove any restrictions.

3. Visually check main bearing bores for any spun bearing damage or for discoloration from overheating. *Line boring is mandatory if either has happened.*

4. Install main bearing caps and torque to specifications. Measure across each main bearing housing bore at several points to determine size and roundness. Record the measurements. *NOTE: A housing bore should never be smaller at the parting line surface.*

5. If main bearing housing bore measurements are okay, measure and record the crankshaft main journals for size. Measure the crown wall thickness of a main bearing insert, and add the thickness of two inserts plus the main journal diameter. Subtract the sum from the main bearing housing bore. This figure is the main bearing oil clearance. Refer to Appendix E for a listing of main bearing housing bore diameters.

6. If all measurements are good at this point, measure connecting rod bores and crankshaft rod journals using the same procedure as for the mains. Determine rod bearing oil clearance using the crown wall method.

NOTE: As a general rule, main and rod bearing oil clearance should not exceed .001" per inch of journal

511

FIGURE 25-1

Prior to assembling the engine, clean out all oil holes and check for restriction (Courtesy of Imperial Clevite, Inc., Engine Parts Division).

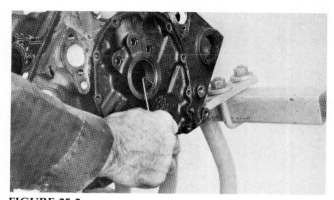

FIGURE 25-2

Install new camshaft bearings and check oil hole alignment (Courtesy of Imperial Clevite, Inc., Engine Parts Division).

Refer to Chapter 22 for detailed information regarding cam bearing installation.

12. Check to make sure the camshaft is correct for the engine application. Install camshaft—lubricate lobes and bearing journals thoroughly before installation. Some engines use mushroom type lifters. These lifters must be installed before the camshaft.

13. Study the rear main seal for proper installation position, and install upper half. Use sealer when required. *The time to prevent a rear main oil leak is now.*

14. Place main bearing inserts in the block and in the caps (Figures 25-3 and 25-4). Make certain the thrust bearing is installed in the right location. Also, make sure oil holes in the bearing shells are placed on the block if each insert pair only has a hole in one shell.

15. Lubricate the rear main seal and bearings thoroughly. Gently lay crankshaft into position on the main bearings (Figure 25-5). Hold crankshaft parallel to the bores to prevent damage to bearings or rear main seal.

16. Install rear main bearing cap and seal first. Then install the remaining main bearing caps, and tighten all cap bolts or nuts finger tight. Observe the position and number marks on the caps (Figure 25-6). While rotating the crankshaft, tap each cap so it registers into its natural position.

17. Pry crankshaft back and forth on its axis to properly align the thrust bearing.

18. Tighten main bearing caps to specifications. After tightening each one, check if the

diameter, or be less than .0005" per inch of journal diameter.

7. Once connecting rod bearing clearance has been determined as correct, scrutinize the block for cleanliness. Look under the pan rails, around the main bearing webs, and at all hidden areas for any material that should be removed.

8. Break any sharp edges at the main cap parting surfaces with a fine file. Use fine emery cloth to remove any burrs within the main bearing housing bores.

9. Doublecheck block, tools, and area where block is to be assembled for cleanliness.

10. Install all oil plugs using a hardening type sealer. Where applicable, make certain drive bushings and shaft plugs have been installed in the block.

11. Make certain you have the right cam bearings. Draw or drive in each cam bearing straight to its proper location. All oil holes must be aligned properly (Figure 25-2). Where applicable, be sure rocker arm assembly feed holes are open. This can be checked by blowing air from the top of the block through oil holes in the cam bearings.

FIGURE 25-3
Install upper main bearing halves (Courtesy of Imperial Clevite, Inc., Engine Parts Division).

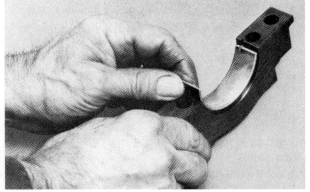

FIGURE 25-4
Assemble lower main bearing halves in caps (Courtesy of Imperial Clevite, Inc., Engine Parts Division).

FIGURE 25-5
Carefully position crankshaft in crankcase (Courtesy of Imperial Clevite, Inc., Engine Parts Division).

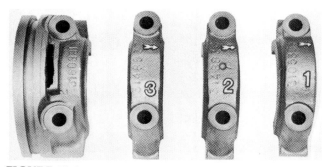

FIGURE 25-6
Typical main bearing cap identification (Courtesy of American Motors Corporation).

FIGURE 25-7
Check crankshaft turning resistance after tightening each bearing cap (Courtesy of Imperial Clevite, Inc., Engine Parts Division).

crankshaft still turns freely by hand (Figure 25-7). *If it binds, stop and find the reason.*

19. Measure crankshaft end play.

20. Install valve timing components. Check cam lobes to be sure your timing procedure is correct. On most engines, correct relationship between the camshaft and crankshaft is achieved when the intake and exhaust lobes are split (equidistant) from the lifters while the piston of that cylinder is at TDC.

21. The pistons and rods should be balanced for top performance. The next paragraph explains one procedure for this.

First, weigh the big-ends of each rod (Figure 25-8). Then grind the rod cap balance pad(s) until all big-ends are of equal weight. Next, determine the total weight of each rod and arrange them in progressive sequence from lightest to heaviest. Then determine the total weight of each piston (including the pin) and arrange them with the rods as a unit (Figure 25-9). Grind the rod small-

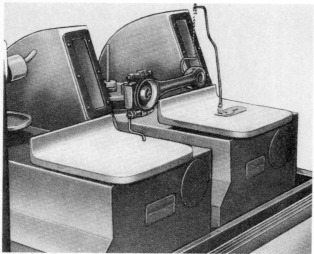

FIGURE 25-8

Connecting rod balancing (Courtesy of Ford Parts and Service Division).

end balance pad on the heavier units until all units match weight. Assemble each rod and piston unit together in the correct relationship.

22. Install the bottom or oil ring first. If the oil ring is segmented, install the top rail before the bottom rail.

   Observe the second ring carefully. Check for an *O* or *T* mark. These marks always face toward the piston top. If there are no ring markings, a step on the outer circumference should be installed down. A step on the inner circumference should be installed up.

   Before installing the top ring, check its end gap in the cylinder. *As a general rule, minimum end gap should not be less than .003" per inch of cylinder diameter.*

   Check the top ring for markings. Marks are always installed upward. If there are no markings, install the ring with the inside bevel up.

23. Place lower bearing halves in the connecting rod caps (Figure 25-10). Wipe off the bearing backs and bearing seat on the cap with a clean cloth. Snap the bearing into place, observing that the bearing tang does not bottom in the indentation into which it fits.

   Place upper bearing halves in the connecting rod (Figure 25-11). Follow the same procedure as used for the lower halves. Lubricate all bearing halves.

24. Stagger ring gaps. Place gaps away from piston thrust faces. Dip the entire piston assembly above the piston pin in oil. Use a good ring compressor to install the piston assemblies.

FIGURE 25-9

Suggested layout for weighing engine parts. A total weight spread of less than one gram should be attained if possible. A 10-gram (about $\frac{1}{3}$ of an ounce) weight differential becomes 60 pounds at 6000 rpm.

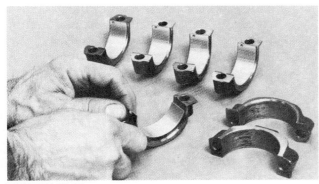

FIGURE 25-10

Assemble lower connecting rod bearing halves (Courtesy of Imperial Clevite, Inc., Engine Parts Division).

FIGURE 25-11

Assemble upper connecting rod bearing halves (Courtesy of Imperial Clevite, Inc., Engine Parts Division).

FIGURE 25-12

Install piston and rod assemblies (Courtesy of Ford Parts and Service Division).

FIGURE 25-13

Check rod reference marks to be sure each assembly is in its proper location.

25. Position crankshaft rod journal at BDC when installing the piston assembly. Check reference marks to be sure the piston is facing in the right direction, and that the rod has been installed correctly.

   Protect the crankshaft from nicks by using boots over both rod bolts.

   With the rings properly squeezed, lightly bump the piston head with a wooden hammer handle until the top ring is just inside the cylinder (Figure 25-12).

26. After the rod is seated on the crankshaft, check to see that the upper bearing half is still in position. Install the rod cap, making sure it faces in the right direction (Figure 25-13). Do not forget to lubricate the bearing.

27. Turn rod cap nuts down finger tight. Tap the cap to seat it in its proper position. Proceed with all other piston assemblies in the same manner.

28. Tighten *all* rod nuts alternately with a torque wrench to the specified limit. Rotate engine and measure connecting rods for side clearance (Figure 25-14).

FIGURE 25-14
Check connecting rod side clearance.

FIGURE 25-15
Assemble other parts as required and install new oil filter (Courtesy of Imperial Clevite, Inc., Engine Parts Division).

29. Install oil pump. Turn engine upright and install new oil filter (Figure 25-15). Add other items as required.

30. Paint the block and protect the assembly with a cover. *NOTE: Many professional racers paint their blocks, rocker covers, and oil pans a flat black color. They do not polish and chrome various engine parts. This is done for cooling efficiency.*

It is a fact that a polished surface will emit less heat by radiation than a black surface.

Engineering handbooks indicate that a black surface has 10 times more heat radiating ability than a plain cast surface. Have you ever wondered why radiators and starters are painted black? This is the reason.

SUGGESTED PARTS REPLACEMENT

A substantial investment is made in rebuilding an engine. The replacement of certain component parts is well worth the additional cost, and usually entails little, if any, extra labor expense when installed with the rebuilt engine. The items listed on Chart 25-1 are vital to the successful operation and long life of an engine. When you rebuild an engine it may not be necesssary to replace all the items. However, when the part is not rebuilt or replaced, good judgment must be applied to prevent a comeback.

INSTALLATION OF NEW WATER PUMP

It is very important to follow these rules when a new water pump is being installed. Failure to do so can cause problems.

1. Make certain the cooling system is free of rust and scale and has no leaks.

2. Check condition of hoses, clamps, belt, thermostat, and radiator cap—replace if necessary.

3. Check engine mounts, making sure that rubber blocks are not worn or split, and that all mounting bolts are tight. Loose or worn engine mounts allow the engine to pivot excessively and could result in the fan hitting either the radiator or shroud.

   Mounts can be checked while the engine is mounted in the chassis with a ground level check. Set parking brake, start engine, and place transmission in drive. Press down on brake pedal while gradually accelerating engine. If left motor mount shows separation, it needs replacement. Place transmission in reverse and repeat procedure for right motor mount.

4. Inspect the new water pump and, if applicable, tighten back plate bolts (Figure 25-16). Turn pump shaft by hand to make sure it rotates freely.

CHART 25-1

Suggested Parts Replacement List	Possible Results if Not Replaced with Engine
Engine oil	Bearing failure, short engine life
Oil filter	Bearing failure, short engine life
Upper radiator hose	Overheating, major damage, short engine life
Lower radiator hose	Overheating, major damage, short engine life
Fan clutch	Overheating, major damage, short engine life
Heater hose	Loss of coolant, overheating, short engine life
Thermostat	Overheating, slow warmup, short engine life
Heat riser valve	Overheating, slow warmup, short engine life
Spark plugs	Misfiring, piston and cylinder damage
Ignition kit	Misfiring, erratic operation
Spark plug wire set	Misfiring, erratic operation
Distributor cap	Misfiring, erratic operation
PCV valve	Excessive crankcase pressure, oil leaks, high oil consumption
Oil fill cap	Excessive crankcase pressure, oil leaks, high oil consumption
Air filter	Incorrect air-fuel ratio, detonation, worn cylinders, short engine life
Carburetor or kit	Flooding, damage to pistons and cylinders, short engine life
Fuel filter	Incorrect fuel flow, stalling, flooding
Engine mounts	Vibration, erratic operation of engine and transmission
Fan belts	Overheating, inoperative alternator and accessories, short engine life
Water pump	Overheating, major damage, short engine life

5. Check fan clutch (if so equipped). *COM-MENT: The automatic viscous fan clutch unit is a major cause of water pump failure.*

- Replace if there is any indication of silicone fluid leakage.

FIGURE 25-16
Some water pumps use a back plate. Coolant leakage can occur from a defective gasket seal, or from loose bolts (Courtesy of Dana Corporation—Engine Parts Group).

- Replace if a noisy or rough feeling can be noticed when the clutch unit is turned by hand.

- Replace if bearings or bushings are worn out or have excessive wobble. Rocking motion of opposing fan blade tips should not exceed more than 3/16" from front to rear.

6. Check fan assembly. Relatively minor defects can amount to substantial loads on the water pump shaft at higher rpm. This can result in severe whipping action and a sudden shaft fracture.

- Lay fan assembly on a flat surface with the leading edge facing down. Replace if there is more than .090" total clearance between opposing blade tips and the flat surface. *Never bend or straighten a fan.*

- Check and replace fan assembly if there are cracks, bends, loose rivets, or broken welds.

7. Adjust fan belt tension. Use any commercially available tension gauge and adjust to OEM specification (Figure 25-17).

If a fan belt is overtightened, the water

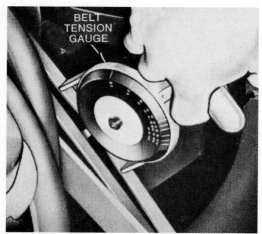

FIGURE 25-17

Checking belt tension (Courtesy of Ford Parts and Service Division).

pump shaft deflects from its true center of rotation. This can cause excessive vibratory forces and result in shaft breakage.

INSTALLATION OF CARBURETOR

For proper performance and to prevent carburetor or engine damage, the following instructions for installation and adjustment must be followed.

1. Remove *all* of the old base gasket. Install a new gasket (and steel shim if so equipped) on intake manifold. Check carburetor base-to-manifold match (Figure 25-18).

2. Mount carburetor on manifold. Set base nuts loosely.

3. A good fuel filter should be installed in the line at the carburetor to prevent dirt from entering.

4. To avoid stripped or crossed threads, first tighten fuel and vacuum lines by hand before using a wrench. Make sure there is no tension on the lines, as this may result in a broken carburetor body.

5. Tighten base nuts and lines evenly. Hookup choke assembly and throttle linkage. Make sure linkage does not bind and throttle goes from full closed to full open.

6. Set idle speed screw(s) on throttle so engine will not stall or race when started. Check position of idle mixture screw(s).

VALVE ADJUSTMENT PROCEDURE

After the cylinder head assembly is installed, perform a valve lash (clearance) adjustment. The following procedure is for an eight-cylinder engine with adjustable hydraulic valve lifters.

1. Make three chalk marks on the vibration damper (90° apart) so that with the timing mark, the damper is divided into four equal sections. See Figure 25-19.

2. Position the no. 1 piston at TDC on the compression stroke. Now, the intake and exhaust valve clearance for the no. 1 cylinder can be adjusted.

   First, loosen the rocker arm stud nut until there is end clearance between the push rod and rocker arm. Then, tighten the nut to just

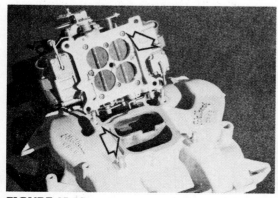
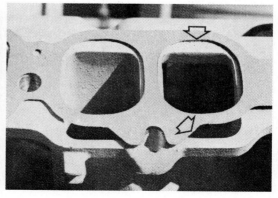

FIGURE 25-18

Air leaks (vacuum leaks) and poor performance can result from gasket mismatch and holes not covered (Courtesy of Edelbrock Equipment Company).

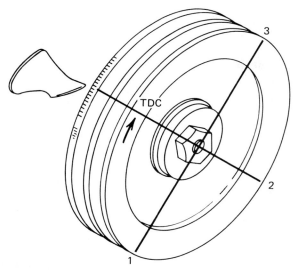

FIGURE 25-19
Valve clearance adjustment points for an eight-cylinder engine.

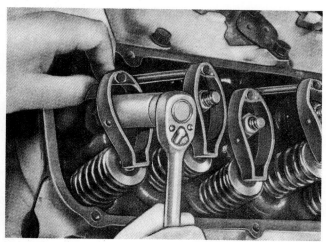

FIGURE 25-20
Determining zero lash position (Courtesy of Ford Parts and Service Division).

remove all the clearance (this is called *zero lash position*). Zero lash position can be easily determined by spinning the push rod with your fingers as the rocker arm stud nut is tightened (Figure 25-20).

When the push rod-to-rocker arm clearance is zero, tighten the stud nut one-half turn. This will place the hydraulic valve lifter plunger at or near the desired midpoint operating position.

3. Turn the crankshaft 90° (one-quarter turn in the direction of rotation) to the next mark on the vibration damper. Adjust the valves for

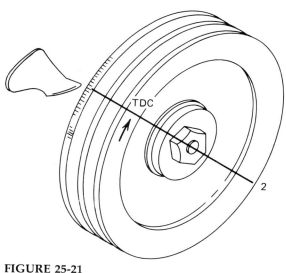

FIGURE 25-21
Valve clearance adjustment points for a four-cylinder engine.

the cylinder that is next in the firing order sequence.

4. Repeat this one-quarter turn procedure until all valve sets are adjusted. When finished, the crankshaft will have made exactly two revolutions (720°).

The procedure just described can also be used on a four-cylinder and on a six-cylinder engine, provided the vibration damper is chalked accordingly.

For a four-cylinder engine, make a chalk mark on the damper directly opposite TDC. The damper is now divided into two equal sections 180° apart. See Figure 25-21.

For most six-cylinder engines, make a chalk mark 120° on each side of TDC. The damper is now divided into three equal sections. See Figure 25-22.

When working with most solid lifter setups, adjust to the specified clearance between the valve stem tip and the rocker arm (Figure 25-23). *On certain OHC engines (Toyota, Datsun, and Pinto), valve lash is set by adjusting the clearance between the cam lobes and rocker arms.* Check the factory manual if in doubt.

SETTING STATIC IGNITION TIMING

1. Position the no. 1 piston at TDC on the compression stroke. Rotate the crankshaft just

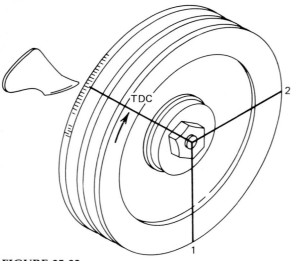

FIGURE 25-22
Valve clearance adjustment points for most six-cylinder engines.

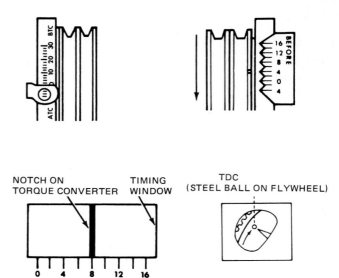

NOTCH ON TORQUE CONVERTER   TIMING WINDOW   TDC (STEEL BALL ON FLYWHEEL)

0   4   8   12   16
TIMING MARKS ON BELL HOUSING
FIGURE 25-24
Examples of timing marks (Copyright, Champion Spark Plug Company, 1980. Used with Permission).

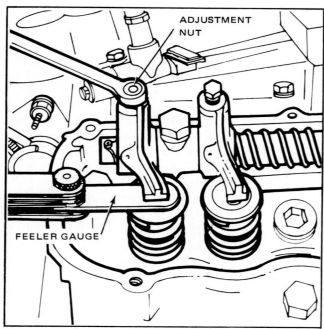

ADJUSTMENT NUT

FEELER GAUGE

FIGURE 25-23
Adjusting valve clearance on Ford 2.8L V-6 engine. Note placement of feeler gauge. Do not insert feeler gauge at outboard edge of valve and move toward carburetor. This method will result in an erroneous "feel" that will produce excessively tight valves (Courtesy of Ford Parts and Service Division).

until the timing marks are aligned according to specifications (Figure 25-24).

2. Install the distributor in the engine with the rotor tip pointing to the no. 1 spark plug wire tower on the cap.

3. Turn the distributor so the rotor tip is halfway to the last cap tower that fired. This is accomplished by slightly turning the distributor housing (with the cap attached) *in the direction of rotation.*

4. Remove the no. 1 plug wire from the spark plug and hold the end close to a good ground.

5. Turn the key switch *on.* Turn the distributor housing *against the direction of rotation* just until a spark occurs at the no. 1 plug wire end.

6. Tighten down the distributor at the exact position it was in when the spark occurred.

7. Turn the key switch *off* and replace the no. 1 plug wire.

BREAK-IN PROCEDURE

1. Use the original manufacturer's specified weight oil for your engine. Only use oil that equals or exceeds the OEM service rating. It

is permissible to use multiviscosity oil, such as 20-50 weight oil.

2. Be sure the radiator is in good condition; otherwise the engine will be damaged. A rebuilt engine will operate at a higher temperature until the break-in period is passed.

3. Make sure all emission control components are clean, open, and in operating condition. Replace filter elements where applicable. Incorrect breathing by an engine will create excessive crankcase pressure and result in oil leaks or higher than normal oil consumption.

4. Hook up a muffler in the exhaust system so any abnormal noise is not masked.

5. Be sure oil pressure is achieved immediately on starting engine. Always use an oil pressure gauge to check pressure. *Do not rely on the dash light.*

6. After first starting the engine, let it run at a fast idle (1500 rpm) for about 20 minutes without a load. This will cause extra oil to be thrown onto the cylinder walls for added lubrication.

   While the engine is running, check for the following.

- Liquid leaks (coolant, fuel, and oil).
- Unusual noises (if necessary, use a stethoscope to locate the source).
- Exhaust color and smell (Is the mixture rich? Is oil being burned? Is water present?).
- Exhaust sound (Is the tailpipe ''pulse beat'' smooth? Does it have a motor boat sound?).
- Exhaust leaks.
- Compression leaks.
- Vacuum leaks.
- Head and block temperature (check for any ''hot spots'' by touching casting surfaces).
- Throttle response (Is there any hesitation or flat spot?).
- Fuel pull-over (Is gasoline dripping from the carburetor main nozzle?).
- Oil pressure.

7. Recheck the ignition timing, carburetor settings, and valve lash adjustment (if applicable) after engine is up to temperature. Then take the vehicle out and subject it to moderate acceleration from 20 to 50 mph (ob-serving speed laws) for at least 10 times. This will aid in seating the rings.

For the first 100 miles, drive at varying speeds to 55 mph. After 100 miles the engine may be operated normally, but do not sustain a steady speed. Vary speeds for the first 1000 miles and avoid extremely high speeds. Follow the applicable speed laws. Never overload the engine during the break-in period.

Cooling System Maintenance and Electrolysis Control

Certain engine parts are subject to electrolyis attack (freeze plugs and aluminum heads in particular). If electrolysis is not stopped, parts will erode and eventually fail. Electrolysis can be prevented. It is the automobile owner's responsibility to do so.

- Maintain a 50% ratio of ethylene glycol in the cooling system.
- Drain, flush, and renew coolant at least once each year.
- Correct cooling system leaks immediately.
- Keep the proper temperature thermostat in the cooling system.
- Maintain the proper level of coolant at all times.
- The engine must be securely grounded by means of a strap or cable to the body or frame.

500 Mile Service and Inspection

To receive top value from a rebuilt engine, it is extremely important to bring the car back to the installer after 500 miles. The following service and inspection items should be taken care of at this time.

1. Change oil and filter.
2. Unless a no-retorque head gasket has been used, retorque the cylinder head bolts. This should be done after the engine has reached operating temperature and has been allowed to cool.
3. Retorque intake manifold bolts.
4. Readjust solid lifters.
5. Check ease of starting with engine cold and at operating temperature.
6. Check idle smoothness.
7. Take care of any customer complaints.

TRUE CASE HISTORY

Vehicle	Ford Pickup Truck
Engine	8 cylinder, 360 CID
Mileage	Rebuilt engine just installed
Problem	1. Lack of power. The customer said, ''This engine doesn't have near the power of my old one.'' 2. No automatic transmission kickdown.
Cause	The carburetor throttle plates were opening only halfway. The engine installer had mounted the accelerator rod bracket incorrectly. This changed the length of the accelerator rod and prevented wide open throttle (WOT). With no WOT, there would be no kickdown.

TRUE CASE HISTORY

Vehicle	Volkswagen Beetle
Engine	4 cylinder, 1200 cc
Mileage	800 miles on a just installed rebuilt engine
Problem	Vehicle caught fire while driving down the road. The two passengers in the back seat suffered burns that required medical attention.
Cause	The rebuilder had installed a larger size battery than original under the back seat. The battery box and cover were removed in the conversion. The upholstery caught fire when the weight of the two passengers in the back caused the seat springs to sag and short across the top of the battery.

TRUE CASE HISTORY

Vehicle	Ford
Engine	8 cylinder, 302 CID
Mileage	11,000 miles on an overhauled engine
Problem	1. Hard starting and rough idle shortly after engine was overhauled.
	2. Burned points. Four sets of contacts had been installed since the engine overhaul.
Cause	A worn distributor breaker plate (Figure 25-25). Notice the egg-shaped wear in the brass bushing on the upper plate. This allowed the point gap to change beyond specifications.

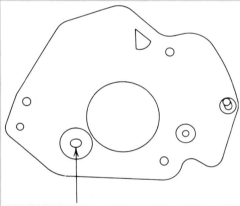

FIGURE 25-25
Worn breaker plate.

Comment   When installing new points, always inspect the breaker plate for wear. This can be done by checking the plate for side movement. Do not forget to lubricate the bushing and pivot areas when installing a new plate.

TRUE CASE HISTORY

Vehicle   Ford Pinto

Engine   4 cylinder, 2300 cc

Mileage   Overhaul recently performed on engine at 77,000 original miles

Problem
1. After about 10 miles of driving on a hot day with the air conditioning on, the engine would quit running.
2. The engine would not restart until allowed to set for at least 30 minutes.

Cause   The ignition control module (Figure 25-26) was defective. After the overhaul, engine operating temperature was increased just enough to upset electrical values in the original module on a hot day.

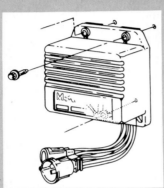

FIGURE 25-26
Dura Spark I and II control module (Courtesy of Ford Parts and Service Division).

Cure   Installing a new module.

Comment   High underhood operating temperatures can sometimes result in intermittent malfunction of an ignition control module. This can be diagnosed by heating the module with a 250-watt heat lamp for about 10 minutes with the engine running (Figure 25-27). Hold the lamp approximately one inch from the module surface. *Do not allow module temperature to exceed 212° F.* If this procedure results in ignition malfunction, replace the module.

FIGURE 25-27
Heating module for test purposes (Courtesy of Ford Parts and Service Division).

TRUE CASE HISTORY

Vehicle	Ford
Engine	8 cylinder, 289 CID
Mileage	Zero miles on a just overhauled engine
Problem	Engine would crank, but would not start. Fuel delivery to carburetor, valve timing, and spark timing were okay.
Cause	The plate for the PCV valve that fits under the carburetor was installed upside down. As a result, there was a complete loss of intake manifold vaccum.

TRUE CASE HISTORY

Vehicle	Saab
Engine	4 cylinder, 1985 cc
Mileage	75 miles on a just overhauled engine
Problem	1. Engine quit running. 2. Starter unable to crank engine.
Cause	The engine was mechanically locked from chunks of hard carbon that had wedged between the top ring lands and the cylinder walls.
Comment	This particular vehicle has the purge hose from the carbon canister routed directly into the filtered side of the air cleaner. Inside the canister, there is a perforated diaphragm that covers the carbon granules. This diaphragm became torn and allowed carbon to be drawn into the engine while driving.

TRUE CASE HISTORY

Vehicle	Chevrolet
Engine	8 cylinder, 307 CID
Mileage	7000 miles on a newly rebuilt engine
Problem	1. Engine overheating.
	2. Lack of power.
	3. Poor gas mileage.
Cause	A collapsed exhaust pipe (Fig. 25-28). The inner wall of the pipe had folded inward creating a restriction that prevented the engine from properly breathing.

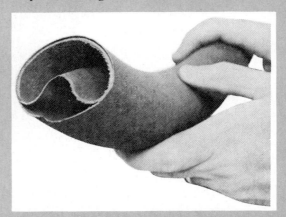

FIGURE 25-28
Collapsed exhaust pipe.

Comment	This kind of problem is often diagnosed with a vacuum gauge. Accelerate the engine and allow it to quickly return to idle. A restricted exhaust will cause a sluggish rise in the return of the needle when the throttle closes. A better way to diagnose this problem is by listening. Increase engine rpm to half-throttle and listen to the sound of the engine at the carburetor air horn. A very distinct "roar" will be heard if the exhaust is restricted.

CHAPTER 25 SELF-TEST

1. What is the main enemy of an engine?
2. A main bearing bore shows discoloration from overheating. What should be done?
3. What is the procedure for determining bearing oil clearance using crown wall thickness?
4. As a general rule, main and rod bearing oil clearance should not exceed what limits?
5. How can cam bearing oil hole alignment be checked?
6. A pair of main bearings only has a hole in one shell. Where should the bearing with the hole be placed?
7. On most engines, valve timing is correct when the intake and exhaust lobes are split from the lifters while the piston of that cylinder is at _____ .
8. Outline the procedure given for balancing pistons and rods.
9. A step on the outer circumference of the second compression ring should face what direction when installed?
10. If the oil ring is segmented, which rail should be installed first?
11. What is the minimum allowable piston ring end gap?
12. A top compression ring has no markings. However, there is a bevel on the inner circumference. What direction should the ring face when installed?
13. Should a bearing tang bottom in the indentation into which it fits?
14. Should ring end gaps be placed away from piston thrust faces?
15. Where should the crankshaft be positioned when installing the piston assembly?
16. Why should rod caps be tapped with a hammer before final tightening?
17. What is the advantage of painting an oil pan flat black?
18. What procedure can be used to check motor mounts while the engine is mounted in the chassis?
19. What is a major cause of water pump failure?
20. What can happen if a fan belt is overtightened?
21. Why should fuel lines be first tightened by hand before using a wrench?
22. An eight-cylinder engine with adjustable hydraulic valve lifters has just been assembled. Outline the valve adjustment procedure.
23. Name several OHC engines that require the valve lash to be adjusted between the cam lobes and rocker arms.
24. Outline the procedure for setting static ignition timing.
25. After first starting a rebuilt engine, how should it be run?
26. List 10 items to check for when a rebuilt engine is first started.
27. For the first 100 miles, how should a rebuilt engine be operated?
28. What steps can be taken to help prevent electrolysis?
29. After 500 miles on a rebuilt engine, what are some service and inspection items that should be checked?

FINAL EXAMINATION QUESTIONS

The following test will help to prepare you for the Automotive Service Excellence (ASE) Engine Repair Certification Examination.

These questions are based on information from this book, and are very similar in content and style to those on the actual certification test.

1. Mechanic A says to replace the stem seals when doing a valve job.

    Mechanic B says to replace the spring retainers when doing a valve job.

    Who is right?

    a. Mechanic A
    b. Mechanic B
    c. Both A and B
    d. Neither A nor B

2. What measurement check is the mechanic performing in this figure?
   a. Piston clearance
   b. Cylinder out-of-round
   c. Cylinder taper
   d. Piston ring end gap

3. What is the mechanic doing in this figure?
   a. Measuring crankshaft thrust play
   b. Measuring connecting rod side clearance
   c. Checking the fillet radius
   d. Checking the oil clearance

4. This given measurement is greater than specification. To correct, what should be done?
   a. Grind the valve stem.
   b. Add shims.
   c. Sink the valve seat.
   d. Replace the spring.

5. What is the mechanic doing in this figure?
   a. Determining main bearing clearance
   b. Determining rod bearing clearance
   c. Measuring crankshaft end play
   d. Measuring journal out-of-roundness

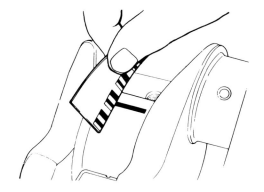

6. A customer is having the third new water pump installed on a car. The two previous pumps failed within a few thousand miles of driving (see figure). What could be causing this problem?
   a. A defective viscous fan clutch assembly
   b. An overtightened belt
   c. A bent fan blade
   d. Any of the above

7. You are replacing rod bearings on an engine. The crankpin diameter measures 2.016". Standard crankpin diameter measures 2.046". What size bearings would you order from parts?
   a. U/S–.015
   b. U/S–.030
   c. Standard size
   d. O/S–.030

8. Mechanic A says a cracked cylinder can cause ethylene glycol to seep into the oil.
   Mechanic B says a cracked cylinder head can cause engine overheating.
   Who is right?
   a. Mechanic A
   b. Mechanic B
   c. Both A and B
   d. Neither A nor B

9. You are rebuilding a typical V-8 engine. The most normal main bearing clearance specification would be
   a. .300"
   b. .030"
   c. .003"
   d. .0003"

10. After honing or deglazing, cylinder walls should be washed with
    a. oil soaked rags
    b. soap and water
    c. unleaded gasoline
    d. cleaning solvent

11. The pattern shown is observed on the skirt of a piston. What is indicated?
    a. A twisted rod
    b. A collapsed piston
    c. An out-of-round bushing
    d. The piston was installed with the offset to the wrong side of the rod.

12. A mechanic has just completed a valve job on a V-8 engine that is equipped with non-adjustable hydraulic lifters. The engine cannot be started. A compression test is made, and six cylinders read zero. What should be done?
    a. Install two head gaskets on each bank.
    b. Grind material off the bottom of each rocker stand.
    c. Grind the valve tips.
    d. Install longer push rods.

13. You make a compression test and obtain a low reading in two adjacent cylinders. This would *most likely* indicate a
    a. leaking intake manifold gasket
    b. leaking exhaust manifold gasket
    c. leaking cylinder head gasket
    d. burned valve

14. Mechanic A says that this rear main seal must be installed with lip **X** facing the flywheel.

    Mechanic B says to dip the seal in oil before installing.

    Who is right?
    a. Mechanic A
    b. Mechanic B
    c. Both A and B
    d. Neither A nor B

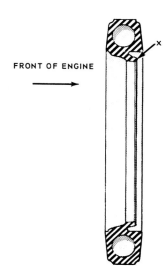

FRONT OF ENGINE

15. A vehicle with a ruptured oil filter is towed into a shop.

    Mechanic A says an oil filter often ruptures when the filter in a full-flow system clogs up.

    Mechanic B says pieces of nylon from a worn timing gear may have entered into the pressure relief valve of the oil pump, causing extreme high pressure.

    Who is right?
    a. Mechanic A
    b. Mechanic B
    c. Both A and B
    d. Neither A nor B

16. What would be the major problem if the parts shown in the figure were left out when rebuilding a typical V-8 engine?

   a. The engine would have no oil pressure.
   b. In the event of a clogged oil filter, there would be no bypass.
   c. The engine would have excessively high oil pressure.
   d. The camshaft would have excessive end thrust.

17. The no. 1 piston has just fired in the above four-cylinder in-line engine. What are the two possible firing orders?

   a. 1-2-3-4 and 1-3-4-2
   b. 1-3-4-2 and 1-2-4-3
   c. 1-4-3-2 and 1-2-3-4
   d. 1-2-4-3 and 1-4-3-2

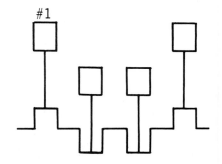

18. A car owner complains of erratic performance. A mechanic checking the timing notices that the timing marks do not return to the same spot when the rpm is raised, then dropped. What could be causing this problem?

   a. The weight and cam assembly needs lubrication.
   b. A worn breaker plate
   c. Both a and b
   d. Neither a nor b

9. The crankshaft shown
   a. has five main bearing journals
   b. is from a six-cylinder engine
   c. both a and b
   d. neither a nor b

20. A cylinder leakage test is being performed. When air is introduced into the cylinder, bubbles are seen in the radiator at the filler opening. What does this possibly indicate?
   a. A cracked cylinder block
   b. A blown head gasket
   c. Either a or b
   d. Neither a nor b

21. The setup shown is for
   a. Knurling
   b. Cleaning
   c. Honing
   d. None of the above

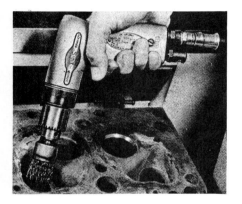

22. How is valve lash adjusted on this valve train?
   a. Add shims at point W.
   b. Grind material from part X.
   c. Turn nut Y.
   d. Turn nut Z.

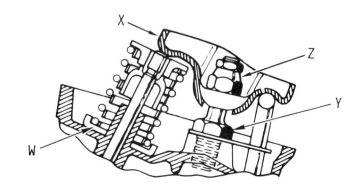

23. Jumper cables are going to be used to start a disabled vehicle.

    Mechanic A says the negative jumper cable should be connected to the engine in the disabled vehicle.

    Mechanic B says the positive jumper cable should be connected to the positive battery terminal in the disabled vehicle.

    Who is right?
    a. Mechanic A
    b. Mechanic B
    c. Both A and B
    d. Neither A nor B

24. All of the lower main bearing shells in an engine show this distress pattern.

    Mechanic A says a lack of oil could be the cause.

    Mechanic B says lugging could be the cause.

    Who is right?
    a. Mechanic A
    b. Mechanic B
    c. Both A and B
    d. Neither A nor B

25. What is this tool used for?
    a. Removing the ridge
    b. Installing a liner
    c. Pulling a wet sleeve
    d. Boring the cylinder

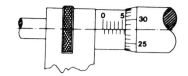

26. What is the reading on the metric micrometer shown?
    a. 6.28 mm
    b. 5.78 mm
    c. 5.53 mm
    d. 5.28 mm

27. Oil is dripping from the bell housing of an engine.
    Mechanic A says this could be from a leaky rear main seal.
    Mechanic B says this could be from a leaky oil gallery plug.
    Who is right?
    a. Mechanic A
    b. Mechanic B
    c. Both A and B
    d. Neither A nor B

28. An engine is being repaired because of a badly spun rod bearing.
    Mechanic A says to replace the connecting rod.
    Mechanic B says to replace the oil cooler core.
    Who is right?
    a. Mechanic A
    b. Mechanic B
    c. Both A and B
    d. Neither A nor B

29. A mechanic removes the piston assemblies from an engine and notices that all the ring faces are covered with dull gray vertical scratches (see figure). What could have caused this?
    a. Failure to run an air filter
    b. A worn carburetor throttle shaft
    c. Either a or b
    d. Neither a nor b

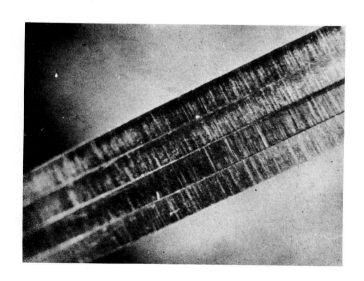

30. What would cause a rod bearing assembly to buckle as shown in the figure?
    a. Drawing down the cap too tightly
    b. Excessive crush
    c. Either a or b
    d. Neither a nor b

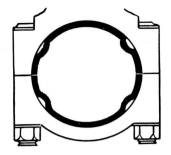

31. A set of engine bearings is shown.
    Mechanic A says these are the upper-half main bearings.
    Mechanic B says bearing X controls crankshaft end play.
    Who is right?
    a. Mechanic A
    b. Mechanic B
    c. Both A and B
    d. Neither A nor B

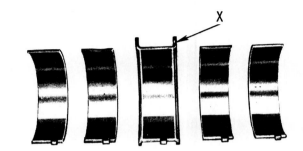

32. When installing a new camshaft in a vehicle, you accidentally knock out part X (shown in figure). What would result when the engine was started?
    a. Engine coolant would get into the oil
    b. Oil pressure loss would occur
    c. Both a and b
    d. Neither a nor b

33. A late model automobile is towed into the repair shop. The mechanic attempts to start the car. The engine cranks very slowly and will not start. Also, the headlights dim considerably during cranking. What could be the problem?
    a. A spun connecting bearing
    b. Oil contaminated with ethylene glycol
    c. Either a or b
    d. Neither a nor b

34. The starter spins and makes a loud grinding noise but does not turn the engine.

    Mechanic A says this could be caused by a worn flywheel ring gear.

    Mechanic B says this could be caused by a worn crankshaft thrust bearing.

    Who is right?
    a. Mechanic A
    b. Mechanic B
    c. Both A and B
    d. Neither A nor B

35. An engine will not start after being rebuilt. What could be the reason?
    a. Valve adjustment
    b. Camshaft timing
    c. Either a or b
    d. Neither a nor b

36. After surfacing both cylinder heads on a V-8 OHV engine, the intake manifold mounting holes will not line up.

    Mechanic A says to enlarge the mounting holes.

    Mechanic B says to use a thick intake manifold gasket set and extra sealer.

    Who is right?
    a. Mechanic A
    b. Mechanic B
    c. Both A and B
    d. Neither A nor B

37. Refer to the valve timing diagram shown. The intake valve is off its seat for how many degrees of crankshaft rotation?
    a. 68°
    b. 102°
    c. 248°
    d. 282°

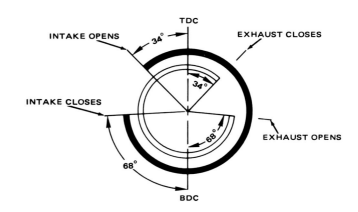

38. An engine is equipped with valve rotators. Which wear pattern shown would indicate that the rotator is functioning properly?
    a. Figure 1
    b. Figure 2
    c. Either Figure 1 or Figure 2
    d. Neither Figure 1 nor Figure 2

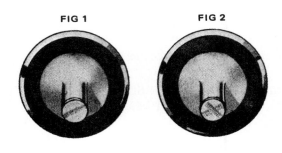

39. While assembling a cylinder head after someone else has ground the valves and seats, you discover that several exhaust valves have seat contact as shown in the figure. What needs to be done to correct this situation?
    a. Grind the seat with a 45° stone, and then narrow with a 60° cutter.
    b. Widen the seat with a 45° stone.
    c. Grind the seat with a 45° stone, and then widen with a 15° cutter.
    d. Widen the seat with a 30° stone.

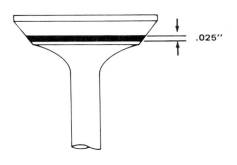

40. Mechanic A says valve seals on an OHV engine can be replaced without taking off the head.

    Mechanic B says the head must be removed when replacing the valve springs on an OHV engine.

    Who is right?
    a. Mechanic A
    b. Mechanic B
    c. Both A and B
    d. Neither A nor B

41. What is the mechanic doing in the illustration?
    a. Cleaning a compression ring groove
    b. Cleaning an oil ring groove
    c. Remachining a worn ring groove
    d. None of the above

42. Crankshaft bearings are being installed in an engine.

    Mechanic A says to lubricate the back of the bearing.

    Mechanic B says to lubricate the front of the bearing.

    Who is right?
    a. Mechanic A
    b. Mechanic B
    c. Both A and B
    d. Neither A nor B

43. The main bearing bores in an engine are stretched (out-of-round).

    Mechanic A says the problem can be corrected by line boring.

    Mechanic B says the problem can be corrected by filing the caps.

    Who is right?
    a. Mechanic A
    b. Mechanic B
    c. Both A and B
    d. Neither A nor B

44. A wet sleeve is fitted too loose in a cylinder block bore.

    Mechanic A says this can cause cavitation.

    Mechanic B says this can cause rapid piston ring wear.

    Who is right?
    a. Mechanic A
    b. Mechanic B
    c. Both A and B
    d. Neither A nor B

45. Based on the picture shown, which of the following statements is true?
    a. Use a new ring when making this check.
    b. A measurement of .020″ would be excessive.
    c. Both a and b
    d. Neither a nor b

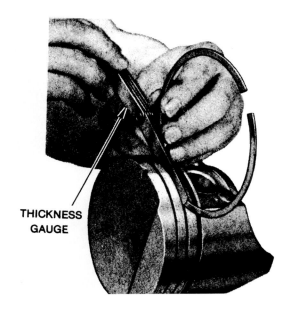

THICKNESS
GAUGE

46. Mechanic A says to turn the crankshaft a full revolution when using Plastigage.

Mechanic B says to span the full width of the bearing when using Plastigage.

Who is right?

   a.   Mechanic A

   b.   Mechanic B

   c.   Both A and B

   d.   Neither A nor B

47. A spark plug has molten chunks of aluminum imbedded between the electrode and the insulator (see figure).

Mechanic A says this can be caused by preignition.

Mechanic B says this can be caused by using low grade gasoline.

Who is right?

   a.   Mechanic A

   b.   Mechanic B

   c.   Both A and B

   d.   Neither A nor B

48. Mechanic A says that if gear X turns 1500 rpm, gear Y turns 750 rpm.

Mechanic B says that gear X and gear Y are timed with a light.

Who is right?

   a.   Mechanic A

   b.   Mechanic B

   c.   Both A and B

   d.   Neither A nor B

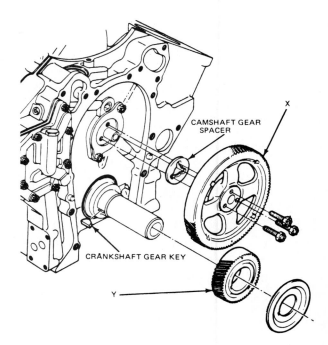

CAMSHAFT GEAR SPACER

CRANKSHAFT GEAR KEY

49. Mechanic A says a broken crankshaft can result if the vibration damper is defective.

    Mechanic B says a sheared woodruff key can result if the vibration damper fits loose.

    Who is right?
    a. Mechanic A
    b. Mechanic B
    c. Both A and B
    d. Neither A nor B

50. While the engine shown is running, a mechanic pinches hose **X** closed. There is a 50 rpm drop.

    Mechanic A says to replace hose **Y**.

    Mechanic B says to replace hose **X**.

    Who is right?
    a. Mechanic A
    b. Mechanic B
    c. Both A and B
    d. Neither A nor B

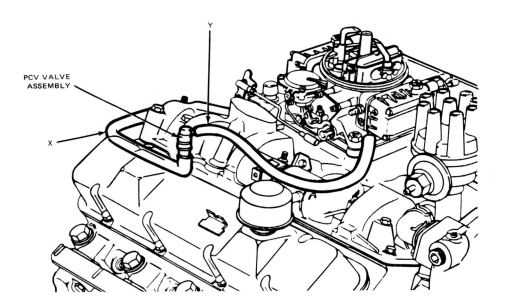

51. Mechanic A says oil can reach the combustion chamber from the crankcase through worn valve guides.

    Mechanic B says oil can reach the combustion chamber from the crankcase through a torn intake manifold gasket.

    Who is right?
    a. Mechanic A
    b. Mechanic B
    c. Both A and B
    d. Neither A nor B

52. What could have caused this camshaft damage?

    a. Incorrect thrust clearance
    b. Contact between lobe and adjacent lifter
    c. Either a or b
    d. Neither a nor b

53. All of these valve grinding stones have been dressed at a 45° angle. However, only one fits the seat correctly. Which one should be used?

    a. Stone A
    b. Stone B
    c. Stone C
    d. Stone D

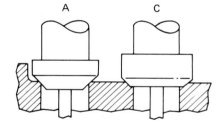
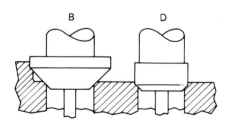

54. A mechanic primes the lubrication system of a rebuilt engine just before the initial starting. Why?

    a. To prevent a dry start
    b. To prevent oil pump damage
    c. To prevent main bearing damage
    d. For all of the above reasons

55. Mechanic A says the part shown contains a bypass valve.

Mechanic B says the part shown bolts to the oil pump.

Who is right?

   a. Mechanic A
   b. Mechanic B
   c. Both A and B
   d. Neither A nor B

Final Examination Answers Appear on Page 590.

APPENDIX A

DECIMAL
EQUIVALENTS

DECIMAL EQUIVALENTS

Fraction	Decimal		Fraction	Decimal
1/64	.015625		33/64	.515625
1/32	.03125		17/32	.53125
3/64	.046875		35/64	.546875
1/16	.0625		9/16	.5625
5/64	.078125		37/64	.578125
3/32	.09375		19/32	.59375
7/64	.109375		39/64	.609375
1/8	.125		5/8	.625
9/64	.140625		41/64	.640625
5/32	.15625		21/32	.65625
11/64	.171875		43/64	.671875
3/16	.1875		11/16	.6875
13/64	.203125		45/64	.703125
7/32	.21875		23/32	.71875
15/64	.234375		47/64	.734375
1/4	.250		3/4	.750
17/64	.265625		49/64	.765625
9/32	.28125		25/32	.78125
19/64	.296875		51/64	.796875
5/16	.3125		13/16	.8125
21/64	.328125		53/64	.828125
11/32	.34375		27/32	.84375
23/64	.359375		55/64	.859375
3/8	.375		7/8	.875
25/64	.390625		57/64	.890625
13/32	.40625		29/32	.90625
27/64	.421875		59/64	.921875
7/16	.4375		15/16	.9375
29/64	.453125		61/64	.953125
15/32	.46875		31/32	.96875
31/64	.484375		63/64	.984375
1/2	.500			

MILLIMETER EQUIVALENTS

MM.	IN.	MM.	IN.	MM.	IN.	MM.	IN.	MM.	IN.
1	0.03937	21	0.82677	41	1.61417	61	2.40157	81	3.18897
2	0.07874	22	0.86614	42	1.65354	62	2.44094	82	3.22834
3	0.11811	23	0.90551	43	1.69291	63	2.48031	83	3.26771
4	0.15748	24	0.94488	44	1.73228	64	2.51968	84	3.30708
5	0.19685	25	0.98425	45	1.77165	65	2.55905	85	3.34645
6	0.23622	26	1.02362	46	1.81102	66	2.59842	86	3.38582
7	0.27559	27	1.06299	47	1.85039	67	2.63779	87	3.42519
8	0.31496	28	1.10236	48	1.88976	68	2.67716	88	3.46456
9	0.35433	29	1.14173	49	1.92913	69	2.71653	89	3.50393
10	0.39370	30	1.18110	50	1.96850	70	2.75590	90	3.54330
11	0.43307	31	1.22047	51	2.00787	71	2.79527	91	3.58267
12	0.47244	32	1.25984	52	2.04724	72	2.83464	92	3.62204
13	0.51181	33	1.29921	53	2.08661	73	2.87401	93	3.66141
14	0.55118	34	1.33858	54	2.12598	74	2.91338	94	3.70078
15	0.59055	35	1.37795	55	2.16535	75	2.95275	95	3.74015
16	0.62992	36	1.41732	56	2.20472	76	2.99212	96	3.77952
17	0.66929	37	1.45669	57	2.24409	77	3.03149	97	3.81889
18	0.70866	38	1.49606	58	2.28346	78	3.07086	98	3.85826
19	0.74803	39	1.53543	59	2.32283	79	3.11023	99	3.89763
20	0.78740	40	1.57480	60	2.36220	80	3.14960	100	3.93700

1/4 MM — .00984   1/2 MM — .01968   3/4 MM — .02953

DECIMAL EQUIVALENTS AND TAP DRILL SIZES

DRILL SIZE	DECIMAL	TAP SIZE
1/64	.0156	
1/32	.0312	
60	.0400	
59	.0410	
58	.0420	
57	.0430	
56	.0465	
3/64	.0469	0-80
55	.0520	
54	.0550	1-56
53	.0595	1-64, 72
1/16	.0625	
52	.0635	
51	.0670	
50	.0700	2-56, 64
49	.0730	
48	.0760	
5/64	.0781	
47	.0785	3-48
46	.0810	
45	.0820	3-56, 4-32
44	.0860	4-36
43	.0890	4-40
42	.0935	4-48
3/32	.0937	
41	.0960	
40	.0980	
39	.0995	
38	.1015	5-40
37	.1040	5-44
36	.1065	6-32
7/64	.1093	
35	.1100	
34	.1110	6-36
33	.1130	6-40
32	.1160	
31	.1200	
1/8	.1250	
30	.1285	
29	.1360	8-32, 36
28	.1405	8-40
9/64	.1406	
27	.1440	
26	.1470	
25	.1495	10-24
24	.1520	
23	.1540	
5/32	.1562	
22	.1570	10-30
21	.1590	10-32
20	.1610	
19	.1660	
18	.1695	
11/64	.1719	

DRILL SIZE	DECIMAL	TAP SIZE
17	.1730	
16	.1770	12-24
15	.1800	
14	.1820	12-28
13	.1850	12-32
3/16	.1875	
12	.1890	
11	.1910	
10	.1935	
9	.1960	
8	.1990	
7	.2010	1/4-20
13/64	.2031	
6	.2040	
5	.2055	
4	.2090	
3	.2130	1/4-28
7/32	.2187	
2	.2210	
1	.2280	
A	.2340	
15/64	.2344	
B	.2380	
C	.2420	
D	.2460	
E, 1/4	.2500	
F	.2570	5/16-18
G	.2610	
17/64	.2656	
H	.2660	
I	.2720	5/16-24
J	.2770	
K	.2810	
9/32	.2812	
L	.2900	
M	.2950	
19/64	.2968	
N	.3020	
5/16	.3125	3/8-16
O	.3160	
P	.3230	
21/64	.3281	

DRILL SIZE	DECIMAL	TAP SIZE
Q	.3320	3/8-24
R	.3390	
11/32	.3437	
S	.3480	
T	.3580	
23/64	.3594	
U	.3680	7/16-14
3/8	.3750	
V	.3770	
W	.3860	
25/64	.3906	7/16-20
X	.3970	
Y	.4040	
13/32	.4062	
Z	.4130	
27/64	.4219	1/2-13
7/16	.4375	
29/64	.4531	1/2-20
15/32	.4687	
31/64	.4844	9/16-12
1/2	.5000	
33/64	.5156	9/16-18
17/32	.5312	5/8-11
35/64	.5469	
9/16	.5625	
37/64	.5781	5/8-18
19/32	.5937	11/16-11
39/64	.6094	
5/8	.6250	11/16-16
41/64	.6406	
21/32	.6562	3/4-10
43/64	.6719	
11/16	.6875	3/4-16
45/64	.7031	
23/32	.7187	
47/64	.7344	
3/4	.7500	
49/64	.7656	7/8-9
25/32	.7812	
51/64	.7969	
13/16	.8125	7/8-14
53/64	.8281	
27/32	.8437	
55/64	.8594	
7/8	.8750	1-8
57/64	.8906	
29/32	.9062	
59/64	.9219	
15/16	.9375	1-12, 14
61/64	.9531	
31/32	.9687	
63/64	.9844	
1	1.000	

PIPE THREAD SIZES

THREAD	DRILL	THREAD	DRILL
1/8-27	R	1 1/2-11 1/2	1 47/64
1/4-18	7/16	2-11 1/2	2 7/32
3/8-18	37/64	2 1/2-8	2 5/8
1/2-14	23/32	3-8	3 1/4
3/4-14	59/64	3 1/2-8	3 3/4
1-11 1/2	1 5/32	4-8	4 1/4
1 1/4-11 1/2	1 1/2		

DECIMAL EQUIVALENTS OF NUMBER SIZE DRILLS

No.	Size of Drill in Inches	No.	Size of Drill in Inches	No.	Size of Drill in Inches	No.	Size of Drill in Inches
1	.2280	21	.1590	41	.0960	61	.0390
2	.2210	22	.1570	42	.0935	62	.0380
3	.2130	23	.1540	43	.0890	63	.0370
4	.2090	24	.1520	44	.0860	64	.0360
5	.2055	25	.1495	45	.0820	65	.0350
6	.2040	26	.1470	46	.0810	66	.0330
7	.2010	27	.1440	47	.0785	67	.0320
8	.1990	28	.1405	48	.0760	68	.0310
9	.1960	29	.1360	49	.0730	69	.0292
10	.1935	30	.1285	50	.0700	70	.0280
11	.1910	31	.1200	51	.0670	71	.0260
12	.1890	32	.1160	52	.0635	72	.0250
13	.1850	33	.1130	53	.0595	73	.0240
14	.1820	34	.1110	54	.0550	74	.0225
15	.1800	35	.1100	55	.0520	75	.0210
16	.1770	36	.1065	56	.0465	76	.0200
17	.1730	37	.1040	57	.0430	77	.0180
18	.1695	38	.1015	58	.0420	78	.0160
19	.1660	39	.0995	59	.0410	79	.0145
20	.1610	40	.0980	60	.0400	80	.0135

DECIMAL EQUIVALENTS OF LETTER SIZE DRILLS

Letter	Size of Drill in Inches	Letter	Size of Drill in Inches
A	.234	N	.302
B	.238	O	.316
C	.242	P	.323
D	.246	Q	.332
E	.250	R	.339
F	.257	S	.348
G	.261	T	.358
H	.266	U	.368
I	.272	V	.377
J	.277	W	.386
K	.281	X	.397
L	.290	Y	.404
M	.295	Z	.413

APPENDIX B

ENGINE PARTS

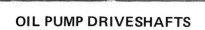

OIL PUMP DRIVESHAFTS

Make & Year	Cyl.	Application	Mfg. No.
CHEVROLET			
1955-63	6	Tube Delivery 7-3/16'' Delivery Length	3705400
1959-79	8	Nylon Sleeve - Oil Pump Shaft Sleeve	3764554
1965-79	8	Driveshaft: 1965-75 Pass. 396, 400, 427 & 454 (Exc. SP/H/P) 1966-74 Corvette (Exc. SP/H/P) 1970 - 396 Truck, 1971-73 - 400 & 454 Truck, 6-33/64'' Length	3998289 3860365
1965-71	8	1965 Pass. & Corvette 396, 427 & 427 w/SP. H/Perf. 1967-68 Pass. & Corvette 427 w/SP: H/Perf. 1968-71 - 396, 427, 454, 6-53/64'' Length	3865886
1959-79	8	1959-69 Corvette 2nd Design 283, 327 1970-77 - 307, 350 & 400 Pass., 1971-74 - 307 & 350 Truck, 5.77'' Length	3998287 3951565
1958-59	8	Driveshaft: 283 1st Design, 5-31/32'' Length	3739826
1959-65	8	1959-61 - 348 2nd Design Pass., 1961-65 - 409 Pass., 6-31/64'' Length	3764553
1968-79	8	Driveshaft: 366 & 427 Serial 50-60, 6-21/64'' Length	3928934
1966	8	Driveshaft: 366 Series 60-80, 6-19/64'' Length	3860469
1969-71	8	350 Eng. - Serial 50, 5-9/16'' Length	3942598
FORD PRODUCTS			
1972-74	4	122 Eng., Pinto, Capri, 6mm Hex 2.31'' Length	D1FZ-6A618-A
1974-79	4	140 Eng., Pinto, Capri, 5/16'' Hex, 3'' Length	D4FZ-6A618-A
1969-79	6	250 Eng., 5/16'' Hex Shaft, 4-9/16'' Length	C9DZ-6A618-A
1965-79	6	1965-72, 240 & 300 Eng. Pass., 1965-77, 240 & 300 L/D & H/D Truck 5/16'' Intermediate Hex Driveshaft, 4.52'' Length	C5AZ-6A618-A
1964-65	6	170 & 200 Eng. - 5/16'' Hex Shaft - 4-1/2'' Long	C4DZ-6A618-A
1961-63	6	144, 170 & 200 Eng. - 1/4'' Hex Shaft - 4-15/32'' Long	C1DE-6A618-A
1961-63	6	223 Eng. - 1/4'' Hex Shaft - 3-61/64'' Long	C2AZ-6A618-A
1960	6	144, 170 Eng. - 1/4'' Hex Shaft - 2-15/16'' Long	CODZ-6A618-A
1955-61	6	223 Eng. - 1/4'' Hex Shaft - 3-27/32'' Long	B9A-6A618-B
1970-79	8	351C, 400 Eng. - 5/16'' Hex Shaft - 8-7/16'' Long	DOAZ-6A618-A

OIL PUMP DRIVESHAFTS — Cont'd

Make & Year	Cyl.	Application	Mfg. No.
FORD PRODUCTS — Cont'd.			
1968-78	8	429, 460 & 460 P.C., 1973-75, 460 Truck, 5/16" Hex, 9-11/32" Length	C8SZ-6A618-A
1969-79	8	351W Eng. - 5/16" Hex Shaft - 8-3/16" Long	C9OZ-6A618-A
1963-79	8	221, 260, 289 & 302 Eng. - 1/4" Hex Shaft - 7.44" Long	C2OZ-6A618-A
1964-78	8	330 H.D., 361 & 391 Eng. - 5/16" Hex Shaft - 8.53" Long	C4TZ-6A618-A
1958-78	8	332, 352, 390 & 406 Eng. - Pass. 330 M.D., 360 & 390 Eng. - Truck - 1/4" Hex Shaft - 8-9/16" Long	B8AZ-6A618-A
1955-64	8	272, 292 & 312 Eng. - 1963-64 Truck only 1/4" Hex Shaft - 8-1/16" Long	B9A-6A618-A C1AE-6A618-A
1955-60	8	279, 302, 317 & 332 Eng. Truck - 1/4" Hex Shaft - 9-5/32" Long	B9T-6A618-A
GMC			
1960-72		305 & 351 Eng. Intermediate 401M, 478M 5/16" Hex Shaft - 11-35/64" Long	2378038
OLDSMOBILE			
1961-79	8	All (Except F-85) 5/16" Hex Shaft - 8-9/64" Long	579237
PONTIAC			
1955-75	8	All Models — Intermediate Oil Pump Driveshaft	518022
1968-79	8	350, 400 & 455 Engine	9794305

ROD BOLTS

Make & Year	Cyl.	Application	Size	Mfg. No.
CHEVROLET				
1978-80	4	151 and X-151 Engine	11/32″	10000569
1976-79	4	1.4, 1.6 Engine	9mm	359158
1978-80	V6-8	196, 231, 350-X Engine	3/8″	1246266
1963-79	6	All w/292 Engine Knurled Type — 3/8″ - 24 x 1-29/32″	3/8″	3789421 360182
1957-76	6-8	All w/194, 230, 250 & 283 Engine Knurled Type — 11/32″-24 x 1-7/8″	11/32″	3892670 3703530 334021 460405
1967-79	8	All w/262, 302, 305, 307, 327 & 350 Engine Knurled Type — 3/8″-24 x 1-59/64″	3/8″	3916399 461372
1970-79	8	All w/400 Engine w/2 B.C. Knurled Type — 3/8″-24 x 1-23/32″	3/8″	3951631
1965-79	8	All w/396, 427 & 454 Engine	3/8″	3862720
CHRYSLER PRODUCTS				
1959-80	6-8	All w/198, 225, 273, 318, 340 & 360 Engine	3/8″	3830614 1944865
FORD				
1974-79	4	140 Engine (2300cc)	9mm	D4FZ-6214-A
1974-80	6	171 Engine (2800cc)	8mm	D4ZZ-6214-A
1960-80	6-8	All w/144, 170, 200, 221, 250, 260, 289 & 302 Engine — 5/16″-24 x 2.08″	5/16″	C3A2-6214-A
1958-68	8	All w/352, 390 Engine — 3/8″-24 x 2-7/32″	3/8″	C1AZ-6214-A
1972-79	8	All w/351W; 429, 460 Engine 3/8″ - 24 x 2-3/16″	3/8″	C8SZ-6214-A
INTERNATIONAL				
	4-8	All w/152, 196, 220, 240, 264 265, 266, 304 & 345 Engine	3/8″	70013-R1
	6	All w/372, 406, 450 and 501 Engine	7/16″	118704-H
	8	392 Engine up to Serial No. 125000	3/8″	309290-C1 332924-C1
	8	461, 478 & 549 Engine	7/16″	133525-R3
	6	Shim Type Washer - Used		348170-C1
	4-8	Shim Type Washer - Used		328278-C1

ROD NUTS

Size	Description	Application	Mfg. No.
8mm	Connecting Rod Nut	Ford	D2RY-6212-A
9mm	Connecting Rod Nut	Chevrolet Ford	361970 D4FZ-6212-A
5/16"-24	Connecting Rod Nuts	Ford	C3DZ-6212-A
5/16"-24	Marsden Connecting Rod Nuts	Ford	Special
11/32"-24	Connecting Rod Nuts	Chevrolet	8703531
11/32"-24	Marsden Connecting Rod Nuts	Chevrolet	Special
11/32"-24	Special Self-Locking Rod Nuts —Hi. Perf.	Chevrolet	Special
3/8"-24	Connecting Rod Nuts	Ford Chevrolet Chrysler	C9AZ-6212-A 354128-S 839103 1737695
3/8"-24	Marsden Connecting Rod Nuts	Ford	C1TE-6212-A
3/8"-24	12 Point Connecting Rod Nuts	Buick Oldsmobile	1174201 395921
7/16"-20	Connecting Rod Nuts	Ford Chevrolet	370664-S
7/16"-20	Marsden Connecting Rod Nuts	Ford	
3/8"	Special Lockwasher	Chrysler	668555

HEAD BOLTS

Make & Year	Cyl.	Application	Size	Mfg. No.
CHEVROLET				
1963-79	4-6	151, 250 and 292 Engine	1/2''-13 x 3-37/64''	3788417
1955-79	6-8	200, 265, 327, 302, 305, 307, 350 & 400 Engine	7/16'' - 14 x 3-3/4''	3734594
1955-79	6-8	200, 265, 327, 302, 305 307, 350 & 400 Engine	7/16'' - 14 x 1-3/4''	3704796
1955-79	6-8	200, 265, 327, 302, 305 307, 350 & 400 Engine	7/16'' - 14 x 3''	3767468
1978-80	V-6	196 & 231 Engine	7/16'' - 14 x 2.69	1243262
1978-80	V-6	196 & 231 Engine	7/16'' - 14 x 3-1/2''	1253539
CHRYSLER				
1937-59	6	All	7/16'' - 14 x 2-13/16''	66014
1960-79	6	All	7/16'' - 14 x 2-13/16''	1947687
FORD				
1974-79	6	171 Engine - 2800cc Also for Main Bearing Cap		D1FZ-6345-A
1965-79	6-8	240, 289, 300 Engine	7/16'' - 14 x 3-29/32''	C2OZ-6065-B
1969-79	8	302 Engine	7/16'' - 14 x 3-29/32''	C2OZ-6065-B
1961-79	8	330, 352, 360, 361, 390 & 391 Engine	1/2'' - 13 x 4-29/64''	C1AE-6065-A
1961-79	8	330, 352, 360, 361, 390 & 391 Engine	1/2'' - 13 x 2-13/16''	C1AE-6065-C
1961-79	8	330, 352, 360, 361, 390 & 391 Engine	1/2'' - 13 x 3-1/2''	C4TZ-6065-A

APPENDIX C

SPECIFICATION TABLE FOR MAIN AND ROD HOUSING SIZE

Engine Description	Main Bearing Housing Size	Rod Bearing Housing Size
AMC, 6 cylinders 196 CID	2.625"/2.626"	2.208"/2.2085"
AMC, 6 cylinders 232 CID	2.691"/2.692"	2.208"/2.2085"
AMC V-8 304, 360 CID	2.941"/2.942"	2.208"/2.2085"
AMC V-8 327 CID	2.691"/2.692"	2.3745"/2.375"
AMC V-8 390, 401	2.941"/2.942"	2.3745"/2.375"
Buick V-6 225 CID	2.687"/2.688"	2.1247"/2.1252"
Buick V-8 300 CID	2.687"/2.688"	2.1247"/2.1252"
Buick V-8 340, 350 CID	3.188"/3.189"	2.1247"/2.1252"
Buick V-8 400, 430, 455 CID	3.438"/3.439"	2.374"/2.3745"
Cadillac V-8 429 CID	3.1883"/3.189"	2.374"/2.3745"
Cadillac V-8 472, 500 CID	3.438"/3.439"	2.6243"/2.625"

Engine Description	Main Bearing Housing Size	Rod Bearing Housing Size
Fiat, 4 cylinders 767, 843, 847 cc	2.1459"/2.1464"	1.7188"/1.7193"
Fiat, 4 cylinders 1089, 1222 cc	1.7991"/1.7996"	1.7188"/1.7193"
Fiat, 4 cylinders 1116, 1197, 1438 cc	2.1459"/2.1465"	1.9144"/1.9152"
Fiat, 4 cylinders 1295, 1393, 1401, 1481, 1795, 1901, 2054, 2279 cc	2.625"/2.6255"	2.233"/2.2335"
Fiat, 4 cylinders 1608 cc	2.1459"/2.1465"	2.021"/2.0215"
General Motors LUV 110.8 CID	2.322"/2.3228"	2.0465"/2.0473"
General Motors Vega 140 CID	2.4906"/2.4916"	2.1247"/2.1252"
General Motors 153 CID	2.4906"/2.4916"	2.1247"/2.1252"
General Motors Corvair 145, 164 CID	2.288"/2.289"	1.9247"/1.9252"
General Motors Chevrolet 6 194, 230, 250 CID	2.4906"/2.4916"	2.1247"/2.1252"

Engine Description	Main Bearing Housing Size	Rod Bearing Housing Size
General Motors Chevrolet 6 235, 261 CID	2.877" 2.9075" 2.9385" 2.9695" (front to rear)	2.4367"/2.4372"
General Motors Chevrolet V-8 283 CID	2.4906"/2.4916"	2.1247"/2.1252"
General Motors Chevrolet 6 292 CID	2.4906"/2.4916"	2.2247"/2.2252"
General Motors Chevrolet V-8 307, 327, 350 CID	2.6406"/2.6415"	2.2247"/2.2252"
General Motors Chevrolet V-8 366 CID Truck	2.937"/2.938"	2.3247"/2.3252"
General Motors Chevrolet V-8 396, 427, 454 CID	2.937"/2.938"	2.3247"/2.3252"
Chrysler Corp. 6 cylinder 170, 198 CID	2.9425"/2.943"	2.3125"/2.313"
Chrysler Corp. 6 cylinder 218, 230 CID	2.6565"/2.657"	2.1675"/2.168"
Chrysler Corp. 6 cylinder 225 CID	2.9425"/2.943"	2.3125"/2.313"
Chrysler Corp. 6 cylinder Dodge Truck 251, 265 CID	2.6565"/2.657"	2.250"/2.2505"
Chrysler Corp. 273, 318, 340 CID V-8	2.6925"/2.693"	2.250"/2.2505"
Chrysler Corp. 361, 383 CID V-8	2.8175"/2.818"	2.50"/2.505"
Chrysler Corp. 413, 440 CID V-8	2.9425"/2.943"	2.50"/2.505"
Datsun, 4 cylinders 1171, 1288 cc	2.1120"/2.1129"	1.8898"/1.8902"
Datsun, 4 cylinders 1189, 1299 cc	2.146"/2.1465"	2.021"/2.0215"
Datsun, 4 cylinders 1296, 1428, 1595, 1770 cc	2.3094"/2.3098"	2.0866"/2.087"
Datsun, 4 cylinders 1488, 1595, 1883, 1982 cc	2.5059"/2.5064"	2.1654"/2.1659"
Datsun, 4 cylinders 1982 cc	2.6238"/2.6243"	2.1654"/2.1659"
Datsun, 6 cylinders 1988, 2393 cc	2.3089"/2.3094"	2.0866"/2.0871"
Datsun, 4 cylinders 1991, 2164 cc	2.9520"/2.9528"	2.2042"/2.2047"

Engine Description	Main Bearing Housing Size	Rod Bearing Housing Size
Ford, Pinto, Capri OHC 122 CID (2000 cc)	2.3866"/2.3874"	2.1654"/2.1661"
Ford Mustang II, Pinto, Capri OHC 140 CID (2300 cc)	2.590"/2.591"	2.1718"/2.1728"
Ford 6 144, 170, 200 CID	2.4012"/2.402"	2.239"/2.3998"
Ford 6 223 CID	2.6912"/2.692"	2.423"/2.4238"
Ford 6 240 CID	2.5902"/2.591"	2.275"/2.2758"
Ford 6 250 CID	2.5902"/2.591"	2.239"/2.2398"
Ford 289, 302 CID	2.4412"/2.442"	2.239"/2.2398"
Ford 292 CID	2.6912"/2.692"	2.312"/2.3128"
Ford 6 Truck 300 CID	2.5902"/2.591"	2.275"/2.2758"
Ford 351 CID	2.9417"/2.9425"	2.4361"/2.4369"
Ford 352 CID	2.9412"/2.942"	2.5907"/2.515"
Ford Truck 360, 361 CID	2.9412"/2.942"	2.5907"/2.5915"
Ford 390 CID	2.9412"/2.942"	2.5909"/2.5915"
Ford Truck 391 CID	2.9412"/2.942"	2.5907"/2.5915"
Ford 400 CID	3.1922"/3.193"	2.4361"/2.4369"
Ford 428 CID	2.9412"/2.942"	2.5907"/2.5915"
Ford 429 CID	3.1922"/3.193"	2.6522"/2.653"
GMC V-6 Truck 351, 401 CID	3.325"/3.326"	3.016"/3.0165"
I.H.C. Truck 266, 304, 345, 392 CID V-8	2.941"/2.942"	2.4995"/2.500"
I.H.C. Truck 308 CID 6	2.874"/2.875"	2.2365"/2.237"
I.H.C. Truck 401, 461, 549 CID V-8	3.316"/3.317"	2.7755"/2.776"
I.H.C. Truck 450, 501 CID 6	3.423"/3.424"	2.898"/2.8985"
Jeep Corp. 134 CID (F–head)	2.4995"/2.50"	2.0432"/2.044"
MG, 4 cylinders 1198, 1489, 1622, 1588 cc	2.146"/2.1465"	2.021"/2.0215"
MG, 4 cylinders 1798 cc	2.271"/2.2715"	2.021"/2.0215"
MG, 6 cylinders 2912 cc	2.521"/2.5215"	2.114"/2.1145"
Oldsmobile 330, 350 CID V-8	2.687"/2.688"	2.2495"/2.250"
Oldsmobile 400, 425, 455 CID V-8	3.188"/3.189"	2.6245"/2.625"

Engine Description	Main Bearing Housing Size	Rod Bearing Housing Size
Opel, 4 cylinders 1196, 1205, 1477, 1488, 1680, 1700, 1736 cc	2.4409"/2.4415"	2.0079"/2.0083"
Opel, 4 cylinders 986, 993, 1078 cc	2.2835"/2.2840"	1.8898"/1.8902"
Opel, 4 cylinders 1492, 1584, 1698, 1897 cc	2.4409"/2.4415"	2.1653"/2.1658"
Opel, 6 cylinders 2473, 2586, 2605 cc	2.4409"/2.4415"	2.0079"/2.0083"
Pontiac 350, 389, 400 CID V-8	3.188"/3.189"	2.3745"/2.375"
Pontiac 421, 428, 455 CID V-8	3.438"/3.4385"	2.3745"/2.375"
Toyota, 4 cylinders 993, 1077, 1166 cc	2.126"/2.1265"	1.7717"/1.7722"
Toyota, 4 cylinders 1407, 1588 cc	2.4409"/2.4414"	2.0079"/2.0084"
Toyota, 4 cylinders 1490, early 1587 cc	2.4245"/2.4255"	2.0865"/2.0875"
Toyota, 4 cylinders late 1587 cc	2.4252"/2.4257"	2.0874"/2.0879"
Toyota, 4 cylinders 1591, 1707, 1858, 1968 cc	2.5197"/2.5202"	2.2047"/2.2052"

Engine Description	Main Bearing Housing Size	Rod Bearing Housing Size
Toyota, 4 cylinders 1897 cc	2.5598"/2.5606"	2.2845"/2.2852"
Toyota, 6 cylinders 1988, 2253, 2563 cc	2.5205"/2.521"	2.1661"/2.166"
Toyota, 4 cylinders 1994 cc	2.5197"/2.5606"	2.2842"/2.2847"
Triumph, 4 cylinders 1670, 1850, 1991, 2088, 2138 cc	2.6255"/2.626"	2.2327"/2.2335"
Triumph, 4 cylinders 1147, early 1296 cc	2.146"/2.1465"	1.7705"/1.771"
Triumph, 4 cylinders late 1296, 1493 cc	2.457"/2.4575"	2.021"/2.0215"
Triumph, 6 cylinders 1998, 2498 cc	2.457"/2.4575"	2.021"/2.0215"
Volkswagen 1131, early 1192 cc	1,2,3(2.3622"/2.3629") 4(1.9685"/1.9695")	2.0787"/2.0795"
Volkswagen late 1192, 1285, 1493, 1584, 1679 cc	1,2,3(2.5591"/2.5598") 4(1.9685"/1.9695")	2.2756"/2.2763"
Volvo, 4 cylinders 1590 cc	2.276"/2.277"	1.9995"/2.000"
Volvo, 4 cylinders 1780, 1990 cc	2.656"/2.657"	2.2765"/2.277"
Volvo, 6 cylinders 2979 cc	2.656"/2.657"	2.2765"/2.277"

APPENDIX D

NAMES AND ADDRESSES OF SUPPLIERS

Core Suppliers

A&A Midwest Rebuilders Suppliers, Inc.
3950 South Wentworth Ave.
Chicago, IL 60609

Boles Parts Supply, Inc.
1057 Blvd. S.E.
Atlanta, GA 30312

Eastern Parts Distributors, Inc.
726 North West St.
Raleigh, NC 27611

Fitz Core Supply
24000 Woodinville-Snohomish Rd.
Woodinville, WA 98072

General Automotive Parts
1711 West Fullerton Ave.
Chicago, IL 60614

Harry A. Greenblot Company
3275–77 East 11th Ave.
Columbus, OH 43219

Southwest Engine Core Company
1261 Goodrich Ave.
Wilmington, CA 90744

Vander Haag's Inc.
Hwy. 18 and 71 North
Spencer, IA 51301

Suppliers of Imported Engine Parts

Arenco Inc.
3120 West Lake St.
Melrose Park, IL 60160

Beck/Arnley
548 Broad Hollow Rd.
Melville, NY 11746

CEW Distributing Company
P.O. Box 3056
Winston Salem, NC 27101

Columbia Motor Corporation
140 West 21st St.
New York, NY 10011

Geon Automotive (BAP/GEON)
East Victoria St.
Compton, CA 90221

International Parts Service
1655 East 27th St.
Los Angeles, CA 90011

MPA (Motor Parts Association)
135 West 10th St.
Huntington Station, NY 11746

Worldparts
800 Space Park South
Nashville, TN 37211

**Source of
Permanent Metal Marker**

John P. Nissen Jr. Co.
P.O. Box 188
Glenside, PA 19038

Source of Flex-Hone

Brush Research Manufacturing Company
4642 East Floral Drive
Los Angeles, CA 90022

Trade Organizations

Automotive Engine Rebuilders Association
234 Waukegan Rd.
Glenview, IL 60025

Production Engine Remanufacturers Association
1800 North Argyle, Suite 400 A
Hollywood, CA 90028

**Sources of
Engine Rebuilding Magazines**

Babcox Publications, Inc.
11 South Forge St.
Akron, OH 44304

Peterson Publishing Company
8490 Sunset Blvd.
Los Angeles, CA 90069

RENEWS
707 Lake-Cook Rd.
Deerfield, IL 60015

Companies Providing Tufftriding

American Crankshaft Company
4150 Yancey Rd.
Charlotte, NC 28210

Cal-Doran Metallurgical Services
2830 East Washington Blvd.
Los Angeles, CA 90023

Commercial Steel Treating
6100 Tireman Ave.
Detroit, MI 48204

Cook Heat Treating Company
8316 East Freeway, P.O. Box 9463
Houston, TX 77306

Lindberg Heat Treating Company
650 East Taylor St.
Cincinnati, OH 45216

Xtek Corporation
Townshop Ave. and Chestnut St.
Cincinnati, OH 45216

**Companies Having Literature
And/Or Publications Available**

Black and Decker Manufacturing Company
701 East Joppa Rd.
Towson, MD 21204

D.A.B. Industries Inc., Replacement Parts Division
Elm Hill Pike
Nashville, TN 37210

Dana Corporation
P.O. Box 455
Toledo, OH 43692

Federal-Mogul Corporation
P.O. Box 1966
Detroit, MI 48235

Gould Inc., Engine Parts Division
6161 Halle Drive
Cleveland, OH 44125

Hastings Manufacturing Company
325 North Hanover St.
Hastings, MI 49058

Irontite Products Company, Inc.
9858 Baldwin Place
El Monte, CA 91734

K.O. Lee Company
P.O. Box 970, 200 South Harrison
Aberdeen, SD 57401

Kwik-Way Manufacturing Company
500 57th St.
Marion, IA 52302

McQuay-Norris Manufacturing Company
1201 Macklind Ave.
St. Louis, MO 63110

Muskegon Piston Ring Company
1839 Sixth St.
Muskegon, MI 49443

Parts Craft
P.O. Box 455
Toledo, OH 43692

Repcoparts U.S.A., Inc.
13921 Bettencourt St.
Cerritos, CA 90701

Rottler Manufacturing Company
8029 South 200 St.
Kent, WA 98031

Sealed Power Corporation
100 Terrace Plaza
Muskegon, MI 49443

Sioux Tools Inc.
2801 Floyd Blvd.
Sioux City, IA 51102

Storm-Vulcan, Inc.
2225 Burbank St.
Dallas, TX 75235

Sunnen Products Company
7910 Manchester Ave.
St. Louis, MO 63143

Tobin Arp Manufacturing Company
15200 West 78th St.
Eden Prairie, MN 55343

TRW Replacement Division
8001 East Pleasant Valley Rd.
Cleveland, OH 44131

Van Norman Machine Company
3640 Main St.
Springfield, MA 01107

Sources of Repair Manuals

Chilton Book Company
Chilton Way
Radnor, PA 19809

Mitchell Manuals Inc., A Cordura Company
P.O. Box 26260
San Diego, CA 92126

Motor Manuals, Hearst Books/Business Publishing
Hearst Corporation
224 West 57th St.
New York, NY 10019

APPENDIX E

CASTING AND FORGING IDENTIFICATION NUMBERS

BUICK

Year	CID	Number of Cylinders	Special Data	Block Casting Number	Crank Number	Head Casting Number
80	173	V-6		14003987		
62–63	198	V-6		1349430	1349429	1354391
61–63	215	8	Aluminum	1193724	1193645	1193741
64–67	225	V-6		1358435 1396736	1357898 1378354	1358676 1368767
75–76	231	V-6		1250581	1357898 1378354	
77–78	231	V-6		1255862		1254170 1257051 1257661 1257722
79–80	231	V-6		25506397 1254083 1261785 1261787		1261377 1262098
68–76	250	6	With or without air injection	3877178 3850817 3854036 3921968 3877178	3876802	3824437 3895052 3872708 377133 3864886 3927763

BUICK

Year	CID	Number of Cylinders	Special Data	Block Casting Number	Crank Number	Head Casting Number
73–78	250	6				331184 355795 377133
80	252	V-6		25505554		
64	300	8	Cast iron block only	1357943 1374557	1357899 1378350	1366379
65–67	300	8	Le Sabre and Special	1374871 1375161	1357899 1378350	1376330
77–79	301	8		0000708		01
76–80	305	8		14016381		378450 376450 358741
66	340	8		1374547	1375809	1366379
67	340	8		1381621	1375809	1376330
68–69	350	8		3182201	1375809 1378346	1233472 1382546
70–79	350	8		1242788 1394987	1375809 1378346	1237650 1247716 1242556 1237749 1238145 1241861
77–80	350	8		25504744		544716 554717
60	364	8	Three hole oil filter mount	1173201 1174372	1173341	1196914 1190415
61–66	364	8	Cam not interchangeable with early 364	1196547	1173341	1185485 1190415 1196914
67	400	8		1381624	1379242	1385649
68–76	400	8		1393444	1379242	1385649 1231786
62–66	401	8		1364705 1349046	1185539 1361798	1196914 1374603 1190415 1185485
63	425	8		1354714 1364707	1185539	1185485 1190415 1196914
64–66	425	8		1364705 1354714 1364704	1361798	1185485 1190415 1196914 1374603
67–69	430	8		1383434 1381625	1379242	1231109
70–76	455	8		1242694	1379242	1238148 1238529 1238530 1231786 1237661

CADILLAC

Year	CID	Number of Cylinders	Special Data	Block Casting Number	Crank Number	Head Casting Number
76–80	350	8		557752		
77–80	350	8	Diesel	558306		
60–62	390	8		1473267	1469265	1468025 1473449 1479831
63	390	8		1473267	1469265 1477634	1468025 1473449
77–79	425	8		1609110		1609112
64	429	8	1.625" crank flange hub, except series 75	1481283	1481021	1481395
64–65	429	8	1.700" crank flange hub, except series 75 with air injection	1481283	1481267 1481021	1481395
65	429	8	Except series 75	1481283	1481267	1481395
66–67	429	8	With air injection	3632948	1486340	1484240
68	472	8		1486238	1486424	3633123
69	472	8		145200	1486424	1486250
70	472	8		145200	1486424	3633450
71–74	472	8		3633568	1486424	3633544
70–76	500	8		3633955	1496793 1495094	3633544 3633964

CHEVROLET

Year	CID	Number of Cylinders	Special Data	Block Casting Number	Crank Number	Head Casting Number
77–78	98	4	Chevette	357153 454208		357151
71–75	140	4	With or without air injection	6259393	3994226	6271105
76	140	4			6270024	3987397
62–68	153	4		3788514	3788522	3788519
69–70	153	4		3833035 3833045 3849449	3788522	3824433 3864888
66–67	194	6	Without air injection	3782856	3788424 3820618	3788380 3788414 3824435 3864883
66–67	194	6	With air injection	3782856	3788424 3820618	3895056

CHEVROLET

Year	CID	Number of Cylinders	Special Data	Block Casting Number	Crank Number	Head Casting Number
63–65	230	6		3788406	3788424 3820618	3824437 3864886 3872708 3872710
66–70	230	6	With or without air injection	3850817 3854036 3877178 3921968	3788423 3788424 3820618 3885052	3824437 3864886 3872708 3885052 3927763
60–62	235	6	Front center mount	3738307 3739716 3764476	3701488	3836848
66–76	250	6	Temperature sender unit location varies; without air injection	3877178 3850817 3854036	3876802	3872708 3885052
66–67	250	6	Temperature sender unit location varies; with air injection	3850817 3854036 3877178 358825	3876802	3824437 3864886 3872708 6259693
75–76	262	8		360851	354490	358741
62–64	283	8	Standard and automatic transmission	3789935 3849852 3864812	3851822	3767460 3774682 3774692 3795896 3814480 3836842
65–67	283	8	Standard and automatic transmission	3849852 3849935 3896944 3932388	3825822 3876768 3949847	3767460 3774682 3774692 3795896 3814480 3836842
66–76	292	6	Without air injection; ½" bolt holes in crank flange	3789404 3851659 3921770	0863 3910363	3895052 3927763
66–73	292	6	With air injection; ½" bolt holes in crank flange	3789404 3851659 3921770	3910363	3927763
63–66	292	6	Without air injection; ⁷⁄₁₆" bolt holes in crank flange	3788378 3789412 3855914 3886061	3789412	3799019 3824437 3864980
75–78	292	6		8994256		370896 3895052

CHEVROLET

Year	CID	Number of Cylinders	Special Data	Block Casting Number	Crank Number	Head Casting Number
67	302	8	High performance special		6764 3815822 3977768	3782461 3814482
68–69	302	8	High performance special		1178 1179 3279	3917291 3927186
73–74	305	V-6		698703	2367417 2457325 2476575	2487298 2487299
75–79	305	8				354434 376450
76–80	305	8		364979 460776 460777		
68–73	307	8		3914636 3932373 3970020	391101A 3911001	3927185
65–67	327	8	Standard and automatic transmission	3782870 3789817 3858174 3892657	2680 4577 3734627 3782680	3782461 3817681
68–69	327	8		3914660 3914678 3932386 3955618	391101A 3911001 3941174	3917293 3927185 299
68–76	350	8	70–73 has counterbore for distributor using shorter oil pump drive	3855961 3932388 3955618 6259425	310514 3882442 3932442 3932444	3927185 3923441 3964286 6260856
68–76	350	8	Heavy-duty engine with steel crank and 4-bolt mains	3956618 3970010	1182 2690	3927185 3932441 3964286
77–80	350	8		1400209 14010207 25504744		3932454 376445
68–76	366	8	Chain or gear driven	3904354 6272176	6223 3824553 3863144	3856213 3418915
69–72	396	8		3965449	3824553 3863144 3874874	3964380
70–80	400	8		330817 3951511	3951529	3961598 3998997
70–73	402	8			3856223	3856206
66–72	427	8		3937726 3869942 3904351	3836223	3856206 3933148
73–76	427	8		6272181	350570	336768

CHEVROLET

Year	CID	Number of Cylinders	Special Data	Block Casting Number	Crank Number	Head Casting Number
77–80	427	8				3876875
70–78	454	8		3963512 345014 359070	3967463 3964414	3964290 343783 352625 336781
75–80	454	8	Heavy-duty truck			346236

CHRYSLER CORPORATION

Year	CID	Number of Cylinders	Special Data	Block Casting Number	Crank Number	Head Casting Number
78–79	105	4	Horizon/Omni 1700 cc	049-103-021		49-103-373
68–76	170	6	Standard and automatic transmission; center sump pan may require change to fit vehicle	2202843 2463230	2843933	2128654 2202952 2206035 2843169
70–76	198	6		2806830	2951979 2951265	2202952 2206035 2843169 3671476
60–80	225	6	Inside diameter of hub on rear crank flange varies; must specify forging number	2663440 2806830 2205528 2202857 2128418 530 030 030-5 030-11	2128492 2264479 220554 2268613 2843935	2128654 2202952 2206035 2121476 4027600 4027600-5 4104362
61–64	265	6	Sodium valves	1400229	1400188 718	1120805 1120806 1632430
64–69	273	6		2465330 2463930 2806030	2658278 2128869 2658278 2264182	2465315 2658920 2843675
62–66	318	8		2264320 2264370	2128869 2464230	2268341
67–80	318	8	Steel or cast iron crank; may be equipped with either air or hot water heated intake manifold	2806030 2566080 2566090 2536030 4006730	2128869	1888888 2658920 2843674 3769973 2843675

CHRYSLER CORPORATION

Year	CID	Number of Cylinders	Special Data	Block Casting Number	Crank Number	Head Casting Number
68–73	340	8	Steel or cast iron crank; inside diameter of hub on rear crank flange varies	2951521	2128869	2531902
71–80	360	8	Steel or cast iron crank	4045601 3769447 367163cf 3870552 3870230 34184ew 3418496	3418640	3418915 3671587 4071051 3769974 2027596
62–66	361	8		2120229 2205712 2658930	2206159	2406516 2463200 2128521 2128522 2206324 2406516 2463200
67–73	361	8		2120229	2206157	2899440
65–67	383	8	Alternator mount at front of block	2120329 2120429 2468130	2206159 3436923	2128521 2128522 2206324 2406516 2463200
68–80	383	8		2468130	2206159 3436923	3751213 2843906 3462346 3614476
72–80	400	8		4006530 3698276 3751287 3751287E 3614230 3698630	2206159 3462923	3769910 3671640 3769954 2843906 3462346 3751213 3769975
61–73	413	8	Truck engine is gear driven with 8 hole crank flange; 1851524 has sodium valves	2120529	1851436 1851524 1851527 1855127 2126160 2206158	2202594 2406740 1851524
74–76	413	8		2120529	2206158 2206160 1851436 1851527	2202594 2406740 1851524
65–71	426	8	Specify transmission	2120529 2406730 2951623	2206158 2206160 2780533 V-42	2406732 2406736 2525053 2780557

CHRYSLER CORPORATION

Year	CID	Number of Cylinders	Special Data	Block Casting Number	Crank Number	Head Casting Number
66–80	440	8		3462615 3462344 2536430	2206160 2126160 3512036	2806019 2843906 2806019
72–76	440	8	3698460 is a cast crank; use vibration damper with elliptical weight; 3671242 is a forged crank	3870556E 3870556 3870557 3462615 3462344 4006630	3751889 3751891 2206160 2226160 3512036	3671640 3769910 3769954 2843906 3462346 3751213 3769975

FORD: All Ford block and head casting numbers contain the digits 6015 and 6090, respectively. This just identifies the part as a block or head. In accordance with industry practice, this listing has omitted the 6015 and 6090 digits. For instance, a C4DE-6090-D cylinder head would simply be listed as C4DE-D.

Year	CID	Number of Cylinders	Special Data	Block Casting Number	Crank Number	Head Casting Number
71–73	98	4	1600cc	D1FZ-A	D1FZ	D1FZ
70–74	122	4	2000cc	D1FZB	72HM 70HM-AKA 72HM-JA	D1FZ
74–78	140	4	2300cc	D4FZ6010A	1D	D4FZ6049A D4FZ6049D D42E
60–64	144	6	Oil pump shaft and temperature sender size varies	C1DE C3DE C4DE-A	2½" center counter-balance	C0DE C3DE-G C3DE-F
61–76	170	6	Oil pump shaft and temperature sender size varies; with or without air injection	C1DE C3DE C4DE C4DE-A C5DE-A C3DE-A D0DZ-A	3" center counter-balance	C1DE C3DE-D C5DE-B C4DE-D C5DE-A C6DE-A C7DE-C C8DE-B D0OZ-A
74–79	171	V-6	2800cc	74TM 74TM-TA		75TM-AA 75TM-HA 75TM-KA
66–79	200	6	With or without air injection	C5DE-H C6DE-H C7DE-H D0DZ-B	C-50, 1-H 1H C5OZ, 1-H 1H C5OZ C5DZ-A	C6DE-A C7DE-C C6DE-A C9DZ-B
60–64	223	6	Different head bolt sizes	C1AE C3AE	C1AE-B C1AE-F	C1AE C2AE C3AE

FORD

Year	CID	Number of Cylinders	Special Data	Block Casting Number	Crank Number	Head Casting Number
65–74	240	6	7 main bearings	C5TE	1-LA, 1-L 1-L 1-LA	C5TE C8AE-A C8AE-D
69–79	250	6		C9DF-B	2H	D3DE-AA C90E-M D5DE-BA
62–63	260	8	6″ between motor mount holes	C30E-C	C20Z C30Z 1M 1MA	C30E-B C30E-F
64–65	260	8	7″ between motor mount holes	C40E-D C40E-B	C20Z C30Z 1M 1MA	C40E-A
64	262	6	3″ or 2⁷⁄₁₆″ crank snout	C4TE	C1TE-H	C1TE C3TE
63	289	8	Oil filter hole in timing cover; 5 bolt holes in flywheel mounting	3AE-1 C4AE	C20Z C30Z 1M, 1MA	C3AE C4AE
64	289	8	7″ between motor mount holes	C30E-G C40E-D	C20Z C30Z 1M, 1MA	C40E-A
65–68	289	8	6 hole flywheel housing; without air injection	C5AE-E	C20Z C30Z 1M, 1MA	C5DE-B C6DE-G C70E-D
66–68	289	8	With air injection	C5AE C50E-A	C20Z C30Z 1M, 1MA	C60E-C C60E-E C70E-A C70E-B
68	289	8		C5AE-A	C20Z C30Z	C80E-D
60–64	292	8	Heavy duty engine with steel crank and stellite valves	ECK ECZ EDB	C1TE-F	C0TE C1TE ECR 5750471 5752113
69–76	300	6	With or without air injection	C5TE-B C5TE-C	C5TE-F C6TE-G	C8TE-E
77–80	300	6				D2TE-BA C8TE-T
68–76	302	8	With or without air injection	C80E-A C80E-B	2M	C8AE-F C8AE-J
75–79	302	8		C4AE-B D40E-AA		D80E-DA

FORD

Year	CID	Number of Cylinders	Special Data	Block Casting Number	Crank Number	Head Casting Number
64–76	330	8	May have ½" or 7/16" distributor shaft hole	C4TE-A C4TE-G C4TE-F	C4TE-A C6TE-D C7TE-A 2T C4AE-A	C4TE-A C4TE-C
76–80	330	8				D5AE-A2A
69–80	351C	8	Cleveland	DCAZ-D	4M	D1AE D0AZ-A D0AZ-G D4AE
70–78	351W	8	Windsor	D0AE-J D5AZ D4AE-DA	3M, 7M 1K	D0AZ-A D5AE-CA D5AZ D5AE-DA
75–80	351M	8		D5AZ	1K	D5AE-A3A D5AZ A2A
65–67	352	8	With or without air injection	C5AE	2T	C5AE C6AE
68–73	360	8		C8TE-A	2T	C8AE-A C8AE-B
64–73	361	8	May have sodium valves	C4TE-B C4TE-D	C4TE-A C6TE-D C7TE-A	C4TE-A C7TE
65–67	390	8	With or without air injection	C5AE-A C6AE-A	2U	C1AE C2SE C4AE C6AE-AA
68	390	8		C8AE-C	2U 3U	C8E-A C8E-B
69–76	390	8	With or without air injection	C8AE-C	2U 3U	C6AE-R C8AE-A C8AE-G
64–76	391	8		C4TE-B C4TE-D C4TE-M	C4TE-B C6TE-E C7TE-B	C4TE-D C7TE
71–80	400	8		D1AE-A D1AE-C	5MA	D1AE D0AZ-A D0AZ-G
64–76	401	8	Special type front cover	EDL	EDL C1TE-B	D0TE C3TE
65–68	427	8		C5AE-D C6AE-C C8AE-B	2U C4AE-B C4AE-E C8AE-B	C5AE-C C8AE-H
68–70	428	8	With or without air injection	C6AE-F	1U 1UB	C8AE-A C8AE-B
72–73	429	8		D0SZ-A D0SZ-D	4UA C8SZ-A	D0VE-C C8SZ-B

FORD

Year	CID	Number of Cylinders	Special Data	Block Casting Number	Crank Number	Head Casting Number
61–65	430	8		C1VE	C1VE	C0ME
68–76	430	8		C8VY-A	2Y 2YA,B,C	D0VE

OLDSMOBILE

Year	CID	Number of Cylinders	Special Data	Block Casting Number	Crank Number	Head Casting Number
76	140	4	Astre	6259394 3967094	6270024	359981
77–80	151	4		10009542 100099540		4-98597-1
77–79	196	V-6		1256690 1261439 1261785		
64–65	225	V-6		1396736	1357898	1368767 1358676
75–78	231	V-6		12494432 1261438		1249175 1257051 1257661 1257722
79–80	231	V-6		255506397 25505554 1254083 1261787		1261377 1262098
65–76	250	8	With air injection	3850817 3854036 3877178	3876802	3927763
65–76	250	8	Without air injection	3877178 3921968	3876802	3872708 3895052
75–76	260	8		554965		231882
77–80	260	8		554965-2A		554715
76–79	305	8		364909 361979 14016381 460776 460778		376450 378450
64–67	330	8	Size of crank snout bolt hole varies	381917	381919 388766 381269	381918 384189 230532
68–74	350	8		395558	230376 393654	397742 230341
75–76	350	8		395558		376445
77–78	350	8		3970010 554964-3A		715
63–64	394	8		585786	582746	583832
65	400	8		389298	384722	383821
66–69	400	8	With or without air injection	390925	384722	395056 230369

OLDSMOBILE

Year	CID	Number of Cylinders	Special Data	Block Casting Number	Crank Number	Head Casting Number
73–78	400	8		9799914 9792250	495201 496414	370696
77–78	403	8		554990 557265 557751		
65	425	8		386525	384722	383821
66–67	425	8	Lifter diameter is .842" or .921"	389244D	384722 390370	230342 396055
68–76	455	8	With or without air injection	231386 396021F	397363 400943	230342 383021

PONTIAC

Year	CID	Number of Cylinders	Special Data	Block Casting Number	Crank Number	Head Casting Number
76–77	140	4		3967094		359981 357635
80	151	4		498886 10009542 10009540		
80	173	V-6		14003987		
77–79	196	V-6		1256690 1261439 1261785		
61–65	215	6		1193724 3838019	1193645 3788423 3788424	1358219 9782005
66–67	230	6	OHC	9787437	3833057 3820618	9782229 9778693
75–78	231	V-6		1249432 1261438		1249175 1257051 1257661 1257722
77–78	231	V-6				12547051 1257661 1257722 1261377 1262098
79–80	231	V-6		255506397 25505554 1254083 1261787		1261377 1262098 25504443
68–69	250	6	OHC	9790043	387680	9791194 9791195
70–76	250	6	Crank snout is drilled and threaded	3850817 3854036	3876802	3895052 3962084

PONTIAC

Year	CID	Number of Cylinders	Special Data	Block Casting Number	Crank Number	Head Casting Number
77–78	250	6		377127 328575 328576		14002012 355795 377133
77–78	260	8		557751 557751-2B		
77–80	301	8		0000708		01
76–80	305	8		525934 460776 460778 14016381 361979 355909 525937		358741 376450
71–73	307	8		3914636 3932373 3970020	391101A 3911001	3927185
63–67	326	8		548211 9778155 9778840 9783669 9787840	541585 9773383 9773382 9783786 9782770	543796 543797 9773345 8778776 2092 141
68–69	350	8		9789679 9790079	9793573	17 18
70–76	350	8		483391 487641 491947	9773382 9773383 9782646 9783786	483713 487076 491945
77–80	350	8		4888986 14016379 25504744 554964-3A 550557 550355 488996 491947		554717 715 1241861 1242556 1243452
64–66	389	8		9778155	9773383 9773382	538177 543796 543797 9773345
65	389	8		9778789	9773383 9773382	9778775 9778776 9778777
66	389	8		9778789	9782646 9783786	092
67–70	400	8	67 Pontiac 400 engines use two types heads; each requires a different piston	9786133 9792510 0792250	103427YF 0773524 9773524	142 143 31 14

PONTIAC

Year	CID	Number of Cylinders	Special Data	Block Casting Number	Crank Number	Head Casting Number
71–76	400	8		487640	103427YF	492466
77–78	400	8		9792250		
61–63	421	8		9773157	9770488 9783787	9770716 092
67–69	428	8		9792281	9783787	16
70–76	455	8		487638	9799103	491548

AMERICAN MOTORS

Year	CID	Number of Cylinders	Special Data	Block Casting Number	Crank Number	Head Casting Number
64	196	6	OHV	3166465	3159614 3134394 3134294 3160033F	3144631 3176459 3157609 3151433
64–70	199	6	OHV	3170721 3181884	3173429	3170717
64–76	232	6	OHV	3170721 322745 3181884	3169034 3172218	3170717
71–72	258	6	With rocker arm shaft	8120533	8120508	8120136
71–79	258	6	Without rocker arm shaft	8122263 3227445	8120508	8122375 3224490
70–80	304	8		4487211 3195527	RD51261	4487242 3220502
63–66	327	8		3166463	3144388	3146596
70–80	360	8		3195528	4487237	4488895 3231475
68–70	390	8	Head bolts may be ½" or 7/16" diameter	4486279	3208795	3188558
71–78	401	8		4488874	4488664	8122377 3212993

ACKNOWLEDGMENTS

Thanks to the following firms and individuals who supplied invaluable literature, illustrations, and assistance.

AC Spark Plug Division, General Motors Corporation, Flint, Michigan

Alio, Joe Jr., Alio Chevron Service, Los Angeles, California

American Hammered Automotive Replacement Division, Sealed Power Corporation, Muskegon, Michigan

American Honda Motor Company, Inc., Gardena, California

American Motors Corporation, Detroit, Michigan

Argus Publishers Corporation, Los Angeles, California

Automotive Services Division/3M Company, St. Paul, Minnesota

Babcox Publications, Akron, Ohio

Balandis, Drew, Hyster Company, City of Commerce, California

Ball, John, American Honda Motor Company, Inc., Gardena, California

Barnes, Robert, Rio Hondo College, Whittier, California

Barrow, David, Taylor Engine Rebuilding, Whittier, California

Bear Service Equipment, Applied Power Inc., Automotive Division, Milwaukee, Wisconsin

Black and Decker Inc., Automotive Products Division, Solon, Ohio

Bolin, Mike, Southern California Gas Company Garage

Bonded Motor and Parts Company, Los Angeles, California

Bowman Distribution, Division of Barnes Group Inc., Cleveland, Ohio

Brush Research Manufacturing Company, Inc., Los Angeles, California

Buick Motor Division, General Motors Corporation, Detroit, Michigan

Cadillac Motor Division, General Motors Corporation, Detroit, Michigan

Callahan, Ace, El Monte Head Repair, El Monte, California

Camp, Charlie

Central Tool Company, Division of K-D Tools, Cranston, Rhode Island

Champion Spark Plug Company, Toledo, Ohio

Chevrolet Motor Division, General Motors Corporation, Detroit, Michigan

Chitty, Dick, Toyota Motor Sales, U.S.A., Inc., Torrance, California

577

Christian, Anthony, Fullemans Full Service, Whittier, California

Chrysler Corporation, Detroit, Michigan

Cloyes Gear and Products, Inc., Willoughby, Ohio

Cole, Gino, Burch Ford, La Habra, California

Corbetts, David, 3M Company, St. Paul, Minnesota

Corsillo, Joe, TRW Replacement Division, Cleveland, Ohio

Cottam, Chris, U.S. Grant High School, Van Nuys, California

Cy-Lent Timing Gears Corporation, Elk Grove Village, Illinois

D.A.B. Industries Inc., Replacement Parts Division, Nashville, Tennessee

Dana Corporation, Toledo, Ohio

Davis Sandblasting Machines, Eugene, Oregon

Dorman Products, Cincinnati, Ohio

Dresser Industries, Inc., Franklin Park, Illinois

Durston Manufacturing Company, La Verne, California

Ed Iskenderian Racing Cams, Gardena, California

Edelbrock Equipment Company, El Segundo, California

Edwards, LeRoy, Engine Rebuilding Corporation, Los Angeles, California

Engle, Diedre, Toyota Motor Sales, U.S.A., Inc., Torrance, California

Estrada, Chito, Alpha Beta Distribution Center, La Habra, California

Fausel, Ray, California State College, Los Angeles, California

Federal-Mogul Corporation, Detroit, Michigan
The illustrations provided by Federal-Mogul Corporation appearing in chapters 2, 9, 12, 19 and 22 are copyrighted and reprinted with permission of Federal-Mogul Corporation. All rights reserved.

Fejfar, Leslie J., Stoody Company, City of Industry, California

Fel-Pro Incorporated, Skokie, Illinois

Ford Parts and Service Division, Dearborn, Michigan

Gameros, Lorenzo, General Electric Corporation, Ontario, California

Gettman, Alan Jr.

Godwin, Paul, S&W Auto Repair, Inc., Whittier, California

Graymills, Chicago, Illinois

Guildner, Boyd, Bundy's Garage, Whittier, California

Haeberlein, Lee, Mountain View High School, South El Monte, California

Hahn, David, Southern Utah State College, Cedar City, Utah

Hall-Toledo, Bowling Green, Ohio

Hastings Manufacturing Company, Hastings, Michigan

Havens, G.A. (Geff), Silver Seal Products Company, Inc., Trenton, Michigan

Hearst Corporation, New York, New York

Heli-Coil Products, Division of Mite Corporation, Danbury, Connecticut

Heron, Russ

Houser Engineering and Manufacturing Inc., Markle, Indiana

Hunckler Products Inc., Huntington, Indiana

Imperial Clevite Inc., Engine Parts Division, Cleveland, Ohio

Ingersoll-Rand Company, Proto Tool Division, Fullerton, California

Irontite Products Company, Inc., El Monte, California

Jeep Corporation, American Motors Corporation, Toledo, Ohio

Jenny Division, Homestead Industries Inc., Corapolis, Pennsylvania

John Wiley & Sons, Inc., Publishers, New York, New York

K-D Tools, Lancaster, Pennsylvania

K-Line Industries, Holland, Michigan

K. O. Lee Company, Aberdeen, South Dakota

K & W Products Inc., Whittier, California

Kent-Moore Corporation, Service Tool Division, Warren, Michigan

Kwik-Way Manufacturing Company, Marion, Iowa

Lampman, Owen

Lawson Products, Inc., North Dallas, Texas

Lincoln St. Louis, Division of McNeil Corporation, St. Louis, Missouri

Lipschultz, Tim, M&L Motor Supply Company, St. Paul, Minnesota

Lisle Corporation, Clarinda, Iowa

L. S. Starrett Company, Athol, Massachusetts

Lukes, Lee, Fel-Pro Incorporated, Skokie, Illinois

Lusby, Arlo, Rowland High School, Rowland Heights, California

Mac Tools, Inc., Washington Court House, Ohio

Magnaflux Corporation, Chicago, Illinois

Marcum, Audrey

Martin Wells, Industries, Los Angeles, California

McGee and Franklin Engine Rebuilders, Sante Fe Springs, California

McKay Manufacturing Company, Los Angeles, California

McQuay-Norris Manufacturing Company, St. Louis, Missouri

Melling Tool Company, Jackson, Michigan

Mercado, Jim, O'Shea Chevron, Hacienda Heights, California

Miranda, Ysidro

Moroso Performance Products, Inc., Guilford, Connecticut

Morris, Rich, Fel-Pro Incorporated, Skokie, Illinois

Munden, Erich, Rio Hondo College, Whittier, California

Muskegon Piston Ring Company, Muskegon, Michigan

National Institute for Automotive Service Excellence, Washington, D.C.

Nylen Products Inc., Bridgman, Michigan

Ohly, Gene, Evans Speed Equipment, South El Monte, California

Oldsmobile Motor Division, General Motors Corporation, Lansing, Michigan

Owatonna Tool Company, Owatonna, Minnesota

Peterson Machine Tool Inc., Shawnee Mission, Kansas

Peterson Publishing Company, Los Angeles, California

Peugeot, France

Pioneer/Barnes Group Inc., Meridian, Mississippi

Pontiac Motor Division, General Motors Corporation, Pontiac, Michigan

Premier Industrial Corporation, Cleveland, Ohio

Ramsco Corporation, South Gate, California

Rank Scherr-Tumico, Inc., St. James, Minnesota

Repcoparts U.S.A. Inc., Cerritos, California

Rexnord Specialty Fastener Group, Torrance, California

Rottler Manufacturing Company, Kent, Washington

Safeguard Engine Parts, Inc., Nashville, Tennessee

Sanchez, Chino, LAPD Garage, Los Angeles, California

Sealed Power Corporation, Muskegon, Michigan

Seal-Lock International, Glenside, Pennsylvania

Sibus, Tim

Silver Seal Products Company, Inc., Trenton, Michigan

Sioux Tools Inc., Sioux City, Iowa

Snap-on Tools Corporation, Kenosha, Wisconsin

Society of Automotive Engineers, Warrendale, Pennsylvania

Steen, Bill, Chaffey College, Alta Loma, California

Storm-Vulcan, A Division of the Scranton Corporation, Dallas, Texas

Sun Electric Corporation, Crystal Lake, Illinois
The copyright associated with the picture of the cylinder leakage tester included herein is the property of Sun Electric Corporation and is used with the express permission of Sun Electric Corporation. None of these materials may be reproduced or used without the express permission of Sun Electric Corporation. SUN is the Registered Trademark of Sun Electric Corporation.

Sunderland, Craig, Taylor Engine Rebuilding, Whittier, California

Sunnen Products Company, St. Louis, Missouri

TRW Inc., Replacement Parts Division, Cleveland, Ohio
Illustrations from TRW publications appear in chapters 12, 15, 16, 18, 20, 21, 22, and 23, © 1975, 1978, 1980, by TRW Inc. Reprinted with permission of TRW Inc.

TRW Inc., Valve Division, Cleveland, Ohio

Templin, James H., Automotive Engine Rebuilders Association, Glenview, Illinois

Tobin Arp Manufacturing Company, Edin Prairie, Minnesota

Tomadur Engine Company, City of Industry, California

Toyota Motor Sales, U.S.A., Inc., Torrance, California

Trinchero, Bart, Pierce College, Woodland Hills, California

Tumasian, Manny, College of Alameda, Alameda, California

Vandervell Canada Limited, Mississauga, Ontario, Canada

Van Vliet, Ron, Met-Co-Aire, Fullerton, California

Vezerian, Steve, Angelo's Shell Service, San Gabriel, California

Walsh, Bill, Board Ford, Whittier, California

Webb, Milton, Andy Granatelli's Tune-Up Masters, Woodland Hills, California

Williams, Bob, J-B Chemical Products, Monrovia, California

Winona Van Norman Machine Company, Winona, Minnesota

Worrel, James, Toyota Motor Sales, U.S.A., Inc., Torrance, California

Yamazen U.S.A., Inc., Carson, California

Zero Manufacturing Company, Blast-N-Peen Division, Washington, Missouri

Zollner Corporation, Fort Wayne, Indiana

INDEX

580

FINAL EXAMINATION ANSWERS

1. a
2. d
3. b
4. b
5. a
6. d
7. b
8. c
9. c
10. b
11. a
12. c
13. c
14. b
15. b
16. a
17. b
18. c
19. a
20. c
21. b
22. d
23. c
24. b
25. a
26. b
27. c
28. c

29. c
30. b
31. b
32. b
33. c
34. c
35. c
36. d
37. d
38. d
39. a
40. a
41. b
42. b
43. a
44. c
45. c
46. b
47. c
48. d
49. c
50. d
51. c
52. c
53. a
54. d
55. c